CHIRAL
NUCLEAR
DYNAMICS

CHIRAL NUCLEAR DYNAMICS

Maciej A. Nowak
Institute of Physics, Jagellonian University, Cracow, Poland

Mannque Rho
Service de Physique Théorique, Saclay, France

Ismail Zahed
Department of Physics, SUNY, Stony Brook, USA

World Scientific
Singapore • New Jersey • London • Hong Kong

Published by

World Scientific Publishing Co. Pte. Ltd.

P O Box 128, Farrer Road, Singapore 912805

USA office: Suite 1B, 1060 Main Street, River Edge, NJ 07661

UK office: 57 Shelton Street, Covent Garden, London WC2H 9HE

Library of Congress Cataloging-in-Publication Data
Nowak, M. A. (Maciej A.)
    Chiral nuclear dynamics / by M. A. Nowak, Mannque Rho, I. Zahed.
      552 p., 21.5 cm.
    Includes bibliographical references and index.
    ISBN 9810210000
    1. Particles (Nuclear physics) -- Chirality.   2. Quantum
chromodynamics.   I. Rho, Mannque.   II. Zahed, Ismail.   III. Title.
QA793.3.C54.N69   1996
539.7'548--dc20   96-16455
                                        CIP

British Library Cataloguing-in-Publication Data
A catalogue record for this book is available from the British Library.

Cover design by Alina Mokrzycka-Juruś and Oliver Rho.

This book is printed on acid-free paper.

Printed in Singapore by Uto-Print

To Our Wives
Ewa, Helga, and Adella

PREFACE

The physics of strongly interacting many-body systems known as nuclear physics is a mature discipline which has achieved a remarkably quantitative success. It has explained with an impressive accuracy the properties of nuclei from the deuteron to heavy nuclei containing several hundreds of nucleons. This is the more remarkable when one realizes that in no way did the success depend on the existence of, or knowledge derived from, the fundamental theory of strong interactions now believed to be quantum chromodynamics (QCD). Guided mostly by a wealth of experimental observations and with the help of symmetry patterns, nuclear physicists were able to build models valid in different regimes, such as the shell model, Fermi liquid model, collective model, interacting boson model, etc. and deduce relations between each other. The only "microscopic" information if ever needed was at the level of mesons and baryons and forces mediating between them. Interactions probed in nuclear processes range from long-distance regime, say of a pion Compton wavelength, to a tenth of fermi corresponding to a hard core: Nuclear forces involve the ranges from zero to several fermis while a nucleon has a size of roughly 1 fermi.

QCD dictates that quarks must be confined within a region of about 1 fermi in radius to give a hadron, so one would expect that as two or more nucleons approach each other within a nucleus, quarks and gluons should take over the dynamics and show up in observables. So the puzzle is: How is a nucleus understood without ever invoking explicitly quarks and gluons and furthermore why do quark-gluon signatures not show up in processes involving short-distance encounters?

This monograph is a first attempt to explain how this puzzle can be understood. We should stress, to start with, that this volume is mostly a collection of works and thoughts of the three authors and as such may not represent a broad review of what has been done in the field although we devote a large portion of the volume to the basic notions developed by other workers. It is certainly not a textbook. It is also largely incomplete. First of all, we have no understanding of how quarks are confined. This explains why the volume hardly touches on this topic although when we treat hadrons, confinement is implicitly assumed. Second, we have no clear idea as to how a nucleus is formed of nearly independent "bags" of quarks and gluons, although this issue may be related to the first. Despite these considerable limitations, we find we can make a significant progress towards understanding nuclei from QCD. The key to this understanding is found in chiral symmetry and its spontaneous breaking in the hadronic world. Throughout most of the book, nonperturbative calculation schemes will be used in the form of semi-classical arguments or chiral counting arguments, providing the seeds for a systematic analysis.

In Chapter 1, the central concept of asymptotic freedom in QCD is discussed from the point of view of the external field problem. Despite that QCD is intractable in the infrared by conventional perturbative methods, a number of nonperturbative results are known. They are related to the structure of the QCD anomalies and to the fact that the QCD vacuum is constrained by a number of symmetries. In this chapter, we discuss the Goldstone theorem, the Vafa-Witten theorem for vector symmetries and 't Hooft's anomaly matching conditions. A generic behavior of the QCD Dirac spectrum is also discussed in the context of random matrix theory.

In Chapter 2, we develop the idea that the QCD vacuum consists of a dilute and random system of instantons and anti-instantons. Although this system screens color and hence does not confine it, we show that most of the theorems listed above are actually realized. After introducing the basics of instanton calculus, we go on to characterize the QCD vacuum as a grand canonical ensemble of instantons and anti-instantons. In the long-wavelength approximation the bulk properties as well as the low-lying excitations of the system can be derived from an effective action for a large number of colors $N_c \gg 1$, in agreement with instanton simulations and cooled lattice calculations.

In Chapter 3, the spectrum of two-dimensional QCD in the limit of large number of colors is analyzed. Some key issues related to the 't Hooft equation as well as the bosonization program are discussed. Following the original ideas of 't Hooft and Witten, we outline what is presently understood of four-dimensional QCD in the large N_c limit. In particular, we show the emergence of an effective theory of weakly interacting mesons that embodies chiral symmetry, and from which baryons arise as solitons.

In Chapter 4, we review the historical role of chiral symmetry in the strong-interaction physics in the context of current algebra. After defining the basic physical currents in hadronic physics, we go over to Gell-Mann current algebra and its embodiment in the linear and nonlinear sigma models. A number of low-energy theorems are discussed from the point of view of conventional PCAC.

In Chapter 5, the current algebra and the PCAC hypothesis are systemized into an effective Lagrangian field theory, namely, chiral perturbation theory. In this part, we follow the historical route through the works of Li, Pagels, Weinberg, Dashen and Weinstein. In particular, we analyze in simple terms the effects of chiral loops on processes involving pions and nucleons. Weinberg's seminal proposal for analyzing the S-matrix in the soft-pion region, together with the chiral power counting, is discussed, laying the ground for modern chiral perturbation theory.

In Chapter 6, Weinberg's program for the S-matrix is extended to all Green's functions, following the work of Gasser and Leutwyler. A number of results in chiral perturbation theory in the meson sector are briefly reviewed, with a thorough presentation of chiral anomalies. Chiral perturbation theory exploits the chiral Ward identities around the chiral point, using power counting in the pion mass. As such, it is limited. We briefly explore in this chapter the possibility of exploiting the chiral Ward identities *on-shell*, without recourse to any power counting.

In Chapter 7, we extend the chiral structure of QCD to the basic structure of the baryons, reviving in the modern context Skyrme's brilliant conception of unified structure of the hadrons. Using the large N_c reasoning of Chapter 3, we discuss how baryons emerge as chiral solitons of the Skyrme type from a weakly-coupled meson Lagrangian. Their structure and properties are presented both for two and three flavors. The picture of strange hyperons developed by Callan and Klebanov is discussed here and taken up later in Chapter 9.

In Chapter 8, we introduce the notion of Cheshire Cat Principle in hadrons and use the Cheshire Cat Mechanism to show the equivalence between the chiral soliton description of baryons and the chiral bag model. This may be viewed as a hadronic realization of an approximate duality. A number of interesting quantum phenomena taking place in the chiral bag is reviewed. We stress that the emergence of anomalies and nonabelian Berry phases is the key to understanding the duality between the bag formulation and the soliton formulation in four dimensions.

In Chapter 9, the role of strangeness in the structure of non-strange baryons is discussed. Also discussed is the possibility that the strange quark may be considered in certain circumstances as "heavy." Heavy-flavor QCD is formulated, and an effective action for the soft interactions between heavy and light flavors is discussed in the context of the instanton model of Chapter 2. The model is approximately bosonized and shown to yield the standard heavy-quark effective actions. In the large N_c limit, the effective action describes baryons as heavy solitons. We show that the heavy solitons acquire their quantum numbers through Berry phases, along the line of reasoning given in Chapters 7 and 8.

In Chapter 10, we show how some of the basic concepts developed in the preceding chapters can be exploited in analyzing hadronic matter, both at normal density and at higher density. The strong coupling problem is analyzed using both semi-classical techniques (solitons) and chiral power counting (chiral perturbation theory). We start by showing how the basic structure of the nucleon-nucleon interaction can be recovered from the two-skyrmion system, and how many-nucleon systems may eventually emerge from a multi-skyrmion solution. We go on to show that the basics of chiral perturbation theory can be incorporated into an effective description of the nuclear medium. A number of applications involving few-body and many-body systems (*i.e.*, light and heavy nuclei) is presented. This chapter represents an early and crude attempt in this unexplored domain of physics.

In Chapter 11, chiral symmetry, current algebra and PCAC are used to analyze the bulk structure of QCD at low temperature. A number of low-temperature results in the chiral limit are derived. We discuss the use of finite-temperature sum rules, instanton models, Nambu-Jona-Lasinio models and random matrix formulations for studying the physics taking place in the chiral phase transition region. The phase transition is analyzed using universality arguments and lattice QCD results. A brief account of the nonperturbative aspects of QCD at high temperature is also given, with special emphasis on the space-like correlations. This chapter is meant to make the bridge to the regime where the effective degrees of freedom figuring in this volume cede to the fundamental QCD degrees of freedom.

In the course of writing this volume, we have benefited immeasurably from generous support and encouragement from many institutes and many people. Stony Brook Nuclear Theory Group through the DOE-grant DE-FG-88ER40388, Saclay's Service de Physique Théorique and Theory Division at Jagellonian University through its Zakopane School 1993 provided financial support so that all three authors could get together at their places at various stages of writing. The National Institute for Nuclear Theory (INT) of the University of Washington, through its 1995 Spring session on chiral dynamics in hadrons and nuclei, and the Gesellschaft für Schwerionenforschung (GSI) at Darmstadt and the Humboldt Foundation of Germany, through their generous support, provided the crucial opportunity to get together to finalize the volume. We have benefited from lively discussions on all aspects of the book with the participants of the INT session as well as with the members of the host institutes (at Stony Brook, Saclay, Cracow, Seattle and Darmstadt). The structure of the volume – particularly its philosophy if one is permitted to use such an expression – is greatly influenced by our collaborators, with whom much of the original work was done and with whom we have had extensive discussions. We owe our deep gratitude in particular to Gerry Brown, Fred Goldhaber, Hans Hansson, Andy Jackson, Marek Jeżabek, Kuniharu Kubodera, Chang-Hwan Lee, Hyun Kyu Lee, Paweł Mazur, Dong-Pil Min, Hélène Nadeau, Holger Bech Nielsen, Byung-Yoon Park, Tae-Sun Park, Madappa Prakash, Michał Praszałowicz, Georges Ripka, Dan Olof Riska, Norberto Scoccola, Edward Shuryak, Vicente Vento, Jac Verbaarschot, Andreas Wirzba, and Hidenaga Yamagishi.

Special thanks go to Oliver Rho who spent numerous hours patiently editing the volume with imaginative suggestions for style, correcting as best as he could our poor English and eliminating, most of all, physics jargons employed much too carelessly.

Maciej A. Nowak, Mannque Rho, Ismail Zahed

Cracow – Paris – Stony Brook
December 1995

CONTENTS

CHAPTER 2. INSTANTON VACUUM

CHAPTER 3. LARGE N_c

CHAPTER 4. CURRENT ALGEBRA

CHAPTER 5. EFFECTIVE CHIRAL LAGRANGIANS

CHAPTER 7. CHIRAL SOLITONS

CHAPTER 8. CHIRAL BAGS

CHAPTER 9. STRANGE AND HEAVY BARYONS

CHAPTER 10. BARYONIC MATTER

CHAPTER 11. HADRONS AT FINITE TEMPERATURE

CHIRAL
NUCLEAR
DYNAMICS

CHIRAL
NUCLEAR
DYNAMICS

CHAPTER 1

INTRODUCTION TO CHIRAL DYNAMICS

1.1. Basics of Quantum Chromodynamics

It is now generally accepted that Quantum Chromodynamics (QCD) is the theory of strong interactions. Strong forces are the consequence of the interaction between quarks and gluons. The structure of the interactions comes from the Lagrangian density of the theory, which at the classical level is

$$
\begin{aligned}
\mathcal{L}_{QCD} &= -\frac{1}{4} G^a_{\mu\nu} G^{a\mu\nu} + \frac{\theta}{16\pi^2} \epsilon^{\mu\nu\alpha\beta} G^a_{\mu\nu} G^a_{\alpha\beta} \\
&\quad + \sum_{f=1}^{N_f} \bar{q}^i_f (i D^{ij}_\mu \gamma^\mu - m_f \delta^{ij}) q^j_f.
\end{aligned}
\tag{1.1}
$$

We have used the following notations

$$
\begin{aligned}
G^a_{\mu\nu} &= \partial_\mu A^a_\nu - \partial_\nu A^a_\mu + ig f^{abc} A^b_\mu A^c_\nu, \\
D^{ij}_\mu &= \partial_\mu \delta^{ij} + ig(A^a_\mu T^a)^{ij}.
\end{aligned}
\tag{1.2}
$$

where A^a_ν is the gluon field with color index a and q^i_f is the quark field with flavor index f and color index i. The generators $T^a = \lambda^a/2$ are $SU(3)_c$ valued and normalized, and f^{abc} refer to the $SU(3)_c$ structure constants. The implicit sum over color and Lorentz indices is assumed.

At tree level, QCD is a theory involving six flavored quarks (up, down, strange, charm, bottom and top), each with three colors, that interact strongly with the octet of vector mesons. The strength of this interaction is universal and described by a dimensionless coupling constant g. In many

1

ways, the structure of the quark-gluon interaction in QCD resembles the structure of the electron-photon interaction in QED. If it were not for the induced three and four gluon couplings, QCD would be a mere copy of QED with some extra color factors. All the nontrivial properties of QCD result from the three and four gluon vertices with strength g and g^2 respectively.

The θ term in (1.1) is a total derivative. Contrary to QED, this term *may* generate nontrivial physics, since the classical QCD equations admit nontrivial topological solutions (see below). This term is odd under time reversal and thus breaks CP by the CPT theorem. The strong CP problem in QCD is still a theoretically debated issue[1]. Empirically, however, this term is believed to be negligible from the constraints derived from the neutron electric dipole moment ($\theta \leq 10^{-10}$).

The interactions figuring in the QCD Lagrangian (1.1) follow from symmetry. QCD is invariant under $SU(3)_c$ gauge transformations, that is, local or position-dependent transformations of the color variables. Specifically, (1.1) is invariant under the following set of gauge transformations

$$
\begin{aligned}
q(x) &\rightarrow \Omega(x)q(x), \\
\bar{q}(x) &\rightarrow \bar{q}(x)\Omega(x)^{-1}, \\
A_\mu(x) &\rightarrow \Omega(x)A_\mu(x)\Omega^{-1}(x) - \frac{i}{g}\Omega(x)\partial_\mu\Omega^{-1}(x)
\end{aligned}
\tag{1.3}
$$

where $\Omega(x)$ is $SU(3)_c$ valued. As a result, the gauge structure of QCD is nonabelian in contrast to QED where the gauge structure is abelian. Thus, QCD is a nonabelian gauge theory of quarks and gluons. The gauge freedom amounts to the ability to change color at any point in space-time, thus making the concept of color-fixing meaningless. In QCD color cannot be observed. This is in contrast with QED where charge is observed.

The bare (tree) Lagrangian (1.1) is the starting point of the quantum theory of the strong interactions, defined in perturbation theory through Feynman graphs. The latter diverge and require renormalization. QCD is a renormalizable theory, so only a finite class of Feynman graphs diverge in the ultraviolet regime. This fact is due to the underlying gauge symmetry of the theory much like in QED. However, because of the nonabelian character of the gauge group in QCD there is a substantial difference between the way the coupling constant renormalizes in QED and QCD, as we now discuss.

To discuss the renormalization of the charge in QED, consider two static bare electric charges of the same magnitude and opposite signs, separated by a distance R. Classically, the vacuum is passive and the charges are subjected to only Coulomb's law. Quantum mechanically, the vacuum is active and the charges are affected by its zero-mode fluctuations. These effects can be estimated in perturbation theory, as shown in Fig. 1.1 (one-loop).

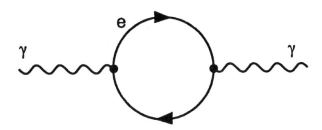

Fig. 1.1. Perturbative contribution to the photon polarization tensor.

The contribution of this diagram to the electrostatic energy is divergent in the ultraviolet. This is a well-known fact in the context of field theories, and because the theory is renormalizable these divergences can be tamed. The machinery for that is the renormalization procedure, whose general exposition is not of interest to us here. Here we will be content with cutting off the one loop integral in the ultraviolet using the cutoff Λ_{co}. Thus, the one-loop contribution to the electrostatic energy of two charges is

$$\mathcal{E} = -\frac{e_0^2}{4\pi R}\left(1 - \frac{2e_0^2}{12\pi^2}\ln(R\Lambda_{co})\right). \tag{1.4}$$

The first term is the familiar Coulomb law and the second term is the vacuum modification. Because of the cutoff, the result is ambiguous or prescription-dependent. This ambiguity may be lifted by *renormalizing* the result to some known, well measured observable (*e.g.* the electric charge measured at the Thomson limit). Thus, renormalization at the scale $1/R_0$ means

$$\mathcal{E}(R_0) = -\frac{e_0^2}{4\pi R_0}\left(1 - \frac{2e_0^2}{12\pi^2}\ln(R_0\Lambda_{co})\right) \equiv -\frac{e(R_0)^2}{4\pi R_0}. \tag{1.5}$$

Now we can express the cutoff in (1.4) through the renormalization condition (1.5)

$$\mathcal{E}(R) = -\frac{e(R_0)^2}{4\pi R}\left(1 - \frac{2e(R_0)^2}{12\pi^2}\ln(R/R_0)\right). \tag{1.6}$$

The dependence on the cutoff scale has disappeared. We may look at (1.6) as a modification of the standard Coulomb's law by the vacuum viewed as a dielectric medium with an inhomogeneous dielectric constant

$$\epsilon(R) = 1 + \frac{2e(R_0)^2}{12\pi^2}\ln(R/R_0). \tag{1.7}$$

Alternatively, we may absorb the R dependence into the charge, defining a "running charge" $e(R) = e(R_0)/\sqrt{\epsilon(R, R_0)}$. In QED the dielectric constant of the vacuum grows with the distance which means that at large separations the electric charge is screened by virtual pairs of electrons and positrons. From (1.6) it follows that the perturbative contribution becomes of order 1 at distance scales on the order of 10^{-62} fermi. Thus, in QED perturbation theory can be used in a very broad range of momenta.

We may characterize the rate at which the charge runs with respect to the change in the scale by introducing the concept of a "beta function"

$$\beta = -R\frac{\partial e(R)}{\partial R}. \tag{1.8}$$

In this language, the lowest order perturbative correction to the charge in QED contributes

$$\beta(e) = +\frac{e^3}{12\pi^2} \tag{1.9}$$

with a positive coefficient. Thus the coupling constant decreases as the momentum decreases. This means that QED is stable in the infrared regime.

Now that we have developed some insights into the quantum problem in QED, let us turn to QCD. By analogy with the QED problem, we will estimate the chromostatic energy at one-loop level of two heavy and colored sources in a color-singlet configuration. The parallel that exists between the quark-gluon interaction and the electron-photon interaction allows us to write immediately the quark contribution to the chromomagnetic energy. For N_f species of quarks, the result is

$$\mathcal{E} = -\frac{1}{N_c}T^aT^a\frac{1}{\epsilon_q}\frac{g^2(R_0)}{4\pi R}. \tag{1.10}$$

The corresponding dielectric constant is

$$\epsilon_q(R) = 1 + \frac{2g^2(R_0)}{12\pi^2} \frac{N_f}{2} \ln(R/R_0). \tag{1.11}$$

The quark contribution in QCD is just a replica of the electron contribution in QED, with extra color and flavor indices. The extra factor of $1/2$ in (1.11) comes from the normalization of the color generators T^a assigned to the external static charges.

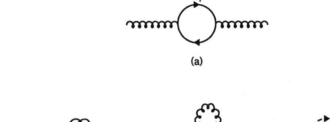

(a)

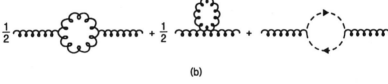

(b)

Fig. 1.2. Perturbative contribution to the gluon polarization tensor: (a) fermionic, (b) gluonic. The wiggly line denotes the gluon and the dashed line the ghost.

The gluon contribution (see Fig. 1.2) modifies the dielectric constant substantially. The result is

$$\epsilon_{q+G}(R) = 1 + \frac{g^2(R_0)}{12\pi^2}(N_f - \frac{33}{2})\ln(R/R_0) \tag{1.12}$$

where the first term in parentheses is the quark contribution given above and the second the gluon contribution. If the number of flavors is less than 17, the gluonic contribution overwhelms the quark contribution and flips the sign of the one-loop correction. This means that the strong coupling is now "anti-screened", that is, the coupling constant becomes smaller when the momentum is increased. Defining the beta function in QCD as we did for QED (cf. (1.8)), we can read off the lowest-order term in the perturbative expansion of the beta function

$$\beta(g) = \frac{g^3}{16\pi^2}(\frac{2}{3}N_f - 11).$$ (1.13)

We see that the first perturbative coefficient is negative for the relevant value of N_f, implying stability in the ultraviolet limit. This phenomenon, known as asymptotic freedom, distinguishes Yang-Mills theory from all other four-dimensional theories.

Asymptotic freedom is a quantum phenomenon, whose interpretation depends on the gauge choice. In Coulomb gauge ($\vec{\partial} \cdot \vec{A} = 0$), the one-loop contributions are shown in Fig. 1.3. Their respective contributions to the beta function are

$$\left(a + b + c + d\right) \sim \left(\frac{2N_f}{3} + 1 - 12 + 0\right).$$ (1.14)

The anti-screening is caused by Fig. 1.3c, which has no imaginary part, and corresponds to *magnetic* gluons produced by an *electric* source. In background covariant gauge with static magnetic fields, one may interpret this result as a consequence of a paramagnetic structure of the gluon modes. In the same gauge, but looking at the fluctuations around the instanton classical solution (see Chapter 2), asymptotic freedom may be understood as the dominance of the 12 instanton zero modes over the one Gaussian hard mode, resulting in the remarkable factor of $-11 = (1 - 12)$ in (1.13).

As we are mostly interested in the long-wavelength limit of QCD, an immediate problem arises: the perturbative estimate indicates that the coupling constant grows with distance, causing the perturbative analysis to break down at about 1 fm. Thus, the basic properties of the QCD vacuum and its hadronic excitations belong to the nonperturbative realm. Before we start our exposition of the nonperturbative approaches to QCD, we would like first to discuss the constraints imposed on the theory by the underlying exact and approximate symmetries.

1.2. Symmetries and Anomalies of QCD

1.2.1. *Symmetries*

We start with the exact symmetries of QCD. At the classical level, the theory is invariant under local $SU(3)_c$ gauge transformations. What

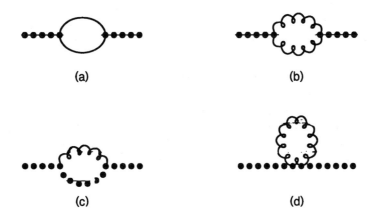

Fig. 1.3. Perturbative contribution to the gluon polarization tensor in Coulomb gauge: (a) fermionic, (b) magnetic, (c) electric-magnetic and (d) tadpole. The solid lines are quarks, the dotted lines scalar gluons and the wiggly lines transverse gluons.

this means is that the classical theory has infinitely many copies related to one another by a gauge transformation. In the process of quantizing the theory only one gauge copy has to be retained to avoid spurious divergences. This procedure is called gauge-fixing. In covariant gauges, naive gauge-fixing induces acausal effects in the theory. To fix this, one is forced to introduce auxiliary, anticommuting but spin-zero fields — the ghosts — whose sole role is to keep track of unitarity in perturbation theory. The gauge-fixed theory with ghosts has still some residual (global) symmetry. The latter is related to the original gauge structure and is usually referred to as the BRST (Becchi-Rouet-Stora-Tyutin) symmetry. Gauge-fixed QCD is BRST-invariant in perturbation theory.

At the non-perturbative level, the BRST gauge-fixing suffers from the Gribov problem, and may not be globally complete[2]. A global gauge fixing requires a good understanding of the fundamental domains in gauge theories [3], and may signal the need for a θ angle in the BRST formalism. Indeed, fundamental domains in a gauge theory allow for a multiple vacuum structure, with the possibility of tunneling through instanton barriers. Such transitions violate CP, and carry a θ dependence as we now discuss.

Due to the nontrivial structure of $SU(3)_c$, classical QCD has a class of topologically distinct configurations (instantons) that minimize the Euclidean action in the limit $N_f \to 0$. Each configuration carries an integral topological charge Q (see Chapter 2). The effective potential $\mathcal{V}[Q]$, viewed as a hypersurface in $Q - Q_\perp$, where $Q_\perp$ are all the degrees of freedom orthogonal to Q, is shown qualitatively in Fig. 1.4. The instanton configuration is just the field configuration that tunnels between two consecutive minima. It represents a large gauge transformation. The presence of the θ term in the Lagrangian tells us that in order to make the theory invariant under large gauge transformations, we have to include in the quantum description all copies of the theory labeled by their topological charge. In Euclidean space, the generating functional (or partition function) is

$$Z_{QCD}[\theta] = \sum_{Q=-\infty}^{+\infty} e^{iQ\theta} \ Z_{QCD}[Q]. \tag{1.15}$$

The θ parameter is dual to the topological charge Q, and plays the role of a complex chemical potential in Euclidean space (time odd). As mentioned above, $\theta \neq 0$ means that the quantum version of QCD breaks CP and thus T *

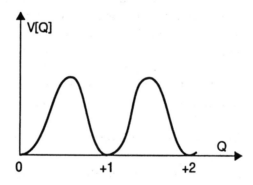

Fig. 1.4. Effective potential in QCD versus the topological charge Q.

The tree Lagrangian of QCD also exhibits a set of "essential approximate symmetries," by which we mean symmetries that are completely restored when one of the parameters of the theory is switched off. Let us look

---

*The conventional form of the BRST quantization yields a version of QCD that does not break CP [4].

at QCD in the *chiral limit*, where the quark masses are set equal to zero. This limit is a relevant limit for such light quarks as the up (u) and down (d) quarks whose masses are small compared with the scale associated with the strong interactions. The same limit will also be applied, whenever appropriate, to the strange (s) quark although its mass is comparable to the fundamental scale of QCD, $\Lambda_{QCD} \approx 100 - 200$ MeV. In this limit, we can rewrite the fermionic part of (1.1) as a sum of two decoupled terms

$$
\begin{aligned}
\mathcal{L} = & \ \bar{q}_R(i\partial_\mu\gamma^\mu - gA_\mu\gamma^\mu)q_R + \\
& \ \bar{q}_L(i\partial_\mu\gamma^\mu - gA_\mu\gamma^\mu)q_L
\end{aligned}
\tag{1.16}
$$

where the left and right chiral fields are defined by

$$
q_R = \frac{1+\gamma_5}{2}q, \qquad q_L = \frac{1-\gamma_5}{2}q.
\tag{1.17}
$$

Thus, the number of left-handed quarks and the number of right-handed quarks are separately conserved in the chiral limit. These conservation laws mean that the Lagrangian (1.16) is invariant under the following global transformations

$$
q_R^i \to A_{ij}q_R^j, \qquad q_L^i \to B_{ij}q_L^j
\tag{1.18}
$$

where A and B are $U(N_F)$ matrices, and N_F is the number of massless flavors. Instead of using the left and right bases, we could use $L+R$ and $L-R$ bases, called the vector and axial bases, respectively. Therefore massless QCD is invariant under global chiral $\left(U_V(1) \times SU_V(N_F)\right) \times \left(U_A(1) \times SU_A(N_F)\right)$ transformations. The corresponding Noether currents are

$$
V_\mu^a = \bar{q}\gamma_\mu t^a q, \qquad A_\mu^a = \bar{q}\gamma_\mu\gamma_5 t^a q
\tag{1.19}
$$

where the index a runs from 0 to N_F, with $t^0 = \mathbf{1}$ and the rest of the t's are Lie algebra-valued.

Massless QCD has no dimensional parameters. Thus the theory is also scale-invariant,

$$
A_\mu \to \lambda A_\mu(\lambda x), \qquad q(x) \to \lambda^{\frac{3}{2}}q(\lambda x).
\tag{1.20}
$$

The corresponding Noether current (dilatational current) reads

$$
j_\mu^D = x^\nu \Theta_{\nu\mu}
\tag{1.21}
$$

where the energy-momentum tensor for QCD without quarks is

$$\Theta_{\mu\nu} = -G_{\mu\alpha}G_\nu^\alpha + \frac{1}{4}g_{\mu\nu}G_{\alpha\beta}G^{\alpha\beta}. \tag{1.22}$$

In the classical theory, the tensor (1.22) is traceless, so the dilatational current is conserved. Scale invariance is part of a broader group of symmetry transformations, called special conformal transformations. The Lagrangian (1.16) is invariant under the following set of special conformal transformations,

$$x_\mu \quad \rightarrow \quad x_\mu^{s.c.} = \frac{x_\mu - c_\mu x^2}{1 - 2c \cdot x + c^2 x^2} \tag{1.23}$$

where c_μ is an arbitrary four-constant. The corresponding Noether currents are

$$j_\mu^{(s.c.)\,\alpha} = x^2 \Theta_\mu^\alpha - 2x^\alpha x^\nu \Theta_{\nu\mu}. \tag{1.24}$$

They are all conserved classically since the energy momentum tensor is traceless.

1.2.2. *Anomalies*

At the quantum level, the situation is more subtle. The classical conservation equations may be upset by quantum effects (loop contributions), giving rise to anomalies. QCD suffers from two anomalies: an abelian axial $(U_A(1))$ anomaly and a trace anomaly. As a result, the axial $U_A(1)$ current and the five conformal currents $j_\mu^D, j_\mu^{(s.c.)\alpha}(\alpha = 0,1,2,3)$ are not conserved at the quantum level.

What causes these anomalies to appear? The answer lies in the necessity to regulate the theory in the ultraviolet. Every regularization procedure introduces a scale. In lattice QCD simulations, the scale is the size of the lattice; in Pauli-Villars' regularization, it is the inverse of the large regulator mass; and in dimensional regularization, it is the scale $1/\mu$ introduced to keep the coupling constant dimensionless in dimensions other than four. Since every loop diagram introduces this scale, it is not surprising that the theory breaks scale invariance through loops. This explains the nonconservation of the dilatational and special conformal currents.

The axial anomaly follows from the conflict between gauge invariance and chiral invariance in the process of regulating the theory at the quantum level. This is manifest for instance in Pauli-Villars' regularization. Since the concept of gauge invariance is held sacred in gauge theories, gauge invariance is imposed in perturbation theory at the expense of chiral invariance. It is actually arbitrary as to which chiral symmetry one chooses to sacrifice: left, right, vector or axial. Since the vector symmetry corresponds to the conservation of the number of all fermions (baryon number) and since we want the latter to be conserved at low energy, the axial symmetry is chosen to be sacrificed in the process of regularization, thus generating the axial anomaly.

We should mention that both the axial and scale anomalies are present in QED for the same reasons. What makes the axial anomaly in QCD special is the presence of instantons making the effects of the anomaly physically manifest in the η' mass problem (Chapter 2). Before discussing these issues, we shall first elaborate on the above general statements.

1.2.3. *Axial anomaly*

The axial anomaly has a simple interpretation in the Hamiltonian formulation of the theory[5]. Consider the Dirac sea of massless quarks in the background of some external field $A^\mu(x)_{ext}$. Axial symmetry forbids the creation of pairs of quarks and anti-quarks of opposite chiralities, but allows a global asymmetric shift in the overall spectrum as shown in Fig. 1.5. Chiral symmetry requires that the change of the chiral charge at the surface of the Dirac sea must be compensated by the corresponding change of the chiral charge at the bottom of the Dirac sea. In continuum theory, the Dirac sea is bottomless, and the *uncompensated* change of the chiral charge in the continuum manifests itself as an anomaly.

Let us see how this scenario works in a quantitative way. For simplicity, we shall consider the massless case, and specialize to a colored magnetic field constant in color and in space, $H^{ai} = \delta^{a8}\delta^{i3}H$. H^{ai} is nonvanishing along the color hypercharge and the z-direction. Now, let us switch on a parallel electric field during the time interval τ,

$$\bar{a}_\mu(t) = \delta_{\mu 2}\bar{a}(t) \begin{cases} 0 & \text{for } t < 0 \\ a_\tau T^8 & \text{for } t > \tau. \end{cases} \qquad (1.25)$$

For $t < 0$, only the magnetic field is present. The quark trajectories fall into Landau orbits[6] in the (x, y) plane, with free motion along the z direction.

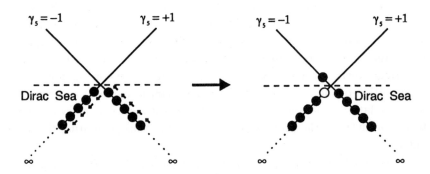

Fig. 1.5.   The asymmetric, global shift of all the energy levels corresponding to definite chirality.

The spectrum is

$$\omega_n = (k_z^2 + (2n+1)gH \pm gH)^{\frac{1}{2}}. \tag{1.26}$$

We note the degeneracy for spin projection $S = \pm 1/2$

$$\omega_n(S = +\frac{1}{2}) = \omega_{n+1}(S = -\frac{1}{2}). \tag{1.27}$$

Clearly the $n = 0$ mode with the quark spin parallel to the magnetic field behaves as a free mode. Any small colored electric field can push this state up or down! The vacuum state at $t < 0$ is the Dirac sea bunched into Landau orbits with infinite degeneracy. At $t > \tau$, under the influence of the electric field, the energies of the positive chirality states decrease by $g\bar{a}_\tau$, and the energies of the negative chirality states increase by $g\bar{a}_\tau$ (*cf.* Fig. 1.5). The associated Dirac wave functions are given in Appendix A_I.

To assess this effect, we need to find out how to count states. For that, we note that due to the matrix structure of the color charge T^8, the above Landau problem is a triplicate of a simpler problem: charged fermions in (1+1)-QED subject to an external *electric* field $\bar{a}$ [†] The triplicate charges are: $\frac{1}{2\sqrt{3}}, \frac{1}{2\sqrt{3}}, -\frac{1}{\sqrt{3}}$. In 1+1-dimensions, the total number of particle-hole pairs induced by the electric field $\bar{a}_\tau$ into or out of the vacuum is equal to the energy shift divided by the spacing between the energy levels, *i.e.*,

$$N_{(1+1)} = \frac{g\bar{a}_\tau}{2\pi/L_z}. \tag{1.28}$$

[†]Modulo the replacement of the four-dimensional Dirac γ_5 by a two-dimensional counterpart, and suitable adjustments for the normalization of the generators.

In 3+1-dimensions, we have to take into account the degeneracy of the Landau orbits, the normalization of the hypercharge generator and the fact that each pair induces a chirality change of -2. Note that due to the degeneracy (1.27) all other states except the discussed one do not contribute to the change of the chiral charge. The number of Landau orbits in the (x, y) plane is restricted by $0 < x_0 = k_y/(gH) < L_x$, so

$$0 < n_y < \frac{L_x L_y}{a_0^2} \tag{1.29}$$

where $a_0 = \sqrt{2\pi}/\omega_c$ is the size of the Landau orbit and $\omega_c = \sqrt{gH}$ is the cyclotron frequency. Therefore the total change in the chiral charge is equal to

$$
\begin{aligned}
\Delta Q_5 = -2N_{(3+1)} &= -2 \cdot \text{Tr}(T^8 T^8) \frac{L_x L_y gH}{2\pi} N_{(1+1)} \\
&= -\frac{1}{2} \frac{g^2}{2\pi^2} L_x L_y L_z H a_\tau.
\end{aligned}
\tag{1.30}
$$

This result can be rewritten in a covariant form using Lorentz and gauge invariances,

$$\Delta Q_5 = \frac{1}{2} \frac{g^2}{8\pi^2} \int d^3 x \int_0^\tau F_{\mu\nu}^{(B)} \tilde{F}^{(B)\mu\nu} \tag{1.31}$$

where $F_{\mu\nu}^{(B)}$ is the field strength associated to the constant magnetic field, and $\tilde{F}_{\mu\nu} = \frac{1}{2}\epsilon_{\mu\nu\alpha\beta}F^{\alpha\beta}$ is its dual. Note that the above calculation gives also the axial anomaly in QED if one replaces the hypercharge matrices by a single charge e

$$\Delta Q_5^{QED} = \frac{e^2}{8\pi^2} \int d^3 x \int_0^\tau F \cdot \tilde{F}. \tag{1.32}$$

By gauge invariance and taking into account the flavor content involved, we can generalize the form of the anomaly to the full nonabelian form

$$\Delta Q_5 = \frac{N_f}{2} \frac{g^2}{8\pi^2} \int d^3 x \int_0^\tau dt G_{\mu\nu}^a \tilde{G}^{a\mu\nu} \tag{1.33}$$

with

$$\tilde{G}_{\mu\nu} = \frac{1}{2}\epsilon_{\mu\nu\alpha\beta}G^{\alpha\beta}.$$

Finally, we can rewrite the above in a covariant form, assuming that three-currents vanish at infinity

$$\partial_\mu j_5^\mu = \frac{N_f}{2} \frac{g^2}{8\pi^2} \tilde{G} \cdot G + \sum_f m_f \bar{q}_f q_f \tag{1.34}$$

where we have restored the explicit breaking of chiral symmetry by non-zero quark masses.

What is the Euclidean interpretation of the above result? In Euclidean space, the anomaly relation could be rewritten[7] (we present the derivation of this result in Chapter 6)

$$\sum_n (\phi_n^\dagger \gamma_5 \phi_n)_{\text{reg}} = \frac{g^2}{32\pi^2} G \cdot \tilde{G} \tag{1.35}$$

where ϕ_n are now the eigenstates of the QCD Dirac operator in Euclidean space and the divergent sum is regularized in some scheme (heat kernel, Pauli-Villars, ...). Let us now integrate both sides of eq.(1.35) over the four-dimensional Euclidean space. Since

$$\sum_n \phi_n^\dagger \gamma_5 \phi_n = \sum_n (\phi_n^{\dagger R} \phi_n^R - \phi_n^{\dagger L} \phi_n^L) \tag{1.36}$$

and γ_5 anticommutes with the Dirac operator $\gamma \cdot D(A)$, the left non-zero modes are exactly compensated by the right non-zero modes, so the sum receives no contribution from the fermionic non-zero modes. This is, however, not the case for the *normalizable* zero modes. The left-hand side of the integrated anomaly relation (1.35) counts the difference between the number of right- and left-handed normalizable zero modes of the Dirac operator. Therefore the right-hand side has to be an integer. We will see in the next chapter that the integrated right-hand side of the anomaly represents the topological charge of the $SU(3)$ gauge field defined by

$$n = \frac{g^2}{32\pi^2} \int d^4x\, G \cdot \tilde{G}. \tag{1.37}$$

The surprisingly simple form of the integrated anomaly is known as the Atiyah-Singer index theorem:[8]

$$n_R - n_L = n. \tag{1.38}$$

It relates the difference of the right- and left-handed normalizable fermionic zero modes of the Dirac operator in some background gauge field A, to the topological charge of this background gauge field.

1.2.4. *Trace anomaly*

Let us now study the anomalous contribution to the dilatational and special conformal currents. Consider for simplicity the Yang-Mills action in $D = 4 - \epsilon$ dimensions with a metric $g^{\mu\nu}$

$$S_{YM} = -\frac{1}{4} \int d^D x \sqrt{-\det g} g^{\mu\sigma} g^{\nu\tau} G_{\mu\nu} G_{\sigma\tau}. \tag{1.39}$$

The gauge-invariant minimal energy-momentum tensor $\Theta^{\mu\nu}$ may be defined by

$$\Theta_{\mu\nu} = -\frac{2}{\sqrt{-\det g}} \frac{\delta S}{\delta g^{\mu\nu}}. \tag{1.40}$$

Since

$$\delta \sqrt{-\det g} = \frac{1}{2} \sqrt{-\det g} g_{\alpha\beta} \delta g^{\alpha\beta}, \tag{1.41}$$

we conclude that

$$\Theta_{\mu\nu} = -g_{\mu\nu} \mathcal{L}_{YM} - G_{\mu\alpha} G_\nu^\alpha \tag{1.42}$$

with its trace given by

$$\Theta_\mu^\mu = \epsilon \mathcal{L}_{YM}. \tag{1.43}$$

Classically, the left-hand side vanishes as $\epsilon \to 0$. Quantum mechanically, the Yang-Mills part needs a proper regularization as it involves fields at the same space-time points which are ill-defined. A careful analysis[9] shows that the renormalized action is

$$\mathcal{L}_{YM} = -\frac{1}{\epsilon} \frac{2\beta(g)}{g} \left(-\frac{1}{4} G_{\mu\nu} G^{\mu\nu} \right)_R \tag{1.44}$$

where $\beta(g)$ is the beta-function. In terms of (1.44) the trace of the energy momentum tensor for pure Yang-Mills field is

$$\Theta_\mu^\mu = \frac{2\beta(g)}{g} \left(\frac{1}{4} G_{\mu\nu} G^{\mu\nu} \right)_R. \tag{1.45}$$

This result can be generalized to the full QCD action. The answer in this case is

$$\Theta_\mu^\mu = \frac{2\beta(g)}{g} \left(\frac{1}{4} G_{\mu\nu} G^{\mu\nu} \right)_R + \sum_f m_f (1 + \gamma_f) \left(\bar{q}_f q_f \right)_R \tag{1.46}$$

where γ_f is the anomalous dimension for the quark of flavor f. This relation is referred to as the trace anomaly. All the quantities involved there are renormalized. Note that the trace anomaly receives contribution from all multi-loop diagrams since $\beta(g)$ represents an infinite series in g^2. It is to be contrasted with the case of the axial anomaly, where there appear to be no n-loop contributions to the anomaly for $n > 1$. However there may be a caveat in this comparison. The proof of the one-loop saturation of the axial anomaly is made at the operator level, whereas the trace anomaly relation (1.46) is obtained for the matrix elements of the operators. For possible complications in the comparison between these two approaches, we refer to the excellent review by Shifman[10]. In the vacuum, Lorentz invariance implies that the zero-point fluctuations sum up to

$$
\begin{aligned}
\mathcal{E}_{vac} &= \frac{1}{4}\langle 0|\Theta_\mu^\mu|0\rangle = \frac{\beta(g)}{8g}\langle 0|G^2|0\rangle \\
&+ \sum_f \frac{1}{4}m_f(1+\gamma_f)\langle 0|\bar{q}_f q_f|0\rangle.
\end{aligned}
\tag{1.47}
$$

The renormalized quantities are implied in the matrix elements, so perturbative subtractions are understood. In the chiral limit, no subtraction is required for $\bar{q}q$. The subtraction in G^2 is understood in the sense of normal ordering around the trivial vacuum. Thus the gluon condensate $\langle 0|G^2|0\rangle$ depends on the choice of the renormalization point. Around $\mu^2 \sim 1$ GeV^2 this dependence is usually *assumed* to be weak.

Asymptotic freedom implies that $\beta(g)$ is negative. Confinement means that the vacuum energy is below zero (nonperturbative). If we were to ignore the effects of the quarks in (1.47), then we would conclude that in the vacuum $\langle 0|G^2|0\rangle = \langle 0|B^2|0\rangle/2 = -\langle 0|E^2|0\rangle/2$ is positive. The vacuum displays an excess of magnetic fluctuations and a deficiency of electric fluctuations.

Thus far we have been discussing only the anomalies associated with the Lagrangian (1.1) of the QCD sector. If we couple external (*e.g.* electroweak) currents to our fields, then new anomalies, abelian and nonabelian, can occur and even the form of the existing anomalies can get modified. We do not discuss these issues now, since some of them will be taken up later in great detail in the course of this book.

Let us finally mention that there is no conclusive evidence that even one of the quarks is massless. Neither is there any reason to conclude that one of them (say, the up quark) cannot be massless at some hadronic scale. We will take the point of view that at the scale we are interested in, chiral symmetries are approximate symmetries, and all chiral currents are not conserved due to the explicit masses of the quarks. The same concerns the trace anomaly which also receives explicit contribution from the mass terms.

1.3. Symmetries of the QCD Vacuum

The specification of the Lagrangian in terms of its symmetries does not completely define the structure of the theory. It can happen that the ground state of the theory does not respect some of the symmetries of the Lagrangian. In such a case we say that the symmetry which is not respected by the vacuum is "spontaneously broken." Looking at the spectrum of hadrons, we immediately come to the conclusion that this is likely to be what happens in strong interactions. First of all, the elementary fields of the theory, *i.e.*, the quarks and gluons are not observed explicitly in nature. Only color singlet ("white") states are observed. There are infinitely many possibilities for the formation of white states. Color singlets could be built from gluons only (glueballs) or from single-quark-single-anti-quark pairs (mesons) or from triplets of quarks (baryons) or from multiples of quark-anti-quark pairs (exotics) or from multiples of quark triplets (composites) or from combinations of pairs and triplets etc. The experimental signals correspond overwhelmingly to baryons and mesons only. The physics behind this superselection rule is not really understood, although the large N_c arguments to be discussed in Chapter 3 provide an interesting suggestion. We also do not understand the physics of how the quarks cluster into the observed meson and baryon states, *i.e.*, the *physical mechanism* of confinement. The commonly accepted belief is that this is an inherently complex nonperturbative mechanism. While many fascinating and ingenious models of confinement have been proposed, its mysteries still remain unexposed. We refer to the work by Greensite for a comprehensive analysis of the various confining mechanisms[11].

One thing the experimental data tell us unequivocally is that the QCD vacuum state is not symmetric under left and right chiral transformations. If it were symmetric, the quarks would more or less retain their current masses and as a consequence, the baryons would be very light, definitely lighter than pions. We would like to understand why the symmetry is "hidden" in the vacuum and which symmetry is thereby spontaneously broken. Here is a list of observations that may help in clarifying these issues.

1.3.1. *Goldstone theorem*

The relation between the spontaneous breakdown of a global continuous symmetry and the presence of massless, spinless particles was first observed by Goldstone[12]. Let us look at the case of complex fields ϕ_i ($i = 1, N$) described by the general Lagrangian density

$$\mathcal{L} = \frac{1}{2} \partial^\mu \phi^{\star i} \partial_\mu \phi^i - \mathcal{V}(\phi^{\star i} \phi^i). \tag{1.48}$$

Let us assume that the Lagrangian is invariant under some continuous global symmetry transformation

$$\vec{\phi} \rightarrow \exp(it^a \omega^a)\vec{\phi} \tag{1.49}$$

where t^a are the hermitian generators of the symmetry with $a = 1, ..., p$. Let us now consider the minimum of the potential $\mathcal{V}$ characterized by $\mathcal{V}'(\vec{\phi}_{vac}) = 0$ where the prime denotes derivative with respect to the field. The spontaneous breakdown of the symmetry means that the state $\vec{\phi}_{vac}$ is not invariant under the full group of continuous symmetries enjoyed by the Lagrangian. In other words, we can split the generators into two classes: those that annihilate the vacuum (respect the symmetries) and those that change the vacuum (break the symmetries):

$$\begin{aligned} t^a \vec{\phi}_{vac} &= 0 \quad (a = 1, ..., k), \\ t^a \vec{\phi}_{vac} &\neq 0 \quad (a = k+1, ..., p). \end{aligned} \tag{1.50}$$

This implies that the condition $\mathcal{V} = $ minimum is fulfilled on a multi-dimensional hypersurface spanned by the $(p - k)$ broken generators. The excitations along this surface will be massless and those orthogonal to this surface will be massive. Therefore the spectrum of the theory has to include $(p - k)$ massless spinless particles, one for each broken generator. Their quantum numbers follow from the quantum charges associated to the corresponding Noether currents. An alternative to the Goldstone theorem is discussed in Appendix A_{II}.

1.3.2. *Vafa-Witten theorem*

Vafa and Witten have argued[13] – modulo some plausible assumptions – that vector-like theories (like QCD) cannot spontaneously break vector symmetries (*e.g.* baryon number and isospin). Let $J_\mu^a(x)$ be an operator with conserved quantum numbers carried by the quarks such as flavor or baryon number. Let us assume that the masses of the lightest quarks (u,d) are equal but *non-zero*. In this case the axial symmetry is explicitly broken due to the non-zero masses of the quarks, but the vector symmetry is maintained. Following Vafa and Witten, we will only consider the flavor-changing currents, Fig. 1.6a and disregard the exchange graphs such as Fig. 1.6b.

Let us consider the Euclidean current-current correlator of the flavor current $J_\mu^a = \bar{q}\gamma_\mu T^a q$ in the vacuum

$$\Pi(x)_{\mu\nu}^{ab} = \langle J_\mu^a(x) J_\nu^b(0) \rangle = -\langle \operatorname{Tr}\left(\gamma_\mu T^a S^A(x,0;m)\gamma_\nu T^b S^A(0,x;m)\right)\rangle \tag{1.51}$$

where $S^A(x,0;m)$ is the propagator from x to 0 of a quark of mass m in a fixed gauge configuration A. Typical contributions to (1.51) are shown in Fig. 1.6. For vanishing vacuum angle θ, (see eq.(1.1)), the averaging involves a positive definite measure in Euclidean space. Viewed as a matrix in space, spin and flavor spaces, the norm of the fermion propagator in a fixed background A is bounded from above. The fermion operator is bounded[‡]

$$|S^A(x,0;m)| \le \alpha e^{-\beta x} \tag{1.52}$$

where α and β are A-independent. Combining (1.51) with (1.52) and making use of the Cauchy-Schwartz inequality, we obtain

$$|\Pi(x)| \le \alpha^2 \operatorname{Tr} T^2 e^{-2\beta x}. \tag{1.53}$$

The exponential fall-off implies the nonexistence of massless particles in the vector correlation functions with conserved fermion quantum numbers. This shows that the vector symmetry is not spontaneously broken.

[‡]For the subtleties in the proof of the boundedness of the fermionic propagator we refer to the original paper.

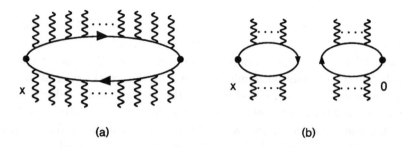

(a)   (b)

Fig. 1.6.　Typical contribution to the current-current correlator with conserved fermion quantum numbers in a fixed background.

1.3.3. *Anomaly matching conditions*

So far, we have been interested in the internal anomalies of QCD, *i.e.*, in the axial $U(1)$ anomaly and the dilatational anomaly. If the quarks couple to external fields, such as the electroweak fields, new forms of anomalies, both abelian and nonabelian, can arise as a consequence of which the structure of the internal anomalies may also be affected. We will postpone the detailed discussion of these anomalies to Chapter 6. For the moment we consider only the most celebrated one, *i.e.*, the one responsible for the decay of the pion into two photons. Let us consider the axial $SU_A(2)$ current corresponding to the π^0

$$A_\mu^3 = \bar{u}\gamma_\mu\gamma_5 u - \bar{d}\gamma_\mu\gamma_5 d. \tag{1.54}$$

This current is anomaly-free in QCD.

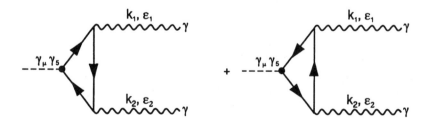

Fig. 1.7.　Triangle anomaly.

If we allow quarks to couple to photons, the matrix element of the divergence of this current between the vacuum state and two photons with momenta k_1, k_2 and polarizations ϵ_1, ϵ_2 is non-zero due to the triangle graph Fig. 1.7, and reads

$$\langle 0|\partial^\mu A_\mu^3|\gamma\gamma\rangle = \frac{\alpha}{2\pi}N_c(Q_u^2 - Q_d^2)\langle 0|F_{\mu\nu}\tilde{F}^{\mu\nu}|k_1, \epsilon_1; k_2, \epsilon_2\rangle$$

$$= -\frac{\alpha}{\pi}N_c(Q_u^2 - Q_d^2)\epsilon^{\mu\nu\alpha\beta}(k_\mu^1\epsilon_\nu^1 - k_\nu^1\epsilon_\mu^1)(k_\alpha^2\epsilon_\beta^2 - k_\beta^2\epsilon_\alpha^2). \qquad (1.55)$$

It is easy to understand the form of eq.(1.55). This is simply the difference between two electromagnetic anomalies (1.32) with the appropriate fractional charges Q of u and d quarks. We could immediately infer from (1.55) the form of the matrix element $\langle 0|A_\mu^3|\gamma\gamma\rangle$

$$\langle 0|A_\mu^3|\gamma\gamma\rangle = -\frac{iq_\mu}{q^2}\frac{4\alpha}{\pi}N_c(Q_u^2 - Q_d^2)\epsilon^{\alpha\beta\gamma\delta}k_\alpha^1\epsilon_\beta^1 k_\gamma^2\epsilon_\delta^2 + \dots \qquad (1.56)$$

where $q = k_1 + k_2$ and the ellipsis stands for less singular terms. Taking a derivative corresponds to multiplying the matrix element (1.56) by iq_μ. We would like to stress that the above result is exact, *i.e.*, no low-energy regime has been assumed. The pole is simply the remnant of the massless quark.

't Hooft has formulated a very powerful condition relating fundamental theories with theories in which particles are bound states of the fundamental constituents. This condition, known as anomaly matching condition, states that a composite particle has to reproduce exactly the anomaly present in the fundamental theory. In other words, the fundamental anomalies and the anomalies in the composite theory must match.

If we now assume confinement, the 't Hooft condition says that in the hadronic (composite) world, there must exist a massless, colorless object which couples to the axial current and photons and reproduces exactly the pole in the anomaly (1.56). See Fig. 1.8.

The anomaly matching condition does not, however, specify the nature of this particle: it could be either a baryon or a meson. Let us consider massless baryons first. In the world with massless baryons and massive pions, the $SU_A(N_F) \times SU_V(N_F)$ symmetry has to be realized linearly, since the Vafa-Witten theorem forbids the vector symmetry to be spontaneously broken. If one assumes that the only contribution to the pole comes from the octet of baryons circulating through the triangle graph Fig. 1.9a, a

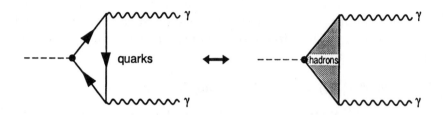

Fig. 1.8. Anomaly matching condition.

direct calculation shows that the pole vanishes[10]. Therefore the baryonic scenario is not realized in nature. The exact proof of this statement should involve calculation of the contributions from *all* baryons, and not just from the "lowest" octet. On the other hand, in the world with massless mesons (pions), the anomaly condition could be easily fulfilled. To ensure this, we do not need to circulate massless pions through the loop. One only needs to assume that there is a relation between the axial current and the pion (see Chapter 4)

$$\langle 0|A_\mu^3|\pi^0\rangle = i f_\pi q_\mu \qquad (1.57)$$

relating the axial current to the decay constant of the pion f_π into muon and neutrino (*cf.* Fig. 1.9(b)). Thus, modulo the restrictions mentioned above for the baryonic contributions to the anomaly, confinement and the anomaly-matching condition lead unambiguously to the world with massless mesons coupled to the axial current.

In summary, the Vafa-Witten argument, the anomaly matching condition, and the Goldstone theorem tell us that chiral symmetry is spontaneously broken to its diagonal subgroup in the QCD vacuum. The triplet (or octet in the case of flavor $SU(3)$) bosons constitute the Goldstone particle multiplet. The QCD vacuum state is characterized by

$$Q_V^a|0\rangle = 0, \quad Q_A^a|0\rangle \neq 0 \qquad (1.58)$$

where Q_V, Q_A are charges corresponding to the vector and axial currents (1.19) respectively.

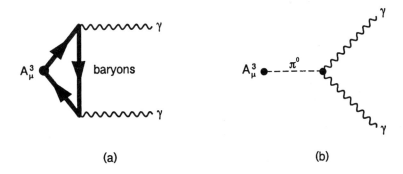

(a) **(b)**

Fig. 1.9. Two possible scenarios for the saturation of the fundamental anomaly - (a) baryonic, (b) mesonic.

1.4. Banks-Casher Relation

The physics by which chiral symmetry is spontaneously broken in the vacuum is still at the level of conjectures. While much about the consequences of spontaneous breakdown of chiral symmetry has been learned as we shall discuss throughout this book, little is known about this mechanism from first principles. This aside, it is largely believed that the spontaneous breakdown of chiral symmetry in the vacuum is followed by a quantitative rearrangement of the Dirac spectrum near zero virtuality.

1.4.1. *Dirac spectral density*

To relate the properties of the Dirac spectrum to the chiral order parameter, consider the quark propagator in an external background field A. In four Euclidean space dimensions,

$$S(x, y, A) = \langle 0|q(x)\bar{q}(y)|0\rangle_A = \langle 0|(-i\not{D}(A) - im)^{-1}|0\rangle_A. \qquad (1.59)$$

The spectral representation for the propagator gives

$$S(x, y, A) = \sum \frac{\phi_n(x)\phi_n^\dagger(y)}{-\lambda_n - im} \qquad (1.60)$$

where ϕ_n, λ_n are eigenvectors and eigenvalues of the Dirac equation

$$i\not{D}(A)\phi_n(x) = \lambda_n\phi_n(x). \qquad (1.61)$$

Thus the fermion condensate in Euclidean space is

$$\langle q^\dagger q \rangle = -\lim_{x \to y} \langle\langle q(x)q^\dagger(y)\rangle\rangle = -\langle\langle \text{Tr}\, S(x,x,A)\rangle\rangle \tag{1.62}$$

where $\langle\langle \ldots \rangle\rangle$ denotes the averaging over the gluonic configurations A using the QCD action. In the limit where the four-volume V_4 goes to infinity, the spectrum becomes dense and we may replace the sum over the states by an integration over the mean spectral density

$$\nu(\lambda) = \langle\langle \sum_n \delta(\lambda - \lambda_n) \rangle\rangle. \tag{1.63}$$

In terms of (1.63) the fermion condensate (1.62) becomes

$$\langle q^\dagger q \rangle = -\langle\langle \text{Tr}(-i\slashed{D}(A) - im)^{-1}\rangle\rangle = \frac{1}{V_4} \int d\lambda \frac{\nu(\lambda)}{\lambda + im}. \tag{1.64}$$

In the chiral limit, $m \to 0$, the Dirac operator $\slashed{D}$ anticommutes with γ_5, so the non-zero eigenvalues come in pairs $(\lambda, -\lambda)$ and the spectral function is symmetric. Inserting

$$\lim_{m \to 0^+} \frac{1}{\lambda \pm im} = \text{P}\frac{1}{\lambda} \mp i\pi\delta(\lambda) \tag{1.65}$$

into (1.64) and using the fact that the principle-value part drops out as the spectrum is λ-even, we obtain

$$\langle q^\dagger q \rangle = -\frac{i\pi}{V_4}\nu(0). \tag{1.66}$$

A Wick rotation to Minkowski space $(q^\dagger, q) \to (i\bar{q}, q)$ gives[§]

$$\langle \bar{q}q \rangle = -\frac{\pi}{V_4}\nu(0). \tag{1.67}$$

This relation was first derived by Banks and Casher[14]. It is very important that the chiral limit $m \to 0$ is taken after the thermodynamical limit $V_4 \to \infty$, for otherwise the result would be zero. The spontaneous breakdown of a continuous symmetry cannot take place in finite volumes, unless the

---

[§]We note that the quark condensate is complex in Euclidean space and real in Minkowski space.

condition $m\langle \bar{q}q \rangle V_4 >> 1$ is fulfilled. The result (1.67) states that in vector-like theories with chiral symmetric (even) spectra, the *quark condensate* is related to the mean spectral density at zero virtuality ($\lambda = 0$).

We could imagine two scenarios for the QCD ground state as depicted in Fig. 1.10. Figure 1.10a exhibits a mean eigenvalue distribution at zero virtuality, thus a nonvanishing quark condensate. This is typical of a ground state with a spontaneous breakdown of chiral symmetry and is likely to be observed in QCD. Figure 1.10b shows a dip at zero virtuality, thus a vanishing quark condensate. This is typical of a state with unbroken chiral symmetry. In fact, the free quark case shows that $\nu(\lambda) \sim \lambda^3$ and vanishes at $\lambda = 0$. This may occur in QCD at high temperature (see Chapter 11). Qualitative spectra of the type shown in Fig. 1.10 have been observed in instanton approaches to the QCD vacuum (see Chapter 2) for high (a) and low (b) instanton densities.

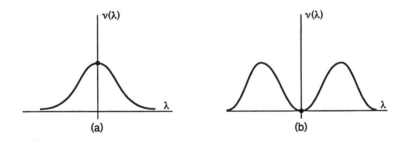

Fig. 1.10. Two possible behaviors of the spectral function of the Dirac operator.

What causes the mean spectral density $\nu(\lambda)$ to be nonvanishing at zero virtuality? A possible answer is delocalization of the quark modes around zero in the thermodynamic limit. The delocalization is caused by strong correlations that randomize the Dirac spectrum near zero, triggering a huge accumulation at zero. As $V_4 \to \infty$ the number of eigenvalues grows as V_4 as opposed to $V_4^{1/4}$ in the free case[15]. A simple way to realize this idea is to use instanton configurations as we will discuss in Chapter 2.

In a way, the Banks-Casher relation (1.67) is reminiscent of the conductivity in metals, where the latter is proportional to the density of states at the Fermi surface. Indeed, it follows from the Kubo formula that the d.c. conductivity σ relates to the density of states at the Fermi level, $\rho(E_F)$, through

$$\sigma = e^2 \, \mathbf{D} \, \rho(E_F) \qquad (1.68)$$

where e is the electron charge and $\mathbf{D}$ the diffusion constant. This result
(1.68) is reminiscent of (1.67) with the identification of $\sigma/e^2\mathbf{D}$ with $\langle\bar{q}q\rangle$,
suggesting that the quarks in the QCD vacuum are in a diffusive state. In
fact, it was also suggested by Banks and Casher[14] that the spontaneous
breaking of chiral symmetry in four dimensions is the result of a diffusive
walk by the quarks that is forced to be quasi-one-dimensional by the strong
gauge forces. This interpretation is natural in the first quantized version of
QCD, and provides an alternative to the one discussed above.

1.4.2. *Vafa-Witten theorem revisited*

It is interesting to discuss Vafa-Witten theorem in light of the above
result. Let u and d be two light quarks with masses m_u and m_d. Vafa
and Witten[13] have argued that in vector-like gauge theories such as QCD,
$\langle u^\dagger u - d^\dagger d\rangle \to 0$ as $m_u \to m_d$. In other words, the QCD ground state does
not spontaneously break isospin symmetry. This can be understood simply
using the Banks-Casher argument, as follows. Using (1.64), we have

$$\langle u^\dagger u - d^\dagger d\rangle = \frac{1}{V_4} \int d\lambda\, \nu(\lambda) \left(\frac{1}{\lambda + im_u} - \frac{1}{\lambda + im_d} \right). \qquad (1.69)$$

Since the maximum of $|1/(\lambda+im)|$ occurs at zero virtuality and furthermore
$\nu(\lambda)$ is positive and normalized to N, we have

$$|\langle u^\dagger u - d^\dagger d\rangle| \le \frac{|m_d - m_u|}{m_u m_d} \frac{1}{V_4} \int d\lambda\nu(\lambda) \le \frac{|m_d - m_u|}{m_u m_d} \frac{N}{V_4}. \qquad (1.70)$$

In the thermodynamic limit, N/V_4 is fixed and so $\langle u^\dagger u - d^\dagger d\rangle \to 0$ as
$m_u \to m_d$, as announced.

Note that in the above argument, the u and d quark masses are equal
but nonzero. What happens when the two masses vanish is not clear.
Indeed, Fig. 1.11 shows a possible scenario of the effective potential along
the $\phi = \langle u^\dagger u - d^\dagger d\rangle$ direction as the quark mass $m_u = m_d = m$ is lowered,
say from 1 MeV, to 1 eV, to 0. Such a behavior would imply the existence of
an accidental symmetry and a spontaneous breakdown of isospin symmetry.
This case was not considered by Vafa and Witten as accidental degeneracies
were ruled out *ab initio*.

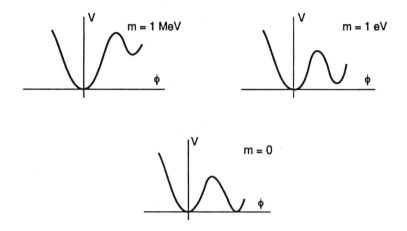

Fig. 1.11. Qualitative behavior of the effective potential versus $\phi = \langle u^\dagger u - d^\dagger d \rangle$.

1.4.3. *Universal properties of the QCD Dirac spectrum*

(i) Results of effective field theory

Is it possible that the spectrum of QCD near zero virtuality follows from symmetries alone? The Banks-Casher relation[14] and newly discovered sum rules by Leutwyler and Smilga[15] suggest so. Indeed, the latter authors have suggested that the QCD spectrum to order m^2 follows from an effective field theory solely based on symmetries in the static limit. Specifically, they have shown that if λ_n denote the set of eigenvalues (1.61), then

$$\frac{1}{V_4^2} \sum_n \frac{1}{\lambda_n^2} = \frac{\Sigma^2}{4N_F} \qquad (1.71)$$

where $\Sigma = \langle \bar{q}q \rangle$. This sum rule and others follow from the static partition function

$$Z^{eff}[m] = \int [dU]\, e^{m V_4 \Sigma \mathrm{Re}\, \mathrm{Tr} U} \langle \prod_F \prod_n \left(\lambda_n^2 + m_F^2 \right) \rangle \qquad (1.72)$$

where the integration $[dU]$ is over the coset G/H which is $SU(N_F)$-valued. The last relation corresponds to the partition function of four-dimensional

Euclidean QCD. The sum rules follow by differentiation with respect to the *current* mass m. As will be discussed in Chapters 5 and 6, (1.72) is none other than the zero-momentum part of the chiral effective action with a mass term in the representation $(N_F, N_F^*) \oplus (N_F^*, N_F)$ of the chiral group.

For a dense spectrum, the sum rule (1.71) can be rewritten in the form

$$\frac{1}{V_4^2} \sum_n \frac{1}{\lambda_n^2} = \int dx \left(\lim_{V_4 \to \infty} \frac{1}{V_4} \nu(\frac{x}{V_4}) \right) = \frac{\Sigma^2}{4N_F} \qquad (1.73)$$

where the integrand involves the microscopic limit $\nu_s(x)$ of the spectral density $\nu(\lambda)$ with $x = \lambda V_4$. The microscopic level density ν_s is a four-volume magnification of the mean spectral density near the origin ($\lambda \to 0, V_4 \to \infty$ with x fixed). It reflects one-level spacing in the spectral distribution. It was postulated by Verbaarschot and Zahed[16] that the microscopic level density in QCD is universal and follows from random matrix theory.

(ii) Quantum spectra and universality

The theory of correlations in quantum spectra[17] has revealed that the microscopic fluctuations can be separated from ordinary variations in the eigenvalues. Using unfolded spectra

$$\Lambda_n = \int_{-\infty}^{\lambda_n} \nu(\lambda) d\lambda, \qquad (1.74)$$

i.e. spectra with average level spacing of 1 as shown in Fig. 1.12, it was shown[17] that the resulting correlations between the eigenvalues are *universal*. In other words, the correlations are solely determined by symmetries provided that the underlying classical system is chaotic. These correlations follow from random matrix theory.

The universality classes in random matrix theory have been discussed by Dyson and others[17]. They fall into three classes: The Gaussian Orthogonal Ensemble (GOE) which is real and reflects chaotic systems with time reversal invariance; the Gaussian Unitary Ensemble (GUE) which is complex and reflects chaotic systems that break time reversal invariance; the Gaussian Symplectic Ensemble (GSE) which is quaternion and real, and reflects chaotic systems with time reversal invariance but broken rotational invariance. In which ensemble does the QCD spectrum near zero virtuality fall?

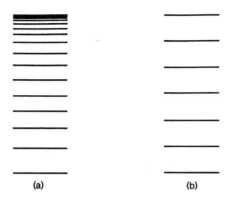

Fig. 1.12. (a) Quantum spectrum; (b) unfolded quantum spectrum.

(iii) Chiral random matrix theory

The pertinent random matrix theory for the QCD spectrum near zero virtuality follows from three assumptions: a) chiral symmetry, $[\gamma_5, \not{V}]_+ = 0$; b) representation of the color group for the quarks (fundamental); and c) number of flavors N_F. In the zero topological charge sector ($Q = 0$), the partition function follows from random matrix theory. Generically the generating functional is[16],

$$Z[m] = \int dT \prod_F^{N_F} \det \begin{pmatrix} im_F & T \\ T^\dagger & im_F \end{pmatrix} e^{-N\Sigma^2 \,\mathrm{Tr}(T^\dagger T)} \tag{1.75}$$

where T is a random $n \times n$ matrix, with $N = 2n$ identified here with the four volume V_4 and Σ the chiral condensate appearing in the Banks-Casher relation (1.67). The integration is over complex matrices. ¶Equation (1.75) will be referred to as the generating function for the chiral Gaussian Unitary Ensemble (GUE). The joint eigenvalue density following from (1.75) reads[16]

$$\nu(\lambda_1, ..., \lambda_n) = C_n \prod_{i \leq j} |\lambda_i^2 - \lambda_j^2|^2 \prod_i (\lambda_i^2 + m_F^2)^{N_F} |\lambda_i| \, e^{-n\Sigma^2 \sum_{l=1}^{n} \lambda_l^2}. \tag{1.76}$$

Integrating (1.76) over $n - 1$-eigenvalues and taking the microscopic (mesoscopic) limit yields the following form for the microscopic spectral density[16]

¶For $Q \neq 0$, T is a complex $n \times m$ matrix with $Q = |n - m|$.

in the chiral limit

$$\nu_s(x) = \frac{x\Sigma^2}{2}\left(J_{N_F}^2(\Sigma x) - J_{N_F-1}(\Sigma x)J_{N_F+1}(\Sigma x)\right) \tag{1.77}$$

which is the *master* formula for all the sum rules discussed by Leutwyler and Smilga[15]. Here J_n are Bessel functions. For instance,

$$\sum_n \frac{1}{V_4^{2p}}\frac{1}{\lambda^{2p}} = \int dx \frac{\nu_s(x)}{x^{2p}} = \left(\frac{\Sigma}{2}\right)^{2p}\frac{\Gamma(2p-1)\Gamma(N_F-p+1)}{\Gamma(p)\Gamma(p+1)\Gamma(N_F+p)}. \tag{1.78}$$

The first equality is a rewriting of the sum over the inverse moments of the eigenvalues in terms of the microscopic spectral density in the thermodynamic limit. The second equality follows from using (1.77) for the microscopic spectral density and performing the integration.

Other sum rules have also been derived by Verbaarschot[18] with the help of Selberg's integral formulae[17]. Also, it was shown by the same author that the GOE ensemble is realized for QCD with two colors, while the GSE ensemble is realized for QCD with adjoint fermions[18]. In the former, the symmetry breaking pattern in the vacuum is assumed to be $SU(2N_F) \to Sp(2N_F)$, while in the latter the breaking pattern is assumed to be $SU(2N_F) \to O(2N_F)$.

The macroscopic spectral density carries the bulk of the dynamical content in the quark spectrum. It is not universal. From the random matrix ensemble discussed above, a qualitative description can be obtained and may be used as a guideline for lattice simulations. Indeed, using the definition (1.63) for $\nu(\lambda)$, where the average is over the joint eigenvalue density (1.76), allows one to rewrite eq. (1.75) in terms of an effective action

$$\begin{aligned}
\mathbf{S}[\nu] = & -\int d\lambda d\lambda'\, \nu(\lambda)\nu(\lambda')\ln|\lambda^2 - \lambda'^2| \\
& -\int d\lambda\, \nu(\lambda)\left(\sum_{F=1}^{N_F}\ln(\lambda^2 + m_F^2) + \ln|\lambda|\right) \\
& +n\Sigma^2\int d\lambda\nu(\lambda)\,\lambda^2 \\
& +\xi\left(\int d\lambda\,\nu(\lambda) - n\right)
\end{aligned} \tag{1.79}$$

where ξ is a Lagrange multiplier. For a dense spectrum, the integration over the eigenvalues λ_n may be replaced by a functional integration over the eigenvalue density $\nu(\lambda)$, modulo a Jacobian. This method is known as the Coulomb gas description in random matrix theory. [17] In the large n limit the Jacobian is one and the extremum of $\mathbf{S}[\nu]$ determines the macroscopic spectral density. Variation of (1.79) with respect to ν and differentiation with respect to λ yield

$$2\mathbf{P} \int \frac{\nu(\lambda')}{\lambda^2 - \lambda'^2} = n\Sigma^2 - \sum_{F=1}^{N_F} \frac{1}{\lambda^2 + m_F^2} - \frac{1}{2\lambda^2} \qquad (1.80)$$

with the normalization condition

$$\int d\lambda \, \nu(\lambda) = n. \qquad (1.81)$$

In (1.80), $\mathbf{P}$ stands for principle-value integral. In the thermodynamical limit $(n \to \infty)$, the fermions drop out from the macroscopic spectral density in the chiral limit. Thus

$$\mathbf{P} \int \frac{\nu(\lambda')}{\lambda^2 - \lambda'^2} = n\frac{\Sigma^2}{2}. \qquad (1.82)$$

This integral equation yields a semi-circular (Wigner) distribution for the macroscopic spectral density

$$\nu(\lambda) = \frac{n\Sigma}{\pi} \sqrt{1 - \frac{\lambda^2 \Sigma^2}{4}}. \qquad (1.83)$$

A comparison with the microscopic spectral density shows that the fermions cause a level repulsion at $\lambda = 0$ (dip), with a width that is $1/n$ in size. The microscopic spectral density magnifies this dip and provides a quantitative description of the level distribution in this $1/n$-window. The macroscopic level density focusses on the rest of the spectrum.

Recent detailed QCD lattice simulations by Kalkreuter [19] using staggered as well as Wilson fermions have provided a first direct insight into the fermionic spectrum of QCD. Figure 1.13 shows a typical behavior of the spectral density for Wilson fermions with different values of hopping parameter $\kappa = (2m + 8)^{-1}$ on 6^4 lattice with $\beta = 1.8$. The dip around the zero is due to the fermionic level repulsion. The detailed analysis of Kalkreuter's staggered fermion distributions has shown that the level correlations (fluctuations in the spectral density as introduced above) are also in agreement with random matrix theory. [20] All this points to the fact that the underlying quark dynamics in the QCD vacuum is random (or chaotic).

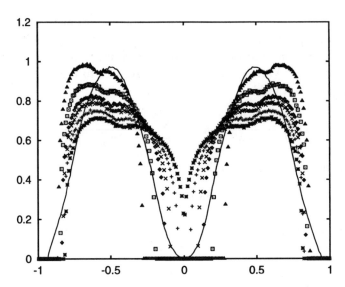

Fig. 1.13. Spectral densities for Wilson fermions at $\beta = 1.8$. The stars (pluses, crosses, diamonds, squares, triangles) correspond to the values of κ equal to 0.25 (0.20, 1/6, 0.15, 0.125, 0.1), respectively.[19]

The microscopic level density is hard to construct from present lattice QCD simulations, as only few eigenvalues are available near $\lambda = 0$. It is however, in good agreement with the microscopic spectral density derived from models with spontaneous symmetry breaking as shown in Fig. 1.14 for the instanton liquid model (Chapter 2). As mentioned above, the microscopic spectral density vanishes at the origin. Is this at odds with the Banks-Casher result? The answer is no, since the size of the magnified dip in Fig. 1.14 is of the order of $1/V_4$ and vanishes in the thermodynamic limit. As noted above, it is important that the chiral limit ($m \to 0$) which characterizes the sampling follows the thermodynamic limit ($V_4 \to \infty$). Some of the agreement between the predicted microscopic spectra and the results of the instanton liquid simulations suggest that the model captures some of the essence of the spontaneous breaking of chiral symmetry. We discuss this model for the QCD vacuum state and its excitations in the next section.

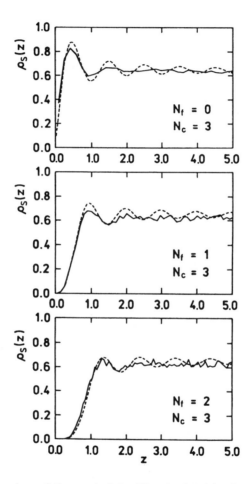

Fig. 1.14. Comparison of the spectral densities simulated in the random instanton model with the exact analytical result. The solid line corresponds to the random system of instantons and the dashed line the exact result from random matrix theory.

Appendix

(A_I) Time-dependent Dirac equation

In the presence of the electric field (1.25), the Dirac equation reads

$$
\begin{aligned}
i\partial_t u(k) &= [\gamma_5(k - g\bar{a}(t)]u(k) \\
i\partial_t v(-k) &= [\gamma_5(k - g\bar{a}(t)]v(-k).
\end{aligned}
\tag{A1.1}
$$

For $t < 0$, the solutions are

$$
\begin{aligned}
u(k,t) &= \exp(-i|k|t)u(k,0) \\
v(-k,t) &= \exp(i|k|t)v(-k,0).
\end{aligned}
\tag{A1.2}
$$

For $t > \tau$, the solutions read

$$
\begin{aligned}
u(k,t) &= \exp\left(-i\chi(k)\int_0^\tau (g\bar{a}_\tau - g\bar{a}(t^{'}))dt^{'}\right) \\
&\quad \times \exp(-i(|k| - \chi(k)g\bar{a}_\tau)t)u(k,0), \\
v(-k,t) &= \exp\left(-i\chi(k)\int_0^\tau (g\bar{a}_\tau - g\bar{a}(t^{'}))dt^{'}\right) \\
&\quad \times \exp(i(|k| - \chi(k)g\bar{a}_\tau)t)v(-k,0),
\end{aligned}
\tag{A1.3}
$$

where $u(k,0), v(-k,0)$ are states with a given chirality $\chi(k)$.

(A_{II}) Alternative to Goldstone theorem

Using the replica trick, McKane and Stone have derived an interesting relation between the chiral condensate and the pseudoscalar-isoscalar correlator at zero momentum.[21] For convenience we will ignore the isospin. Their result reads[21]

$$
\langle \bar{q}q \rangle = m \int d^4x \langle \bar{q}\gamma_5 q(x)\bar{q}\gamma_5 q(0)\rangle.
\tag{A1.4}
$$

Here we will sketch the intuitive diagrammatic proof given by Kogut *et al.*[22] On lattice, the Ward identity (A1.4) can be rewritten

$$
\mathrm{Tr}\Big(G(n,n)\Big) = \sum_{n'} \mathrm{Tr}\Big(|G(n,n')|^2\Big)
\tag{A1.5}
$$

where the sum runs over the lattice sites. A typical graph contributing to
(A1.5) is shown in Fig. 1.15a. The same graph with γ_5 insertions (circles)
is shown in Fig. 1.15b. They follow from one another by sliding one of the
two γ_5's along the quark path. The different graphs alternate in sign, and
cancel pairwise except for the original one. Interpreting the diagrams of
Fig. 1.15 as contributions to $\gamma_5 G(n, n')\gamma_5 G(n', n)$ yields (A1.5), and finally
(A1.4) in the continuum.

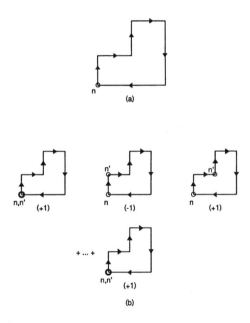

Fig. 1.15. (a) Graph contributing to $G(n, n)$. (b) Same graph with two γ_5's inserted
(circles). Adding all the graphs give back (a).

The Goldstone theorem follows from the observation that the pseudoscalar correlator develops a zero mass *pole* as $m \to 0$, thus a nonvanishing
quark condensate. Alternatively, McKane and Stone[21] have suggested that
(A1.4) is still maintained in the case where the pseudoscalar correlator
does not develop a pole, but diverges as $1/m$ for *all* momenta, not just
zero momentum. In this case there is no Goldstone particle. The fermions
are subject to Anderson localization. In both the Goldstone and Anderson
phases, the density of levels $\nu(\lambda)$ relates to the order parameter (here $\langle \bar{q}q \rangle$)

and *does not* vanish. In the localized or Anderson phase, the conductivity vanishes. The fact that the density of states does not vanish at the mobility edge (substitute for Fermi level) together with (1.68) implies that the diffusion coefficient $\mathbf{D} = 0$ in the localized phase. The observation of light pions in nature rules out the Anderson phase for QCD. We do not know of any field theory case where the localization option is realized.

References

1. S. Samuel, Mod. Phys. Lett. **A7**, 2007 (1992); H. Kikuchi and J. Wudka, Phys. Lett. **B284**, 111 (1992).
2. K. Fujikawa, Tokyo-UT-722-95, HEP-TH-9510111.
3. P. van Baal and N.D. Hari Dass, Nucl. Phys. **B385**, 185 (1992).
4. H. Yamagishi and I. Zahed, SUNY-preprint-NTG-95-29.
5. J. Ambjorn, J. Greensite and C. Petersson, Nucl. Phys. **B221**, 381 (1983).
6. L.D. Landau and E.M. Lifshitz, *Quantum Mechanics* (&112) (Pergamon Press, 1977).
7. K. Fujikawa, Phys. Rev. Lett. **42**, 1195 (1979).
8. M. Atiyah and I. Singer, Ann. Math. **87**, 484 (1968).
9. J. Collins, A. Duncan and S. Joglekar, Phys. Rev. **D16**, 438 (1977).
10. M.A. Shifman, Phys. Rep. **209**, 341 (1991).
11. G. Greensite, Nucl. Phys. **B249**, 263 (1985).
12. J. Goldstone, Nuovo Cimento **19**, 154 (1961).
13. C. Vafa and E. Witten, Nucl. Phys. **B234**, 173 (1984).
14. T. Banks and A. Casher, Nucl. Phys. **B169**, 103 (1980).
15. H. Leutwyler and A. Smilga, Phys. Rev. **D46**, 5607 (1992).
16. J.J.M. Verbaarschot and I. Zahed, Phys. Rev. Lett. **70**, 3852 (1993).
17. M.L. Mehta, *Random Matrices* (Academic Press, N.Y. 1991).
18. J.J.M. Verbaarschot, Phys. Lett. **B329**, 351 (1994).
19. T. Kalkreuter, Berlin preprint, HUB-EP-95-26, hep-lat/9511009.
20. M.A. Halasz and J.J.M. Verbaarschot, Phys. Rev. Lett. **74**, 3920 (1995).
21. A.J. McKane and M. Stone, Ann. Phys. **131**, 36 (1981).
22. J. Kogut, *et al.*, Nucl. Phys. **225** [FS9], 326 (1983).

CHAPTER 2

INSTANTON VACUUM

2.1. Basics of Instantons

Instantons in QCD constitute an important class of inhomogeneous gauge configurations that might be important at distance scales of the order of a fraction of a fermi. They are exact solutions to the classical Yang-Mills equations in Euclidean space. In the instanton approach to the QCD vacuum, it is assumed that the world in the femtometer scale is dominated by instantons and anti-instantons, perhaps in a liquid state, where light (current) quarks are turned heavy by their successive rescattering with the instantons and anti-instantons. This concept of turning heavy is at the origin of the spontaneous breaking of chiral symmetry. The purpose of this section is to provide a short but hopefully comprehensive description of what QCD instantons are, why they might be relevant for the spontaneous breaking of chiral symmetry, and how they organize themselves in the vacuum state. This approach to the QCD ground state will constitute the cornerstone for the development of models of chiral symmetry breaking in the following chapters.* Unless specified otherwise, most of the discussion in this section is understood in Euclidean space, with the conventions given in the Appendix.

The nonabelian character of QCD allows for the existence of classical, finite action, "large gauge configurations," that interpolate (or tunnel) continuously between topologically inequivalent classical vacua in QCD. What

---

*This does not imply, however, that other mechanisms, such as monopole condensation and related mechanisms, cannot do as well in breaking chiral symmetry. In fact, the mechanism based on instantons treated in this chapter may be a generic feature reflecting symmetries rather than specific dynamics.

do we mean by "large gauge configurations" and what are these classical vacua? To answer the first question, consider a gauge configuration A_μ^a of a finite action,

$$S = \frac{1}{4} \int d^4x \, G_{\mu\nu}^a G_{\mu\nu}^a. \tag{2.1}$$

Clearly S is finite if the gauge configuration approaches the vacuum value at infinity modulo a gauge transformation,

$$A_\mu(|x| \to \infty) = \frac{i}{g} \Omega^\dagger \partial_\mu \Omega \tag{2.2}$$

where Ω is $SU(2)$-valued. Equation (2.2) provides a mapping from the sphere S^3 at infinity ($|x| = \infty$) onto $SU(2)$.[†] Since $\Pi_3(SU(2)) = \mathbf{Z}$ (see Appendix in Chapter 6), the gauge configurations (2.2) fall into different topological classes, each of which is characterized by an integer topological number Q,

$$Q = \frac{g^2}{32\pi^2} \int d^4x \, G_{\mu\nu}^a \tilde{G}_{\mu\nu}^a \tag{2.3}$$

where

$$\tilde{G}_{\mu\nu}^a = \frac{1}{2} \epsilon_{\mu\nu\alpha\beta} G_{\alpha\beta}^a. \tag{2.4}$$

Large gauge configurations are characterized by a nonvanishing Q, which is the degree of winding of the map (2.2). Indeed, the integrand in (2.3) is a total divergence $G \cdot \tilde{G} \propto \partial \cdot K$, where K_μ is a gauge-dependent four-current (related to the Chern-Simons current defined later)

$$K_\mu \sim 2\epsilon_{\mu\nu\alpha\beta} \left(A_\nu^a G_{\alpha\beta}^a + \frac{1}{3} g f^{abc} A_\nu^a A_\alpha^b A_\beta^c \right). \tag{2.5}$$

Inserting (2.5) into (2.3) through the four-current, and specializing to $SU(2)$ maps give

$$Q = \frac{i}{24\pi^2} \int_{|x|=\infty} d^3x \epsilon_{ijk} \, \mathrm{Tr}(A_i A_j A_k) \tag{2.6}$$

which is precisely the winding number of (2.2) onto the sphere at infinity.[‡]

[†]Embedding $SU(2)$ into $SU(N_c)$ is described below.
[‡]We have assumed that the Euclidean space has the geometry of : $[0,1] \times S^3$.

The various classical vacua are built on the topologically inequivalent gauge configurations. They are the precursors of the quantum vacua each with a definite charge $Q = n$, where the charge Q plays the role of a superselection rule. The large gauge configurations in QCD interpolate continuously between these classical vacua, by changing the value of Q as shown qualitatively in Fig. 1.4. Gauge transformations do not change the energy, hence the periodicity of $\mathcal{V}(Q)$. The infinite degeneracy of $\mathcal{V}(Q)$ is usually removed by sub-barrier tunneling transitions. In Minkowski space they are transitions conditioned by the uncertainty principle $(\Delta E \Delta t) \sim \hbar$. In Euclidean space they are related to particle motion along the Euclidean time axis with a *finite* transition probability. Thus the name instantons.

2.1.1. *SU(2) instantons*

QCD instantons are classical, self-dual solutions to the Yang-Mills equations in Euclidean space with a nonvanishing topological charge Q. The self-duality condition implies that $G = \tilde{G}$, and guarantees that instantons are classical solutions to the Yang-Mills equations. Indeed,

$$D_\mu G_{\mu\nu} = D_\mu \tilde{G}_{\mu\nu} = \frac{1}{6}\epsilon_{\mu\nu\alpha\beta}\left(D_\mu G_{\alpha\beta} + D_\alpha G_{\beta\mu} + D_\beta G_{\mu\alpha}\right) = 0 \quad (2.7)$$

where we have used the antisymmetric character of $\epsilon_{\mu\nu\alpha\beta}$ and the Bianchi identity. Instantons carry minimal action,

$$S = \frac{1}{4}\int d^4x\, G^a_{\mu\nu}\tilde{G}^a_{\mu\nu} = Q\frac{8\pi^2}{g^2} \quad (2.8)$$

and in this sense saturate the Bogomol'nyi bound.[§] The finite action (2.8) yields a finite transition probability $W \sim e^{-2S} \sim e^{-16\pi^2/g^2}$.

The explicit form of the instanton depends on the gauge choice. A natural and simple choice follows from the boundary condition (2.2) and the self-duality constraint. For the first, we demand a nontrivial map on S^3 at infinity. A simple gauge choice is

$$\Omega(|x| = \infty) = \tau^+_\mu \cdot \hat{x}_\mu, \qquad \tau^\pm_\mu = (\vec{\tau}, \mp i\,\mathbf{1}). \quad (2.9)$$

Here τ's are the usual Pauli matrices, so that Ω is $SU(2)$-valued. Inserting (2.9) into (2.2) gives

$$\Omega^\dagger \partial_\mu \Omega = \frac{1}{2}[\tau^-_\lambda, \tau^+_\sigma]\,\hat{x}_\lambda \partial_\mu \hat{x}_\sigma = i\tau^a \bar{\eta}^a_{\lambda\mu} \frac{x_\lambda}{x^2}. \quad (2.10)$$

[§]For general large gauge configurations we expect $S \geq Q8\pi^2/g^2$, since $G^2 = G\tilde{G} + (G - \tilde{G})^2/2$. The bound on S is usually referred to as the Bogomol'nyi bound.

The above defines $\bar{\eta}$ known as the 't Hooft symbol. Its properties are listed in the Appendix. In the regular gauge, the general form of A_μ^a is

$$A_\mu^a = \frac{2}{g}\bar{\eta}_{\lambda\mu}^a \frac{x_\lambda}{x^2}\, F(x^2) \tag{2.11}$$

with the boundary conditions $F(x^2 \to \infty) = 1$ and $F(x^2 \to 0) = \alpha x^2$ with α a constant to be fixed. The latter guarantees the finiteness of A at the origin. The explicit form of F follows from the self-duality relation $G = \tilde{G}$ which implies that

$$x^2 F' + F(F - 1) = 0 \tag{2.12}$$

where F' is the derivative with respect to x. Using the boundary conditions on F yields A in the form

$$A_\mu^a = \frac{2}{g}\bar{\eta}_{\lambda\mu}^a \frac{x_\lambda}{x^2 + \rho^2} \tag{2.13}$$

for arbitrary instanton size ρ (which is an integration constant). Because of the scale invariance of the classical Yang-Mills equations, the instanton size is not fixed by the classical equations of motion. In fact (2.13) belongs to a family of configurations with the same action $S = 8\pi/g^2$ and charge $Q = 1$. They follow from (2.13) by translation $x_\mu \to x_\mu - R_\mu$, $O(4)$ rotations $A^a \to \mathbf{R}^{ab}A^b$ and dilatation $\rho \to \lambda\rho$, therefore the most general form of the instanton solution in the regular gauge is

$$A_\mu = \frac{2}{g}\bar{\eta}_{\lambda\mu}^a \frac{x_\lambda - R_\lambda}{(x_\mu - R_\mu)^2 + \rho^2} U \frac{T^a}{2} U^\dagger \tag{2.14}$$

where U are $SU(2)$ matrices and $\mathbf{R}_{ab}(U) = \frac{1}{2}\text{Tr}(\tau_a U \tau_b U^\dagger)$.

For completeness, we note that all the above arguments for instantons also hold for anti-instantons except that for the latter, the gauge field is anti-self-dual $G = -\tilde{G}$, carrying the same action $S = 8\pi/g^2$ as the instantons but opposite charge $Q = -1$. Their explicit form follows from (2.13) through the substitution $\bar{\eta} \to \eta$ (see Appendix).

Instantons in $SU(N_c)$ with $N_c \geq 3$, follow from the $SU(2)$ instantons discussed here by simple embeddings. In the case of $SU(N_c)$, the number of independent color components of the instanton is equal to the dimension of the coset manifold $SU(N_c)/(SU(N_c - 2) \times U(1))$, that is, $4N_c - 5$.

2.1.2. *Quark zero modes*

What makes the instantons special in QCD is their effects on light quarks. In an instanton background the Dirac equation admits a normalizable zero-mode solution with specific handedness, namely right-handed for the instanton. The existence of this zero mode is guaranteed by the Atiyah-Singer index theorem. The explicit form of this zero mode follows from the specific choice of the instanton background.

The zero-mode solution to the Dirac equation $\not{D}\psi_0 = 0$ can be sought in the Weyl basis

$$\sigma^{\pm} \cdot \psi_{L,R} = 0 \tag{2.15}$$

where $\sigma_{\mu}^{\pm} = (\vec{\sigma}, \mp i)$ are the spin Pauli matrices, and L, R stand, respectively, for left- and right-handed bispinors. The following identity holds

$$(\sigma^{+} \cdot D)(\sigma^{-} \cdot D) = D \cdot D + \frac{g}{4}\bar{\eta}_{\mu\nu}^{a}\sigma^{a}\tau^{b}G_{\mu\nu}^{b} \tag{2.16}$$

since

$$\sigma_{\mu}^{+}\sigma_{\nu}^{-} = 1_{\mu\nu} + i\bar{\eta}_{\mu\nu}^{a}\sigma^{a} \tag{2.17}$$

which is a hedgehog mapping of color τ onto space x. For instantons, $G_{\mu\nu}^{a} \sim \bar{\eta}_{\mu\nu}^{a}$, and for anti-instantons $G_{\mu\nu}^{a} \sim \eta_{\mu\nu}^{a}$. Therefore, using the properties of the 't Hooft tensors

$$\bar{\eta}_{\mu\nu}^{a}\bar{\eta}_{\mu\nu}^{b} = 4\delta^{ab},$$
$$\bar{\eta}_{\mu\nu}^{a}\eta_{\mu\nu}^{b} = 0$$

we get the following equations in the case of the instanton:

$$D \cdot D \psi_L = 0,$$
$$D \cdot D \psi_R = 4\sigma \cdot \tau \frac{\rho^2}{x^2 + \rho^2}\psi_R.$$

Since $(iD)^2$ is a positive operator, the first equation implies that $\psi_L = 0$, and the second one implies that the right-handed fermions can only occur in the singlet spin-color channel (as opposed to the triplet channel), *i.e.*, $\sigma \cdot \tau = -3$. This means that the right-handed fermionic zero mode is a hedgehog in spin and color, *i.e.*, $\psi \sim \epsilon^{am} \sim [i\tau_2]^{am}$, where a is a color index and m is a spin index. The explicit form of the right-handed color-spin-singlet zero mode in the field of an instanton reads

$$\psi_{R,ai} = \left(\frac{1}{\pi\sqrt{2x^2}} \frac{\rho}{(x^2 + \rho^2)^{3/2}} \right) (x \cdot \gamma)_{ij} \, \Pi_{jb} \, U_{ab} \qquad (2.18)$$

with Π = diagonal $(0, i\tau^2)$ for gauge choice (2.13). Anti-instantons allow for only left-handed zero modes. They follow from (2.18) by parity.

The structure and character of these zero modes are at the origin of flavor mixing in the vacuum. Indeed, in QCD the axial-singlet current is anomalous. For N_f flavors,

$$\partial^\mu j_{5\mu} = \frac{N_f g^2}{16\pi^2} G \cdot \tilde{G} = 2N_f \, \Delta Q(x) \qquad (2.19)$$

which is eq.(1.34) in the chiral limit for simplicity. Integrating both sides of (2.19) and assuming that the vector part of the current is bounded on the sphere at infinity, we obtain

$$\Delta Q_5 = Q_5(+\infty) - Q_5(-\infty) = 2N_f \Delta Q. \qquad (2.20)$$

The change in the axial charge is directly related to the change in the topological charge. Thus the possibility of motion through instantons (Q-direction) means non-conservation of the quark axial charge in a vacuum with instantons. For one instanton, $\Delta Q = 1$, and the net change in the axial charge is $\Delta Q_5 = 2N_f$. Even though the instanton is a configuration in color space, the fact that it breaks parity has an important effect on the chirality of the light quarks. The instanton flips chirality. This mechanism provides the answer to the $U_A(1)$ problem and allows for flavor mixing in the chiral limit.

2.2. Vacuum State

When restricted to the light flavors (u, d, s) with vanishing current masses ($m_u \sim m_d \sim m_s \sim 0$), the QCD Hamiltonian is symmetric under rigid chiral $SU(3)_L \times SU(3)_R$ transformations. The absence of parity doublets in the spectrum suggests that the vacuum spontaneously breaks $L \times R$ symmetry, as discussed in Chapter 1. What this means is that vacuum fluctuations mix handedness, so that

$$\langle 0|Q_L^a|0\rangle = \langle 0|Q_R^a|0\rangle = 0 \qquad \text{and} \qquad \langle 0|Q_L^a Q_R^a|0\rangle \neq 0. \qquad (2.21)$$

Here $Q_{L,R}^a$ are the left and right isovector charges. Instantons and anti-instantons in the vacuum do precisely this. Indeed, a topologically neutral ensemble of instantons and anti-instantons, while globally isospin invariant, may yield (2.21) through statistical fluctuations in the thermodynamic limit. Such an ensemble, when used as a model for the QCD vacuum, provides a natural mechanism for the spontaneous breaking of chiral symmetry.

The idea that the vacuum in QCD may be largely dominated by instanton-anti-instanton fluctuations dates back to the original work of 't Hooft[1] and also that of Callan, Dashen and Gross[2]. Originally, the QCD vacuum was viewed as a non-interacting gas of instantons and anti-instantons in the singular gauge.¶ While very appealing, this idea met rapidly its demise when it was discovered that the relevant physical quantities were infrared sensitive. In the dilute gas approximation, the instanton size distribution scales as[2] ‖

$$d(\rho) \sim \frac{(\rho\Lambda)^{11N_c/3}}{(\rho\Lambda)^5}, \tag{2.22}$$

to one loop. The numerator in (2.22) is the one-instanton contribution to e^{-S} and the denominator is the one-loop pre-exponent in the quenched approximation. Since $d \sim \rho^6$ ($N_c = 3$), *large size* instantons dominate the partition function in the vacuum, causing most calculations to diverge in the infrared.

The infrared problem is most probably an artifact of the non-interacting gas approximation, and may be cured by instanton-anti-instanton interactions in the quenched approximation and instanton-quark-anti-instanton induced interactions in the unquenched approximation. These facts, although always suspected, have never been established convincingly. In this respect, the variational calculations by Diakonov and Petrov[3] in the *quenched* approximation using the sum ansatz in singular gauge for the instanton-anti-instanton configuration are very suggestive. Their calculations show that on the average, the instanton anti-instanton interaction is repulsive due to the dominance of the repulsive color-spin orientations in the statistical sum. Their variational estimate provides an instanton size

---

¶The use of the singular gauge is important. It is only in this gauge that the instanton and anti-instanton configurations are localized.

‖Here Λ refers to the QCD renormalization-group-invariant scale.

distribution that is cut off in the infrared[3],

$$d(\rho) \rightarrow \frac{(\rho\Lambda)^{11N_c/3}}{(\rho\Lambda)^5} \, e^{-\beta \bar{\rho}^2 \rho^2 \Lambda^4} \tag{2.23}$$

where $\bar{\rho}$ is the average instanton size and β is a number of order 1. However, it should be noted that the repulsion depends quantitatively on the form of the ansatz at short distances (here a sum ansatz). This notwithstanding, the same arguments show that in the vacuum, the average instanton size $\bar{\rho}$ is about 0.3 fm. This result is in agreement with an earlier phenomenological estimate made by Shuryak [4] using QCD sum-rule parameters and new lattice results by Chu, Grandy, Huang and Negele[6]. The variational calculations also suggest that on the average $\rho/R \sim 1/3$ where R is a typical instanton-anti-instanton separation. In four dimensions, this gives rise to a small packing fraction $(\rho/R)^4 \sim 0.01$. Thus only a small fraction of space-time is populated by strong gluonic fields (instantons). The variational calculations are compatible with a dilute system. It was also found that the instanton action is large: $S = 8\pi^2/g^2 \sim 10$. Since the quantum corrections go as $1/S \sim 0.1$ the semiclassical approach was in a way justified. How are these results affected by the presence of light dynamical quarks?

2.2.1. Instanton-anti-instanton molecule

To answer this question, we start by considering the schematic problem of an instanton-anti-instanton configuration. This configuration has a trivial topology. In the chiral limit, it does not exhibit rigorous zero modes. To illustrate this, consider a gauge field configuration given by the linear superposition of an instanton and an anti-instanton. When the relative distance R between the pseudoparticles (instantons and anti-instantons) is finite, the zero modes begin to overlap. More specifically, for a sum ansatz (a linear superposition of an instanton and an anti-instanton), the eigenfunctions of the Dirac equation are given by $\phi_\pm = (\phi^R \pm \phi^L)/\sqrt{2}$. The corresponding energies are split about 0 by a gap of

$$\Delta = 2 \int \phi_R^\dagger i\slashed{\partial}\phi_L = 2T. \tag{2.24}$$

The overlap matrix T can be calculated explicitly (see the Appendix). The overlap matrix is proportional to the cosine of a suitably defined relative orientation angle[7] between the instanton and the anti-instanton. The overlap is maximized for the relative orientations of 0 and π (this resembles a dipole-dipole interaction). For asymptotically large values of R, $T \sim R^{-3}$. The dependence is the same as the one found in the propagator of a noninteracting massless quark. For a small interparticle distance, $T \sim R$. Instanton-anti-instanton molecules are likely to bind through light quarks, provided that a sufficiently strong repulsive mechanism is established at short separations.

2.2.2. Instanton-anti-instanton ensemble

In the instanton model for the QCD vacuum, the ground state is described by a statistical ensemble of N_+ instantons and N_- anti-instantons in a Euclidean volume V_4 with a partition function determined by the gauge field action[2,8]. To ensure topological neutrality we require that the expectation value $\langle N_+ - N_- \rangle = 0$. To account for the axial singlet anomaly, the topological susceptibility should be nonvanishing, *i.e.* $\langle (N_+ - N_-)^2 \rangle \neq 0$. This is achieved by the statistical fluctuations inherent in the thermodynamic limit $N_\pm \to N/2$, $N \to \infty$, $V_4 \to \infty$ at a fixed pseudoparticle density N/V_4.

For a more quantitative understanding of the vacuum state, we will truncate the Hilbert space of the fermions to the space of zero modes. This approximation is justified for the long-wavelength properties of the vacuum, that is, scales > 0.3 fm, a typical instanton size. For a topologically neutral system of pseudoparticles interacting with N_f flavors of quarks, the partition function is given by[8]

$$\left\langle \prod_f \det \begin{pmatrix} K_{11} & \cdots & K_{1N} & T_{1\bar{1}} & \cdots & T_{1\bar{N}} \\ \vdots & \ddots & \vdots & \vdots & \ddots & \vdots \\ K_{N1} & \cdots & K_{NN} & T_{N\bar{1}} & \cdots & T_{N\bar{N}} \\ T_{\bar{1}1} & \cdots & T_{\bar{1}N} & K_{\bar{1}\bar{1}} & \cdots & K_{\bar{1}\bar{N}} \\ \vdots & \ddots & \vdots & \vdots & \ddots & \vdots \\ T_{\bar{N}1} & \cdots & T_{\bar{N}N} & K_{\bar{N}\bar{1}} & \cdots & K_{\bar{N}\bar{N}} \end{pmatrix} \right\rangle \tag{2.25}$$

with $N = N_+$ and $\bar{N} = N_-$. The overlap matrices T and K are defined by

$$T_{IJ} = \int d^4x \, \phi_I^\dagger i\gamma_\mu \partial_\mu \phi_J,$$

$$K_{IJ} = im \int d^4x \, \phi_I^\dagger \phi_J. \tag{2.26}$$

Each zero mode as well as mass m carry an implicit flavor index. Note that due to the specific chiral structure of the zero modes, the overlap matrix T connects instantons and anti-instantons, whereas the overlap K connects pseudoparticles of the same kind. The explicit form of the overlap integrals is given in the Appendix. For a dilute system of pseudoparticles, we may approximate $K_{IJ} \sim im\delta_{IJ}$. Brackets denote averaging over the instanton and anti-instanton positions (z_I^μ), orientations (U_I) for a *fixed* size, with the weight provided by the invariant measure and the effective instanton-instanton interaction. Generically $(\Omega = (R_\mu, \rho, U))$

$$Z_{Q=0} = \int \prod_I d\Omega_I \, e^{-S(\Omega)} \prod_f \det \begin{pmatrix} im_f & T(\Omega) \\ T^\dagger(\Omega) & im_f \end{pmatrix} \qquad (2.27)$$

in the $Q = 0$ sector. For an entirely random system, (2.27) reduces to (1.75) with the invariant measure of complex hermitian matrices.

In this approach, the spontaneous breaking of chiral symmetry may be understood as follows. Let us assume that the pseudoparticles (instantons and anti-instantons) are some four-dimensional atoms each with one valence electron (zero mode). In the dilute approximation, the quark zero modes are well localized on the pseudoparticles, and the system resembles an insulator. As the density of pseudoparticles is increased, the zero modes start to overlap. At some critical density, the zero-mode states percolate, causing a delocalization of the flavor quantum numbers. This system resembles a conductor, with the conductivity being the analogue of the quark condensate discussed in Chapter 1. The spontaneous breaking of chiral symmetry is analogous to the metal-insulator (Mott) transition in three-dimensional solids[9].

This analogy can be made more quantitative using (2.25-2.27). Consider a finite cluster (n) of pseudoparticles in the dilute limit. Typically, this cluster will contribute a factor in the partition function of the form

$$\int \prod_{i=1}^{n} d^4 R^I d^4 R^{\bar{I}} \, \text{Tr}(T^\dagger T)^{nN_f} \sim V_4 R^{n(8-6N_f)-4}, \qquad (2.28)$$

where R is a typical distance between neighboring instantons and anti-instantons and integration is over the positions of the pseudoparticles. For three flavors, large-size clusters are suppressed, whereas for one flavor, contributions from clusters of all possible sizes remain important. We see a quantitative difference between these two cases. For three flavors, terms consisting of factors of two-cycles ($n = 1$) are the dominant contribution in the dilute limit. This corresponds to the phase in which instantons and anti-instantons are forming the smallest possible clusters, *i.e.*, a molecule. This molecular phase is shown schematically in Fig. 2.1a. The spectrum

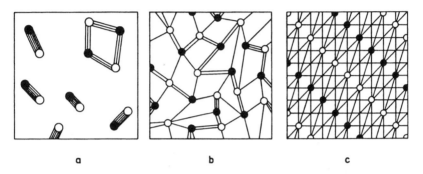

Fig. 2.1. Three phases of the instantonic "matter"[5]. White and black circles represent, respectively, instantons and anti-instantons, straight lines denote zero-mode "bonds". (a) Molecular phase. (b) Random phase. (c) Crystalline phase (projection of a four dimensional *fcc* lattice).

of the Dirac equation in this case resembles the inverted two-humped potential and vanishes at zero virtuality. By the Banks-Casher theorem, the quark condensate is zero and the vacuum is chirally invariant.

For a larger density of pseudoparticles, the quark-anti-quark hopping becomes important causing the delocalization of flavor. This phase is qualitatively shown in Fig. 2.1. The delocalization caused by the random hopping induces a phase coherence in the quark wave functions, leading to an accumulation at zero virtuality. As a result, the vacuum spontaneously breaks chiral symmetry. To see this, consider the case where the T's are randomly distributed with the invariant Haar measure as discussed in Chapter 1. The maximum entropy principle suggests that the eigenvalues of T are Gaussian-distributed with a variance κ. Thus the random matrix theory is of the GUE (Gaussian Unitary Ensemble) type.[10] It follows from (1.79) that the average distribution of eigenvalues is semi-circular. This implies that chiral symmetry is spontaneously broken. Equation (1.83) follows immediately from the Coulomb gas method in random matrix theory, and is exact in the large N_c limit.

In general, simulations with light dynamical quarks have proven to be more difficult in the instanton model of the QCD vacuum, because of the fermion determinant. The latter screens the long-distance dipole-dipole interaction between instantons and anti-instantons in the singular gauge. As a result, the topological charge is screened, thereby resolving the $U_A(1)$ problem. Mean-field arguments suggest that for an instanton density of 1 fm^{-4}, the topological charge is screened over a distance of 0.5 fm[11]. In this phase, the instantons and anti-instantons are in a relatively diluted phase, a point in favor of the mean-field arguments.

2.3. Random Instanton Vacuum

2.3.1. *Quark propagator*

In the instanton approach to the QCD ground state, the vacuum is thought of as an ensemble of instantons and anti-instantons. Let N_+ and N_- be the number of each species respectively. A quark propagating in this medium will be described by

$$S = (-i\slashed{\partial} - \slashed{A} - im)^{-1} \qquad (2.29)$$

where m is the mass matrix and the gauge configuration A is

$$A = \sum_I^{N_+} A_I + \sum_J^{N_-} A_J \qquad (2.30)$$

a schematic form of which is shown in Fig. 2.2. The propagator (2.29) can be reorganized in the language of many-body theory through

$$S = S_0 + \sum_I (S_I - S_0) + \sum_{I \neq J} (S_I - S_0) S_0^{-1} (S_J - S_0) + ... \qquad (2.31)$$

where the sum extends over $1, ..., (N_+ + N_-)$. S_I is the propagator in the field of one instanton: it is the sum of the zero-mode part plus a scattering part. In what follows, we choose to approximate the distorted part by the free propagator $S_0 = (i\slashed{\partial} + im)^{-1}$ so that

$$S_I \sim S_0 + \frac{\phi_I \phi_I^\dagger}{-im}. \qquad (2.32)$$

This approximation is reliable at momenta very high and very low compared to the inverse Compton wavelength of the instanton (for a fixed size). Inserting (2.32) into (2.31) gives

$$S = S_0 + \sum_{I,J} \phi_I \left(T + K - 2im\right)_{IJ}^{-1} \phi_J^\dagger \tag{2.33}$$

where T is the hopping matrix defined in (2.26). For a very dilute system, we can approximate the overlaps K by the diagonal matrix $im\mathbf{1}$. Note that due to the chirality-odd character of the kinetic term, T is off-diagonal in instanton-anti-instanton space, while K is diagonal.

Fig. 2.2. A quark propagating in a random instanton gas.

In what follows, we will restrict ourselves to the case where color orientations and positions of the instantons are distributed according to their invariant measure, where a hard core is assumed. The effect of the fermion determinant is the screening of the long-range dipole-dipole interaction. For a dilute system, the instanton-anti-instanton system behaves as a free gas. Throughout, the instanton size is fixed at $1/3$ fm and the instanton density at 1 fm^{-4}. For a random gas of instantons and anti-instantons with fixed size, the measure in (2.27) becomes

$$\int \prod_I^{N_+ + N_-} d\Omega_I = \prod_I^{N_+ + N_-} \int d^4 R_I dU_I \tag{2.34}$$

where R_I and U_I are, respectively, the position and orientation of the instanton in Euclidean space.

In x-space, the quark propagator reads

$$S(x,m) = S_0(x,m) + S_1(x,m) \tag{2.35}$$

where $S_0(x,m)$ is the free massive quark propagator

$$S_0(x,m) \sim \frac{im}{4\pi^2} \left(\frac{\not x}{x^2} K_2(mx) + \frac{1}{x} K_1(mx) \right) \tag{2.36}$$

and

$$S_1(x,m) = \int \frac{d^4k}{(2\pi)^4} e^{ik\cdot x} \frac{-1}{\not k - i(m - k^2/M_k(m))} \tag{2.37}$$

the interacting part. The running mass $M_k(m)$ is given by (2.53) below. We show in Fig. 2.3 the behavior of the quark propagator in the instanton vacuum. Figure 2.3a shows the scalar part $m \operatorname{Tr} S(x,m)/\operatorname{Tr} S_0(x,m = 0)$ versus x up to 2 fm. The open squares are the simulated results of Verbaarschot and Shuryak[12] with a current mass of 10 MeV, using 128 instantons and 128 anti-instantons in a periodic box $3.36^3 \times 6.72$ fm^4. The solid lines are the mean-field calculations of Kacir, Prakash and Zahed[13] for $m = 5, 10$ MeV respectively. Figure 2.3b shows the behavior of the vector part $\operatorname{Tr} \gamma_4 S(x,m)/\operatorname{Tr} S_0(x,m = 0)$ of the quark propagator. At short distances, the quark propagator is given by the free-field result (2.36), while at large distances, the propagator follows qualitatively (2.36) with the substitution $m \to M_0(m) \sim 350$ MeV for small current quark masses. This substitution, however, should be qualified, as we will discuss below.

Up to approximately 1.5 fm, both the simulated and the mean-field calculations give similar results. The agreement stems from the diluteness of the instanton ensemble in the random approximation. Indeed, the diluteness factor of the ensemble is given by the dimensionless factor $N\rho^4/2V_4N_c \sim 10^{-3}$ for $N_c = 3$. We will consider in the mean-field approximation[14,15] the random instanton system as a dilute gas with specific susceptibilities.

2.3.2. *Effective action*

The vacuum-to-vacuum amplitude in QCD is given by

$$Z[0] = \langle \det(i\not D + im) \rangle \tag{2.38}$$

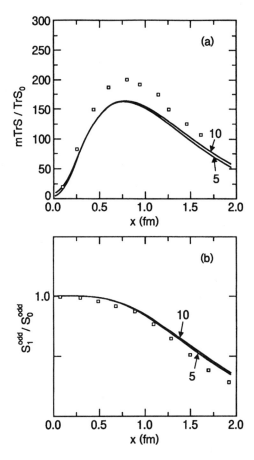

Fig. 2.3.   Behavior of the scalar (a) and vector (b) parts of the quark propagator S in a dilute instanton gas, normalized to the massless quark propagator. The squares are simulated results of Verbaarschot and Shuryak[12] with current quark mass of 10 MeV and the solid lines are the mean-field results of Kacir *et al.*[13]

where the averaging is made over all gauge configurations. When restricted to non-interacting instantons and anti-instantons, this averaging can be considerably simplified. Indeed, the exact determinant in a gauge background splits into $\det_{high} \times \det_{low}$ where "high" and "low" refer to high and low frequencies respectively. The low frequency part becomes

$$\det_{low} = \frac{\det(-S^{-1})}{\det(-S_0^{-1})} \sim \exp\left[\int_m^M i\,dm\,\text{Tr}(S - S_0)\right]. \qquad (2.39)$$

In the zero-mode space, $S - S_0 \sim \phi(T - im)^{-1}\phi^\dagger$, so that**

$$\det_{low} \sim \exp\left[\int_m^M i\,dm\,\text{Tr}\phi_I(T - im)_{IJ}^{-1}\phi_J^\dagger\right] = \frac{\det(T - im)}{\det(T - iM)}. \qquad (2.40)$$

Throughout, we have assumed that the zero modes originating from different instantons are orthogonal. Note that the factor involving M is canceled by the high momentum part of the fermion determinant. Thus

$$Z[0] = \langle\langle\det(-S^{-1})\rangle\rangle = \det(-S_0^{-1})\langle\langle\det(T - im)\rangle\rangle. \qquad (2.41)$$

where $\langle\langle\ldots\rangle\rangle$ denotes averaging over the collective coordinates of the instantons according to their invariant measure.

(i) 't Hooft's determinants

To display the dynamical content of the partition function (2.41), we use a trick due to Ilgenfritz[16] to rewrite the determinants in (2.40) in terms of integrals over Grassmann variables (the explicit summation over pseudoparticles, Dirac, flavor and color indices is omitted in the exponent)

$$\det(-S_0^{-1}) = \int [d\psi][d\psi^\dagger]\exp\int \psi^\dagger(i\partial_\mu\gamma_\mu + im)\psi,$$

$$\det(T - im) = \int [d\chi][d\chi^\dagger]\exp\chi^\dagger(\int \phi^\dagger i\partial_\mu\gamma_\mu\phi - im)\chi. \qquad (2.42)$$

For fixed values of χ and $\chi^\dagger$, a shift in the Grassmann variables ψ and $\psi^\dagger$ by $\psi \to \psi + i\phi\chi$, and $\psi^\dagger \to \psi^\dagger + i\chi^\dagger\phi^\dagger$ yield

$$\langle\langle\det(-S^{-1})\rangle\rangle = \langle\langle\int [d\psi][d\psi^\dagger]\exp(-\int \psi^\dagger S_0^{-1}\psi)[d\chi][d\chi^\dagger](-i)^{-NN_f}$$

$$\times \exp[i\chi^\dagger\int \phi^\dagger S_0^{-1}\psi + \int i\psi^\dagger S_0^{-1}\phi\chi - im\chi^\dagger\chi]\rangle\rangle. \qquad (2.43)$$

** We ignore K for simplicity, *i.e.* we put $K_{IJ} = im\delta_{IJ}$.

Using the chiral properties of zero modes, one can show that in the limit $m \to 0$, $\det S^{-1}$ is invariant under $U_L(N_f) \times U_R(N_f)$ transformations of the fermion field ψ. Now we rewrite the sum over the instanton and anti-instanton indices I and flavor indices f in the exponent of (2.43) as a product over I and f. Due to the Grassmannian character of the variables only the zeroth and the first order terms of the Taylor expansion of the exponent are nonzero. Thus, the $\chi_{If}^\dagger, \chi_{I,f}$-integrals are Gaussian and can be evaluated in a straightforward way. The result reads,

$$
\begin{aligned}
\langle\langle \det(-S^{-1}) \rangle\rangle &= \int [d\psi][d\psi^\dagger] \exp\left(- \int \psi^\dagger S_0^{-1} \psi\right) \\
&\times \left\langle\left\langle \prod_{If} \left(i \int \psi^\dagger S_0^{-1} \phi_I \int \phi_I^\dagger S_0^{-1} \psi \right) \right\rangle\right\rangle.
\end{aligned}
\tag{2.44}
$$

In order to obtain the last term of this equation, we have reversed the order of ψ and $\psi^\dagger$. By using the Grassmannian properties of the fermion fields, (2.44) can be rewritten in terms of a 't Hooft's determinant in flavor space as

$$
\begin{aligned}
\langle\langle \det(-S^{-1}) \rangle\rangle &= \int [d\psi][d\psi^\dagger] \exp\left(- \int \psi^\dagger S_0^{-1} \psi\right) \\
&\times \prod_I \frac{1}{N_f!} \left\langle\left\langle \det_{fg} \left(i \int \psi^{\dagger f} S_0^{-1} \phi_I \int \phi_I^\dagger S_0^{-1} \psi^g \right) \right\rangle\right\rangle.
\end{aligned}
\tag{2.45}
$$

Since the zero modes have definite chirality, the determinantal factor in (2.45) explicitly breaks the $U_A(1)$ symmetry and induces nonlocal interactions between fermions on the scale of the size of the instantons. At the mean-field level they also induce the spontaneous breakdown of chiral symmetry.

(ii) Case $N_f = 1$

To demonstrate how fermionic zero modes cause the spontaneous breakdown of chiral symmetry, we will outline the mean-field calculation for one flavor. We note, however, that this case is only academic since for $N_f = 1$, chiral symmetry is always broken by the anomaly. Nevertheless, the exercise will be useful for the generalization to $N_f = 2, 3$.

For $N_f = 1$, integration over the color group can be easily performed with the help of the formula

$$
\int dU \, U_i^a U_b^{\dagger j} = \frac{1}{N_c} \delta_b^a \delta_i^j.
\tag{2.46}
$$

We obtain

$$Z = \int [d\psi][d\psi^\dagger] \exp\left[\int d^4x \psi^\dagger i\partial\!\!\!/\psi\right] \Theta_+^{N_+} \Theta_-^{N_-}, \qquad (2.47)$$

where

$$\Theta_\pm = -\frac{1}{N_c V_4} \int \frac{d^4k}{(2\pi)^4} k^2 \varphi'^2(k)\, \psi_\alpha^\dagger(k)\gamma_\mp \psi_\alpha(k). \qquad (2.48)$$

Here φ' is the Fourier transform of the 't Hooft zero modes and $\gamma_\pm = (1 \pm \gamma_5)/2$. If we notice that the Θ terms in (2.48) play the role of "fugacities" in an otherwise conventional partition function, then Z can be analyzed using standard techniques from statistical mechanics. More specifically, with the help of the inverse Laplace transform,

$$\Theta_\pm^{N_\pm} = \frac{N_\pm!}{2i\pi} \int_{\epsilon-i\infty}^{\epsilon+i\infty} d\beta_\pm \frac{1}{\beta_\pm^{N_\pm+1}} \exp(-\beta_\pm \Theta_\pm), \qquad (2.49)$$

we can rewrite (2.47) in terms of quark bilinears only. Since the convergence of the integrals does not enter in the derivation of the mean-field equations, we will omit the range of integrations from now on. The fermionic integrations can then be performed analytically. Up to constant factors we obtain

$$Z = \int d\beta_\pm \exp\left[-(N_+ + 1)\ln\beta_+ - (N_- + 1)\ln\beta_-\right]$$

$$\times \exp\left[V_4 N_c \int \frac{d^4k}{(2\pi)^4} \operatorname{Tr}\ln\left(k\!\!\!/ + i\gamma_\pm \frac{\varphi'^2(k)k^2\beta_\mp}{N_c}\right)\right]. \qquad (2.50)$$

In order to highlight that the saddle point approximation is exact in the thermodynamic limit, we have transformed the variables $\beta_\pm \to V_4 \beta_\pm$. Note that due to the two-point "vertex" generated by the instantons, the fermion determinant in (2.50) contains a constituent mass term. In the thermodynamic limit ($N_\pm$ and $V_4 \to \infty$ with N/V_4 finite), the integration in (2.50) can be performed by the saddle point method. For $N_+ = N_- = N/2$, this leads to the following gap equation:

$$\int \frac{d^4k}{(2\pi)^4} \frac{M_k^2(0)}{k^2 + M_k^2(0)} = \frac{N}{4N_c V_4} \qquad (2.51)$$

where

$$M_k(0) = \frac{\varphi'^2(k)k^2\beta}{N_c} \qquad (2.52)$$

and φ' is the zero-mode profile around an instanton of the size ρ, explicitly defined in the Appendix. Since $\beta_+ = \beta_-$ for $N_+ = N_-$, we have omitted the subscripts $\pm$. Remember that here the pseudoparticle density N/V_4 is proportional to the vacuum gluon condensate. Using the value of the gluon condensate suggested by the QCD sum rules $(N/V_4 \sim (200\,\mathrm{MeV})^4)$, one obtains from eq. (2.51) a constituent quark mass $M_0(0)$ of about $345\,\mathrm{MeV}$ and a quark condensate of about $(-250\,\mathrm{MeV})^3$.

For finite current masses, the gap equation (2.51) takes the general form [13]

$$\frac{N(1 - 2m\lambda(m))}{4N_c V_4} = \int \frac{d^4k}{(2\pi)^4} \frac{(k^2 + m^2)M_k^2(m) - mM_k(m)k^2}{k^4 - 2mM_k(m)k^2 + (k^2 + m^2)M_k^2(m)} \quad (2.53)$$

where we have set $M_k(m)/M_k(0) = \lambda(m)/\lambda(0)$. For $k \to 0$, $M_0(m) \to \lambda(m)(N/(V_4 N_c))(2\pi\rho)^2$, while for $k \to \infty$, $M_k(m) \to 1/k^6$. The sensitivity of $\lambda(m)$ to m for light quark masses is small. We note that at low momentum, $M_k(m)$ acquires a non-analytical contribution [13]

$$M_k(m) \sim M_0(m)\left(1 + 3z^2\log\frac{z}{2}\,e^{c+\frac{1}{2}}\right)_{z=k\rho/2} \quad (2.54)$$

where $c = 0.577$ is Euler's constant. The running quark mass does not show up as a simple pole. What this means is that as $k \to 0$, the quark modes become tachyonic in the long-wavelength limit. To the extent that long-wavelength quarks are unphysical, this should be a priori of no real concern. The problem, though, is that the model does not confine. This implies that the instanton vacuum as a state is unstable. How this can be fixed is not known. It is our hope that this will have mild consequences on the remainder of our discussion.

(iii) Case $N_f = 2, 3$

Consider now the case of two or more flavors. In this case, the partition function involves both quartic and sextic fermionic operators around the single instanton (anti-instanton). In other words, the instanton (anti-instanton) can trap N_f quarks (anti-quarks) with different flavors. The group averaging is equivalent to projecting onto a color singlet. Proceeding as in the one-flavor case, we obtain the following expression for the determinant in the partition function (2.45):

$$\frac{1}{2}\langle\langle\det{}_{fg}\rangle\rangle_{N_f=2} = \frac{1}{4(N_c^2 - 1)}k^4\varphi'^4[(1 - \frac{1}{2N_c})(\psi^\dagger\gamma_+\psi)^2$$

$$+\frac{1}{8N_c}(\psi^\dagger\gamma_+\sigma_{\mu\nu}\psi)^2] + (\gamma_5 \leftrightarrow -\gamma_5). \quad (2.55)$$

Here, ψ is an isodoublet of the u and d quark fields. We have used the convention $\sigma_{\mu\nu} = \frac{1}{2}[\gamma_\mu, \gamma_\nu]$. We recognize the expected 't Hooft interaction, but here the interactions are smeared over the size of an instanton. The case of three flavors is more complicated. The effective sextic vertex for three flavors and for two or three colors reads

$$\frac{1}{3!}\langle\langle\det{}_{fg}\rangle\rangle_{N_f=3} = \frac{1}{3!}\frac{1}{N_c(N_c^2-1)}\int dR^3_{u,d,s}\epsilon_{f_1f_2f_3}\epsilon_{g_1g_2g_3}$$

$$\times\left[(1 - \frac{1}{2(N_c+2)})(q^\dagger_{f_1}\gamma_+q_{g_1})(q^\dagger_{f_2}\gamma_+q_{g_2})(q^\dagger_{f_3}\gamma_+q_{g_3})\right.$$

$$\left.+\frac{3}{8(N_c+2)}(q^\dagger_{f_1}\gamma_+q_{g_1})(q^\dagger_{f_2}\gamma_+\sigma_{\mu\nu}q_{g_2})(q^\dagger_{f_3}\gamma_+\sigma_{\mu\nu}q_{g_3})\right]$$

$$+(\gamma_5 \leftrightarrow -\gamma_5). \quad (2.56)$$

The measure is defined as

$$dR^{N_f}_{u_1,\cdots,u_{N_f}} = (2\pi)^4\delta^{(4)}(\sum_{i=1}^{N_f}q_{u_i} - \sum_{i=1}^{N_f}k_{u_i})$$

$$\times\prod_{i=1}^{N_f}\frac{d^4k_{u_i}}{(2\pi)^4}\frac{d^4q_{u_i}}{(2\pi)^4}\varphi'(k_{u_i})k_{u_i}\varphi'(q_{u_i})q_{u_i}. \quad (2.57)$$

The first term, leading in N_c, has the 't Hooft determinant structure. It is surprising that the subleading term has such a simple form in the case of three flavors.

It can be easily demonstrated that for an arbitrary number of flavors and in leading order in N_c, the partition function (2.47) reduces to the partition function composed of 't Hooft determinants. To prove this, we evaluate the average over the color orientations using a cumulant expansion,

$$\langle\langle\exp(i\chi^\dagger\int\phi^\dagger S_0^{-1}\psi + i\int\psi^\dagger S_0^{-1}\phi\chi))\rangle\rangle_U =$$

$$\exp\langle\langle-\chi^\dagger\int\phi^\dagger S_0^{-1}\psi\int\psi^\dagger S_0^{-1}\phi\chi\rangle\rangle_U, \quad (2.58)$$

where the brackets $\langle\langle\cdots\rangle\rangle_U$ denote averaging over the color orientations for a fixed instanton size. To leading order in N_c, the higher-order cumulants do not contribute. In the integrand, the $\chi^\dagger, \chi$-variables occur as the exponent of a quadratic form, diagonal in the pseudoparticles and nondiagonal in the flavor indices. Performing this integral yields

$$\langle\langle\det(-S^{-1})\rangle\rangle = \int [d\psi][d\psi^\dagger]\exp(-\int\psi^\dagger S_0^{-1}\psi)$$
$$\times \prod_I \int dz_I \det(i\langle\langle\int\psi^\dagger S_0^{-1}\phi_I\int\phi_I^\dagger S_0^{-1}\psi\rangle\rangle_U). \qquad (2.59)$$

The determinant in this equation is over the flavor indices and can be identified as a smeared 't Hooft determinant. One could generalize this procedure to the case of massive current quarks. In this case, the cumulant expansion gives

$$Z[0] = \int [d\psi][d\psi^\dagger]\exp(-\int\psi^\dagger S_0^{-1}\psi)\prod_I\int dz_I$$
$$\times \det\left(m\rho_I - \frac{\rho_I}{2i}\langle\langle\int\psi^\dagger S_0^{-1}\phi_I\int\phi_I^\dagger S_0^{-1}\psi\rangle\rangle_U\right). \qquad (2.60)$$

The determinant is over flavor and the averaging is over color orientations. In a totally random instanton vacuum, the effect of the instantons and anti-instantons is to induce effective interactions among the light quarks as shown in Fig. 2.4.

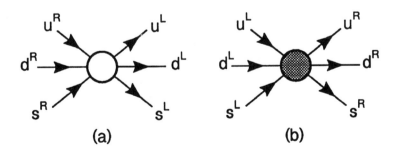

Fig. 2.4. Induced effect from an instanton (a) and an anti-instanton (b) on three light flavors.

(iv) Coarse graining

For a sufficiently random system, it is best to trade the averaging over the instanton positions for the averaging over the instanton $(n_+(x))$ and anti-instanton $(n_-(x))$ densities in Euclidean space. This procedure consists in coarse-graining the system as shown in Fig. 2.5. We choose to rewrite the integral over the positions of the instantons as an integral over the density of the instantons by allowing the number of pseudoparticles to change. This corresponds to the partition of the total Euclidean volume V_4 into blocks (hypercubes) ΔV_i, each of which contains $n^+(x_i)$ instantons and $n^-(x_i)$ anti-instantons. Here x_i denotes the position of the block. Specifically this amounts to rewriting

$$\int \prod_I d^4 R_I d\rho_I \mathbf{F}_I \;=\; \int \prod_{\Delta_i} \Delta^+(x_i)\Delta^-(x_i)\mathbf{F}_+^{n^+(x_i)}\mathbf{F}_-^{n^-(x_i)}\mu_\pm$$

$$=\; \int \prod_x dn^+(x)dn^-(x)\mathbf{F}_+^{n^+(x)}\mathbf{F}_-^{n^-(x)}\mu_\pm \quad (2.61)$$

where the induced vertices are

$$\mathbf{F}_\pm = \det_\pm = \frac{1}{N_f!}\det\left(m\rho_I - \rho_I\langle\frac{1}{2i}\int \psi^\dagger S_0^{-1}\phi_I\phi_I^\dagger S_0^{-1}\psi\rangle_U\right). \quad (2.62)$$

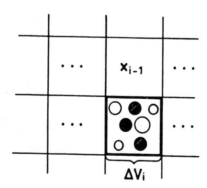

Fig. 2.5. Coarse graining of the random instanton phase.

The measure $\mu_\pm$ reflects the bulk properties of the random system. In the totally random case, $\mu_\pm = 1$. In general, it is unknown. However, if we were to ensure that the ensemble is neutral and has the correct susceptibility χ and compressibility σ^2, then we would expect the measure $\mu_\pm$ to fulfill

$$\langle n^+ \rangle = \langle n^- \rangle = \frac{1}{2}\frac{N}{V_4} = \frac{n}{2},$$

$$\chi = \frac{N}{V_4} = \int d^4x \langle (n^+ - n^-)^2 \rangle,$$

$$\sigma^2 = \int d^4x \langle (n^+ + n^- - n)^2 \rangle. \tag{2.63}$$

Combining (2.63) with (2.61), we have

$$
\begin{aligned}
\mathcal{S} = \ &+ \frac{1}{2n} \int d^4x (n^+ - n^-)^2 \\
&+ \frac{n}{\sigma^2} \int d^4x (n^+ + n^-)(\ln\left(\frac{n^+ + n^-}{n}\right) - 1) \\
&+ \int d^4x \left(\psi^\dagger S_0^{-1} \psi - n^+ \ln \det_+ - n^- \ln \det_-\right)
\end{aligned}
\tag{2.64}
$$

which is the minimal effective action in Euclidean space for the quarks in the instanton vacuum. The number difference between instantons and anti-instantons is Gaussian and is in agreement with the susceptibility χ. The number sum between the instanton and anti-instanton is consistent with the expression of the compressibility in the Gaussian approximation.

2.3.3. *Anomalies*

The coarse-grained effective action (2.64) reproduces both the axial and scale anomalies. Indeed, under a local $U_A(1)$ transformation

$$\psi_L \to e^{i\alpha}\psi_L, \qquad \psi_R \to e^{-i\alpha}\psi_R \tag{2.65}$$

the determinants in (2.64) transform as follows

$$\ln \det_\pm \to \ln\left(e^{\mp 2iN_f\alpha}\det_\pm\right). \tag{2.66}$$

As a result, the axial singlet current $j_\mu^5 = \psi^\dagger \gamma_\mu \gamma^5 \psi$ is not conserved, even in the chiral limit,

$$\partial \cdot j^5 = 2N_f(n_+ - n_-) + 2i\sum_f m_f \psi^\dagger \gamma_5 \psi. \tag{2.67}$$

Under a scale transformation, we expect

$$\rho \to \lambda^{-1}\rho, \qquad \psi(x) \to \lambda^{3/2}\psi(\lambda x), \qquad n^{\pm}(x) \to \lambda^4 n^{\pm}(\lambda x). \qquad (2.68)$$

The first two terms in (2.64) are scale non-invariant. Hence, the dilatational current $J_\mu = x^\sigma \Theta_{\sigma\mu}$, with $\Theta_{\mu\nu}$ the energy-momentum tensor, is not conserved. Specifically

$$\partial \cdot J = \left(\frac{\delta S}{\delta \lambda}\right)_{\lambda=0} = \frac{4n}{\sigma^2}(n^+ + n^-) + \frac{2}{n}(n_+ - n_-)^2 + \mathcal{O}(N_f). \qquad (2.69)$$

Since $\partial \cdot J = \Theta^\mu_\mu$, a comparison with the trace anomaly (see Chapter 1) in the limit $N_f \to 0$, gives

$$\sigma^2 = \frac{12}{11N_c}\left(\frac{N}{V_4}\right). \qquad (2.70)$$

The compressibility of the instanton vacuum $\sigma^2 \sim N_c^0$ for $n \sim N_c$. It vanishes in case $n \sim N_c^0$.

2.3.4. Glueballs

To describe the long-wavelength glueball excitations in the instanton vacuum, we can define two interpolating fields:

$$H(x) = \frac{1}{2}(n^+(x) + n^-(x)),$$

$$Q(x) = \frac{1}{2}(n^+(x) - n^-(x)). \qquad (2.71)$$

H refers to the scalar glueballs and Q to the pseudoscalar glueballs. It is straightforward to rewrite (2.64) in terms of the effective glueball fields and the constituent fermion fields

$$\mathcal{S} = \int [d\psi][d\psi^\dagger] e^{-\int \psi^\dagger S_0^{-1} \psi}$$

$$\times \int [dH]\,[dQ]\,W(H,Q)\,(\det_+ \det_-)^H (\det_+ /\det_-)^Q. \qquad (2.72)$$

The weight W is *Gaussian* in H and *logarithmic* in Q, and follows from (2.64). It is worth noting that the scalar glueballs couple to the $U_A(1)$-invariant fermionic combinations, while the pseudoscalar glueballs couple to the $U_A(1)$-variant fermionic combinations. Both combinations are explicitly invariant under $SU_L(N_f) \times SU_R(N_f)$. The field H describes fluctuations in the total number of instantons and anti-instantons in the vacuum, while the field Q describes fluctuations in the difference in the number of instantons and anti-instantons in the vacuum. A detailed discussion of these issues is given in ref. [13].

2.3.5. *Mesons*

To discuss the mesonic excitations in the instanton vacuum, we will rely on an approximate bosonization scheme of $Z[0]$ with the effective action (2.64). For that, consider the set of mesonic correlators,

$$\mathbf{C}_\gamma(x) = \langle T^* \psi^\dagger \gamma \psi(x)\, \psi^\dagger \gamma \psi(0) \rangle \tag{2.73}$$

where $\gamma = (1, \gamma_5, \gamma_\mu, ...) \otimes (1, T^a)$. To describe the long-wavelength fluctuations, let us introduce auxiliary fields through the identity

$$1 = \int d\pi^\pm\, dP^\pm\, e^{\int P^\pm (\pi^\pm - \langle \psi^\dagger S_0^{-1} \phi_\pm \phi_\pm^\dagger S_0^{-1} \psi \rangle)}. \tag{2.74}$$

Here $P^\pm$ are complex $N_f \times N_f$ matrices, of which a general parametrization is given by

$$P^\pm = e^{\pm i\kappa/2} \sigma e^{\pm i\kappa/2} \tag{2.75}$$

where matrix κ is $U(N_f)$ valued, and σ a real $N_f \times N_f$ matrix. Inserting (2.75) into (2.64), integrating over the fermionic fields and eliminating the auxiliary fields $\pi^\pm$ via the saddle point approximation, yields

$$\mathbf{C}_\gamma(x) = \int dP^\pm\, e^{-\mathbf{S}_{\text{eff}}^{-1}[P^\pm]} \{ + \text{Tr}(\mathbf{S}[x,0,P]\gamma \mathbf{S}[0,x,P]\gamma)$$
$$- \text{Tr}(\mathbf{S}[x,x,P]\gamma \mathbf{S}[0,0,P]\gamma) \} \tag{2.76}$$

where the effective action is given by

$$\mathbf{S}_{\text{eff}}[P^\pm] = -N_c \text{Tr}\left(\ln \mathbf{S}^{-1} \right)$$
$$+ \frac{n}{2} \left(\text{lndet} \frac{4P^+}{n\rho} + \text{logdet} \frac{4P^-}{n\rho} \right)$$
$$- 2 \int d^4 z \, \text{Tr}_f m(P^+(z) + P^-(z)). \tag{2.77}$$

The trace Tr is over flavor and Dirac indices, as well as four momenta, the trace Tr_f is over flavor indices, and the determinant det is over flavor indices as well as position space. The propagator $\mathbf{S}$ is in the P-background. In momentum space,

$$\mathbf{S}^{-1}[k, l, P] = +\slashed{k} - im$$

$$-iP^+(k-l)\left(1 - \frac{im\slashed{k}}{k^2}\right)\gamma_5^+\left(1 - \frac{im\slashed{l}}{l^2}\right)\sqrt{M_k(m)M_l(m)}$$

$$-iP^-(k-l)\left(1 - \frac{im\slashed{k}}{k^2}\right)\gamma_5^-\left(1 - \frac{im\slashed{l}}{l^2}\right)\sqrt{M_k(m)M_l(m)}.$$

$$(2.78)$$

In the representation (2.76), the mesonic correlator is given by the diagrams shown in Fig. 2.6.

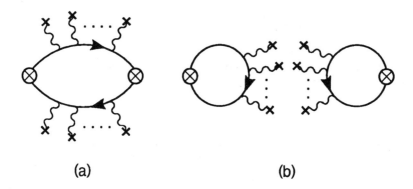

(a)   (b)

Fig. 2.6. Diagrams contributing to the mesonic correlator. The wiggly lines are exchange of P mesons as described in the text.

The running mass $M_k(m)$ in (2.78) is defined by eq. (2.53). The parameter β in (2.52) follows from the saddle point equation with vacuum conditions in (2.76). The latter are specified by $\kappa = 0$ (isospin and parity symmetric vacuum), and $\delta\mathbf{S}^{-1}/\delta\sigma = 0$ with σ diagonal. The use of the saddle point approximation in eliminating the $\pi^\pm$ field is exact in the limit where the number of colors becomes large. In this limit, the two diagrams shown in Fig. 2.6 are dominant.

Figure 2.7a shows the result for the pion correlation function normalized to the free and massless correlator from the work of Kacir, Prakash and Zahed[13]. The current mass used is $m = 5$ MeV. The dotted line is the contribution from the connected diagram, the dashed line is the contribution from the disconnected diagram, and the full line is the sum. The connected contribution is spurious as it yields a two-quark cut in the imaginary part. This contribution is a reminder that the model does not confine. Figure 2.7b shows the raw pion correlator multiplied by $x^{3/2}$ on a semi-log plot. The asymptotic form of the correlator sets in at about 2.5 fm and is consistent with the formation of a pion mass. The two curves are for a current mass of 5 and 10 MeV. The 5 MeV curve is consistent with a pion mass of 158 MeV. Figure 2.7c compares the result shown in Fig. 2.7a with the simulated instanton calculations of Verbaarschot and Shuryak[17] (squares) and the quenched lattice simulations by Chu, Grandy, Huang and Negele[18] (solid circles). The instanton simulations were carried out with a current mass of 10 MeV. Similar results hold for the pseudoscalar nonet.

Figure 2.8 shows the result for the ρ correlator. The symbols are the same as the one for Fig. 2.7c. We note that in the mean-field calculation only the connected diagram contributes to the correlation function. The entire effects in the mean-field calculation come from the spurious exchange of two constituent quarks and the result is in agreement with the instanton simulation. The same remarks apply to the other vector correlators as well as axial-vector (transverse part) correlators.

Are the results in the pseudoscalar channels consistent with PCAC? To answer this question, we will further parametrize κ as

$$\kappa = \sum_{a=0}^{8} \kappa_a T^a \tag{2.79}$$

in eqs. (2.76-2.78). Through a local chiral rotation, we can transfer the phases in the P-fields into the kinetic term. This phase redefinition will affect the residue structure of the correlation function but not its singularities. In the long-wavelength limit, the induced action for these phases is, in momentum representation and physical basis,

$$\mathcal{S}(p) = \frac{1}{2} f^2 \, p^2 (\vec{\pi}^2 + \vec{K}^2 + \eta^2 + \eta'^2) + \cdots. \tag{2.80}$$

Here f is the pseudoscalar decay constant in the chiral limit

$$f^2 = \lim_{q^2 \to 0} \frac{4N_c}{q^2} \int \frac{d^4 k}{(2\pi)^4} \frac{(M_{k_+}(m)k_- - M_{k_-}(m)k_+)^2}{(k_+^2 + M_{k_+}(m)^2)(k_-^2 + M_{k_-}(m)^2)} \tag{2.81}$$

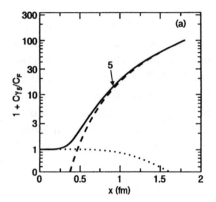
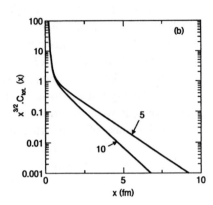

Pseudoscalar channel (pion)

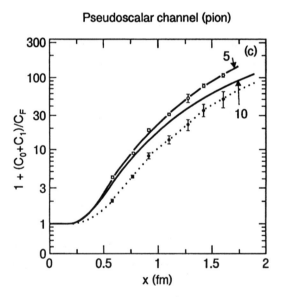

Fig. 2.7. (a) Pion correlator in the mean-field approximation normalized to the free massless correlator. (b) Raw pion correlator times $\sqrt{x^3}$. (c) Same as (a) in comparison to instanton simulations (open circles) and lattice simulations (solid circles). The mean-field calculations were done with the current mass of 5 and 10 MeV.

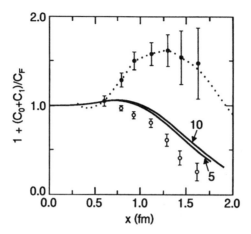

Fig. 2.8. The ρ correlator in the mean-field approximation (full lines), from instanton simulations (open circles) and lattice simulations (solid circles) normalized to the free massless correlator.

with $k_\pm = k \pm q/2$. The meson masses in (2.80) are found to be [tt]

$$m_\pi^2 = -2m_u \frac{\langle \bar{\psi}\psi \rangle}{f^2},$$

$$m_K^2 = -(m_u + m_s) \frac{\langle \bar{\psi}\psi \rangle}{f^2} \qquad (2.82)$$

where $\langle \bar{\psi}\psi \rangle$ is the quark condensate (in Minkowski space) in the chiral limit in the instanton liquid

$$\langle \bar{\psi}\psi \rangle = -4N_c \left(\frac{n}{2N_c} \lambda(0) - \int \frac{d^4k}{(2\pi)^4} \frac{M_k(0)}{k^2 + M_k(0)^2} \right). \qquad (2.83)$$

The relations (2.82) are the Gell-Mann-Oakes-Renner (GMOR) relations. This shows that the light pseudoscalar excitations in the instanton liquid are indeed Goldstone bosons. The pseudoscalar excitations are consistent with PCAC.

[tt]Strictly speaking, the expansion in m_s is unjustified. It turns out, however, that the asymptotic fall off of the strange pseudoscalar correlators is consistent with the GMOR result [13]. All the non-linearities in m_s drop out.

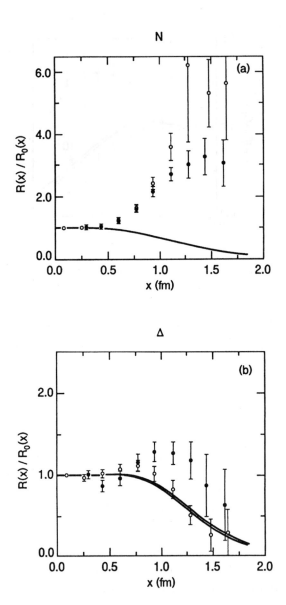

Fig. 2.9. Nucleon (a) and Δ (b) correlator in the mean-field approximation (full line) in comparison with instanton simulation (open circles) and lattice simulations (solid circles). The curves are normalized by the free massless correlator.

The mass of the η' in (2.80) follows from the admixture of the pseudoscalars with the gluons as discussed above. It is not a Goldstone boson. The relevant part in the induced effective action (2.64) is

$$S_{\eta'G} = \frac{1}{2n} \int d^4 z(n^+ - n^-)^2 \; - i\sqrt{2N_f} \int d^4 z(n^+ - n^-)\, \eta'. \quad (2.84)$$

The mass of the η' is obtained by integrating over the instanton and anti-instanton densities. This clearly shows that a vacuum state that correctly reproduces the topological susceptibility yields a reasonable prediction for the η' mass. In our case, $m_{\eta'} = 1172$ MeV.

A similar analysis can be extended to baryons. Here we will only quote a few results, details can be found in the original work. Figure 2.9a shows the behavior of the nucleon correlator[13] (full lines) assumed to be made of three constituent quarks vs. the instanton simulated correlator[19] (open circles) and the lattice simulated correlator[18] (solid circles). Clearly, the mean-field analysis with three constituent quarks ignores the attraction in the spin-isospin zero diquark channel, which is subleading in N_c (see Chapter 3). The attraction, however, is approximately 25 times weaker in the nucleon as compared to the pion shown in Fig. 2.7c. Figure 2.9b shows the behavior of the Δ-correlator in the mean-field analysis (full lines) in comparison with instanton simulations (open circles) and lattice simulations (solid circles). The simulations using instantons are entirely consistent with the mean-field analysis which is composed of only three constituent quarks! There is no binding in the Δ-channel. Aside from generating a constituent mass, the instantons do not affect the correlators in the decuplet. Other baryonic channels can be investigated similarly.

The use of the mean-field approximation seems to be justified by the dilute distribution of the instantons in the vacuum and the use of the large N_c limit. What is the rationale behind the large N_c argument? The answer to this question will be addressed in the following chapter.

Appendix

(A$_I$) Euclidean space conventions

We present here the formulae that allow the transcription from Minkowski to Euclidean space. Throughout $x_4^E = ix_0$, with all spatial coordinates unchanged.

Vector potentials:

$$A_4^E = -iA_0, \quad A_i^E = -A_i \quad (i = 1, 2, 3). \tag{A2.1}$$

Covariant differentiation:

$$D_4^E = -iD_0, \quad D_i^E = -D_i \quad (i = 1, 2, 3) \tag{A2.2}$$

hence

$$D_\mu^E = \frac{\partial}{\partial x_\mu^E} - igA_\mu^E. \tag{A2.3}$$

Stress tensor:

$$G_{\mu\nu}^{a\,E} = \frac{\partial}{\partial x_\mu^E} A_\nu^{aE} - \frac{\partial}{\partial x_\nu^E} A_\mu^{aE} + gf^{abc} A_\mu^{b\,E} A_\nu^{c\,E}. \tag{A2.4}$$

Gamma matrices:

$$\gamma_4^E = \gamma_0 \quad , \gamma_i^E = -i\gamma_i \quad (i = 1, 2, 3) \tag{A2.5}$$

$$\{\gamma_\mu^E, \gamma_\nu^E\} = 2\delta_{\mu\nu}. \tag{A2:6}$$

Fermi fields:

$$\psi^E = \psi, \quad \psi^{\dagger\,E} = i\overline{\psi}. \tag{A2.7}$$

Note the convention relating $\psi^{\dagger\,E}$ and $\overline{\psi}^M$. In this way we keep track of the transformation properties under Lorentz group. For example, $\psi^\dagger\psi$ is a *scalar* in Euclidean space.

Action:

$$S^E = -iS \tag{A2.8}$$

$$S = \int d^4x \left(-\frac{1}{4} F_{\mu\nu}^a F_{\mu\nu}^a + \overline{\psi}(i\gamma_\mu D_\mu - M)\psi \right)$$

$$S^E = \int d^4x^E \left(\frac{1}{4} F_{\mu\nu}^{a\,E} F_{\mu\nu}^{a\,E} + \psi^{\dagger\,E}(-i\gamma_\mu^E D_\mu^E - iM)\psi^E \right).$$

Unless explicitly mentioned, all formulae in this chapter are given in Euclidean space. The index E is usually omitted.

(A_{II}) Zero modes and overlap integrals

The explicit form of the zero mode density matrices are
Instanton - Instanton :

$$\phi_I(x)_{i\alpha}\phi_{I'}^\dagger(y)_{j\beta} = \frac{1}{8}\varphi_I(x)\varphi_{I'}(y)[\not{x}\gamma_\mu\gamma_\nu\not{y}\frac{1-\gamma_5}{2}]_{ij} \otimes (U_I\tau_\mu^-\tau_\nu^+U_{I'}^\dagger)_{\alpha\beta}.$$
(A2.9)

Instanton - Anti-instanton :

$$\phi_I(x)_{i\alpha}\phi_{\overline{I'}}^\dagger(y)_{j\beta} = -i\frac{1}{2}\varphi_I(x)\varphi_{\overline{I'}}(y)[\not{x}\gamma_\mu\not{y}\frac{1-\gamma_5}{2}]_{ij} \otimes (U_I\tau_\mu^-U_{\overline{I'}}^\dagger)_{\alpha\beta},$$
(A2.10)

where

$$\varphi(x) = \frac{\rho_I}{\pi\sqrt{x^2}(x^2+\rho_I^2)^{\frac{3}{2}}}.$$
(A2.11)

The overlap integrals are explicitly given by

$$T_{\overline{I}I} = \int d^4x\phi_{\overline{I}}^{\dagger i\alpha}(x-R_{\overline{I}})i\not\partial\phi^{i\alpha}(x-R_I)$$

$$= \mathrm{Tr}(U_I\tau_\mu^-U_{\overline{I}}^\dagger)r_\mu\frac{1}{2\pi^2}\frac{1}{r}\frac{d}{dr}\mathcal{M}(r),$$
(A2.12)

$$K_{II'} = im\int d^4x\phi_I^{\dagger i\alpha}(x-R_I)\phi_{I'}^{i\alpha}(x-R_{I'})$$

$$= im\,\mathrm{Tr}(U_I^\dagger U_{I'}^\dagger)\frac{1}{2\pi^2}\mathcal{M}(r)$$
(A2.13)

where $r_\mu = R_\mu^I - R_\mu^{I'}$

$$\mathcal{M}(r) = \frac{1}{r}\int_0^\infty dp p^2|\varphi(p)|^2 J_1(pr)$$
(A2.14)

and the Fourier transformations of the zero-mode profile (A-11) is explicitly given by

$$\varphi(p) = \pi\rho^2\frac{d}{dx}(I_0(x)K_0(x) - I_1(x)K_1(x))|_{x=\frac{|p|\rho}{2}}.$$
(A2.15)

(A_{III}) 't Hooft symbols and their properties

't Hooft symbols are defined by the relations

$$\tau_\mu^+ \tau_\nu^- = \delta_{\mu\nu} + i\eta_{\mu\nu}^a \tau^a,$$
$$\tau_\mu^- \tau_\nu^+ = \delta_{\mu\nu} + i\bar{\eta}_{\mu\nu}^a \tau^a \tag{A2.16}$$

or, explicitly

$$\eta_{\mu\nu}^a = \begin{cases} \epsilon_{a\mu\nu} & \mu,\nu = 1,2,3 \\ -\delta_{a\nu} & \mu = 4 \\ +\delta_{a\mu} & \nu = 4 \\ 0 & \text{otherwise} \end{cases} \tag{A2.17}$$

$\bar{\eta}$ follows by flipping the signs of the Kronecker deltas. Some useful properties of the 't Hooft symbols are

$$\eta_{\mu\nu}^a = \frac{1}{2}\epsilon_{\mu\nu\alpha\beta}\eta_{\alpha\beta}^a$$
$$\eta_{\mu\nu}^a \eta_{\mu\nu}^b = 4\delta_{ab}$$
$$\eta_{\mu\nu}^a \eta_{\mu\lambda}^a = 3\delta_{\nu\lambda}$$
$$\eta_{\mu\nu}^a \eta_{\gamma\lambda}^a = \delta_{\mu\gamma}\delta_{\nu\lambda} - \delta_{\mu\lambda}\delta_{\nu\gamma} + \epsilon_{\mu\nu\gamma\lambda}$$
$$\epsilon_{\mu\nu\lambda\sigma}\eta_{\gamma\sigma}^a = \delta_{\gamma\mu}\eta_{\nu\lambda}^a - \delta_{\gamma\nu}\eta_{\mu\lambda}^a + \delta_{\gamma\lambda}\eta_{\mu\nu}^a$$
$$\eta_{\mu\nu}^a \eta_{\mu\lambda}^b = \delta_{ab}\delta_{\nu\lambda} + \epsilon_{abc}\eta_{\nu\lambda}^c$$
$$\epsilon_{abc}\eta_{\mu\nu}^b \eta_{\gamma\lambda}^c = \delta_{\mu\gamma}\eta_{\nu\lambda}^a - \delta_{\mu\lambda}\eta_{\nu\gamma}^a - \delta_{\nu\gamma}\eta_{\mu\lambda}^a + \delta_{\nu\lambda}\eta_{\mu\gamma}^a$$
$$\eta_{\mu\nu}^a \bar{\eta}_{\mu\nu}^b = 0$$
$$\eta_{\gamma\mu}^a \bar{\eta}_{\gamma\lambda}^b = \eta_{\gamma\lambda}^a \bar{\eta}_{\gamma\mu}^b . \tag{A2.18}$$

The relations for the $\bar{\eta}$'s are obtained by exchanging $\eta_{\mu\nu}^a \leftrightarrow \bar{\eta}_{\mu\nu}^a$ and flipping the sign of $\epsilon_{\mu\nu\alpha\beta}$.

References

1. G. 't Hooft, Phys. Rev. **D14**, 3432 (1976).
2. C. G. Callan, R. F. Dashen and D. J. Gross, Phys. Rev. **D17**, 2717 (1978).
3. D. I. Diakonov and Yu. V. Petrov, Nucl. Phys. **B245**, 259 (1984).
4. E. V. Shuryak, Nucl. Phys. **B203**, 116, 140, 237 (1982).
5. M. A. Nowak, Acta Phys. Pol. **B22**, 697 (1991).
6. M.-C. Chu, J.M. Grandy, S. Huang and J. Negele, Nucl. Phys. **B34** [Proc. Suppl.], 170 (1994).

7. E. V. Shuryak, Nucl. Phys. **B302**, 574 (1988).
8. E. V. Shuryak, Phys. Rep. **115**, 151 (1984).
9. J. M. Ziman, *Models of Disorder* (Cambridge University Press, 1979).
10. M.L. Mehta, *Random Matrices* (Academic Press, N.Y. 1991).
11. I. Zahed, Stony Brook preprint, SUNY-NTG-94-22.
12. J.J.M. Verbaarschot and E.V. Shuryak, Nucl. Phys. **410**, 37 (1993).
13. M. Kacir, M. Prakash and I. Zahed, Stony Brook preprint, SUNY-NTG-95-5.
14. D.I. Diakonov and V.Y. Petrov, Leningrad preprint, LNPI-1153 (1986).
15. M.A. Nowak, J.J.M. Verbaarschot and I. Zahed, Nucl. Phys. **B324**, 1 (1989).
16. E. M. Ilgenfritz, Habilitation, Leipzig 1988.
17. J.J.M. Verbaarschot and E. Shuryak, Nucl. Phys. **B410**, 37 (1993).
18. M.-C. Chu, J.M. Grandy, S. Huang and J. Negele, Phys. Rev. **D48**, 3340 (1993).
19. T. Schäfer, J.J.M. Verbaarschot and E. Shuryak, Nucl. Phys. **B412**, 143 (1994).

CHAPTER 3

LARGE N_c

3.1. Generalities

The use of the instanton vacuum model has led to interesting insights on the long-wavelength excitations of the QCD vacuum. The model, as was discussed in Chapter 2, is well motivated from QCD. The mean-field approximation we made use of allowed us to trade the induced effective action in terms of constituent quarks and gluons for one in terms of bosonic variables. The procedure relied on saddle-point approximations and is exact in the limit that the number of colors approaches infinity. Similar constructions have been extensively used in the literature with various other models. We believe that an accurate description of the QCD ground state, starting from QCD, should ultimately yield an effective theory in the long-wavelength approximation much along the same line as for the instanton-liquid model. This approach, which is quite similar in many aspects to the Ginzburg-Landau approach in superconductors, will constitute the principal thesis of this monograph.

In broad terms, since QCD confines, it is clear that the genuine QCD degrees of freedom are not explicit at low energy. Thus it is logical to ask whether we can formulate QCD entirely in terms of effective degrees of freedom. An answer to this question was first given by 't Hooft[1] for mesons in two dimensions. It was later extended to baryons by Witten.[2]

't Hooft observed that the number of colors in QCD, denoted N_c, could be used to simplify the theory. By considering a fictitious world made up of a large number of colors and assuming that confinement and chiral symmetry breaking are still in effect in this limit, 't Hooft found that the meson spectrum in this world consisted of narrow resonances, with Regge behavior. Meson-meson scattering amplitudes are suppressed by powers of $1/\sqrt{N_c}$. However the baryon sector in large N_c QCD turns out to be much more subtle since naive power counting is not applicable in a straightforward way. We first discuss mesons and baryons in two dimensions (QCD$_2$), where most of the results can be obtained with rigorous methods. Then we move to QCD in four dimensions. Finally, we discuss some exact results valid in the limit of infinite number of colors.

3.2. Two-Dimensional QCD in Large N_c

3.2.1. *Mesons*

In this subsection we discuss 't Hooft's original proposal for large N_c in the meson sector. The issue of baryons will also be addressed and some difficulties will be pointed out. Following 't Hooft we will consider QCD in light-cone gauge. Define light-cone variables as

$$x^\pm = \frac{1}{\sqrt{2}}(x^0 \pm x^1) \quad \gamma^\pm = \frac{1}{\sqrt{2}}(\gamma^0 \pm \gamma^1) \tag{3.1}$$
$$g^{+-} = g^{-+} = 1 \quad g^{++} = g^{--} = 0$$

and the light-cone gauge condition $A_- = A^+ = 0$.[*] In the light-cone gauge, the QCD Lagrangian simplifies to

$$\mathcal{L} = -\frac{1}{2}\mathrm{Tr}(\partial_- A_+)^2 + \bar{q}_f(i\partial\!\!\!/ - m_f - g\gamma_- A_+)q_f \tag{3.2}$$

where the flavor index f in the quark field q is summed over. The Feynman rules for (3.2) are shown in Fig. 3.1. Since the quark gluon-vertex is proportional to γ_- and since $\gamma_-^2 = 0$, only the γ_+ part in the fermion propagator contributes. In the strings of the gamma matrices only $\gamma_-\gamma_+$ combinations survive, so after the redefinition of the vertices, the gamma matrices can be dropped from the Feynman rules. The gluon propagator has only one nonvanishing component, so the Lorentz structure can also be simplified.

Now consider the limit where $N_c \to \infty$ with $g^2 N_c$ fixed. Obviously, in this limit $g \sim 1/\sqrt{N_c}$, so only those subclasses of diagrams that have

---

[*]In two dimensional axial gauges (to which light-cone gauge belongs), the gluons do not interact.

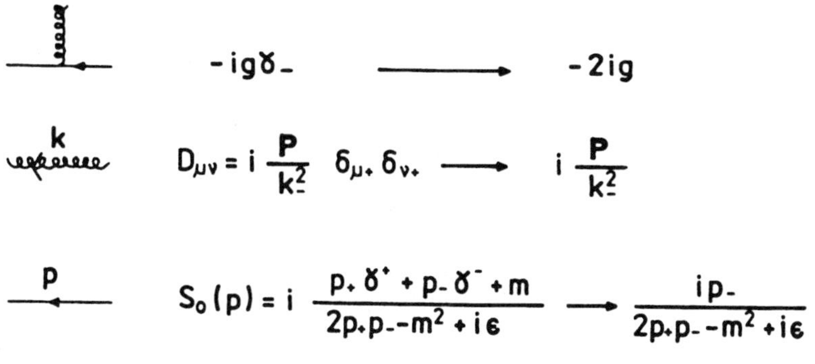

$$-ig\gamma_- \qquad\qquad \longrightarrow \qquad -2ig$$

$$D_{\mu\nu} = i\,\frac{\mathbf{P}}{k_-^2}\,\delta_{\mu_+}\delta_{\nu_+} \qquad \longrightarrow \qquad i\,\frac{\mathbf{P}}{k_-^2}$$

$$S_0(p) = i\,\frac{p_+\gamma^+ + p_-\gamma^- + m}{2p_+p_- - m^2 + i\epsilon} \qquad \longrightarrow \qquad \frac{ip_-}{2p_+p_- - m^2 + i\epsilon}$$

Fig. 3.1. Feynman rules in the light-cone gauge. Here **P** is the principle-value operator.

large enough combinatorial color factors could survive the limit. In this limit, $SU(N_c) \sim U(N_c)$ and only the planar diagrams contribute. Since the gauge fields do not self-interact in this gauge, only non-crossing ladders with fermion self-energy insertions ("rainbow diagrams") contribute. The internal quark loops, suppressed by $1/N_c$, will be ignored.

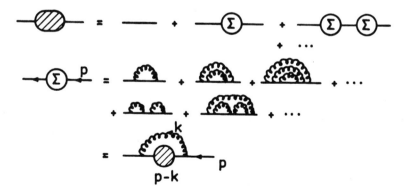

Fig. 3.2. Quark propagator in large N_c.

The fermion propagator in the large N_c world is given by the diagrams of Fig. 3.2. The gap equation for the self-energy is given by [†]

$$i\Sigma = 4g^2 N_c \int \frac{dk_+}{2\pi} \frac{dk_-}{2\pi} S(p-k)D(k_-) \tag{3.3}$$

or explicitly

$$\Sigma(p) = +\frac{g^2 N_c}{\pi^2} \int dk_+ dk_-$$
$$\times \frac{1}{k_-^2} \frac{-i(p_- - k_-)}{m^2 - (2(p_+ - k_+) - \Sigma(p-k))(p_- - k_- + i\epsilon)}. \tag{3.4}$$

The integrand is invariant under the shift $p_+ - k_+ \to k_+$. The integral over k_+ is ultraviolet divergent. Since the singularity is logarithmic, it drops out if one uses a symmetric cutoff. Thus the result for the self-energy is ultraviolet-finite and reads

$$\Sigma(p_-) = \frac{g^2 N_c}{2\pi} \int \frac{dk_-}{k_-^2} \operatorname{sgn}(p_- - k_-). \tag{3.5}$$

The pole $1/k_-^2$ shows that the self-energy is infrared sensitive. If λ is an infrared regulator in momentum space, then the regulated result is

$$\Sigma(p_-) = \frac{g^2 N_c}{\pi} \left(\frac{\operatorname{sgn}(p_-)}{\lambda} - \frac{1}{p_-} \right). \tag{3.6}$$

In fact, the infrared sensitivity in (3.6) is a gauge artifact. To see this, notice that if we were to give the gluons a mass term $-\kappa^2 \operatorname{Tr}(A_+^2)/2$, then the expression for the self-energy would be

$$\Sigma(p_-) = \frac{g^2 N_c}{2\pi} \frac{2}{\kappa} \arctan\frac{p_-}{\kappa} \tag{3.7}$$

which is the result (3.6) if we choose $\kappa = \lambda\pi/2$. This point was first made by Einhorn[3] using different arguments.

In the light-cone gauge, the fermion propagator in large N_c becomes

$$S(p_-) = ip_- \left(-M^2 + 2p_+ p_- - \frac{g^2 N_c |p_-|}{\pi\lambda} + i\epsilon \right)^{-1} \tag{3.8}$$

with the renormalized mass $M^2 = m^2 - g^2 N_c/\pi$. The pole here is gauge-dependent. This shows clearly that the concept of a constituent quark mass,

---

[†]Note the N_c factor multiplying g^2 that differs from the convention of 't Hooft.

a phenomenologically successful notion, is meaningless in a strict sense and is at best only suggestive in a fixed gauge. Such sensitivity to the gauge choice can also be shown in the context of the Schwinger model.[4] A similar remark can be made on the notion of constituent quark masses in four-dimensional QCD. The gauge-dependence disappears in gauge-invariant combinations as we now show.

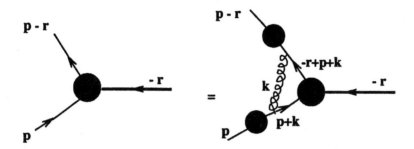

Fig. 3.3. Meson Bethe-Salpeter equation in large N_c.

The meson spectrum in the large N_c limit can be obtained by solving the Bethe-Salpeter equation shown in Fig. 3.3. If we denote by $\psi(p, -r)$ the meson wave function (or vertex function), then from Fig. 3.3, we have

$$\psi(p, -r) = \frac{4g^2 N_c}{i(2\pi)^2} \int S_1(p) S_2(p-r) \frac{d^2 k}{k_-^2} \psi(p+k, -r). \qquad (3.9)$$

The reduced wave function $\phi(p_-, -r) = \int dp_+ \psi(p_+, p_-, -r)$, after integration over k_+, satisfies

$$\begin{aligned}
\phi(p_-, -r) = {}& -\frac{g^2 N_c}{(2\pi)^2 i} \int dp_+ \left(p_+ - \frac{M_1^2}{2p_-} - \left(\frac{g^2 N_c}{2\pi\lambda} - i\epsilon \right) \mathrm{sgn}(p_-) \right)^{-1} \\
& \times \left(p_+ - r_+ - \frac{M_2^2}{2(p_- - r_-)} - \left(\frac{g^2 N_c}{2\pi\lambda} - i\epsilon \right) \mathrm{sgn}(p_- - r_-) \right)^{-1} \\
& \times \int \frac{dk_-}{k_-^2} \phi(p_- + k_-, -r).
\end{aligned} \qquad (3.10)$$

The integration over p_+ is nonzero only if the integration contour runs between the poles. With this in mind $(r_- > 0)$

$$\phi(p_-, -r) = \frac{g^2 N_c}{2\pi} \frac{\theta(p_-)\,\theta(r_- - p_-)}{\left(\frac{M_1^2}{2p_-} + \frac{M_2^2}{2(r_- - p_-)} + \frac{g^2 N_c}{\pi\lambda} - r_+\right)}$$

$$\times \int \frac{dk_-}{k_-^2} \phi(p_- + k_-, -r). \qquad (3.11)$$

The integration over the gluon propagator is infrared-sensitive (or equivalently gauge-dependent). Using the same infrared regulator as for the quark propagator (which is simply the principle-value prescription) yields

$$r_+ \phi(p_-, -r) = \left(\frac{M_1^2}{2p_-} + \frac{M_2^2}{2(r_- - p_-)}\right) \phi(p_-, -r)$$

$$- \frac{g^2 N_c}{2\pi} \theta(p_-)\,\theta(r_- - p_-) \mathbf{P} \int \frac{dk_-}{k_-^2} \phi(p_- + k_-, -r). \qquad (3.12)$$

Note that the gauge dependence has dropped out. What happened is in fact the following: the gauge dependence in the constituent quark mass is canceled by the gauge dependence in the gluon propagator. We believe this cancellation to be of paramount importance in gauge theories. The argument presented here shows the intricate relationship between the gauge dependence of the constituent quark mass and the infrared sensitivity of the gluon propagator. Only a consistent treatment of the two effects, here in the context of leading N_c, leads to an answer that is gauge-invariant and finite. We believe this point to be of relevance also to QCD in four dimensions, although most of the calculations performed so far that we are aware of (*i.e.*, mean-field, Bethe-Salpeter, semi-classical instanton simulations etc.) do not seem to contain this mechanism.

In terms of the "parton momentum fraction" given by $x = p_-/r_-$ and the invariant energy $2r_+r_- = \mu^2$, (3.12) simplifies to

$$\mu^2 \phi(x) = \left(\frac{M_1^2}{x} + \frac{M_2^2}{1 - x}\right) \phi(x) - \frac{g^2 N_c}{\pi} \mathbf{P} \int_0^1 \frac{dy}{(y - x)^2} \phi(y) \qquad (3.13)$$

which is the 't Hooft equation for the reduced meson wave function in the light-cone frame. Solutions to this equation can be obtained numerically. The spectrum is shown in Fig. 3.4. Note that while the "masses" of the quarks could even be tachyonic, the bound states (*i.e.*, the mesons) never are. For the lowest-lying meson (in the chiral limit), the energy stored in

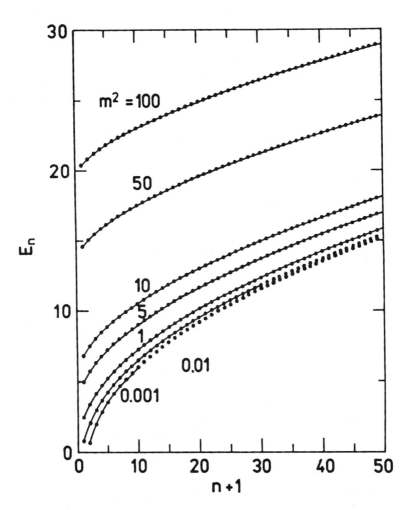

Fig. 3.4. Meson spectrum of QCD_2 in large N_c. The dotted line represents the exact solution of the 't Hooft equation and the solid line the exact semi-classical calculation.[5]

the flux connecting the quark and the anti-quark balances exactly against the quark self-energies. The result is a zero mass meson in the chiral limit. However, this is not a Goldstone boson, as we will show below. For higher-lying states, the energy of the flux is always dominant, hence a *positive* bound state mass.

Asymptotically, the spectrum is linear, a consequence of the relativistic description of the quarks. Indeed, if we set by hand the confining potential to be $V = \sigma x$ with $\sigma = g^2 N_c/2$ and consider two relativistic quarks, each with momentum p, then the total energy would be $\mu = 2p + \sigma x$. Using the Bohr-Sommerfeld quantization condition (for large n)

$$2 \int_0^{x_r} p \, dx = n\pi \tag{3.14}$$

(x_r denotes the point of return) gives the semiclassical spectrum

$$\mu^2 \sim 2\pi\sigma n. \tag{3.15}$$

A more careful analysis of (3.13) using semiclassical techniques yields[1,5]

$$\mu^2 \sim (2n+1)\pi\sigma + 2M^2 \ln\left(\frac{\mu^2}{M^2}\right) \tag{3.16}$$

for $M_1 = M_2 = M$ and integer n ($\sigma = g^2 N_c/2$). The mesonic states asymptotically display a Regge-like behavior.

3.2.2. *Baryons*

(i) Feynman graphs

Baryons in large N_c QCD can be thought of as color singlets of N_c quarks. The planar diagrams shown in Fig. 3.3 contribute to the baryon wave function. However, they are not leading in $1/N_c$ as pointed out by Durgut.[6] The fact that they are subleading can be immediately seen from the diagram of Fig. 3.5 for a diquark. Whereas each iteration of the quark-anti-quark exchange supplies a factor of N_c, in the case of the quark-quark interaction, the N_c behavior is unchanged during the iteration.[‡] What is interesting, though, is that the sum of the planar diagrams shown in Fig. 3.3

[‡] This fact was used in the large N_c analysis of the baryon correlators in the preceding chapter.

is gauge-invariant. Durgut has explicitly shown that the gauge dependence in the quark self-energies cancels precisely against the gauge dependence in the gluon propagator, resulting in a generalized version of the 't Hooft equation for the baryonic wave function in the light-cone frame. Solutions to these equations are not known. The form of the equation is identical to the equation for three quarks connected by strings in the Δ (triangle) configuration. Rossi and Veneziano[7] have suggested that Durgut's construction may be recovered from a dual approximation of QCD.

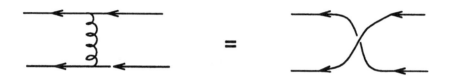

Fig. 3.5. Gluon-exchange in a diquark state.

Witten has argued[2] that 't Hooft's diagrammatic construction is more subtle for baryons. To appreciate this problem, consider the lowest perturbative corrections to the free propagation of an N_c-quark state as shown in Fig. 3.6. Since the coupling constant scales like $1/\sqrt{N_c}$, summing over all pair exchanges in Fig. 3.6(b) gives a weight of the order of $N_c(N_c-1)/2N_c \sim N_c$. The second correction scales like N_c^2, and so on. The perturbative expansion is increasingly divergent in N_c. The reason is that the baryon mass is in fact of order N_c. Indeed, the perturbative expansion for the propagator of a baryon of mass $M \sim N_c\Lambda(1 + g^2 N_c)$ is

$$e^{-itM} = e^{-itN_c\Lambda}\left(1 - it\Lambda N_c^2 g^2 - \frac{1}{2}\Lambda^2 t^2 N_c^4 g^4 + ...\right). \qquad (3.17)$$

The diagrammatic expansion in g^2 is increasingly divergent in N_c. The conclusion is that a large N_c limit exists for baryons but the diagrammatic expansion may not be a convenient way to show it. Instead, Hamiltonian or path integral methods using bosonization techniques may be more appropriate.

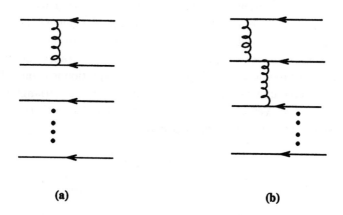

Fig. 3.6. First (a) and second (b) correction to a free propagation of N_c quarks.

(ii) Bosonization

In his original work, Witten[2] relied on a specific bosonization of QCD_2 in axial gauge. Here we will follow later suggestions by a number of authors [8] based on exact two-dimensional nonabelian bosonization also introduced by Witten[9]. Consider first the free theory of N_f massless free Dirac fermions described by

$$S_{fermion} = \sum_{i=1}^{N_f} \bar{q}_i i \slashed{\partial} q_i. \qquad (3.18)$$

Witten[9] has shown how to bosonize such a theory. The bosonized version is the nonlinear $U(N_f)$ sigma model with a topological term (Wess-Zumino term):

$$
\begin{aligned}
S_{boson} \; \equiv \; S_B(U) &= \frac{1}{8\pi} \int d^2x \, \mathrm{Tr} \partial_\mu U \partial^\mu U^\dagger \\
&+ \frac{1}{12\pi} \int_D d^3y \epsilon^{ijk} \, \mathrm{Tr}(U^\dagger \partial_i U)(U^\dagger \partial_j U)(U^\dagger \partial_k U) \quad (3.19)
\end{aligned}
$$

where U is a $U(N_f)$ matrix and the fields in the Wess-Zumino term are defined on a three-dimensional disc D whose boundary corresponds to the original two-dimensional space. A simple dictionary between the fermionic and bosonic degrees of freedom can then be established:

$$J_{R(ij)} =: \bar{q}_R^i q_R^j : \qquad \leftrightarrow \qquad \frac{i}{2\pi}(U^\dagger \partial_+ U)_{ij}$$

$$J_{L(ij)} =: \bar{q}_L^i q_L^j : \qquad \leftrightarrow \qquad \frac{i}{2\pi}(U \partial_- U^\dagger)_{ij}$$

$$m_q \sum_{i=1}^{N_f} \bar{q}_i q_i \qquad \leftrightarrow \qquad m_q c\mu : \mathrm{Tr}(U + U^\dagger) : \qquad (3.20)$$

where normal-ordering is indicated. The last relation shows how to introduce mass terms, with μ an overall normal-ordering scale. For simplicity,
we have set all the quark masses equal. The vector current can be written
as

$$V_\mu = \frac{i}{4\pi}\left((U^\dagger \partial_\mu U + U \partial_\mu U^\dagger) - \epsilon_{\mu\nu}(U^\dagger \partial_\nu U - U \partial_\nu U^\dagger)\right). \qquad (3.21)$$

Adding a color index to our bosonization transcription (3.20) yields

$$S = N_c S(U) + N_f S(H) \qquad (3.22)$$

where H is the analog of U in color space. Equation (3.22) involves *colored*
bosons. The color indices are independent of the flavor indices, hence the
factors N_c and N_f in (3.22). The addition of the gluons (not bosonized)
to (3.22) may be sought by a minimal substitution to the color analog of
(3.21), as H is the only colored field. Thus,

$$V_\mu^C = \frac{i}{4\pi}\left((H^\dagger D_\mu H + H D_\mu H^\dagger) - \epsilon_{\mu\nu}(H^\dagger D_\nu H - H D_\nu H^\dagger)\right). \qquad (3.23)$$

Since $V_\mu^C = \delta S / \delta A_\mu$, the form of the gauge-invariant action S is

$$\begin{aligned} S(H, A_\pm) &= S(H) \\ &+ \frac{i}{2\pi}\int d^2x \, \mathrm{Tr}\left(A_+ H \partial_- H^\dagger + A_- H^\dagger \partial_+ H\right. \\ &\left. + i A_+ H A_- H^\dagger - i A_- A_+\right). \end{aligned} \qquad (3.24)$$

Adding a mass term and specializing to the light-cone gauge gives

$$\begin{aligned} S_B(QCD_2) &= N_f S(H) + \frac{i}{2\pi} N_f \int d^2x \, \mathrm{Tr}_c(A_+ H \partial_- H^\dagger) \\ &- \frac{1}{2g^2}\int d^2x \, \mathrm{Tr}_c(\partial_- A_+)^2 + N_c S(U) \\ &+ m_q c\mu \int d^2x : \mathrm{Tr}_c(H \, \mathrm{Tr}_f U + H^\dagger \, \mathrm{Tr}_f U^\dagger) : \end{aligned} \qquad (3.25)$$

for massive QCD_2. The first two terms are N_f copies of QCD_2 with N_c colors, the third term is the free Yang-Mills term in light-cone gauge, the fourth term is the bosonized free flavor action, and the fifth term is the bosonized mass term. The indices in the traces denote color (c) or flavor (f). The gauge choice allows for an integration over the gauge field. Indeed, defining $\partial_- G = iH\partial_- H^\dagger$ with boundary $G(\pm\infty, x_-) = 0$ reduces the gauge integration to a Gaussian. The result is

$$S = N_f S(H) + N_c S(U) - \frac{1}{2}\left(\frac{N_f g}{\sqrt{2\pi}}\right)^2 \int d^2x \, \text{Tr}_c G(x)^2$$
$$+ \text{``mass terms''}. \tag{3.26}$$

So far (3.26) is an exact transcription of massive QCD_2 with N_f flavors and N_c colors. Unfortunately, the action is too involved to be tractable analytically. It can be solved, however, in the strong coupling limit.

(iii) Strong coupling

In the strong coupling limit, $g/m_q \to \infty$ so that $m_G = (gN_f/\sqrt{2\pi}) \to \infty$. Hence, the G-particle in (3.26) becomes infinitely heavy and decouples. This corresponds to setting $G = 0$, thus $H \to 1_{N_c}$, with $\text{Tr}H = N_c$ except for a renormalization of the mass scale. In describing the Green functions with the external light particles, the mass-scale renormalization amounts to taking into account the effect of the internal loops due to the G particles. Each of these loops brings a logarithm. The cumulative effect of these logarithms brings about an anomalous dimension γ for the mass operator. The final result is

$$S = \frac{1}{8\pi}\int d^2x \, \text{Tr}\partial_\mu U \partial^\mu U^\dagger$$
$$+ \frac{1}{12\pi}\int_D d^3y \epsilon^{ijk}\, \text{Tr}(U^\dagger \partial_i U)(U^\dagger \partial_j U)(U^\dagger \partial_k U)$$
$$+ m^2 \int d^2x \, \text{Tr} : (U + U^\dagger) : \tag{3.27}$$

where

$$m^2 = [N_c c m_q (m_G)^\gamma]^{\frac{2}{1+\gamma}} \tag{3.28}$$

following a rescaling $\hbar/N_c \to \hbar$ in the functional integral. The normal ordering is set at the mass scale m and the anomalous dimension reads

$$\gamma = \frac{N_c^2 - 1}{N_c^2 + N_c N_f}.$$ (3.29)

We note that in the large N_c limit, $\gamma \sim N_c^0$, so that $m \sim \sqrt{N_c}$. This scaling with N_c is *a priori* surprising, since the bare meson mass is usually assigned a canonical scaling of N_c^0. This point will be clarified below. The reader familiar with the phenomenological model of Skyrme[11] (called skyrmion in the literature) will immediately recognize that (3.27) is the two-dimensional analogue of the minimal four-dimensional Skyrme effective action (without higher order terms). In the strong-coupling regime, QCD_2 *is equivalent* to the sigma model with the Wess-Zumino term.

(iv) Mesons revisited

Is the bosonized version of QCD_2 in the strong coupling regime compatible with 't Hooft's bound state description? For mesons, this can be checked by considering (3.27) in the topologically trivial sector. Define

$$U = \exp(i\sqrt{2\pi/N_c} T^a \pi^a)$$ (3.30)

where the π's are mesonic fields. To leading order in $1/N_c$, (3.30) reads

$$S = \int d^2x (\frac{1}{2}\partial_\mu \pi^a \partial_\mu \pi^a - \frac{1}{2}m_\pi^2 \pi^a \pi^a) + \mathcal{O}(1/\sqrt{N_c})$$ (3.31)

with the "pion" mass

$$m_\pi = \sqrt{4\pi} c m_G m_q.$$ (3.32)

In the large N_c limit, the theory becomes a theory of noninteracting scalars with a mass independent of the number of colors. The chiral limit brings the mass to zero, in agreement with 't Hooft's result. Also, since $m_G \sim g$, for fixed $g\sqrt{N_c}$, the pionic mass scales as $m_\pi \sim g\sqrt{N_c}m_q/\sqrt{N_c} \sim m_q/\sqrt{N_c}$ and vanishes in the large N_c limit. This signals quasi-long-range order in QCD_2 at large N_c as discussed in the Appendix. The strong coupling limit, however, may be inappropriate for probing the Regge trajectories.

(v) Solitons

We will now see that (3.27) allows us to find an analytic solution corresponding to a baryon in two dimensions. For static U_s configurations, the Wess-Zumino term vanishes since it is time-odd. The simplest static ansatz for the U matrix is diagonal in flavor space,

$$U_s(x) = \text{diag}\big(\exp(+i\sqrt{4\pi/N_c}\ \phi(x)), 1, ..., 1\big) \tag{3.33}$$

where we have chosen one particular flavor to be nontrivial. The corresponding baryonic current

$$J_\mu^B = -\frac{i}{2\pi}\epsilon_{\mu\nu}\ \text{Tr}(U_s^\dagger \partial^\nu U_s) \tag{3.34}$$

is topologically conserved (that is, conserved independently of the equations of motion), and the baryonic charge is equal to N_c, with boundary conditions: $\sqrt{4\pi/N_c}\phi$ equal to 2π at $+\infty$ and zero at $-\infty$. With the ansatz (3.33), the action (3.27) reduces to the exactly soluble Sine-Gordon model

$$S(U_s) = -\int dx \left[\frac{1}{2}\left(\frac{d\phi}{dx}\right)^2 - 2m^2(\cos\sqrt{4\pi/N_c}\phi - 1)\right] \tag{3.35}$$

with the solution

$$\phi(x) = \sqrt{4N_c/\pi}\ \arctan[\exp(\sqrt{8\pi/N_c}mx)] \tag{3.36}$$

and the classical energy

$$M_s = 4m\sqrt{\frac{2N_c}{\pi}}. \tag{3.37}$$

Thus, a baryon in the strong coupling version of two-dimensional QCD corresponds to a Sine-Gordon soliton. In the large N_c limit, $\gamma \sim N_c^0$ (see (3.29)) and $m \sim \sqrt{N_c}$, so that $M_s \sim N_c$. In the large N_c limit the soliton is infinitely heavy, in agreement with the qualitative discussion given by Witten[2] (see above). We remark that the chiral limit ($m_q \to 0$) and the large N_c limit do not commute. This is explicit in (3.28).

Is the nonlocal character of the soliton field (3.33) consistent with the locality of the fields in QCD_2? From the definition of the baryonic (solitonic) current (3.34), it follows that the baryon charge is

$$B = \frac{1}{2\pi}\Big(\phi(+\infty) - \phi(-\infty)\Big). \tag{3.38}$$

Any field operator Ψ carrying baryon number B, $[B, \Psi(x)] \neq 0$, is nonlocal with respect to ϕ since

$$[\phi(+\infty), \Psi(x)] \neq 0 \qquad \text{or} \qquad [\phi(-\infty), \Psi(x)] \neq 0. \qquad (3.39)$$

Thus locality is naively upset. However, we note that Ψ is in fact local with respect to $e^{\pm i\phi}$, which is identified with the local mesonic field $\bar{q}q$. So the baryonic and mesonic sources of the bosonized version obey local commutation relations as expected from QCD_2. This issue may be subtle in four dimensions [10].

The present discussion is instructive in many ways. It shows that two-dimensional QCD in the strong coupling limit is a nonlinear sigma model with a Wess-Zumino term. The latter admits classical and static two-dimensional solitons that are heavy. The analogy of these results with the Skyrme model or the skyrmion [11] is striking (see Chapter 7). It should, however, be stressed that similarities with two dimensions are sometimes misleading, since the two-dimensional world has its own theorems. We refer to the Appendix for an example. To what extent the above two-dimensional analysis carries quantitatively over to four dimensions is not clear. Below, we give some insights into how this can be achieved qualitatively.

3.3. Four-Dimensional QCD in Large N_c

3.3.1. *Confinement and chiral symmetry*

The two fundamental assumptions in large N_c QCD arguments are that confinement and the spontaneous breaking of chiral symmetry are valid from $N_c = 2$ to $N_c = \infty$. These are, however, conjectures that so far remain unproven. There may be several mechanisms for confinement and spontaneous chiral symmetry breaking for various values of N_c.

This notwithstanding, Coleman and Witten[12] have suggested that under *reasonable* assumptions the case for the spontaneous breaking of chiral symmetry in large N_c can be made. Indeed, assuming that (1) the large N_c limit exists, (2) QCD confines in the large N_c limit, (3) the vacuum state can be reached by variational arguments, (4) the vacuum does not exhibit accidental degeneracies, and (5) the vacuum order-parameter M is bilinear in quark field $M_{ij} \sim \langle \bar{q}_i(1+\gamma_5)q_j \rangle$, they concluded that the vacuum state of large N_c QCD with N_f massless quarks breaks $U(N_f) \times U(N_f)$ spontaneously, resulting in diagonal $U(N_f)$.

To prove the above, note that the order parameter M transforms as $U M V^\dagger$ under the flavor $U(N_f) \times U(N_f)$ group. Thus M can always be rotated to its diagonal non-negative form. The squares of the diagonal form are the eigenvalues of $MM^\dagger$. Because of the symmetry, the induced effective potential $\mathcal{V}$ can only depend on the combination $MM^\dagger$. If we denote by λ_i the eigenvalues of $MM^\dagger$, then to leading order in $1/N_c$

$$\mathcal{V} = N_c \, \mathrm{Tr} F(MM^\dagger) = N_c \sum_i F(\lambda_i) \qquad (3.40)$$

where F is an arbitrary function. Since the eigenvalues are independent variables, the minima of $\mathcal{V}$ occur when all the eigenvalues sit at the minimum of F. Thus the eigenvalues are either all zero (no symmetry breaking) or all equal and nonzero (symmetry breaking to $U(N_f)$).

The first alternative is ruled out by noting that the flavor current $j_\mu = \bar{q}T(1 + \gamma_5)\gamma_\mu q$, with T Lie algebra-valued in $U(N_f)$, yields an anomalous 3-point function: the AAA-triangle anomaly. This means that the Fourier transform of $\langle jjj \rangle$ is not an analytic function of the invariant momenta p, q and $r = -(p+q)$ at the infrared point $p = q = r = 0$. On the other hand, large N_c arguments show that $\langle jjj \rangle$ has poles in p^2, q^2 and/or r^2. Thus the current j must create at least one massless particle when applied to the vacuum. The flavor symmetry is therefore spontaneously broken.

3.3.2. *Diagrammatics*

In four dimensions, the large N_c counting rules obey a systematics that we will now outline. In Fig. 3.7, we have summarized the single and double line technique for quarks and gluons as originally discussed by 't Hooft[1]. The quarks are in the fundamental representation of both flavor (dashed) and color (full). The gluons are in the adjoint representation of color. The double-line notation for gluons explicitly shows the flow of the color with $A^i_{\mu\,j} = \sum_a A^a_\mu (T^a)^i_j$. The advantage of this notation is that it allows us to read off the symmetry factors from the Feynman graphs. Examples are displayed in Fig. 3.8 with their respective weight factors.

One can derive a number of convenient rules useful for extracting the symmetry factors. Instead of trying to derive them, we will just state them with examples. We refer to 't Hooft's original analysis for the general proofs[1].

Rule 1 : Planar gluon insertions give rise only to factors of $g^2 N_c \sim N_c^0$

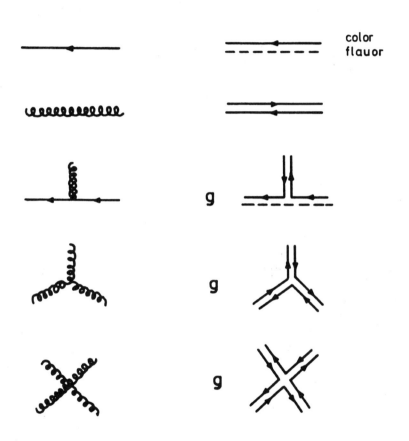

Fig. 3.7. Large N_c rules.

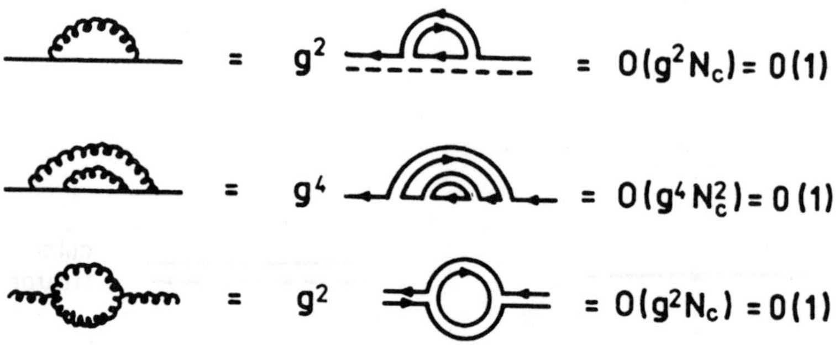

Fig. 3.8. Double-line counting examples.

independent of the number of lines inserted. A few examples of this rule are shown in Fig. 3.9a.

Rule 2 : A quark loop contributes $g^2 \sim 1/N_c$. Some examples of this rule are shown in Fig. 3.9b.

Rule 3 : A non-planar gluon loop contributes $g^2/N_c \sim N_c^{-2}$. An example is shown in Fig. 3.9c.

An arbitrary diagram with L_F quark loops *inserted* and L_G gluon handles is down by $1/N_c^{L_F + 2L_G}$ compared to the diagram without.

3.3.3. *Mesons*

To understand the structure of mesons in the large N_c world, consider the correlation function of two scalar electromagnetic currents $j(Q)$ in the vacuum. To leading order in $1/N_c$, the correlation function is the sum of *all* planar Feynman diagrams. A typical contribution is shown in Fig. 3.10. A cut through the diagram ("cut-diagram") displays neither multi-meson nor meson-gluon intermediate states. The correlation function is thereby dominated by the exchange of single mesons of mass M

$$\langle j(Q)j(-Q)\rangle = \sum_n \frac{\lambda_n^2}{Q^2 + M_n^2}. \tag{3.41}$$

Power counting shows that (3.41) scales like N_c times a smooth function of Q^2. Thus $\lambda_n \sim \langle 0|j|n\rangle \sim \sqrt{N_c}$ and $M_n \sim N_c^0$.

The subleading correction to Fig. 3.10 is shown in Fig. 3.11. It is of order N_c^0 because of the fermion loop (Rule 2). The cut-diagram displays a

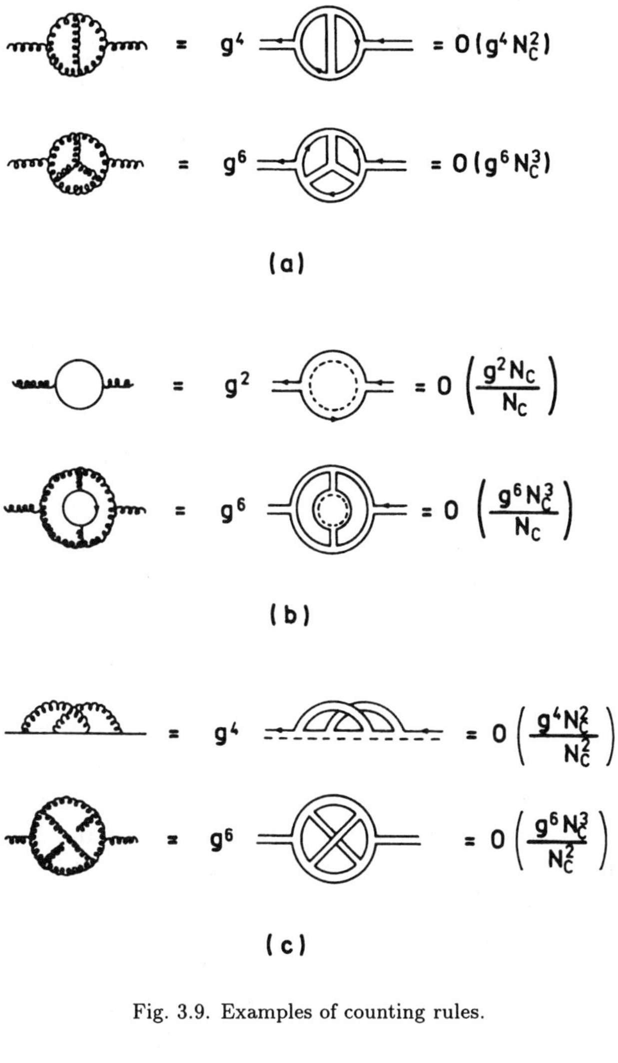

$$\text{(a)}$$

$$\text{(b)}$$

$$\text{(c)}$$

Fig. 3.9. Examples of counting rules.

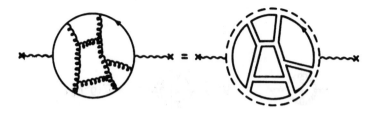

Fig. 3.10. A typical leading planar contribution to (3.42).

two-meson intermediate state. From the optical theorem we conclude that

$$\text{CUT} \ [\text{Fig. (3.11)}] = \lambda_n^2 \Gamma_{M \to 2M}^2 \sim N_c^0 \tag{3.42}$$

where $\Gamma \sim 1/\sqrt{N_c}$ is the width for a meson (M) to decay into two mesons $(2M)$.

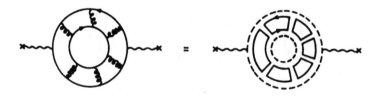

Fig. 3.11. A typical subleading planar contribution to (3.42).

    The N_c-counting for meson-meson interactions can be derived by considering n-point correlation functions (with $n \geq 3$) as shown in Fig. 3.12. The dominant contributions to all correlators are planar and of order N_c. Since the sources scale like $1/\sqrt{N_c}$, we conclude that the 3-meson interaction scales like $1/\sqrt{N_c}$, the 4-meson interaction like $1/N_c$, and so on. Meson-meson interactions vanish in the large N_c world.

    Not only are mesons free (no interactions) for large N_c, they are also stable (zero width) and light (mass of order N_c^0). Glueballs, on the other hand, are heavy (mass of order N_c) but decoupled. These results provide an interesting explanation for why the mesons in nature are usually light, and why they appear as sharp resonances. In this limit, exotics such as $\bar{q}^2 q^2$, $G\bar{q}q$, ... are all suppressed by $1/\sqrt{N_c}$, $1/N_c$, ... and so on. Also the Zweig suppression rule is nicely explained. Indeed, from Fig. 3.12, we can conclude that the decay $\phi \to \pi\rho$ is suppressed by $1/N_c$ relative to the decay $\phi \to \overline{K}K$. In nature, the ratio of these decays is $15 : 83$.

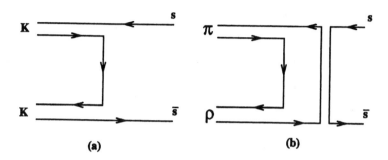

Fig. 3.12. $\phi \to \overline{K}K$ (a) and $\phi \to \pi\rho$ (b) decays in large N_c.

3.3.4. Baryons

We have already encountered the diagrammatic analysis of the baryons in QCD_2 in the large-N_c limit. As the number of colors grows, not only do the color factors proliferate, but also the number of external (valence) quark lines increases. The baryon is an N_c quark configuration produced by a *local* color singlet source $(\epsilon^{i_1 i_2 \cdots i_N} q_{i_1} q_{i_2} \cdots q_{i_N})$ where all the quarks act coherently. In four dimensions, there is no known *exact* bosonization scheme such as the one developed in two dimensions that allows us to transcribe QCD_4 by a chain of equalities. Using the nonrelativistic quark description in large N_c, Witten[2] has made suggestive arguments based on a semiclassical analysis of what causes the baryons to be heavy, $M_B \sim N_c$. This simple result has far-reaching consequences as we will see in later chapters. We may rewrite this condition simply as

$$M_B \sim N_c \sim \frac{1}{g^2} \tag{3.43}$$

suggesting, by analogy with many nonperturbative solutions of simpler models, that the baryon is a nonperturbative, soliton or monopole-like solution of the theory of weakly interacting mesons and glueballs.

(i) Pion-baryon scattering

Can anything generic be said of meson-baryon dynamics taking the above considerations into account? In nonrelativistic quark model the nucleon axial charge is $g_A = (N_c+2)/3$, so that $g_A \sim N_c$ in the large N_c limit. Since $M_B \sim N_c$ and $f_\pi \sim \sqrt{N_c}$ it follows from the Goldberger-Treiman relation (Chapter 4) that the pion-baryon coupling $g_{\pi BB} \sim \sqrt{N_c^3}$ is strong. Generically, the pion-baryon vertex may be written in the form,

$$\langle B|\bar{q}\gamma^i\gamma_5\tau^a q|B\rangle = g(B)N_c\langle B|\mathbf{X}^{ia}|B\rangle \tag{3.44}$$

where $g(B)$ is a number of order 1. The mixed generators $\mathbf{X}^{ia} \sim \sigma^i\tau^a$ form a 4×4 matrix in the nucleon basis $(\uparrow, \downarrow) \otimes (p, n)$.

Since the baryons are heavy, the process $\pi B \to \pi B$ follows from the Born graphs shown in Fig. 3.13 to leading order in $1/N_c$. The seagull term has been omitted as it is subleading in $1/N_c$. Since the recoil effects are subleading in $1/N_c$, the Born amplitude of Fig. 3.13 is proportional to

$$\mathcal{A}_B \sim N_c \left[\mathbf{X}^{ia}, \mathbf{X}^{jb}\right]. \tag{3.45}$$

This result is at odds with unitarity unless the commutator in (3.45) vanishes faster than $1/N_c$. The vanishing of the leading term in the Born amplitude results from the existence of a tower of degenerate baryonic states (intermediate states in the Born graph) as well as the fact that the generators $\mathbf{X}^{ia}$ obey a contracted $SU(4)$ algebra[13],

$$[\mathbf{X}^{ia}, \mathbf{X}^{jb}] = 0 \tag{3.46}$$

to order N_c^0. It is important to stress that (3.46) does not follow from chiral symmetry, but rather from unitarity and is therefore a strong-coupling property[13,14].

Consider now the next process $\pi B \to \pi\pi B$ at low energy. To leading order in N_c the Born graphs contribute as shown in Fig. 3.14. Again the seagull graphs are $1/N_c$ suppressed compared to the leading term. The pion-nucleon amplitude from the diagrams of Fig. 3.14 is proportional to [15]

$$\mathcal{A}_B \sim \sqrt{N_c^3}\,[\mathbf{X}^{ia}, [\mathbf{X}^{jb}, \mathbf{X}^{kc}]] \tag{3.47}$$

and again at odds with unitarity unless the double commutator (3.47) vanishes faster than $1/\sqrt{N_c^3}$. A solution to this problem may be sought by

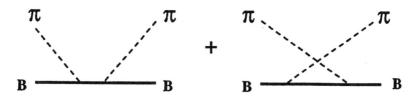

Fig. 3.13. $\pi B \to \pi B$ to leading order in $1/N_c$.

expanding the generators $\mathbf{X}$ in a series in $1/N_c$[§]

$$\mathbf{X} = \mathbf{X}_0 + \frac{1}{N_c}\mathbf{X}_1 + \frac{1}{N_c^2}\mathbf{X}_2 + \cdots . \tag{3.48}$$

Substituting (3.48) into (3.47) and recalling (3.46) to leading order yields the following constraint

$$[\mathbf{X}_0^{ia}, [\mathbf{X}_0^{jb}, \mathbf{X}_1^{kc}]] + [\mathbf{X}_0^{ia}, [\mathbf{X}_1^{jb}, \mathbf{X}_0^{kc}]] = 0. \tag{3.49}$$

This is fulfilled if and only if $\mathbf{X}_1 \sim \mathbf{X}_0$. Thus[16]

$$g\,\mathbf{X} \to g\left(\mathbf{X}_0 + \mathcal{O}(1/N_c^2)\right). \tag{3.50}$$

In this way, unitarity forces all the meson-baryon vertices to be interrelated.

(ii) Dashen-Manohar relations

What are the relations between the various meson-baryon vertices? To answer this question, we will assume that in the large N_c limit baryons are hedgehogs in isospin-spin space

$$|B\rangle \sim |uu...u\ dd...d\rangle \otimes |\uparrow\uparrow ... \uparrow \downarrow\downarrow ... \downarrow\rangle \tag{3.51}$$

with spin-isospin assignment $I = J$ (see Chapter 7). Using the Wigner-Eckart theorem, it follows that

$$\langle J'm'\alpha'|\mathbf{X}^{ia}|Jm\alpha\rangle = X(J,J')\sqrt{\frac{2J+1}{2J'+1}}\begin{pmatrix} J & 1 & J' \\ \alpha & a & \alpha' \end{pmatrix}\begin{pmatrix} J & 1 & J' \\ m & i & m' \end{pmatrix} \tag{3.52}$$

[§]The implicit assumption here is that the expansion is in $1/N_c$ and not some other power!

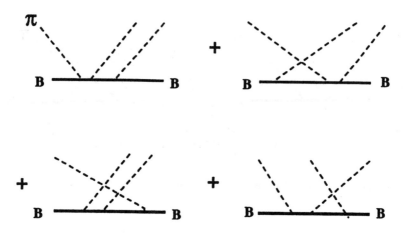

Fig. 3.14. $\pi B \to \pi\pi B$ to leading order in $1/N_c$.

where $X(J, J')$ is a reduced matrix element. It is a tensor with spin–one isospin–one structure, and therefore can only connect states with $\Delta J = \pm 1$. Straightforward Clebsch-Gordaning gives generically

$$\langle J|[\mathbf{X}, \mathbf{X}]|J+1\rangle \ \sim \ X(J, J)X(J, J+1) + X(J, J+1)X(J+1, J+1)$$
$$\sim \ 0 \tag{3.53}$$

for the off-diagonal terms, and

$$\langle J|[\mathbf{X}, \mathbf{X}]|J\rangle \sim X(J, J-1)^2 + X(J, J)^2 + X(J, J+1)^2 \sim 0 \tag{3.54}$$

for the diagonal terms, modulo some numerical factors. These relations show that the meson-baryon reduced vertices obey recursion relations that are dictated by unitarity. These relations were first derived by Dashen and Manohar[15]. So, given $X(J, J)$ and $X(J-1, J)$, the others follow from (3.53) and (3.54). Thus $X(1/2, 1/2) = 1$ and $X(-1/2, 1/2) = 0$ entirely determine the recursion relations. These relations are satisfied by the nonrelativistic quark model and the Skyrme model (discussed in Chapter 7).

(iii) Contracted $SU(4)$

The commutation relations

$$\left[\mathbf{X}^{ia}, \mathbf{X}^{jb} \right] = 0,$$

$$\left[\mathbf{I}^{a}, \mathbf{X}^{ib} \right] = i\epsilon^{abc} \mathbf{X}^{ic},$$

$$\left[\mathbf{J}^{i}, \mathbf{X}^{jb} \right] = i\epsilon^{ijk} \mathbf{X}^{ka}, \tag{3.55}$$

where $\mathbf{I}$ and $\mathbf{J}$ are the isospin and spin generators, constitute a contracted $SU(4)$ algebra. Indeed, if we were to denote by $\mathbf{1} \otimes \mathbf{I}$, $\mathbf{J} \otimes \mathbf{1}$ and $\mathbf{J} \otimes \mathbf{I}$, the generators of $SU(4)$, then the algebra of large N_c QCD follows by taking the limit[13]

$$\mathbf{X}^{ia} = \lim_{N_c \to \infty} \frac{(\mathbf{J}^i \otimes \mathbf{I}^a)}{N_c}. \tag{3.56}$$

This algebra is satisfied by the constituent quark model, with

$$\mathbf{X}^{ia} = \frac{1}{N_c} \sum_{\alpha} q_{\alpha}^{\dagger} \sigma^i \tau^a q_{\alpha} \tag{3.57}$$

where the sum is over color indices. Indeed,

$$
\begin{aligned}
\left[\mathbf{X}^{ia}, \mathbf{X}^{jb} \right] &= \frac{1}{N_c^2} \sum_{\alpha, \beta} \left[q_{\alpha}^{\dagger} \sigma^i \tau^a q_{\alpha}, q_{\beta}^{\dagger} \sigma^j \tau^b q_{\beta} \right] \\
&= \frac{1}{N_c^2} \sum_{\alpha} \left[q_{\alpha}^{\dagger} \sigma^i \tau^a q_{\alpha}, q_{\alpha}^{\dagger} \sigma^j \tau^b q_{\alpha} \right] \sim \frac{N_c}{N_c^2} \sim 0
\end{aligned} \tag{3.58}
$$

and vanishes in the large N_c limit. It is also satisfied by the Skyrme model, with the identification $\mathbf{X}^{ia} = \mathrm{Tr}(A \tau^i A^{\dagger} \tau^a)$, where the collective variables A are $SU(2)$ valued and commute with each other.

The contracted $SU(4)$ algebra is what is required to unitarize the pion-baryon scattering amplitudes to all orders since the pion-baryon interaction is strong. It would be interesting to check whether similar constraints could be derived in models where scalar and vector mesons are also included. In addition, since the baryon-baryon interaction is of order N_c, thus strong, we suspect that similar cancellations are also operative to enforce unitarity for large N_c. In the meson-exchange approach to the baryon-baryon interaction, this is almost self-evident from the above arguments. This point, however, has to be clarified since the pions are now off-shell and there are additional contact interactions.

Baryons form an irreducible representation of the contracted algebra. The contracted representations form infinite towers of states. Generically

$$
\begin{aligned}
K &= 0 &:&\quad N, \Delta, \ldots \\
K &= \frac{1}{2} &:&\quad \Lambda, \Sigma, \Sigma^*, \ldots. \\
K &= 1 &:&\quad \Xi, \Xi^*, \ldots \\
K &= \frac{3}{2} &:&\quad \Omega, \ldots
\end{aligned}
$$

where $\mathbf{K} = \mathbf{I} + \mathbf{J}$. These states fall into the N_c-multiplet shown in Fig. 3.15. Let $g(K)$ be the overall coupling in a given tower. From (3.50), it follows that for fixed K the ratios of the coupling constants is one to order $1/N_c^2$. In particular,

$$
\frac{\Sigma^+ \to \Sigma^0 \pi^+}{\Sigma^+ \to \Lambda \pi^+} = 1 + \mathcal{O}\left(\frac{1}{N_c^2}\right). \tag{3.59}
$$

For $N_c = 3$, (3.59) is $\sqrt{3}F/D$, giving $F/D \sim 0.58$ which is to be compared with $2/3$ in the nonrelativistic quark model.

(iv) Mass relations

The above arguments suggest a constraint on the baryon masses. Indeed, unitarity again implies that the baryon mass operator M should satisfy the constraint $[\mathbf{X}, [\mathbf{X}, M]] \leq 1/N_c$, which is explicit from the Born diagram of Fig. 3.13 with the recoil hyperfine effects taken into account in the intermediate state. Thus, M takes the generic form

$$
M = M_1 \, N_c + M_0 \, N_c^0 + M_{-1} \frac{1}{N_c} \left(\mathbf{J}^2 + \mathbf{I}^2 + \mathbf{K}^2 \right) + \ldots . \tag{3.60}
$$

Hyperfine mass splittings are of order $1/N_c$ [17] (see Chapter 9). We note that the order N_c^0 contribution to the meson-nucleon scattering amplitude comes from the mass splittings after the strong coupling cancellation [14]. A number of mass relations between baryons follow in the large N_c limit [17]. In particular,

$$
(\Sigma^* - \Sigma) = (\Xi^* - \Xi)
$$

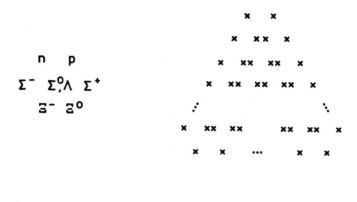

Fig. 3.15. Baryonic multiplet for $N_c = 3$ (a) and $N_c = \infty$ (b).

$$\frac{3}{4}(\Lambda + \Sigma) - \frac{1}{2}(N + \Xi) = \frac{1}{4}(\Sigma^* - \Delta) - \frac{1}{4}(\Omega - \Xi^*)$$
$$\frac{1}{2}(\Sigma^* - \Delta) + \frac{1}{2}(\Omega - \Xi^*) = (\Xi^* - \Sigma^*) \tag{3.61}$$

where the corrections are of order $1/N_c^2$. These are to be compared with the $SU(3)$ mass relations

$$\frac{1}{4}(3\Lambda + \Sigma) = \frac{1}{2}(N + \Xi)$$
$$(\Omega - \Xi^*) = (\Sigma^8 - \Delta)$$
$$\frac{1}{2}(\Sigma^* - \Delta) + \frac{1}{2}(\Omega - \Xi^*) = (\Xi^* - \Sigma^*) \tag{3.62}$$

where the corrections are of order $m_s^{3/2}$.

Appendix

(A_I) Berezinski-Kosterlitz-Thouless alternative

In four dimensions, the well-known Goldstone theorem[18] states that the nonvanishing expectation value of the scalar field is incompatible with the invariance of the vacuum under the continuous group of transformations. Thus Goldstone's dichotomy results: either the expectation value of the scalar field vanishes or the charge of the continuous transformation does not annihilate the vacuum — that is, the vacuum is not invariant under this continuous transformation. For the latter case, the vacuum is said to spontaneously break the symmetry concerned, with a massless (Nambu-Goldstone) boson carrier of long-range order,

$$\lim_{|x|\to\infty} \langle \phi^*(x)\phi(0)\rangle = \text{const.} \tag{A3.1}$$

This is to be compared with the symmetric (Wigner-Weyl) vacuum where the scalar correlations are suppressed exponentially,

$$\lim_{|x|\to\infty} \langle \phi^*(x)\phi(0)\rangle \sim e^{-m|x|}. \tag{A3.2}$$

In statistical mechanics, Mermin and Wagner[19] have shown that continuous symmetries cannot be spontaneously broken unless the dimension is three or higher. Severe infrared divergences in lower dimensions forbid the occurrence of any long-range order. These results were extended to field theory by Coleman, who has shown that the spontaneous breakdown of chiral symmetry cannot occur in $1+1$ dimensions. The Mermin-Wagner-Coleman theorem rules out the possibility of *any* spontaneous breaking of a continuous symmetry in $1 + 1$ dimensions.

Why is it that massless QCD_2 displays a massless meson state in large N_c? The answer to this question was suggested by Witten.[20] He noted, following Berezinski and Kosterlitz and Thouless (BKT in short)[21], that in two-dimensional systems, there can exist, in addition to the Wigner-Weyl phase (A3.2), another phase which we shall call a BKT phase characterized by quasi-long-range order

$$\lim_{|x|\to\infty} \langle \phi^*(x)\phi(0)\rangle = |x|^{-\alpha}. \tag{A3.3}$$

Now if the exponent α is very small, the correlations are almost infinite-ranged. Imagine now that the exponent α is proportional to the *inverse* of the number of colors. This, as shown by Witten, happens in the $SU(N)$ Thirring model, which also seems to violate the Coleman theorem. If we take the limit $N_c \to \infty$ *before* the limit $|x| \to \infty$, the correlator mimics the Goldstone phase. If, on the contrary, we keep the number of colors fixed, then the long-distance correlations will always be suppressed by a power-law behavior (A3.3). However, the larger the number of colors, the weaker the suppression.

(A_{II}) Witten-Veneziano relation

The requirement of a spontaneously broken $U_L(3) \times U_R(3)$ symmetry leads to a nonet of Goldstone bosons. We know, however, that the $U_A(1)$ axial current is anomalous, and due to the presence of instantons, this anomaly has a physical effect. Therefore, the $U_A(1)$ current is not conserved, and the η' is not a Goldstone boson. Since the instantons have a physical effect, one may wonder if gluodynamics should not be modified by adding an extra term to the pure QCD (Yang-Mills) Lagrangian, *i.e.*, the θ term. The Yang-Mills action reads

$$S = \int d^4x \, \text{Tr} \left(-\frac{1}{4} G \cdot G + \frac{\theta g^2}{16\pi^2} G \cdot \tilde{G} \right). \tag{A3.4}$$

Now, if we were to assume that the θ term survives in the large N_c limit, we encounter a paradox[24]. Indeed, on the one hand, the θ angle can be dialed to zero for massless QCD through a chiral singlet rotation of the quark field (thus an induced $U_A(1)$ anomaly) for any N_c. On the other hand, the fermionic loops are suppressed in large N_c massless QCD. How could a subleading effect cancel a leading effect?

To analyze the resolution of this paradox, let us look at the topological susceptibility defined by

$$\chi \equiv \left(\frac{d^2 \ln Z}{d\theta^2} \right)_{\theta=0} = \left(\frac{g^2}{16\pi^2} \right)^2 \int d^4x \langle T^*(G \cdot \tilde{G}(x) G \cdot \tilde{G}(0)) \rangle$$

$$= \left(\frac{g^2}{16\pi^2} \right)^2 \lim_{k \to 0} \chi(k) \tag{A3.5}$$

where in the last line we have expressed the topological susceptibility in terms of the glueball correlation function $\chi(k)$

$$\chi(k) = \int d^4x e^{ik\cdot x} \langle T^*(G\cdot\tilde{G}(x)G\cdot\tilde{G}(0))\rangle. \tag{A3.6}$$

We will assume with Witten that (A3.6) is non-zero.[¶] To leading order in $1/N_c$, the correlation function is saturated by singlet mesons and glueballs,

$$\begin{aligned}
\chi(k) &\sim \chi(k)^{glue} + \chi(k)^{mes} \\
&\sim \left(C(k^2) + \sum_{glueballs} \frac{N_c^2 a_n^2}{k^2 - M_{n,glue}^2}\right) + \sum_{mesons} \frac{N_c b_n^2}{k^2 - M_{n,mes}^2}
\end{aligned} \tag{A3.7}$$

where $C(k^2)$ is an arbitrary polynomial, since $G\tilde{G}$ is a composite operator that requires subtractions. The large N_c counting rules discussed above imply that $a_n \sim b_n \sim N_c^0$. Using (A3.7), the paradox can be stated as follows: the gluonic term is *naively* dominant and θ-dependent, yet the total correlation function (A3.7) is θ-independent in massless QCD. The resolution of this paradox leads to Witten's suggestion: there must exist a meson, whose mass squared goes as $1/N_c$ and not N_c^0 so that at $k = 0$, the glueball contribution is balanced by this meson contribution which is now dominant. For the cancellation to occur, it is important that $C(0) \sim N_c^2$ and positive. Thus

$$\chi^{glue}(0) = \frac{N_c b_{\eta'}^2}{m_{\eta'}^2} \tag{A3.8}$$

with $m_{\eta'}^2 \sim 1/N_c$. We may identify this meson with the η', as it is the lowest flavor-singlet pseudoscalar meson that could contribute to the correlation function $\chi(k)$. The η' is massless in the large N_c and chiral limit of QCD, a feature also shared by the other Goldstone bosons. In this limit, the axial singlet current is conserved (and hence no anomaly).

One can rewrite the relation (A3.8) using the definition of the topological susceptibility and soft-meson theorems for the axial current. The result is the Witten-Veneziano relation[24,25]

$$m_{\eta'}^2 = \frac{4N_f}{f_\pi^2}\chi_{glue} \tag{A3.9}$$

[¶]This assumption may be at odds with the fact that $G\cdot\tilde{G}$ is a total divergence [23]. This point merits to be clarified.

where the topological susceptibility χ_{glue} follows from (A3.5) in a world without light quarks. χ_{glue} is a non-negative constant, thanks to the subtraction constant in (A3.7). Since $\chi_{glue} \sim N_c^0$ and $f_\pi \sim \sqrt{N_c}$, it follows that $m_{\eta'}^2 \sim 1/N_c$ as expected.

An interesting question arises as to why the mass of the η' is so large in nature ($\sim 1\mathrm{GeV}$), while being so small, parametrically, in the $1/N_c$ counting. An explanation has been suggested by Novikov *et al.*[26] The mass of the η' comes from mixing with the glue states (glueballs) which are expected to be heavy ($m_G^2 \sim 15\,\mathrm{GeV}^2$). Therefore it may be more appropriate to say that the mass of the η' is actually *light* compared to the typical mass in this channel.

References

1. G. 't Hooft, Nucl. Phys. **B75**, 461 (1974).
2. E. Witten, Nucl. Phys. **B160**, 57 (1979).
3. M. Einhorn, Phys. Rev. **D14**, 3451 (1976).
4. A. Casher, J. Kogut and L. Susskind, Phys. Rev. **D10**, 732 (1974).
5. M. Prakash, M. Prakash and I. Zahed, Ann. Phys. (NY) **221**, 71 (1993).
6. M. Durgut, Nucl. Phys. **B116**, 233 (1976).
7. R.C. Rossi and G. Veneziano, Nucl. Phys. **B123**, 507 (1977).
8. *E.g.*, D. Gepner, Nucl. Phys. **B252**, 481 (1985); D. Gonzales and A.N. Redlich, Nucl. Phys. **B256**, 621 (1985); G.D. Date, Y. Frishman and J. Sonnenschein, Nucl. Phys. **B283**, 365 (1987).
9. E. Witten, Comm. Math. Phys. **92**, 455 (1984).
10. H. Yamagishi and I. Zahed, SUNY preprint, SUNY-NTG-94-12, unpublished.
11. T.H. Skyrme, Proc. Roy. Soc. **A260** (1961) 2127; Nucl. Phys. **31**, 556 (1962).
12. S. Coleman and E. Witten, Phys. Rev. Lett. **45**, 100 (1980).
13. J.L. Gervais and B. Sakita, Phys. Rev. Lett. **52**, 87 (1984); Phys. Rev. **D30**, 1795 (1984).
14. H. Yamagishi and I. Zahed, Mod. Phys. Lett. **7**, 1105 (1992).
15. R. Dashen and A. Manohar, Phys. Lett. **B315**, 425 (1993).
16. R. Dashen and A. Manohar, Phys. Lett. **B315**, 438 (1993).
17. E. Jenkins, Phys. Lett. **B315**, 441 (1993).
18. J. Goldstone, A. Salam and S. Weinberg, Phys. Rev. **127**, 965 (1962).
19. N.D. Mermin and H. Wagner, Phys. Rev. Lett. **17**, 1133 (1966).
20. E. Witten, Nucl. Phys. **B145**, 110 (1978).
21. V.L. Berezinski, JETP **32**, 493 (1970); J.M. Kosterlitz and D.J. Thouless, J. Phys. **C6**, 1186 (1973).
22. S. Coleman, Comm. Math. Phys. **31**, 259 (1973).
23. H. Yamagishi and I. Zahed, SUNY preprint, SUNY-NTG-95-29, unpublished.
24. E. Witten, Nucl. Phys. **B156**, 269 (1979).

25. G. Veneziano, Nucl. Phys. **B159**, 213 (1979).
26. V.A. Novikov, M.A. Shifman, A.I. Vainshtein and V.I. Zakharov, Nucl. Phys. **B191**, 301 (1981).

CHAPTER 4

CURRENT ALGEBRA

4.1. Introduction

As suggested in Chapter 3, QCD in the large N_c limit transforms into a weakly interacting theory of infinitely many mesons and glueballs in which baryons emerge as solitons. We will see how this picture fares with nature in later chapters. In the real world, however, $N_c = 3$. Is 1/3 small enough to warrant a $1/N_c$ expansion? The rationale provided by the large N_c arguments prompts us to ask if there is a reliable mesonic formulation of QCD for $N_c = 3$. A possible answer may be found in the concepts of current algebra and their subsequent formulations in terms of effective Lagrangians.

Current algebra is a subject that predates QCD. The idea was originally due to Gell-Mann, the technique was elaborated by Fubini and Furlan, and the first important result was obtained by Adler and Weisberger — all within the course of a few years[1,2]. The idea emphasizes the important role of chiral symmetry in fundamental dynamics. The purpose of this chapter is to give a short overview of the approach and a brief account of some of the successful predictions in the form of soft-pion theorems. Our exposition will be mostly historical, with a few comments on some recent developments at the end.

4.2. Physical Currents

We can obtain information on strongly interacting particles in two different ways: (1) through purely hadronic processes, which are usually amenable to the study of hadronic S-matrix elements of events such as

pp or πp scattering; (2) through semi-leptonic processes which reduce to the study of hadronic matrix elements of various physical currents such as $\pi^0 \to \gamma\gamma$, $\pi^- \to \mu^-\bar{\nu}_\mu$, and $e^+e^- \to$ hadrons. The physical currents are the electroweak currents of the standard $SU(2) \times U(1)$ model.

4.2.1. *Electromagnetic current*

The best known physical current is the electromagnetic current described by the interaction Lagrangian

$$\mathcal{L}^I_{em} = -e \sum_f \bar{\psi}_f Q_f \gamma^\mu A_\mu \psi_f \equiv -e A_\mu J^\mu_{em} \tag{4.1}$$

in Maxwell theory with a photon field A_μ. Here Q_f is equal to -1 for electrons, muons and τ mesons, $Q_f = 2/3$ for u, c, and t quarks and $Q_f = -1/3$ for d, s, and b quarks. Perturbation theory applied to (4.1) in the context of Maxwell theory gives a very good description of the electromagnetic processes. The hadronic contribution to the electromagnetic current J^μ_{em} can be organized using $SU(3)$ flavor quantum numbers. The electromagnetic interaction in fact breaks $SU(3)$ symmetry (since $m_{\pi^+} \neq m_{\pi^0}$ for instance). From the Gell-Mann-Nishijima relation, we have

$$Q = I_3 + \frac{1}{2}Y. \tag{4.2}$$

The hypercharge (Y) is the sum of the baryon number (B) and the strangeness (S). Thus

$$J^\mu_{em} = V^\mu_{0,0,0} + V^\mu_{0,1,0} \tag{4.3}$$

with the notation V_{Y,I,I_3}. Here the first term transforms as a singlet under isospin and the second term as a triplet. Both charges corresponding to the currents (4.3) are separately conserved. This isospin decomposition is consistent with the $SU(3)$ assignment in the quark model since

$$Q = I_3 + \frac{1}{2}Y = \frac{\lambda^3}{2} + \frac{1}{\sqrt{3}}\frac{\lambda^8}{2} = \begin{pmatrix} \frac{2}{3} & 0 & 0 \\ 0 & -\frac{1}{3} & 0 \\ 0 & 0 & -\frac{1}{3} \end{pmatrix}. \tag{4.4}$$

Examples of processes measuring hadronic matrix elements of the electromagnetic current are shown in Fig. 4.1 where (a) is a typical photo-emission process $h \to h' + \gamma$ such as

$$\text{Amp}[\omega \to \pi^0\gamma] \sim e A_\mu \langle \pi^0 | J^\mu_{em} | \omega \rangle$$

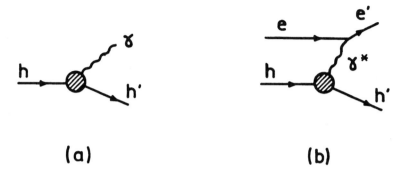

Fig. 4.1.   Processes involving the electromagnetic current: (a) photoemission, (b) electron scattering.

and (b) is a typical scattering process $eh \to eh'$ with amplitude

$$e^2 \overline{u}(p_2) \gamma_\mu u(p_1) \frac{1}{k^2} \langle h' | J_{em}^\mu | h \rangle$$

where all the information about the strong interaction is embedded in purely hadronic states $|h\rangle$.

4.2.2. *Weak currents*

Weak processes can be classified as leptonic, semi-leptonic and non-leptonic. Typical cases (denoting leptons by l and hadrons by h with specific examples in parentheses) are:

Leptonic:

$$l \to l' + \nu_l + \overline{\nu}_{l'}$$
$$(\mu^- \to e^- + \nu_\mu + \overline{\nu}_e) \tag{4.5}$$

Semi-Leptonic:

$$h \to l + \overline{\nu}_l$$
$$(\pi \to \mu + \overline{\nu}_\mu)$$
$$h \to h' + l + \overline{\nu}_l$$
$$(K^- \to K^0 + e^- + \overline{\nu}_e) \tag{4.6}$$

Non-Leptonic:

$$h \to h' + h''$$
$$(K^+ \to \pi^+ + \pi^0) \tag{4.7}$$

Prior to the Standard Model, the above processes were described by a local Fermi interaction

$$\mathcal{L}_F = \frac{G}{\sqrt{2}} \left(J_\mu^\dagger J^\mu + \text{h.c.} \right). \tag{4.8}$$

Here G is the Fermi constant fixed from the decay rate of the muon, $G \sim 10^{-5} m_N^{-2}$, with $m_N = 940$ MeV being the nucleon mass, and J^μ is the sum of the leptonic and hadronic currents $J^\mu = J_l^\mu + J_h^\mu$. The leptonic current is given by the V-A charged current plus the neutral current. The V-A charged current is of the form

$$J_l^\mu = \sum_l \bar{\nu}_l \gamma^\mu \frac{1 - \gamma_5}{2} l \tag{4.9}$$

where $l = e, \mu, \tau$.

The standard model tells us specifically that the Fermi interaction (4.8) is only an effective interaction and that the weak processes are mediated by heavy W bosons as shown in Fig. 4.2 for leptonic, semi-leptonic and non-leptonic processes, respectively. Indeed, the standard-model Lagrangian (the current part) reads

$$\mathcal{L}_{WSG} = -\frac{g}{\sqrt{2}} \sum_{i=l,q} \bar{\Psi}_i \gamma^\mu (\tau_+ W_\mu^+ + \tau_- W_\mu^-) \Psi_i$$
$$- \frac{g}{2 \cos \theta_w} \sum_j \bar{\Psi}_j \gamma^\mu ((I_3^{(j)} - 2Q_f \sin^2 \theta_w) - I_3^{(j)} \gamma_5) \Psi_j Z_\mu^0. \tag{4.10}$$

Here Ψ denotes left-handed doublets of quarks and leptons, θ_w denotes the Weinberg angle, and the sum in the second line runs over all flavors of quarks and leptons. The first line describes *charged currents*, the second line *neutral currents*. A direct comparison of the above two formulations yields the relation

$$\frac{G_F}{\sqrt{2}} = \frac{g^2}{8 M_W^2}. \tag{4.11}$$

Since the decay rates calculated from the above Lagrangians differ by the ratio of the lepton or quark mass to the W or Z boson mass for energies much lower than the W mass, the effective Fermi interaction works rather well.

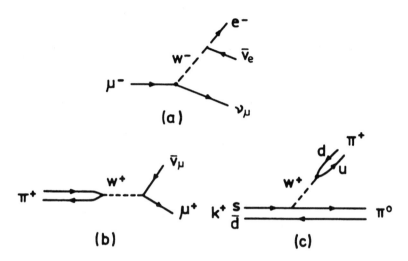

Fig. 4.2. Weak processes in the standard model: (a) leptonic, (b) semi-leptonic, and (c) non-leptonic.

For the hadronic part of the weak current, we decompose the current in terms of good quantum numbers, namely hypercharge (Y) and isospin (I). In general, the hadronic current J_h is the sum of a vector part V and an axial part A, $J_h^\mu = V^\mu + A^\mu$. To get an idea as to how this comes about, we cite here a few examples. In kaon decays $K^- \rightarrow K^0 + e + \bar{\nu}_e$, only V contributes. In pion decays $\pi^+ \rightarrow \mu^+ + \nu_\mu$, only A contributes. Finally, in neutron decays $n \rightarrow p + e^- + \bar{\nu}_e$, both V and A contribute. All these processes conserve the baryon number. It can be noted that the strangeness conserving processes ($\Delta S = 0$) involve $I = 1$ and $I_3 = \pm 1$ such as $\pi^+ \rightarrow \mu^+ + \nu_\mu$, and the strangeness changing processes ($\Delta S = 1$) involve $I = 1/2$ and $I_3 = \pm 1/2$. Thus the general form of the vector and axial-vector hadronic weak currents is

$$V^\mu = \cos\theta_c V_{0,1,I_3}^\mu + \sin\theta_c V_{1,\frac{1}{2},I_3}^\mu$$
$$A^\mu = \cos\theta_c A_{0,1,I_3}^\mu + \sin\theta_c A_{1,\frac{1}{2},I_3}^\mu, \tag{4.12}$$

where we have used the notations V_{Y,I,I_3} and A_{Y,I,I_3}. Note that

$$\Delta Q = \Delta I_3 + \frac{1}{2}\Delta Y = \Delta I_3 + \frac{1}{2}\Delta S$$

since $\Delta B = 0$. For exact $SU(3)$ symmetry, the Cabibbo angle $\theta_c \sim 0.23$. It measures the amount of strangeness-changing current in both the axial and vector parts of the hadronic part of the weak current.

From the point of view of the standard model, the Cabibbo angle is one of the four angles describing the mixing of the $Q = -1/3$ quarks via the Kobayashi-Maskawa unitary matrix. Restricted to d, s quarks only, this matrix reduces within a very good approximation to the planar rotation, parametrized by the Cabibbo angle.

4.3. CVC Hypothesis

The basic idea behind the conserved vector current (CVC) hypothesis put forward by Gell-Mann and Feynman is that the hypercharge zero part of the vector current $(V_{0,1,+1}, V_{0,1,0}, V_{0,1,-1})$ forms an isotriplet of conserved vector currents. The corresponding Noether charges generate the observed $SU(2)$ isospin symmetry in strong interaction processes. In QCD, these currents are the isotriplet flavor currents

$$V_{0,1,a}^{\mu} = \bar{q}\gamma^{\mu}T^a q. \tag{4.13}$$

The CVC hypothesis sets the scale for the weak interaction through charge and isospin normalizations,

$$\langle \pi^{\pm}|V_{0,1,0}|\pi^{\pm}\rangle = \pm 1$$
$$\langle \pi^{+}|V_{0,1,1}|\pi^{0}\rangle = 1.$$

The CVC hypothesis was extended by Cabibbo to the full octet of the vector and axial currents as depicted in Fig. 4.3. This hypothesis simplifies the calculations in strong interaction physics tremendously. With this hypothesis, this would mean that only two matrix elements (one for the octet vector and the other for the octet axial) are needed. The rest of the matrix elements follow from $SU(3)$ symmetry by Clebsch-Gordan coefficients.

In nature, $SU(3)$ is badly broken. Kaons are about three times heavier than pions. Also, the proton mass and the neutron mass are not identical and neither are the π^+ and π^- masses. In other words, even $SU(2)$ is

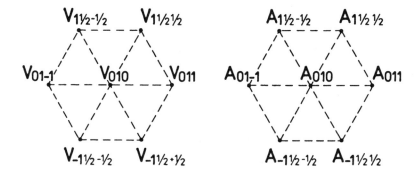

Fig. 4.3. Octet vector currents (left), Octet axial currents (right).

broken. In QCD language, the breaking is primarily due to nonvanishing current quark masses. The strange quark mass is an order of magnitude larger than the up and down masses. The CVC hypothesis is nevertheless interesting as it paves the road to the concept of current algebra.

4.4. Gell-Mann Current Algebra

4.4.1. *Basics*

To illustrate the motivation behind the current-algebra formulation in strong interactions, consider the Fourier transform of the axial-axial correlation function for scattering of a pion off a hadronic target at low momentum. Specifically,

$$T_{\mu\nu}^{ab}(q) = \int d^4x e^{iq\cdot x} \langle h(p_2)|T\left(A_\mu^a(x)A_\nu^b(0)\right)|h(p_1)\rangle. \qquad (4.14)$$

Since the pion is light, at low momentum, the axial current can be used as an interpolating field between the vacuum and a pion state. This is the PCAC hypothesis discussed in Chapter 2. Using the LSZ reduction formulae which relate the physical amplitudes to the residues of the Green functions calculated on-shell and PCAC, we can relate the low momentum part of (4.14) to the pion-hadron scattering amplitude,

$$T_{\mu\nu}^{ab}(q \approx 0) \sim \frac{q_\mu q_\nu}{q^4} f_\pi^2 \langle h(p_1)\pi^a|h(p_2)\pi^b\rangle. \qquad (4.15)$$

Notice that the divergence of the axial-axial correlator (4.14) is given as $q_\mu \to 0$ by (ignoring pion mass terms)

$$q^\mu T^{ab}_{\mu\nu}(q) \sim i \int d^4x \, e^{iq \cdot x} \delta(x^0) \langle h(p_2)|[A^a_0(x), A^b_\nu(0)]|h(p_1)\rangle. \quad (4.16)$$

The equal time commutator in (4.16) involves the axial charge densities. Current algebra assumes, as we will discuss below, that

$$[A^a_0(x), A^b_\nu(0)] = if^{abc}V^c_\nu(0)\delta^3(x) \quad (4.17)$$

where f^{abc} is the $SU(3)$ structure constant. Inserting (4.17) into (4.16) gives

$$q^\mu T^{ab}_{\mu\nu}(q) \sim -f^{abc}\langle h(p_2)|V^c_\nu(0)|h(p_1)\rangle. \quad (4.18)$$

Inserting (4.15) into (4.18) yields at threshold (*i.e.*, $q = (m_\pi, \vec{0})$ and $p_1 = p_2 = (m_h, \vec{0})$)

$$\langle h\pi^a|h\pi^b\rangle \sim -\frac{m_\pi}{f^2_\pi}f^{abc}\langle h|V^c_0(0)|h\rangle. \quad (4.19)$$

The pion-hadron scattering amplitude at threshold is totally given by the isospin content of the hadron and two chiral parameters - namely the pion mass m_π and the pion decay constant f_π. The knowledge of the equal time commutator (4.17) has allowed us to derive a constraint on the pion-hadron scattering amplitude at threshold, without reference to pion-hadron dynamics. This is the essence of the current algebra approach, and the constraint is an example of a soft-pion theorem. This is actually the Weinberg-Tomozawa result as we will discuss more thoroughly below.

4.4.2. *Connection to sum rules*

The idea described above is reminiscent of sum rules in quantum mechanics. Indeed, consider the probability A_{IF} for an atom in state I to move into state F via photon emission,

$$A_{IF} = -\frac{2m}{\hbar^2}(E_F - E_I)|\langle F|X|I\rangle|^2 \quad (4.20)$$

which is just the dipole formula with m the nucleon mass. X is the position operator conjugate to the momentum operator P,

$$[X, P] = i\hbar. \quad (4.21)$$

The sum of all the transitions is given by

$$\sum_F A_{IF} = -\frac{m}{\hbar^2} \langle I | [[H, X], X] | I \rangle$$

$$= -\frac{1}{2\hbar^2} \langle I | [[P^2, X], X] | I \rangle = 1, \qquad (4.22)$$

where $H \sim P^2/2m$ is the free part of the Hamiltonian. The probabilities sum up to one as they should. The morale of this simple story is that the knowledge of the "algebra" (4.21) allows us to derive a "theorem" (4.22) without knowledge of the details of the dynamics of the transition. This theorem is the celebrated Thomas-Reiche-Kuhn sum rule, relating the total cross-section for $E1$ photoabsorption on the atom Z to the fine structure constant

$$\frac{1}{2\pi^2} \int_0^\infty d\omega \sigma_{E1}(\omega) = Z \frac{\alpha}{m_e}. \qquad (4.23)$$

In a way, this is what the current-algebra approach is all about.

4.4.3. *Hypothesis*

In 1961 Gell-Mann postulated the following equal time commutation relations for the vector and axial-vector currents ($x^0 = y^0$)

$$[V_0^a(x), V_\mu^b(y)] = i f^{abc} V_\mu^c(x) \delta^3(x - y)$$
$$[V_0^a(x), A_\mu^b(y)] = i f^{abc} A_\mu^c(x) \delta^3(x - y)$$
$$[A_0^a(x), A_\mu^b(y)] = i f^{abc} V_\mu^c(x) \delta^3(x - y)$$
$$[A_0^a(x), V_\mu^b(y)] = i f^{abc} A_\mu^c(x) \delta^3(x - y) \qquad (4.24)$$

with Schwinger terms explicitly dropped. The Schwinger terms and other matters are discussed in the Appendix. The Gell-Mann hypothesis was that the above equal time commutation relations hold regardless of whether or not the currents are exactly conserved. These relations were derived from a quark model and then assumed to hold for the real world.

4.5. Symmetries and Algebra

Gell-Mann current algebra, while motivated by the increasing empirical facts from electroweak processes with hadronic participants, in fact reflected a possible underlying symmetry of the strong interactions as already hinted at in the CVC hypothesis.

Consider a general Lagrangian $\mathcal{L}(\phi, \partial_\mu \phi)$ for a general field ϕ_i with $i = 1, 2, \ldots$ an isospin index, which under infinitesimal transformations behaves as

$$\delta_a \phi_i = -i\varepsilon T_{ij}^a \phi_j. \tag{4.25}$$

We will think of ϕ_i either as an isotriplet in the adjoint representation, $\phi_i = \pi_i$ (pion), in which case $T_{ij}^a = \epsilon_{ij}^a$, or as a triplet in the fundamental representation, $\phi_i = q_i$ (quark), in which case $T_{ij}^a = \lambda_{ij}^a/2$, where the λ's are the Gell-Mann matrices. The T's in (4.25) generate a Lie algebra

$$[T^a, T^b] = i f^{abc} T^c \tag{4.26}$$

which is $SO(3)$ for the triplet pion and $SU(3)$ for the triplet quark. Under (4.25) the Lagrangian transforms as $\delta_a \mathcal{L} = \varepsilon \partial^\mu J_\mu^a$, where J_μ^a is the associated Noether current

$$J_\mu^a = -i \frac{\partial \mathcal{L}}{\partial(\partial_\mu \phi_i)} T_{ij}^a \phi_j. \tag{4.27}$$

If $\delta_a \mathcal{L} = 0$, the current (4.27) is classically conserved, $\partial \cdot J^a = 0$. If we denote by $\Pi_i = \partial \mathcal{L}/\partial \dot{\phi}_i$, then the classical charge

$$Q^a = \int d^3x J_0^a(x) = \int d^3x \left(-i\Pi_i T_{ij}^a \phi_j \right) \tag{4.28}$$

is conserved under the assumption that the currents are bounded on the three-sphere at infinity. The above arguments are classical. Their extension to the quantum world is somewhat more subtle.

Since the current J_μ^a is a composite operator of the fundamental fields ϕ_i, all defined at the same space-time point, it requires regularization. The occurrence of infinities is usually unavoidable and anomalies arise, *i.e.* the current is no longer conserved quantum mechanically in general. We have already discussed this matter in the context of QCD currents in Chapter 2, and we will discuss it again in Chapter 6 in the context of nonabelian anomalies. Now barring anomalies,

$$J_0^a(x) = -\frac{i}{2} T_{ij}^a \left(\Pi_i \phi_j \pm \phi_j \Pi_i \right) \tag{4.29}$$

yields a conserved quantum charge Q^a. Here the upper sign $(+)$ refers to bosonic fields and the lower sign $(-)$ to fermionic fields. The ordering in (4.29) makes the charge density operator hermitian. If the T's are field-dependent, hermiticity can be enforced via Weyl-ordering. The fields ϕ_i and Π_j are canonical and satisfy the equal-time commutation relation

$$[\phi_i(x), \Pi_j(y)]_{x^0=y^0} = i\delta_{ij}\delta^3(x - y). \tag{4.30}$$

It is clear that the global algebra of the generators (4.26) manifests itself in the form of a current algebra on the charge densities. Indeed, using (4.29) and (4.30) gives

$$[J_0^a(x), J_0^b(y)]_{x^0=y^0} = -i[T^a, T^b]_{ij}\frac{1}{2}(\Pi_i\phi_j \pm \phi_j\Pi_i)(x)\delta^3(x - y). \tag{4.31}$$

Since the T's fulfill (4.26), it follows that

$$[J_0^a(x), J_0^b(y)]_{x^0=y^0} = if^{abc}J_0^c(x)\delta^3(x - y). \tag{4.32}$$

This relation depends solely on (4.26), (4.29) and (4.30) and holds independent of whether the Noether current J_μ^a is conserved or not. The algebra (4.26) and the canonical commutation relations are all that are needed to arrive at (4.32).

The Gell-Mann current algebra may now be viewed as a mere reflection of a possible approximate isospin symmetry of the underlying strong-interaction Hamiltonian on the structure of the currents. For the relations (4.32) to hold, the symmetry need not be exact, and that constitutes the whole beauty of the approach.

4.6. Current Algebra from Models

Well before the advent of QCD and asymptotic freedom, models were proposed by Gell-Mann and followers that embodied the current commutation relations (4.24). In this section, we discuss two popular ones: the linear σ model and the free quark model. These models should be viewed as toy models that reflect the general physical content contained in current algebra, and serve to illustrate the emergence of current-current commutation relations from local formulations. Some important predictions follow from (4.24) without recourse to any model, as we will discuss below.

4.6.1. *The σ model*

The linear sigma model embodies the tenets of current algebra. It was originally formulated by Gell-Mann and Lévy. Let $\phi_i = \pi_i$ be the isotriplet pion field ($i = 1, 2, 3$) and $\phi_4 = \sigma$ a fictitious scalar field. The dynamics of the π-σ fields is given by the Lagrangian density

$$\mathcal{L} = \frac{1}{2}(\partial_\mu \vec{\pi})^2 + \frac{1}{2}(\partial_\mu \sigma)^2 - \mathcal{V}(\sigma^2 + \vec{\pi}^2) + \kappa\,\sigma. \qquad (4.33)$$

Aside from the last term, the Lagrangian (4.33) is invariant under $SO(4) \sim SU_A(2) \times SU_V(2)$ transformations. The infinitesimal form of these transformations is:

$$\delta\sigma = -\vec{\alpha} \cdot \vec{\pi},$$
$$\delta\vec{\pi} = +\vec{\alpha}\sigma + \vec{\beta} \times \vec{\pi}, \qquad (4.34)$$

with $\vec{\alpha}$ and $\vec{\beta}$ the infinitesimal parameters associated to $SU_A(2)$ and $SU_V(2)$ respectively. The Noether currents associated to (4.34) can be readily constructed. They are given by

$$V_\mu^a = (\vec{\pi} \times \partial_\mu \vec{\pi})^a,$$
$$A_\mu^a = (\sigma\partial_\mu\pi^a - \pi^a\partial_\mu\sigma). \qquad (4.35)$$

The vector current is conserved while the axial current is not because of the linear term in (4.33),

$$\partial \cdot V^a = 0, \qquad\qquad \partial \cdot A^a = -\kappa\sigma. \qquad (4.36)$$

The quantum charge densities are free of ordering (that is, the fields are all independent) and read

$$V_0^a(x) = (\vec{\pi} \times \dot{\vec{\pi}})^a(x),$$
$$A_0^a(x) = (\sigma\dot{\vec{\pi}} - \vec{\pi}^a\dot{\sigma})(x). \qquad (4.37)$$

The dotted fields are to be understood as conjugate momenta. It is straightforward to check that the charge densities (4.37) fulfill the current-algebra relations (4.24) using the canonical commutation rules between the fields,

$$[\sigma(x), \dot{\sigma}(y)]_{x^0=y^0} = i\delta^3(x-y), \quad [\dot{\pi}^a(x), \dot{\pi}^b(y)]_{x^0=y^0} = i\delta^{ab}\delta^3(x-y). \,(4.38)$$

4.6.2. *The quark model*

The quark model was also originally proposed by Gell-Mann to address the issue of $SU(3)$ symmetry in hadronic interactions. Here limiting ourselves to flavor $SU(2)$, let $\phi_i = q_i$ be the doublet quark field of mass m, with a free Lagrangian

$$\mathcal{L} = \bar{q}(i\partial\!\!\!/ - m)q. \qquad (4.39)$$

Under the $SU_A(2) \times SU_V(2)$ transformations

$$\delta_A q = -i\gamma_5 \vec{\alpha} \cdot \vec{T}q, \qquad \delta_V q = -i\vec{\beta} \cdot \vec{T}q \qquad (4.40)$$

with $\vec{T} = \vec{\tau}/2$. (For flavor $SU(3)$, we have $T^a = \lambda^a/2$, the Gell-Mann matrices.) The Noether currents following from (4.40) are

$$\begin{aligned} V_\mu^a &= \bar{q}\gamma_\mu T^a q, \\ A_\mu^a &= \bar{q}\gamma_\mu \gamma_5 T^a q. \end{aligned} \qquad (4.41)$$

The vector current V is conserved but the axial current A is not, due to the mass term in (4.39)

$$\partial \cdot V^a = 0, \qquad \partial \cdot A^a = m\bar{q}q. \qquad (4.42)$$

The charge densities read

$$\begin{aligned} V_0^a &= \bar{q}\gamma_0 T^a q, \\ A_0^a &= \bar{q}\gamma_0 \gamma_5 T^a q \end{aligned} \qquad (4.43)$$

and obey the Gell-Mann current algebra (4.24) thanks to the canonical equal-time anticommutators

$$\{q_i(x), \bar{q}_j(y)\}_{x^0=y^0} = i\delta_{ij}\,\delta^3(x-y). \qquad (4.44)$$

The free quark model as originally discussed by Gell-Mann was the precursor of QCD. Thus, QCD should also fulfill the current-algebra relations. This is the case modulo Schwinger terms (described in the Appendix) and anomalies. We shall address some of these issues below in the context of the $\pi^0 \to \gamma\gamma$ decay.

4.7. Low-Energy Theorems

The great achievement of current algebra was its ability to generate a useful set of constraints on hadronic amplitudes near threshold just by assuming that the pion was an unusually light hadron, and this without a theory of strong interactions. These constraints are usually referred to as low-energy theorems.[*] These theorems were found to agree rather well with experiments and today represent still a benchmark for many theoretical descriptions of strong interaction physics. In this section we will give an outline of some of these results. Before we move to the low-energy theorems in hadronic physics, we will start with the more familiar case of electrodynamics.

4.7.1. *Two soft photons*

Let us consider Compton scattering on some general target diagrammatically represented in Fig. 4.4 and described by the amplitude

$$T = e^2 \epsilon_1^{\mu*} \epsilon_2^{\nu} T_{\mu\nu},$$
$$T_{\mu\nu} = i \int d^4x d^4y \exp(ik_2 \cdot x - ik_1 \cdot y) \langle p_2 | T(j_\mu(x) j_\nu(y)) | p_1 \rangle. \quad (4.45)$$

As independent kinematical variables, we choose k_1, k_2, and $P \equiv (p_1 + p_2)/2$. The Lorentz and discrete symmetries of QED allow a general tensor decomposition of the amplitude $T_{\mu\nu}$:

$$T_{\mu\nu} = A g_{\mu\nu} + B P_\mu P_\nu + C(P_\mu k_{2\nu} + P_\nu k_{1\mu}) + D k_{1\mu} k_{2\nu} \quad (4.46)$$

where the amplitudes A, B, C, D are functions of the Lorentz invariants $(P \cdot k_1)$ and $t = (k_1 - k_2)^2$. Gauge invariance ($k_i^\mu T_{\mu\nu} = 0$) implies additional conditions

$$A + (P \cdot k_2)C + (k_1 \cdot k_2)D = 0,$$
$$(P \cdot k_2)B + (k_1 \cdot k_2)C = 0. \quad (4.47)$$

The explicit expression for $T_{\mu\nu}(x, y = 0)$ reads

$$T_{\mu\nu} = -i(2\pi)^3 \sum_n \langle p_2 | j_\mu(0) | n \rangle \langle n | j_\nu(0) | p_1 \rangle \frac{\delta^{(3)}(p_n - p_1 - k_2)}{E_n - E_1 - k_2^0 - i\epsilon}$$

---

[*]Strictly speaking these theorems hold exactly at the soft point, and approximately at threshold. A reanalysis of these theorems around the chiral point was carried out recently in the context of heavy baryon chiral perturbation theory [3].

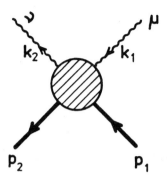

Fig. 4.4. Compton scattering on hadronic target.

$$-i(2\pi)^3 \sum_n \langle p_2|j_\nu(0)|n\rangle\langle n|j_\mu(0)|p_1\rangle \frac{\delta^{(3)}(p_n - p_2 + k_2)}{E_n - E_2 + k_2^0 - i\epsilon}$$

$$(4.48)$$

where, as usual, we have used the definition of the chronological product, Lorentz invariance of the current matrix elements, and completeness.

We would now like to study the Compton process in the *soft* limit when both momenta of the photons tend to zero. First, let us note that in this limit, both terms in (4.48) develop poles, the first term when $E_n = E_1$, the second term when $E_n = E_2$. Current conservation implies

$$\langle p_2|j_\mu|p_1\rangle = \bar{u}(p_2)\left(eF_1(q^2)\gamma_\mu + i\frac{e}{2m}F_2(q^2)\sigma_{\mu\nu}q_\nu\right)u(p_1) \quad (4.49)$$

where F_1 and F_2 are the usual Dirac and Pauli form factors and $q_\mu = p_2 - p_1$. In the low momentum limit,

$$\langle p_2|j_0|p_1\rangle = 2em + \frac{e}{m}\vec{p}_1 \cdot \vec{p}_2 + Q_{ij}(p_1 - p_2)_i(p_1 - p_2)_j$$

$$+ i(2\mu/J - e/m)(\vec{p}_2 \times \vec{p}_1) \cdot \vec{J} + O(p^4),$$

$$\langle p_2|j_i|p_1\rangle = e(p_1 + p_2)_i + (2im\mu/J)\left(\vec{J} \times (\vec{p}_2 - \vec{p}_1)\right)_i + O(p^3).$$

$$(4.50)$$

The charge density is an even function of the momenta and the current density is an odd function of the momenta. The lowest-order terms are fixed by the normalization of the current whereas the higher order terms have traditional multipole interpretation. For example, the real coefficients e, μ, and Q_{ij} correspond to the electric charge, the magnetic moment, and a combination of the mean-square charge radius and quadrupole electric moment, respectively. J is the target angular momentum. Inserting the expansion (4.50) into (4.48), we get

$$
\begin{aligned}
T(k_1, k_2 \to 0) = &-2e^2(\vec{\epsilon}_2^* \cdot \vec{\epsilon}_1) - 4iek_2^0\left(\mu/J - e/(2m)\right)\left((\vec{\epsilon}_2^* \times \vec{\epsilon}_1) \cdot \vec{J}\right) \\
&-\frac{2ie\mu}{Jk_2^0}(\vec{\epsilon}_2^* \cdot \vec{k}_2)\left(\vec{\epsilon}_1 \cdot (\vec{J} \times \vec{k}_2)\right) \\
&+\frac{2ie\mu}{Jk_2^0}(\vec{\epsilon}_1 \cdot \vec{k}_1)\left(\vec{\epsilon}_2^* \cdot (\vec{J} \times \vec{k}_1)\right) \\
&-\frac{2\mu^2}{J^2 k_2^0}[\vec{\epsilon}_2^* \cdot (\vec{J} \times \vec{k}_1), \vec{\epsilon}_1 \cdot (\vec{J} \times \vec{k}_2)] + O(k^2).
\end{aligned} \tag{4.51}
$$

We will work in Coulomb gauge and consider the forward Compton amplitude by restricting the discussion to the following kinematical range

$$
\begin{aligned}
\vec{k}_1 = \vec{k}_2 &= (0, 0, \omega), \\
\epsilon_\mu^\pm &= (0, 1, \pm i, 0)/\sqrt{2}.
\end{aligned} \tag{4.52}
$$

With this in mind, the helicity amplitudes are given by

$$
T^\pm = -2e^2 \pm 2J_3\omega m(\mu/J - e/m)^2. \tag{4.53}
$$

Comparison of eq.(4.53) with eq.(4.46) gives

$$
A = A(0, 0) = 2 \tag{4.54}
$$

and the cross section in the soft photon limit ($\omega \to 0$) reads

$$
\frac{d\sigma_{soft}}{d\Omega} = \frac{\alpha^2}{4m^2}A^2 = \alpha^2/m^2. \tag{4.55}
$$

We stress again that the above low-energy formula was obtained without any reference to a perturbative expansion.

The results obtained above allow us to write down the sum rule we want. Let us define the odd amplitude $T^{odd} = T^+ - T^-$ and assume that for the analytic function $T^{odd}(\omega)$, one could write down the unsubtracted dispersion relation

$$
T^{odd}(\omega^2) = \frac{1}{\pi}\int_0^\infty \frac{\text{Im } T^{odd}(\omega'^2)}{\omega'^2 - \omega^2 - i\epsilon}d\omega'^2. \tag{4.56}
$$

Relating the cuts to the total cross section by the optical theorem and expressing the magnetic moment by the Landé factor $\mu = geJ/2m$, we obtain the well-known Gerasimov-Drell-Hearn sum rule

$$\frac{4m^2}{e^2\pi|J_3|}\int_0^\infty (\sigma(+) - \sigma(-))\frac{d\omega}{\omega} = (g-2)^2 \qquad (4.57)$$

relating the difference of the integrated total cross sections for opposite helicities to the deviation of the gyromagnetic factor for the target particle from its value at tree level. The present sum rule is a prototype of sum rules to be discussed in the context of current algebra.

4.7.2. *One soft pion*

We now turn to soft-pion processes by considering the physics of one soft pion. We start with the PCAC hypothesis, then we derive the Goldberger-Treiman relation, and finally we discuss the Adler consistency condition.

(i) PCAC hypothesis

The Gell-Mann current algebra is merely a statement on the algebraic structure satisfied by the axial- and vector-current densities. We have already noted that this local algebra reflects underlying symmetries of the theory. A striking fact in the hadronic spectrum is the lightness of the pion. From our previous discussions, it is clear that the pion is a Goldstone particle. In the context of strong interactions, the pion therefore is a very special hadron. At low energy, the assumption is that it dominates the scattering amplitude even though it has a mass. What this means is that the physics of interacting hadrons at threshold is predominantly the physics of pions.

To be able to quantify the role of the pion in hadronic matrix elements, one defines an interpolating pion field. For that, we consider the weak decay of the pion, $\pi^- \to e^-\bar{\nu}_e$. The transition amplitude is given by

$$\text{Amp}[\pi \to e\bar{\nu}] = i\frac{G}{\sqrt{2}}\cos(\theta_c)\langle 0|A_\mu^{1+i2}|\pi^-\rangle\bar{u}(e)\gamma^\mu(1-\gamma_5)u(\bar{\nu}) \quad (4.58)$$

where θ_c is the Cabibbo angle. The matrix element of the axial current between the vacuum and the pion state is just the pion decay constant

$$\langle 0|A_\mu^i|\pi^j(q)\rangle = i\delta^{ij}f_\pi q_\mu \qquad (4.59)$$

where $\pi^\pm = (\pi^1 \pm i\pi^2)/\sqrt{2}$. Equation (4.59) involves an interpolation between the vacuum and the pion on shell. The PCAC hypothesis corresponds to assuming an interpolating pion field of the form

$$\pi^a(x) \sim \frac{1}{m_\pi^2 f_\pi} \partial^\mu A_\mu^a(x) \tag{4.60}$$

which extrapolates the content of (4.59) off-shell. This is meaningful only if *the interpolation is smooth*, i.e., the hadronic amplitudes do not vary rapidly near threshold. The relation (4.60) implies that the axial current is a Partially Conserved Axial Current (PCAC) since the pion mass is small.

We stress that the prescription (4.60) is not unique, since one can add any term to it that vanishes on-shell. In fact, more generally, the S-matrix – which is what one measures in experiments – is invariant under certain field transformations consistent with the symmetries of the theory. PCAC is useful for approximating physical amplitudes by those amplitudes with soft kinematics for which rigorous theorems can be established. In the present case, (4.60) satisfies this condition because the divergence of the axial current is pion-dominated at threshold and the pion is nearly massless. (A similar interpolation for the kaon field by the divergence of the axial current is problematic since the threshold point and the soft point are not close — the mass of the kaon is not small.) Indeed the hadronic process shown in Fig. 4.5 is given by

$$\langle\beta|\partial \cdot A^a|\alpha\rangle = \frac{f_\pi m_\pi^2}{-q^2 + m_\pi^2} \ \text{Amp} \ (\alpha \to \beta + \pi^a)$$
$$+ \ \text{regular terms.} \tag{4.61}$$

The regular terms are assumed to be smooth at threshold $q^2 = m_\pi^2$. The interpolating pion field (4.60) can be used to create and annihilate pions. Thus current correlation functions are amenable to pion-dominated amplitudes.

(ii) Goldberger-Treiman relation

Consider the neutron β-decay $n \to p + e^- + \bar{\nu}_e$ as shown in Fig. 4.6. The decay involves both hadronic vector and axial currents. Consider the axial-current part. Covariance and parity imply

$$\langle p|A_\mu^{1+i2}|n\rangle = \bar{u}_p \left(G_1(q^2)\gamma_\mu\gamma_5 + G_2(q^2)q_\mu\gamma_5\right) u_n \tag{4.62}$$

Fig. 4.5. Axial current in a hadronic target.

with $q = p_p - p_n$ where $G_1(q^2)$ and $G_2(q^2)$ are, respectively, the pseudovec-tor (axial) and pseudoscalar form factors. Note that since the hadrons are on-shell,

$$\langle p|\partial^\mu A_\mu^{1+i2}|n\rangle = iq^\mu \langle p|A_\mu^{1+i2}|n\rangle$$
$$= \left(2m_N G_1(q^2) + q^2 G_2(q^2)\right) i\bar{u}_p \gamma_5 u_n. \qquad (4.63)$$

In nuclear β-decays, q^2 involved is rather small (typically a few keV), so

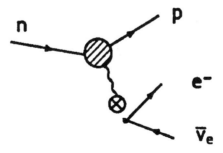

Fig. 4.6. β-decay of the neutron (for a large W^- mass) and pion dominance.

the decay process in which a neutron turns into a proton by emitting an electron and a neutrino occurs close to the mass-shell. Thus by PCAC,

$$\langle p|\partial^\mu A_\mu^{1+i2}|n\rangle = \sqrt{2}m_\pi^2 f_\pi \langle p|(\pi^-)^\dagger|n\rangle$$

$$= 2m_\pi^2 f_\pi \frac{g_{\pi NN}(q^2)}{-q^2 + m_\pi^2} i\bar{u}_p \gamma_5 u_n. \qquad (4.64)$$

Comparing (4.64) with (4.63) as $q \to 0$ gives

$$f_\pi = \frac{m_N g_A}{g_{\pi NN}(0)} \qquad (4.65)$$

where $g_A = G_1(0)$. The smoothness requirement implicit in the PCAC hypothesis is that $g_{\pi NN}(0) \approx g_{\pi NN}(m_\pi^2) \approx g_{\pi NN}$ which is the pion nucleon coupling constant. Inserting this into (4.65) yields

$$f_\pi g_{\pi NN} = m_N g_A \qquad (4.66)$$

which is the Goldberger-Treiman relation. The measured $g_A \approx 1.26$ and $g_{\pi NN}^2/4\pi \approx 14$ imply $f_\pi \approx 89$ MeV, close to the experimental value of 93 MeV. Note that the form factor for the pseudoscalar coupling $G_2(q^2)$ is also measurable, since

$$2m_N G_1(q^2) + q^2 G_2(q^2) = 2m_\pi^2 f_\pi \frac{g_{\pi NN}(q^2)}{-q^2 + m_\pi^2}$$

$$= 2f_\pi g_{\pi NN}(q^2) + q^2 \frac{2f_\pi g_{\pi NN}(q^2)}{-q^2 + m_\pi^2} \qquad (4.67)$$

so that

$$h_A \equiv G_2(0) \approx \frac{2m_N g_A}{m_\pi^2}. \qquad (4.68)$$

(iii) Adler consistency condition

Let us consider the matrix element in the chiral limit ($m_\pi \to 0$)

$$T_\mu^a = \langle f|A_\mu^a|i\rangle. \qquad (4.69)$$

Let us now split the above matrix element into a zero-mass pion pole contribution and a part regular at $q^2 = 0$

$$T_\mu^a = i\frac{f_\pi q_\mu}{q^2} T_{pole}^a + \bar{T}_\mu^a \qquad (4.70)$$

where T_{pole}, by LSZ reduction, is the decay amplitude $\langle f|\pi|i\rangle$. The conservation of the current (in the chiral limit), $q_\mu T^{\mu a} = 0$, implies

$$T^a_{pole} = \frac{iq_\mu \bar{T}^{\mu a}}{f_\pi}. \tag{4.71}$$

If in the *soft* limit ($q_\mu \to 0$) the process does not develop poles from the graphs in which the axial current couples to the external lines of the $i \to f$ subprocess, then eq.(4.71) says that the decay amplitude vanishes with q_μ. This is known as the Adler consistency condition. This condition can also be obtained using PCAC rather than CAC as is done above.

As a specific example, we consider the original construction by Adler, *i.e.*, we take $|i\rangle = |N(p_1)\rangle$ and $|f\rangle = |\pi(q)N(p_2)\rangle$. Then

$$T^{ab}_\mu = \int d^4x \exp(ik \cdot x)\langle \pi^a(q)N(p_2)|A^b_\mu|N(p_1)\rangle. \tag{4.72}$$

The scattering amplitude can be decomposed as

$$T^{ba}(\nu, \nu_B) = \bar{u}(p_2)\left(\mathbf{A}^{ba}(\nu, \nu_B) + \frac{\gamma \cdot (k+q)}{2}\mathbf{B}^{ba}(\nu, \nu_B)\right)u(p_1) \tag{4.73}$$

where $\mathbf{A}$ is the scalar part of T and $\mathbf{B}$ the vector part, both of which are functions of the variables

$$\nu = \frac{(p_1 + p_2) \cdot q}{2m_N}, \qquad \nu_B = -\frac{q \cdot k}{2m_N}. \tag{4.74}$$

They can be related to the Mandelstam variables s and t. Near threshold, the contributions of the isobar Δ, the Roper and other baryon resonances can be ignored. Thus the amplitude involves a nucleon recoiling in the intermediate state as shown in Fig. 4.7. Specifically, near threshold, $\mathbf{B}$ is dominated by the nucleon pole term

$$\mathbf{B}^{ba} \sim \frac{g^2_{\pi NN}}{2m_N}\left(\frac{\tau^b\tau^a}{\nu_B - \nu} - \frac{\tau^a\tau^b}{\nu_B + \nu}\right). \tag{4.75}$$

Using PCAC, we can relate (4.73) to an off-shell amplitude through

$$T^{ba} = \lim_{q^2 \to m^2_\pi}\left(\frac{-q^2 + m^2_\pi}{f_\pi m^2_\pi}\right)iq^\lambda\langle p_2|A^b_\lambda|p_1, ka\rangle. \tag{4.76}$$

The axial current carries momentum q and is off-shell. Now, consider the right-hand side of (4.76) without taking the limit. Because of the explicit

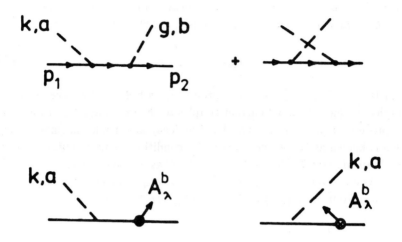

Fig. 4.7. Pion-nucleon amplitude at threshold (upper); pion-nucleon amplitude at the unphysical point (lower).

q-dependence, the axial matrix element is expected to develop a pole as $q \to 0$. The contribution to this off-shell amplitude is shown in Fig. 4.7:

$$\lim_{q \to 0} T^{ba} = \frac{g_A g_{\pi NN}}{f_\pi} \overline{u}(p_2) \left(\delta^{ab} + \frac{\gamma \cdot (k+q)}{4} \left[\frac{\tau^b \tau^a}{\nu_B - \nu} - \frac{\tau^a \tau^b}{\nu_B + \nu} \right] \right) u(p_1).$$
$$(4.77)$$

Comparing (4.73) to (4.77) suggests that at the unphysical point $\nu = 0, \nu_B = 0, q^2 = 0$, the amplitudes **A** and **B** viewed as analytic functions of ν, ν_B and q^2 satisfy

$$\mathbf{A}^{ab}(\nu = 0, \nu_B = 0, q^2 = 0) \sim \delta^{ab} \frac{g_A g_{\pi NN}}{f_\pi} = \delta^{ab} \frac{g_{\pi NN}^2}{m_N} \qquad (4.78)$$

and

$$(\gamma \cdot q \mathbf{B}^{ab})(\nu = 0, \nu_B = 0, q^2 = 0) \sim -\frac{g_A g_{\pi NN}}{f_\pi} \delta^{ab} = -\frac{g_{\pi NN}^2}{m_N} \delta^{ab}. \qquad (4.79)$$

The condition (4.79) is satisfied through the Born term (4.75) using the Goldberger-Treiman relation (4.68). The condition (4.78) is the Adler consistency relation that follows from $\lim_{q_\mu \to 0} T_{ba} = 0$.

The relations (4.78) and (4.79) put a constraint on the S-wave part of the scattering amplitude at threshold. Indeed, if we decompose the amplitude in the isospin space

$$T^{ab} = T^+ \mathbf{1}^{ab} + \frac{1}{2}[\tau^b, \tau^a]T^- \qquad (4.80)$$

then at threshold, *i.e.* $\nu = m_\pi$, $\nu_B = -m_\pi^2/2m_N$ and $q^2 = m_\pi^2$,

$$T^+ = \mathbf{A}^+ + m_\pi \mathbf{B}^+ \sim \frac{g_{\pi NN}^2}{m_N} - \frac{g_{\pi NN}^2}{m_N} \sim \mathcal{O}\left(\frac{m_\pi}{m_N}\right). \qquad (4.81)$$

The even part of the pion-nucleon scattering amplitude vanishes at threshold in the chiral limit. At threshold, the even part of the scattering amplitude is related to the S-wave scattering lengths ($T_{th} = 8\pi(M_N + m_\pi)a$). Thus

$$T^+ \sim \frac{1}{3}(2a_{3/2} + a_{1/2}) = \mathcal{O}\left(\frac{m_\pi}{m_N}\right) \qquad (4.82)$$

where a_I is the isospin-I S-wave scattering length. Empirically, $a_{1/2} \approx 0.17 m_\pi^{-1}$ and $a_{3/2} \approx -0.0881 m_\pi^{-1}$, in agreement with (4.82).

4.7.3. *Two soft pions (Tomozawa-Weinberg relation)*

Let us consider the pion-hadron scattering

$$\pi^a(q) + \alpha(p_1) \to \pi^b(k) + \beta(p_2)$$

where $\pi^a(q)$ is the pion of isospin index a with four-momentum q and $\alpha(p)$ is a hadron labeled by α with four-momentum p. We are interested in the scattering amplitude

$$\begin{aligned} S_{\beta,\alpha}^{ba} = \lim_{q^2, k^2 \to m_\pi^2} (q^2 - m_\pi^2)(k^2 - m_\pi^2) \\ \int d^4x\, d^4y\, e^{ik\cdot x} e^{-iq\cdot y} \langle \beta|T(\pi^b(x)\pi^a(y))|\alpha\rangle \end{aligned} \qquad (4.83)$$

in the soft-pion limit $q_\mu = k_\mu \to 0$. We shall follow Weinberg[4] and decompose the axial current into a "normal" part A_N (with zero divergence) and a "pion" part

$$A^{a,\mu} = A_N^{a,\mu} + f_\pi \partial^\mu \pi^a. \qquad (4.84)$$

Using the following relations

$$\partial_\mu^x T\left(A^{b,\mu}(x)A^{a,\nu}(y)\right) = m_\pi^2 f_\pi T\left(\pi^{b,\mu}(x)A^{a,\nu}(y)\right)$$
$$+\delta(x^0 - y^0)[A^{b,0}(x), A^{a,\nu}(y)] \qquad (4.85)$$

and

$$\partial_\mu^x T\left(A^{b,\mu}(x)\pi^a(y)\right) = m_\pi^2 f_\pi T\left(\pi^{b,\mu}(x)\pi^a(y)\right)$$
$$+\delta(x^0 - y^0)[A^{b,0}(x), \pi^a(y)], \qquad (4.86)$$

replacing $(-q^2 + m_\pi^2)$ by $\Box + m_\pi^2$ in the integrand of (4.83) and setting $q^2 = k^2 = m_\pi^2$, we obtain

$$S_{\beta,\alpha}^{ba} = -\int d^4x\, d^4y\, e^{ik\cdot x} e^{-iq\cdot y} \langle\beta| \left(\frac{k_\mu q_\nu}{f_\pi^2} T(A^{b,\mu}(x)A^{a,\nu}(y))_N\right.$$
$$+\frac{m_\pi^2}{f_\pi}\delta(x^0 - y^0)[A^{b,0}(x), \pi^a(y)]$$
$$\left.-\frac{iq_\nu}{f_\pi^2}\delta(x^0 - y^0)[A^{b,0}(x), A^{a,\nu}(y)]\right)|\alpha\rangle. \qquad (4.87)$$

The term $(...)_N$ has no pion pole. By translational invariance, we may set $x \to 0$ and $y \to y - x = z$ and have

$$S_{\beta,\alpha}^{ba} = i(2\pi)^4\delta^4(p_1 + q - p_2 - k)\, T_{\beta,\alpha}^{ba} \qquad (4.88)$$

with

$$T_{\beta,\alpha}^{ba} = -\int d^4z\, e^{-iq\cdot z} \langle\beta| \left(\frac{k_\mu q_\nu}{f_\pi^2} T(A^{b,\mu}(0)A^{a,\nu}(z))_N\right.$$
$$+\frac{m_\pi^2}{f_\pi}\delta(z^0)[A^{b,0}(0), \pi^a(z)]$$
$$\left.-\frac{q_\nu}{f_\pi^2}\delta(z^0)[A^{b,0}(0), A^{a,\nu}(z)]\right)|\alpha\rangle. \qquad (4.89)$$

No approximation has been made so far and hence (4.89) is exact. Since the spacing between the negative-parity states for even-parity baryonic targets is large compared to the pion mass m_π, we can ignore the contribution of the pion-subtracted (first) term in (4.89). The second term is related to the π-baryon (B) sigma term and contributes only to the isospin-symmetric amplitude. Thus for the isospin-antisymmetric amplitude only the last term contributes in the soft-pion limit. Using current algebra, we have

$$if^{bac}\langle\beta|V_\mu^{c,0}(0)|\alpha\rangle = if^{bac}2p_\mu[T_B^c]_{\beta\alpha} = -2[T_\pi^c]^{ba}[T_B^c]_{\beta\alpha}p_\mu \tag{4.90}$$

where we use $[T_\pi^c]_{ba} = -if^{bac}$ in the vector representation for the pion. At threshold for πN scattering, $p_1 = p_2 = (m_N, \vec{0})$ and $q = k = (m_\pi, \vec{0})$, the scattering lengths a^{ba} are related to T_{th}^{ba} by

$$a^{ba} = \frac{1}{8\pi(m_\pi + m_N)}T_{th}^{ba} = -\frac{1}{4\pi(m_\pi + m_N)}\frac{m_N m_\pi}{f_\pi^2}[T_\pi^c]_{ba}[T_N^c]_{\beta\alpha}. \tag{4.91}$$

Since

$$\vec{T}_\pi \cdot \vec{T}_N = \frac{1}{2}\left(T(T+1) - T_N(T_N+1) - 1(1+1)\right),$$

it follows that

$$a_T \approx -L(1 + \frac{m_\pi}{M_N})^{-1}\left(T(T+1) - T_N(T_N+1) - 2\right) \tag{4.92}$$

where T is the total isospin of the pion-target system, and $L = m_\pi/(8\pi f_\pi^2)$ is a universal length independent of the target. This relation for the scattering length was first derived, independently, by Weinberg and Tomozawa.

For a nucleon target, $T_N = 1/2$ so that $T = 1/2$ or $T = 3/2$. Since $f_\pi \sim 0.68 m_\pi$, we have from (4.92) :

$$m_\pi a_{3/2} \approx -0.075, \qquad m_\pi a_{1/2} \approx +0.15 \tag{4.93}$$

which are in good agreement with the empirical values

$$m_\pi a_{3/2} = (-0.088 \pm 0.004), \qquad m_\pi a_{1/2} = (+0.171 \pm 0.005). \tag{4.94}$$

4.7.4. *One photon and one pion (photoproduction)*

We now consider the generic case of interactions between one soft pion and one soft photon, *i.e.*, the low-energy theorem for electroproduction of pions. The scheme is analogous to the case of two soft pions considered above. Let us specify the target to be nucleons. We start, as usual, with

$$T_\mu^a = i\epsilon^\nu \int d^4x \exp(iq \cdot x)\langle p_2|T(A_\mu^a(x)j_\nu(0))|p_1\rangle. \tag{4.95}$$

Integrating by parts, we obtain

$$q^\mu T^a_\mu = -\int d^4x \exp(iq\cdot x)\langle p_2|[A^a_0(x), j_\nu(0)]|p_1\rangle\delta(x_0)\epsilon^\nu. \quad (4.96)$$

Let us now perform the soft momentum limit in an asymmetric way, *i.e.*, first we put $\vec{q}=0$, then we let q_0 go to zero. In this limit, we get

$$\lim_{q_0\to 0,\vec{q}=0} q^\mu T^a_\mu = -\langle p_2|[Q^a_A, j_\nu]|p_1\rangle\epsilon^\nu. \quad (4.97)$$

The physical amplitude for photoproduction is

$$if_\pi T^a[\gamma P \to \pi P](q^2, \nu = 0) = \langle p_2|[Q^a_A, j_\mu]|p_1\rangle\epsilon^\mu + \lim_{q_\mu\to 0}(q^\mu \bar{T}^a_\mu) \quad (4.98)$$

where $\nu = q\cdot(p_1 + p_2)/2m_N$. Using the current-algebra commutation relations, we get

$$if_\pi T^a[\gamma P \to \pi P](q^2, \nu = 0) = if^{a3b}\langle p_2|A^b_\nu|p_1\rangle\epsilon^\nu + \dots \quad (4.99)$$

where the ellipsis denotes terms regular in the soft-pion limit. The axial-current matrix element can be written as before,

$$\langle p_2|A^a_\mu|p_1\rangle = \bar{u}(p_2)(\gamma_\mu\gamma_5 G_1(t) + (p_2 - p_1)_\mu\gamma_5 G_2(t))\frac{\tau^a}{2}u(p_1) \quad (4.100)$$

so the photoproduction amplitude is

$$if_\pi T^a[\gamma P \to \pi P](q^2, \nu = 0) = if^{a3b}\bar{u}(p_2)\bigg(\gamma_\mu\gamma_5 G_1(t)$$
$$+ (p_2 - p_1)_\mu\gamma_5 G_2(t)\bigg)\frac{\tau^b}{2}u(p_1)\epsilon^\mu$$
$$+\dots. \quad (4.101)$$

Alternatively, we can also make the general tensor decomposition

$$\epsilon^\mu T^a_\mu = \bar{u}(p_2)\left(\gamma_5(\gamma\cdot\epsilon)T^a_1 + \gamma_5(P\cdot\epsilon)T^a_2\right)u(p_1), \quad (4.102)$$

where T_1 and T_2 are the amplitudes related by gauge invariance ($k^\mu T^a_\mu = 0$). Indeed, $T^a_1 = \nu T^a_2$. A further decomposition of the invariant amplitudes in isospin space gives

$$T^a_i = \frac{1}{2}[\tau^a, \tau^3]T^-_i + \delta_{a3}T^+_i + \tau^a T^0_i \quad (i = 1, 2). \quad (4.103)$$

After simplifying eq.(4.102) with the convenient choice of the polarization $(P \cdot \epsilon) = 0$, the direct comparison of eqs.(4.101), (4.102) and (4.103) at threshold $(p_1 = p_2)$ leads – modulo corrections of order m_π – to

$$T_1^- \sim -\frac{G_1(0)}{2f_\pi} = -\frac{g_A}{2f_\pi} = -\frac{g_{\pi NN}}{2M_N},$$

$$T_1^{+,0} \sim 0. \qquad (4.104)$$

These threshold relations are known as the Kroll-Ruderman relations in the chiral limit. As a result, pion photo-production from nucleons obeys the constraint

$$R \equiv \frac{d\sigma[\gamma n \to p\pi^-]}{d\sigma[\gamma p \to n\pi^+]} = \left(\frac{T^- - T^0}{T^- + T^0}\right)^2 = 1. \qquad (4.105)$$

This relation may be improved upon by taking into account the pion mass in the $T_1^{+,0}$ amplitudes (4.104). The analysis follows the Weinberg treatment of the π-nucleon scattering length. The result is

$$T_1^{+,0} = \frac{m_\pi g_{\pi NN}}{4m_N^2}. \qquad (4.106)$$

Thus

$$R = \left(\frac{2m_N + m_\pi}{2m_N - m_\pi}\right)^2 = 1.34, \qquad (4.107)$$

which is to be compared with the experimental value $R = 1.265 \pm 0.075$.

4.7.5. *Low-energy theorems for* $\rho \to \pi\pi$

A soft-pion theorem for the decay of $\rho(p) \to \pi(q)\pi(k)$ follows from the amplitude

$$T_{\mu\nu}^{abc} = i \int d^4x \exp(ik \cdot x)\langle 0|T(A_\nu^a(x)A_\mu^b(0))|\rho^c(p)\rangle. \qquad (4.108)$$

Arguments identical to those of the previous section yield

$$k^\mu T_{\mu\nu}^{abc} = -if^{abd}\langle 0|V_\nu^d|\rho^c\rangle \qquad (4.109)$$

where the commutation relation for two axial currents has been used. Defining f_ρ through the relation

$$\langle 0|V_\mu^a|\rho^b(p)\rangle = f_\rho \epsilon_\mu \delta^{ab}, \qquad (4.110)$$

we can rewrite eq.(4.109) as

$$k^\mu T_{\mu\nu}^{abc} = -if^{abc} f_\rho \epsilon_\nu,$$
$$q^\nu k^\mu T_{\mu\nu}^{abc} = -if^{abc} f_\rho (q \cdot \epsilon). \qquad (4.111)$$

Let us now perform the soft limit $k_\mu, q_\mu \to 0$. The reduction theorem and PCAC relate the decay amplitude to the residue of the Green function, eq.(4.108):

$$\frac{q_\nu k_\mu}{q^2 k^2} f_\pi^2 T^{abc}[\rho \to \pi\pi] = T_{\mu\nu}^{abc}. \qquad (4.112)$$

From the general parity arguments,

$$T^{abc}[\rho \to \pi\pi] = if^{abc}(k - q)_\mu \epsilon^\mu g_{\rho\pi\pi}. \qquad (4.113)$$

Comparison of eq.(4.111) and eq.(4.112), supplemented by the current conservation condition (equivalent here to $P \cdot \epsilon = 0$) predicts, in the soft-pion limit,

$$f_\rho = 2f_\pi^2 g_{\rho\pi\pi}. \qquad (4.114)$$

This soft-pion theorem is usually expressed in a more general form using universality arguments. For that, consider the soft limit of the general hadronic matrix element

$$\langle H_2(p_2)|V_\mu^3|H_1(p_1)\rangle = \tau^3 (p_1 + p_2)_\mu G_E((p_1 - p_2)^2) + \cdots \quad (4.115)$$

where $G_E(t) = F_1(t) + [t/4m^2]F_2(t)$ is the Sachs electric form factor. The normalization of the vector current imposes $G_E(0) = 1$ (*i.e.*, total isovector charge conservation). Assuming that one can write down an unsubtracted dispersion relation for $G_E(t)$,

$$G_E(t) = \frac{g_{\rho H_1 H_2} f_\rho}{m_\rho^2 - t} + \text{continuum} \qquad (4.116)$$

and also that the ρ pole dominates over the continuum as explicitly shown in eq.(4.116), then at $t = 0$ one finds

$$g_{\rho H_1 H_2} = \frac{m_\rho^2}{f_\rho}. \qquad (4.117)$$

This relation is *universal* in the sense that the right-hand side is *independent* of the hadrons H_1 and H_2. Combining eq.(4.117) and eq.(4.114) leads to the relation

$$m_\rho^2 = 2g_{\rho\pi\pi}^2 f_\pi^2 \tag{4.118}$$

known as the KSRF (Kawarabayashi-Suzuki-Riazuddin-Fayazzuddin) formula.

The derivation given here involves a large extrapolation between the soft point and the vector-meson mass, which is difficult to justify. It turns, however, that the KSRF relation emerges quite generally independently of current algebras in effective chiral Lagrangian field theories developed in later chapters. This relation will surface again when we apply the effective Lagrangian concept to nuclear many-body systems.

4.7.6. *Anomalies and current algebra* : $\pi^0 \to \gamma\gamma$

We consider the amplitude $T(q^2)$ for the decay of π^0 to two photons. According to the reduction formula, the amplitude is given as a residue of the relevant Green function. In our case

$$T[\pi^0 \to \gamma\gamma] = \lim_{q^2 \equiv (k_1+k_2)^2 \to m_\pi^2} e^2 \epsilon_1^{*\mu} \epsilon_2^{*\nu} T_{\mu\nu}(k_1, k_2) \tag{4.119}$$

where

$$T_{\mu\nu}(k_1, k_2) = \int d^4x d^4y \langle 0|Tj_\mu(x)j_\nu(y)\pi(0)|0\rangle \exp(ik_1 \cdot x + ik_2 \cdot y). \tag{4.120}$$

Here ϵ_i, k_i denote the polarizations and momenta of the photons.

Now, consider the auxiliary amplitude

$$\begin{aligned}
T_\mu \equiv \langle\gamma\gamma|A_\mu^3|0\rangle = &(k_1 + k_2)_\mu \varepsilon_{\delta\nu\alpha\beta} \epsilon_1^{*\delta} \epsilon_2^{*\nu} k_1^\alpha k_2^\beta A_1(q^2) \\
&+ [(k_1 \cdot k_2)\varepsilon_{\mu\nu\alpha\beta} \epsilon_1^{*\nu} \epsilon_2^{\alpha*}(k_1 - k_2)^\beta \\
&+ (k_1 \cdot \epsilon_2^*)\varepsilon_{\mu\nu\alpha\beta} \epsilon_1^{\nu*} k_1^\alpha k_2^\beta \\
&- (k_2 \cdot \epsilon_1^*)\varepsilon_{\mu\nu\alpha\beta} \epsilon_2^{\nu*} k_1^\alpha k_2^\beta] A_2(q^2).
\end{aligned} \tag{4.121}$$

The R.H.S. of eq.(4.121) is dictated by general Lorentz and discrete symmetries. Taking the divergence, we get

$$\partial^\mu T_\mu = \langle\gamma\gamma|\partial^\mu A_\mu^3|0\rangle = iq^2\varepsilon_{\mu\nu\alpha\beta} \epsilon_1^{*\mu} \epsilon_2^{\nu*} k_1^\alpha k_2^\beta [A_1(q^2) - A_2(q^2)]. \tag{4.122}$$

On the other hand, we recall the anomaly equation (see Chapter 1)

$$\partial_\mu A_\mu^3 = m_\pi^2 f_\pi \pi^3 - \left(\sum_a I_a^{(3)} Q_a^2\right) \frac{\alpha}{2\pi} \varepsilon_{\mu\nu\alpha\beta} F^{\mu\nu} F^{\alpha\beta}$$

$$\equiv m_\pi^2 f_\pi \pi^3 - \lambda F \cdot \tilde{F} \tag{4.123}$$

where the sum is over the charge and third components of the isospin of the fundamental fermions. In the last line, we have introduced a compact notation. Inserting the anomaly equation (4.123) into the L.H.S. of eq.(4.122) leads to

$$\langle \gamma\gamma | \partial_\mu A_\mu^3 | 0 \rangle = m_\pi^2 f_\pi \langle \gamma\gamma | \pi^0 | 0 \rangle - \lambda \langle \gamma\gamma | F \cdot \tilde{F} | 0 \rangle$$

$$= m_\pi^2 f_\pi \varepsilon_{\mu\nu\alpha\beta}\, \epsilon_1^{*\mu} \epsilon_2^{\nu*} k_1^\alpha k_2^\beta \frac{H(q^2)}{m_\pi^2 - q^2} - 4\lambda \varepsilon_{\mu\nu\alpha\beta}\, \epsilon_1^{*\mu} \epsilon_2^{\nu*} k_1^\alpha k_2^\beta \tag{4.124}$$

where we have redefined $T(q^2)$ by pulling out the trivial kinematical factors as

$$T(q^2) = \varepsilon_{\mu\nu\alpha\beta}\, \epsilon_1^{*\mu} \epsilon_2^{\nu*} k_1^\alpha k_2^\beta H(q^2). \tag{4.125}$$

We are now in the position to apply the smoothness hypothesis for $q_\mu \to 0$. From $q_\mu = k_1^\mu + k_2^\mu$, we see that $q^2 \to 0$ requires that k_1 and k_2 be proportional to each other. Expanding eq.(4.125) in this limit, we get

$$\langle \gamma\gamma | \partial^\mu A_\mu^3 | 0 \rangle \to \varepsilon_{\mu\nu\alpha\beta}\, \epsilon_1^{*\mu} \epsilon_2^{\nu*} k_1^\alpha k_2^\beta (f_\pi H(q^2 = 0) - 4\lambda) \sim O(q^2). \tag{4.126}$$

On the other hand, eq.(4.122) in the same limit leads to

$$\langle \gamma\gamma | \partial^\mu A_\mu^3 | 0 \rangle = iq^2 \varepsilon_{\mu\nu\alpha\beta}\, \epsilon_1^{*\mu} \epsilon_2^{\nu*} k_1^\alpha k_2^\beta [A_1(0) - A_2(0)] \sim O(q^4). \tag{4.127}$$

We see that both limits, eq.(4.126) and eq.(4.127), are compatible only if the q^2 term in eq.(4.126) vanishes, *i.e.*,

$$H(q^2 = 0) = \frac{4}{f_\pi} \lambda = \frac{2\alpha}{\pi} \frac{1}{f_\pi} \sum_a I_a^{(3)} Q_a^2. \tag{4.128}$$

This consistency condition leads to the $\pi^0 \to \gamma\gamma$ decay rate

$$\Gamma = \left(\sum_a I_a^{(3)} Q_a^2\right)^2 \frac{\alpha^2 m_\pi^2}{16\pi^3 f_\pi^2}. \tag{4.129}$$

Note that had we ignored the anomaly, the decay of the pion would have been strongly suppressed. This observation is known as the Sutherland-Veltman paradox. It is the anomaly which allows the pion to decay into two photons.

The formula (4.129) is interesting from another point of view. The sum in eq.(4.129) runs over the elementary fermions of the theory. In the naive quark model, the prediction based on eq.(4.129) leads to a value 9 times smaller than the experimentally observed 7.6 eV. This factor 9 is none other than the square of the number of colors of the quarks, ignored in the naive quark model. In QCD, the fermionic loop saturating the anomaly counts colors correctly, leading to a full agreement with the experimental value. This is usually regarded as empirical evidence for $N_c = 3$.

4.8. Hadronic Sum Rules

In this section, we study the sum rules resulting from the current-algebra commutation relations. As with the toy model in Subsection 4.4.2, where the powerful sum rule emerged as a consequence of the simple canonical commutation relations for the momenta and positions, the commutation relations of current algebra constitute sufficient conditions for the hierarchy of sum rules. We start by discussing the generic current-algebra sum rule followed by a review of a few historically significant results.

4.8.1. *Generic sum rule in current algebra*

Consider the time-time current-algebra commutation relation

$$[J_0^a(\vec{x}), J_0^b(\vec{y})] = i f^{abc} J_0^c(\vec{x}) \delta^{(3)}(\vec{x} - \vec{y}) \tag{4.130}$$

where J_0 is the time component of some generic (vector or axial) current. In momentum space, the above relation reads

$$[J_0^a(\vec{q}_2), J_0^b(\vec{q}_1)] = i f^{abc} J_0^c(\vec{q}_1 - \vec{q}_2). \tag{4.131}$$

Taking the expectation value of eq.(4.131) between arbitrary states and saturating the sum with the complete set of states, we get

$$i f^{abc} \langle p | J_0^c | p \rangle = \sum_n (2\pi)^3 \delta^{(3)}(\vec{p} + \vec{q}_2 - \vec{p}_n) \langle p | J_0^a | n \rangle \langle n | J_0^b | p \rangle$$

$$- \sum_n (2\pi)^3 \delta^{(3)}(\vec{p} - \vec{q}_1 - \vec{p}_n) \langle p | J_0^b | n \rangle \langle n | J_0^a | p \rangle. \tag{4.132}$$

Introducing the Fourier representation of the δ function

$$\delta(x_0) = \frac{1}{2\pi} \int dq_0 \exp(iq_0 x_0) \qquad (4.133)$$

allows us to rewrite eq.(4.132) in the form

$$if^{abc}\langle p|J_0^c|p\rangle = \int dq_0 t_{00}^{ab} \qquad (4.134)$$

with

$$t_{\mu\nu}^{ab} = \frac{1}{2\pi} \int d^4x \exp(iq \cdot x) \langle p|[J_\mu^a(x/2), J_\nu^b(-x/2)]|p\rangle. \qquad (4.135)$$

Equation (4.134) is the starting point for our generic formula. The amplitude $t_{\mu\nu}^{ab}$ is related to the absorptive part of the time-ordered product of the currents

$$t_{\mu\nu} = \frac{1}{\pi}\text{Abs } T_{\mu\nu} \equiv \frac{1}{2\pi i}\left(T_{\mu\nu}(q_0 + i\epsilon) - T_{\mu\nu}(q_0 - i\epsilon)\right) \qquad (4.136)$$

where

$$T_{\mu\nu} = i \int d^4x \exp(iq \cdot x)\langle p|T[J_\mu(x/2)J_\nu(-x/2)]|p\rangle. \qquad (4.137)$$

Lorentz covariance dictates the tensor structure of $T_{\mu\nu}^{ab}$ (the subtleties of the covariance of the T product are discussed in the Appendix):

$$T_{\mu\nu}^{ab} = B^{ab}g_{\mu\nu} + A^{ab}p_\mu p_\nu + C^{ab}(p_\mu q_\nu + q_\mu p_\nu) + D^{ab}q_\mu q_\nu \qquad (4.138)$$

where the amplitudes A, B, C, D are functions of Lorentz invariants $2p \cdot q$ and q^2. Let us choose now $\vec{p} \cdot \vec{q} = 0$. Introducing $\nu = p \cdot q/m$ (where m is the target, *e.g.*, nucleon mass), we can rewrite our basic relation (4.134) as

$$\frac{m}{\pi p_0^2} \int d\nu (\text{Im}B^{ab} + \text{Im}A^{ab}p_0^2 + 2C^{ab}\nu m + D^{ab}\frac{m^2\nu^2}{p_0^2})$$
$$= \frac{1}{p_0}if^{abc}\langle p|J_0^c|p\rangle. \qquad (4.139)$$

We still have the freedom of choosing the Lorentz frame. Let us choose the "infinite momentum frame" ($p_0 \to \infty$). Assuming that we could commute

the integration and the limiting procedure, we arrive at the generic sum
rule relation

$$\frac{1}{\pi} \int d\nu \, \text{Im} A^{ab} = \lim_{p_0 \to \infty} \frac{1}{m p_0} \langle p | J_0^c | p \rangle. \tag{4.140}$$

The R.H.S. is always finite due to the normalization of the states. For the
vector current and spinor states for the target, we get

$$\lim_{p_o \to \infty} \frac{1}{p_0} \langle p | V_\mu^a | p \rangle = \lim_{p_o \to \infty} \frac{1}{p_0} \bar{u}(p) \gamma_\mu F_1^a(0) u(p),$$

$$\lim_{p_o \to \infty} \frac{1}{p_0} (2 p_0) \chi^\dagger \chi F_1^a(0) = 2 \chi^\dagger \chi F_1^a(0) \tag{4.141}$$

where χ is a Pauli spinor.

The above procedure is readily generalized to arbitrary matrix ele-
ments of the operators (*i.e.*, $\langle p_1 | J_\mu | p_2 \rangle$ for $|p_1\rangle \neq |p_2\rangle$),

$$\frac{1}{2} \int d\nu \, a^{ab}(\nu, t, -\vec{q_1}^{\,2}, \vec{q_2}^{\,2}) = i f^{abc} F_1^c(t) \tag{4.142}$$

where a is the imaginary part of the invariant amplitude A corresponding
to the tensor $p_\mu p_\nu$ in the tensor decomposition of $T_{\mu\nu}$, eq.(4.138).

4.8.2. *Examples of current-algebra sum rules*

(i) Cabibbo-Radicati sum rule

Let us choose $J_0^a = V_0^{1+i2}$ and $J_0^b = V_0^{1-i2}$. Consider a proton target
and denote its mass by m_N. The commutation relation is

$$[V_0^{1+i2}(\vec{x}), V_0^{1-i2}(\vec{y})] = 2 V_0^3(\vec{x}) \delta^{(3)}(\vec{x} - \vec{y}). \tag{4.143}$$

Now the generic sum rule takes the simple form

$$\frac{1}{2} \int_{-\infty}^{\infty} a^V(\nu) d\nu = \frac{1}{2} \int_0^\infty d\nu \, a^V(\nu, t) + \frac{1}{2} \int_0^\infty d\nu \, a^V(-\nu, t) = 1. \tag{4.144}$$

Due to the analytic properties of the amplitudes, the amplitude $a(-\nu, t)$
corresponds to the crossed process described by $a(\nu, t)$. The currents $V^{1\pm i2}$
correspond to unphysical, charged photons, constituting, together with the
physical V^3, the isotriplet of the conserved vector current (by CVC, they
are the physical charged-vector current components in weak interactions).

Of course by isospin rotation we can later relate them to physical quantities. In eq.(4.144), the process $\gamma^+p \rightarrow \gamma^+p$ starts at threshold while the crossed process $\gamma^-p \rightarrow \gamma^-p$ develops a pole before threshold, due to the possibility of having the intermediate state $\gamma^-p \rightarrow n \rightarrow \gamma^-p$. The explicit contribution from the pole is easily extracted from

$$t_{\mu\nu}^{ab} = (2\pi)^3 \delta^4(p + q - q_n)\langle p|J_\mu^a|n\rangle\langle n|J_\nu^b|p\rangle$$
$$- (a \leftrightarrow b, \mu \leftrightarrow \nu). \qquad (4.145)$$

Using the isospin symmetries of the currents, we can express the matrix elements $\langle p|V^{1+i2}|n\rangle$ in terms of the Dirac and Pauli vector form factors $F_{1,2}^V$. Isolating the pole contribution and decomposing the amplitudes in isospin space as $T^{ab} = T^+\delta^{ab} + T^-\frac{1}{2}[\tau^a, \tau^b]$, the following sum rule is obtained

$$(F_1^V(t))^2 - (t/4m_N^2)(F_2^V(t))^2 + \frac{1}{2}\int_{\text{continuum}} d\nu \, a^V(\nu, t) = 1. \qquad (4.146)$$

Current conservation has not yet been exploited. Since $q^\mu t_{\mu\nu} = 0$, the amplitudes $a \equiv \text{Im } A$ and b ($b = \text{Im } B$) are related to each other through

$$\frac{\partial a(\nu, t)}{\partial t}\Big|_{q=0} = -b(\nu, 0)/m_N^2\nu^2. \qquad (4.147)$$

Differentiating eq.(4.146) with respect to t and using eq.(4.147), we get

$$2F_1^{V'}(0) - (F_2^V(0)/2m_N)^2 - \frac{1}{2}\int_{\text{continuum}} \frac{d\nu}{\nu^2 m_N^2}b^V(\nu, 0) = 0 \qquad (4.148)$$

where the prime denotes the derivative with respect to t. Finally, relating the imaginary part of the amplitude to the total cross-section for photo-production via the optical theorem and then relating via Clebsch-Gordan coefficients the cross sections for the "charged photons" to the cross-sections for the isospin $I = 1/2, 3/2$ states,

$$\sigma(\gamma^+p) - \sigma(\gamma^-p) = 2\sigma_{3/2}^V - 4\sigma_{1/2}^V, \qquad (4.149)$$

gives the Cabibbo-Radicati sum rule:

$$\frac{1}{3}R_E^2 = \frac{\mu_V^2}{4m_N^2} + \frac{1}{2\pi^2\alpha}\int_{\nu_0}^\infty \frac{d\nu}{\nu}(2\sigma_{1/2} - \sigma_{3/2}) \qquad (4.150)$$

where R_E is the proton charge radius and $F_2^V(0) = \mu_V$. Its agreement with experiments confirms the locality assumption that enters in the commutation relations of the currents.

(ii) Adler-Weisberger sum rule

We now choose $J_0^a = A_0^{1+i2}$ and $J_0^b = A_0^{1-i2}$. The commutation relation is

$$[A_0^{1+i2}(\vec{x}), A_0^{1-i2}(\vec{y})] = 2V_0^3(\vec{x})\delta^{(3)}(\vec{x} - \vec{y}). \tag{4.151}$$

Our generic sum rule takes the simple form

$$\frac{1}{2}\int_{-\infty}^{\infty} a^A d\nu = \frac{1}{2}\int_0^{\infty} d\nu\, a^A(\nu, t) + \frac{1}{2}\int_0^{\infty} d\nu\, a^A(-\nu, t) = 1. \tag{4.152}$$

We can immediately write down the analog of eq.(4.146), taking into account pions by PCAC in place of the charged photons. Consider the matrix elements of the *axial* current contributing to the pole in the process $\pi^- p \to n \to \pi^- p$. Since the axial nucleonic form factors are defined by

$$\langle N(p_2)|A_\mu^a(0)|N(p_1)\rangle = \bar{u}(p_2)\left[G_1(q^2)\gamma_\mu + \frac{G_2(q^2)}{2M_N}q_\mu\right]\gamma_5 \frac{\tau^a}{2}u(p_1), \tag{4.153}$$

the sum rule reads

$$G_1(0)^2 + \frac{1}{\pi}f_\pi^2 \int_{\nu_0}^{\infty} \frac{d\nu}{\nu^2}(\mathrm{Im}\, T_{\pi^- p} - \mathrm{Im}\, T_{\pi^+ p}(\nu, 0)) = 1. \tag{4.154}$$

Since $G_1(0) \equiv g_A$, and the difference under the integral can be related via optical theorem to the total cross section, we arrive at the formula

$$\frac{1}{g_A^2} = 1 + \frac{2M_N^2}{\pi g_{\pi NN}^2}\int \frac{d\nu}{\nu^2}\sqrt{\nu^2 - m_\pi^2}\left(\sigma^{\pi^- p}(\nu) - \sigma^{\pi^+ p}(\nu)\right) \tag{4.155}$$

first obtained by Adler and Weisberger. It is interesting to note that a very similar sum rule was written down many years before the advent of current algebras by Goldberger, Miyazawa and Oehme (GMO), on the basis of more general assumptions. For comparison, we discuss in the Appendix the main arguments leading to the GMO sum rule.

(iii) Weinberg sum rules

Weinberg's first sum rule is based on the observation that the vector-vector and axial-axial commutators are equal.

$$\langle 0|[V_0^a(\vec{x}), V_k^b(\vec{0})]|0\rangle = \langle 0|[A_0^a(\vec{x}), A_k^b(\vec{0})]|0\rangle. \tag{4.156}$$

The commutators of space and time components develop Schwinger terms, but the vector-vector and axial-axial Schwinger terms are equal, so eq.(4.156) is not spoiled by their explicit presence. Let us consider the Fourier transform of eq.(4.156)

$$\int d^3x \exp(-i\vec{q}\cdot\vec{x})\langle 0|[V_0^a(\vec{x}), V_k^b(\vec{0})]|0\rangle \,,$$

$$\int d^3x \exp(-i\vec{q}\cdot\vec{x})\langle 0|[A_0^a(\vec{x}), A_k^b(\vec{0})]|0\rangle \,. \tag{4.157}$$

As in the preceding subsection, we introduce the spectral decomposition

$$\delta^{ab}t_{\mu\nu}^J = (2\pi)^3 \sum_n \delta(q-p_n)\langle 0|J_\mu^a|n\rangle\langle n|J_\nu^b|0\rangle$$

$$= \delta^{ab}(a^J(t)g_{\mu\nu} + b^J(t)q_\mu q_\nu) \tag{4.158}$$

where $J = V, A$ and the amplitudes in the Lorentz decomposition are functions of $t = -(\vec{q})^2$. Since our commutation relations eq.(4.156) involve time-space components only, we get the sum rule

$$\int_0^\infty b^V(t)dt = \int_0^\infty b^A(t)dt \tag{4.159}$$

known as the first Weinberg sum rule.

A sum rule for the invariant amplitudes $a(t)$ can be readily derived with the help of the identity

$$\langle 0|[\partial_0 V_k^a - \partial_k V_0^a, V_i^b]|0\rangle = \langle 0|[\partial_0 A_k^a - \partial_k A_0^a, A_i^b]|0\rangle. \tag{4.160}$$

For free fields, the above identity follows from the Cartan-Maurer equation

$$\partial_\mu V_\nu^a - \partial_\nu V_\mu^a + if^{abc}V_\mu^b V_\nu^c = 0. \tag{4.161}$$

Note that the relation eq.(4.160) holds true despite the explicit presence of the Schwinger terms. The immediate consequence of eq.(4.160) and the spectral decomposition eq.(4.158) is the second Weinberg sum rule

$$\int_0^\infty a^V(t)dt = \int_0^\infty a^A(t)dt. \tag{4.162}$$

Saturating the Weinberg sum rules is easy due to the fact that they involve vacuum expectation values. Neglecting the continua, we saturate both sides of the sum rules with single-particle states, *i.e.*, with a pion and an a_1 for the axial current and a ρ for the vector current. Then the first and second sum rules give, respectively,

$$\frac{f_\rho^2}{m_\rho^2} = \frac{f_{a_1}^2}{m_{a_1}^2} + f_\pi^2$$
$$f_\rho^2 = f_{a_1}^2 . \tag{4.163}$$

Then, using KSRF and universality, we find

$$m_{a_1} = \sqrt{2} m_\rho , \tag{4.164}$$

which is well confirmed by experimental data.

(iv) Photoproduction sum rules

The analogy in treating two soft pions (*i.e.*, the Tomozawa-Weinberg formula) and one soft pion and one soft photon (*i.e.*, the Kroll-Ruderman relation) can be extended to sum rules. In this case, we should expect sum rules for photoproduction analogous to the Adler-Weisberger relation for two axial currents. We omit the details, referring the interested reader to the original paper by Fubini, Furlan and Rosetti [5]. The results are

$$F_2^V(0) = \frac{8m_N^2}{g_{\pi NN}} \int_{\nu_0} \frac{d\nu}{\nu} Im A_1^+(\nu) ,$$
$$F_2^S(0) = \frac{8m_N^2}{g_{\pi NN}} \int_{\nu_0} \frac{d\nu}{\nu} Im A_1^0(\nu) \tag{4.165}$$

where A_1 is the invariant amplitude for electroproduction of the massless pions as defined in subsection 4.7.6.

4.9. Further Developments

Most of our discussion in this chapter has dealt with what is now the standard lore of current algebra. The current-algebra results, discussed so far, are exact at the soft point and approximate at threshold. A generalization of the current-algebra approach with non-trivial constraints imposed on the *on-shell* amplitudes has recently been formulated [7], making a direct contact with the non-perturbative sector of QCD. This matter will be taken up in Chapter 6.

The current-algebra results can be transcribed to Feynman graphs, paving the way to chiral perturbation theory, as we discuss in the next chapter. In this way, some systematics will be introduced in the threshold analysis in the form of power counting around the chiral limit[3,6]. One can also expand around the physical point (*i.e.*, the threshold point) in the spirit of a semi-classical expansion. Such an approach is in the process of being developed.[7]

Appendix

(A_I) *Schwinger terms, seagulls, T^*-product*

Let us consider the consistency of the conditions imposed on current commutators by Lorentz invariance. First of all, recall that the 10 generators of the Poincaré group are given by

$$M_{\mu\nu} = -\int d^3x (x_\mu \Theta_{\nu 0} - x_\nu \Theta_{\mu 0}) ,$$

$$P_\mu = \int d^3x \Theta_{\mu 0} \tag{A4.1}$$

where $\Theta_{\mu\nu}$ is the canonical energy-momentum tensor of the theory

$$\Theta_{\mu\nu} = -\mathcal{L}g_{\mu\nu} + \frac{\partial \mathcal{L}}{\partial(\partial^\mu \phi_a)}\partial_\nu \phi_a. \tag{A4.2}$$

In the form given here, the boost generators are

$$K_l \equiv M_{l0} = x_0 P_l - \int d^3x\, x_l \Theta_{00}. \tag{A4.3}$$

We start with the fundamental commutation relation of a conserved hadronic current $j_\mu^a(x)$ with a conserved charge Q^a:

$$[Q^a(0), j_0^b(0)] = if^{abc}j_0^c(0). \tag{A4.4}$$

Commuting both sides of eq.(A4.4) with the boost generator K_l, using the Jacobi identity and the vector transformation properties (that is, $[M_{\alpha\beta}, j_\mu^a(x)]$

$$= i(g_{\alpha\mu}j_\beta^a - g_{\beta\mu}j_\alpha^a) + i(x_\alpha\partial_\beta - x_\beta\partial_\alpha)j_\mu^a(x)), \text{ we have}$$

$$[Q^a(0), j_l^b(0)] = if^{abc}j_l^c(0) + \int d^3x\, x_l[j_0^a(0), \partial_\mu j^{\mu b}(x)]. \qquad (A4.5)$$

Provided that the current is conserved, the relation eq.(A4.4) must hold in any reference frame.

Let us now check the consistency of the *local* commutation relation

$$[j_0^a(\vec{x}), j_0^b(0)] = if^{abc}j_0^c(0)\delta^{(3)}(\vec{x}). \qquad (A4.6)$$

To get the strongest restrictions on eq.(A4.6), we impose the commutator relations with Θ_{00} itself rather than with the integrated densities K_l. An elementary calculation, exploiting the relations

$$[\Theta_{00}, j_0^a(\vec{x})] = -i(\partial_k(j_a^k(\vec{x})\delta(\vec{x})) + \partial_0 j_0^a(\vec{x})\delta(\vec{x})),$$
$$[\Theta_{00}, j_k^a(\vec{x})] = -i(\partial_0 j_k^a(\vec{x})\delta(\vec{x}) - j_0^a(\vec{x})\partial_k\delta_k(\vec{x})) \qquad (A4.7)$$

leads to

$$[j_0^a(\vec{y}), j_k^b(\vec{z})]\partial^k\delta(\vec{x} - \vec{z}) + [j_0^b(\vec{x}), j_k^a(\vec{z})]\partial^k\delta(\vec{z} - \vec{y})$$
$$= if^{abc}\delta(\vec{x} - \vec{y})j_k^c(\vec{z})\partial^k\delta(\vec{z} - \vec{y}) \qquad (A4.8)$$

with the most general solution

$$[j_0^a(\vec{x}), j_l^b(0)] = if^{abc}j_l^c(0)\delta(\vec{x}) + R_{lm}^{ab}\partial^m\delta(\vec{x}) \qquad (A4.9)$$

where the tensor R could be a c-number or q-number. Note that first, the consistency of the time-time (TT) commutation relation, eq.(A4.6), imposes the conditions on the space-time (ST) commutation relations, and second, the most general form of the ST commutation relations involves gradients of the Dirac delta distribution. One could apply a similar analysis to the case of the space-space (SS) commutation relations, with the most general solution

$$[j_i^a(\vec{x}), j_j^b(0)] = S_{ij}^{ab}\delta(\vec{x}) + i[K_k, R_{ij}^{ab}(\vec{x})]\partial^k\delta(\vec{x}). \qquad (A4.10)$$

Again, the structure of the SS commutation relations depends on the structure of the ST commutation relations. In particular, if $R_{lm}^{ab}(\vec{x})$ is a c-number, the SS commutation relations *do not* involve the derivatives of the delta distributions.

In a seminal paper, Schwinger discovered the necessity of derivatives of delta functions in the ST commutation relations in QED, *i.e.*, in the simpler case where $f^{abc} = 0$. Let us consider the expectation value of the commutator $\langle 0|[j_0(\vec{x}), j_i(\vec{y})]|0\rangle$. Naively, $\langle 0|[j_0(\vec{x}), j_i(\vec{y})]|0\rangle = 0$, since

$$[j_0(\vec{x}), j_i(\vec{y})] = [\psi^\dagger(x,t)\psi(x,t), \psi^\dagger(y,t)\gamma_0\gamma_i\psi(y,t)]$$
$$= \delta(\vec{x} - \vec{y})\psi^\dagger(x,t)[1, \gamma_0\gamma_i]\psi(x,t) = 0. \qquad (A4.11)$$

It is easy to see that this result is not consistent with positivity and Lorentz invariance. Let us take the derivative of the vacuum expectation value of the commutation relation, *i.e.*,

$$\langle 0|[j_0(x), \partial_i j_i(y)]|0\rangle = \langle 0|j_0(x), -\partial_0 j_0(y)]|0\rangle$$
$$= \langle 0|[j_0, -i[H, j_0]]|0\rangle = -i\langle 0|j_0(x)Hj_0(y) + j_0(y)Hj_0(x)|0\rangle \,(A4.12)$$

where we have used current conservation, Heisenberg equations of motion and the fact that H annihilates the vacuum. In the limit $x \to y$, the above expression is equal to $-2i\langle 0|j_0Hj_0|0\rangle$, which is always different from zero, since

$$\langle 0|j_0(x)Hj_0(x)|0\rangle = \sum_n \langle 0|j_0(x)|n\rangle\langle n|H|n\rangle\langle n|j_0(x)|0\rangle$$
$$= \sum_n E_n|\langle 0|j_0(x)|n\rangle|^2. \qquad (A4.13)$$

The naive vanishing of the space-time commutator was caused by the careless treatment of the composite operator.

In more complicated cases of the current-algebra commutator relations, we have to keep Schwinger terms as well. The coefficient R in (A4.10) is generally model-dependent (either a c-number or an operator). In the case of the algebra of fields, R is a c-number, and the Schwinger terms appear only in the ST commutation relations.

We should note that the presence of the Schwinger terms spoils the covariance of the time ordered product of two currents. Since the latter (Green function) is related to observable scattering amplitudes or decay rates by the LSZ formulae, we have to find a remedy for this lack of covariance.

The general Lorentz properties of an arbitrary tensor operator $O_{\mu\nu}$ imply

$$[K_\rho, O_{\mu\nu}(x,y)] = -ig_{\mu 0}O_{\rho\nu} + ig_{\mu\rho}O_{0\nu} - ig_{\nu 0}O_{\mu\rho} + ig_{\nu\rho}O_{\mu 0}$$

$$-i(x_0\partial_\rho^x - x_\rho\partial_0^x)O_{\mu\nu} - i(y_0\partial_\rho^y - y_\rho\partial_0^y)O_{\mu\nu}.\text{(A4.14)}$$

The explicit calculation of the commutator of the boost generator K_l and the T product $T_{\mu\nu} = T(j_\mu(x)j_\nu(0))$ yields

$$[K_l, T_{\mu\nu}(x)] = \cdots - ix_l\delta(x_0)[j_\mu(x), j_\nu(0)] \ . \qquad \text{(A4.15)}$$

The ellipsis represents all the terms of the R.H.S. of eq.(A4.14), consistent with the Lorentz covariance requirement. The *extra* term explicitly written down in eq.(A4.15), if not equal to zero, *spoils* the Lorentz covariance of the time-ordered product of the two currents. In particular, the Schwinger terms present in the ST commutation relations invalidate the tensor Lorentz properties of the time-ordered product of two currents. We could formally impose Lorentz properties on $T_{\mu\nu}$, by defining

$$T_{\mu\nu}^*(x) = T_{\mu\nu}(x) + \delta^{(4)}(x)\rho_{\mu\nu} \qquad \text{(A4.16)}$$

where we have added the compensating contact term (called a seagull term) which restores the covariance of the product. Imposing the covariance condition eq.(A4.14) on $T_{\mu\nu}^*$ determines (after some algebra) the form of the seagull term. One finds

$$\rho_{km}(x) = -R_{mk}. \qquad \text{(A4.17)}$$

The seagull term exactly cancels the Schwinger term contribution.

We could therefore use either the T^* *and* the commutation relations with the Schwinger terms, or the naive T products and commutation relations without the Schwinger terms. The last procedure (known as Feynman's hypothesis) gives a correct result due to the cancellation condition (A4.17).

The readers are warned that the analysis given above was based on the implicit assumption that the TT commutation relations, from which we have started our procedure, do not involve the Schwinger terms. The most famous counter-example to the above assumption is the vector and axial charge-density commutation relation

$$[j_0(\vec{x}), j_0^5(\vec{y})] = \frac{e}{4\pi^2}\epsilon^{0jik}F_{ik}\partial_j\delta(\vec{x} - \vec{y}). \qquad \text{(A4.18)}$$

One can find the seagull terms that covariantize the relevant T-products, but their form explicitly breaks gauge invariance of the theory. Preserving gauge invariance leads to the violation of Feynman's hypothesis. The failure of total cancellations between the seagull and the Schwinger term (A4.18) leads to the celebrated case of an axial anomaly in the divergence of the axial current. For a thorough analysis of axial anomalies from this point of view, we refer to the papers by Adler and Jackiw[2].

(A_{II}) Goldberger-Miyazawa-Oehme (GMO) sum rule

The parametrization (4.80) of the pion-nucleon scattering amplitude viewed as an analytic function of the pion energy ν suggests that (for $\nu > 0$)

$$M^i(\nu) = A^i(\nu) + iB^i(\nu) \tag{A4.19}$$

for both $i = even$ and *odd*. Here A is the dispersive part and B is the absorptive part of M respectively. For notational convenience, we have left out ν_B from the argument in (A4.19). As analytic functions, A and B in (A4.19) obey dispersion relations. For the odd part, we have the subtracted dispersion relation[8]

$$A^{odd}(\nu) - \frac{\nu}{m_\pi} A^{odd}(m_\pi) = \frac{2\nu(\nu^2 - m_\pi^2)}{\pi} \int_0^\infty d\nu' \frac{B^{odd}(\nu')}{(\nu'^2 - m_\pi^2)(\nu'^2 - \nu^2)} . \tag{A4.20}$$

Taking $\nu \to \infty$ on both sides of (A4.20) and assuming that $A^{odd}(\nu)/\nu \to 0$ as $\nu \to \infty$,

$$\frac{A^{odd}(m_\pi)}{m_\pi} = \frac{2}{\pi} \int_0^\infty d\nu' \frac{B^{odd}(\nu')}{(\nu'^2 - m_\pi^2)} . \tag{A4.21}$$

At threshold, the L.H.S. of (A4.21) can be written in terms of the scattering lengths (4.83),

$$\frac{A^{odd}(m_\pi)}{m_\pi} = \frac{8\pi}{3} \left(1 + \frac{m_\pi}{M_N}\right) (a_{1/2} - a_{3/2}) . \tag{A4.22}$$

If we denote by $M^\pm$ the coherent forward scattering amplitudes for $\pi^\pm + p \to \pi^\pm + p$, then

$$M^{even} = \frac{1}{2} \left(M^-(\nu) + M^+(\nu)\right) ,$$

$$M^{odd} = \frac{1}{2} \left(M^-(\nu) - M^+(\nu)\right) . \tag{A4.23}$$

Using the optical theorem, we have $(\nu' \geq m_\pi)$

$$B^{odd}(\nu') = 2m_N \sqrt{\nu'^2 - m_\pi^2} \left(\sigma_-(\nu') - \sigma_+(\nu')\right). \qquad (A4.24)$$

Inserting (A4.22)-(A4.23) into (A4.21), we get

$$\begin{aligned}
\frac{a_{1/2} - a_{3/2}}{3m_\pi} =& \frac{1}{8\pi^2} \frac{1}{(m_N + m_\pi)} \int_0^{m_\pi} \frac{(B_-(\nu') - B_+(\nu'))}{\nu'^2 - m_\pi^2} d\nu' \\
&+ \frac{1}{4\pi^2 \left(1 + \frac{m_N}{m_\pi}\right)} \int_{m_\pi}^\infty \frac{(\sigma_-(\nu') - \sigma_+(\nu')) \, d\nu'}{\sqrt{\nu'^2 - m_\pi^2}}. \qquad (A4.25)
\end{aligned}$$

In the kinematical region $0 < \nu' < m_\pi$, only the exchange process $\pi^+ + p \to \pi^+ + p$ contributes, leading to the δ-like singularity for the imaginary part of the amplitude, due to the intermediate neutron state. Specifically,

$$B_+(\nu') = -\pi^2 g_{\pi NN}^2 \frac{\nu'^2 - m_\pi^2}{m_N} \delta\left(\nu' - \frac{m_\pi^2}{2m_N}\right). \qquad (A4.26)$$

Inserting (A4.26) into (A4.25) and using the Goldberger-Treiman relation, we find

$$\begin{aligned}
g_A^2 =& \frac{2f_\pi^2}{3m_\pi} \left(1 + \frac{m_\pi}{m_N}\right)(a_{1/2} - a_{3/2}) \\
&+ \frac{f_\pi^2}{2\pi^2} \int_{m_\pi}^\infty \frac{d\nu'}{\sqrt{\nu'^2 - m_\pi^2}} (\sigma_+(\nu') - \sigma_-(\nu')). \qquad (A4.27)
\end{aligned}$$

This relation was first derived by Goldberger, Miyazawa and Oehme years before the advent of current algebra. It becomes relevant to current algebra when one inserts the values of the scattering lengths as suggested by the Tomozawa-Weinberg relation. The result is the Adler-Weisberger relation. The validity of the Goldberger-Miyazawa-Oehme relation rests solely on the assumption that $M^{odd}(\nu)$ satisfies an unsubtracted dispersion relation, a fact that is confirmed by Regge arguments and experimental data.

References

1. For a comprehensive summary and full list of references, see S.L. Adler and R.F. Dashen, *Current Algebras and Applications to Particle Physics* (W.A. Benjamin, New York, 1968) and V. de Alfaro, S. Fubini, G. Furlan and C. Rossetti, *Currents in Hadron Physics* (North Holland, Amsterdam, 1973).

2. For more recent developments including anomalies, see S.B. Treiman, R. Jackiw, B. Zumino and E. Witten, *Current Algebra and Anomalies* (World Scientific, Singapore, 1985).
3. V. Bernard, N. Kaiser and U.G. Meissner, Int. J. Mod. Phys. **E4**, 193 (1995).
4. S. Weinberg, *Lectures on Elementary Particles and Quantum Field Theory*, 1970 Brandeis Univ. Summer Institute, eds. S. Deser, M. Grisaru and H. Pendleton (MIT Press, 1970).
5. S. Fubini, G. Furlan and C. Rossetti, Nuovo Cim. **40A**, 1171 (1965).
6. For an updated account and references, see J.F. Donoghue, E. Golowich and B.R. Holstein, *Dynamics of the Standard Model* (Cambridge University Press, Cambridge, 1992); B.R. Holstein, "Chiral Perturbation Theory : A Primer," hep-ph/9510344.
7. H. Yamagishi and I. Zahed, Ann. Phys. (1995), in print.
8. M.L. Goldberger, H. Miyazawa and R. Oehme, Phys. Rev. **99**, 986 (1955).

CHAPTER 5

EFFECTIVE CHIRAL LAGRANGIANS

5.1. Introduction

The current-algebra results described in the preceding chapter are not confined to few-pion processes. They can be generalized to amplitudes involving an arbitrary number of soft pions. This extension was achieved in the context of effective Lagrangians. Historically, effective Lagrangians were formulated so as to reproduce the results of current algebra and PCAC at tree level. In these formulations, analyticity, unitarity and symmetries were easier to achieve, with low energy theorems following from Feynman diagrams. The effective Lagrangians were merely convenient alternatives to commutator algebra.

In 1979, Weinberg[1] extended the scope of the effective Lagrangian formulation by postulating that the use of effective Lagrangians can go beyond current algebra. This assertion was based on the observation that for soft-pion processes, chiral Lagrangians offer a powerful parametrization of the S-matrix based on chiral counting arguments and general principles.

In this chapter, we shall give a brief description of Weinberg's initial program in terms of effective chiral Lagrangians and outline the general idea behind power counting in chiral perturbation theory. We will show that simple effective Lagrangians with (broken) chiral symmetry reproduce the tree level current-algebra results. Weinberg's original formulation will be reviewed and illustrated with $\pi\pi$-scattering although the basic idea can be made considerably more general involving other fields. The systematization and extension of Weinberg's program by Gasser and Leutwyler[2] will be discussed in Chapter 6. This latter work has rekindled interest in

Weinberg's approach, and triggered a flurry of activities in recent years.

Before we discuss the role of chiral effective Lagrangians in strong-interaction physics, we should stress one important point. The use of effective Lagrangians beyond tree level as a way to understand the hadronic S-matrix in the soft-pion limit, while well motivated phenomenologically and fully consistent with the present notion of field theory (that is, that all theories to the Planck scale are effective theories), side-steps the basic issues of confinement and broken chiral symmetry. Experimental input data are crucial elements in this approach, whose role is to provide relationships between various processes as dictated by symmetries, analyticity, unitarity and other general principles. This is not an approach with which to *derive* quantities from first principles and is inherently limited in its range of applicability (soft physics) and its scope of predictions (pion mediated processes).

5.2. Chiral Lagrangians

The current-algebra procedure becomes increasingly cumbersome as the number of soft-pion fields emitted or absorbed increases. Being essentially algebraic, the method does not enhance one's intuition beyond the soft-pion limit. To remedy this, Weinberg suggested a prescription based on an effective Lagrangian formulation that, while easier computationally, would bring important physical insights to the algebraic method.

Weinberg's observation starts from the S-matrix for the emission and absorption of any number of soft pions in a general reaction $\alpha \to \beta + n\pi$. Using PCAC, this matrix element can be written, symbolically, in the following form:

$$\langle \alpha | T \left(\pi(x_1)...\pi(x_n) \right) | \beta \rangle \sim$$
$$\left(\frac{1}{f_\pi m_\pi^2} \right)^n \sum_C C^{\mu_1 \cdots \mu_n} \langle \alpha | T \left(A_{\mu_1}(x_1)...A_{\mu_n}(x_n) \right) | \beta \rangle.$$

$$(5.1)$$

The sum is over all possible commutators generated while pulling ∂^μ across the T-ordered product. In the soft-pion limit, the dominant contribution to (5.1) corresponds to the emission and absorption of currents on the external legs of a core process *without* pions. Thus, to recover the current-algebra results, all one needs to do is to use an appropriate Lagrangian that satisfies PCAC and current algebra, and to evaluate the desired soft-pion matrix elements at tree level. Corrections to the soft-pion theorems could be sought by renormalizing the tree level results. So what are the appropriate effective Lagrangians?

5.2.1. *Linear σ model*

The linear σ model discussed in Chapter 4 contains the basic ingredients of chiral symmetry. It is the simplest example of a linear effective chiral Lagrangian. Including explicitly the nucleons (or constituent quarks), we have, for chiral $SU(2) \times SU(2)$ symmetry,

$$
\mathcal{L} = \overline{N}\left(i\partial\!\!\!/ + g\left(\sigma + i\vec{\tau} \cdot \vec{\pi}\gamma_5 \right) \right) N
$$
$$
+ \frac{1}{2}(\partial_\mu \sigma)^2 + \frac{1}{2}(\partial_\mu \vec{\pi})^2 - \frac{\lambda}{4}\left(\sigma^2 + \vec{\pi}^2 - v^2 \right)^2 - m_\pi^2 f_\pi \sigma. \quad (5.2)
$$

The Lagrangian (5.2) is invariant under vector and axial-vector isospin rotations for $m_\pi = 0$. Note that the potential $\mathcal{V}$ has been explicitly fixed to a Mexican hat shape as shown in Fig.5.1. The potential is $O(4)$ invariant. The Noether currents are

$$
V_\mu^a = \left(\vec{\pi} \times \partial_\mu \vec{\pi} \right)^a + \overline{N}\gamma_\mu \frac{\tau^a}{2} N,
$$
$$
A_\mu^a = \left(\sigma \partial_\mu \pi^a - \pi^a \partial_\mu \sigma \right) + \overline{N}\gamma_\mu \gamma_5 \frac{\tau^a}{2} N \quad (5.3)
$$

and satisfy Gell-Mann current algebra. The vector current is conserved $\partial \cdot V^a = 0$, while the axial current is partially conserved $\partial \cdot A^a = m_\pi^2 f_\pi \pi^a$, in agreement with the PCAC hypothesis. The global charges

$$
T^a = \int d^3 x V_0^a(x), \qquad X^a = \int d^3 x A_0^a(x) \quad (5.4)
$$

are the generators of the $SU_V(2) \times SU_A(2)$ rigid algebra, and their linear combinations generate the $SU_L(2) \times SU_R(2)$ rigid algebra,

$$
\left[\frac{1}{2}\left(T^a \pm X^a \right), \frac{1}{2}\left(T^b \pm X^b \right) \right] = i\epsilon^{abc} \frac{1}{2}\left(T^c \pm X^c \right),
$$
$$
\left[\frac{1}{2}\left(T^a \pm X^a \right), \frac{1}{2}\left(T^b \mp X^b \right) \right] = 0. \quad (5.5)
$$

In terms of (5.4) the fields in (5.1) transform linearly. For isospin,

$$\left[T^a, \pi^b\right] = i\epsilon^{abc}\pi^c, \qquad \left[T^a, \sigma\right] = 0, \qquad \left[T^a, N\right] = -\frac{1}{2}\tau^a N \quad (5.6)$$

while for axial isospin,

$$\left[X^a, \pi^b\right] = -i\delta^{ab}\sigma, \qquad \left[X^a, \sigma\right] = i\pi^a, \qquad \left[X^a, N\right] = -\frac{1}{2}\tau^a\gamma_5 N. \quad (5.7)$$

The chiral group is realized linearly.

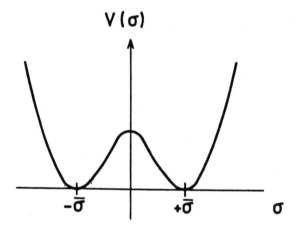

Fig. 5.1. Effective potential $\mathcal{V}$ in the linear σ model.

In the vacuum, chiral symmetry is explicitly broken so that $m_\pi^2 \neq 0$. This term generates tadpoles, *i.e.*, vacuum expectation values for the σ field, or the minima in V as shown in Fig.5.1. These tadpoles are eliminated by redefining the σ field through $\sigma \to \sigma - \overline{\sigma}$, with

$$\lambda\overline{\sigma}(\overline{\sigma}^2 - v^2) = m_\pi^2 f_\pi. \quad (5.8)$$

In the shifted variables, the effective Lagrangian (5.2) becomes

$$\mathcal{L} = \overline{N}\left(i\partial\!\!\!/ - m_N + g_{\pi NN}[\sigma + i\vec{\tau}\cdot\vec{\pi}\gamma_5]\right)N$$
$$+ \frac{1}{2}(\partial_\mu\sigma)^2 - \frac{1}{2}m_\sigma^2\sigma^2 + \frac{1}{2}(\partial_\mu\vec{\pi})^2 - \frac{1}{2}m_\pi^2\vec{\pi}^2$$

$$+\frac{1}{2}g_{\pi NN}\frac{m_\sigma^2 - m_\pi^2}{m_N}\sigma\left(\sigma^2 + \vec{\pi}^2\right)$$

$$-\frac{1}{8}g_{\pi NN}^2\frac{m_\sigma^2 - m_\pi^2}{m_N^2}\left(\sigma^2 + \vec{\pi}^2\right)^2. \tag{5.9}$$

The physical quantities appearing in (5.9) are defined as follows

$$g_{\pi NN} = g, \qquad \overline{\sigma} = \frac{m_N}{g_{\pi NN}}, \qquad m_N = f_\pi g_{\pi NN},$$

$$m_\sigma^2 = 2\lambda\left(\frac{m_N}{g_{\pi NN}}\right)^2 + m_\pi^2. \tag{5.10}$$

Due to chiral symmetry, the σ coupling to the nucleon is the same as the pion coupling to the nucleon. A simple power counting shows that the linear σ model is renormalizable in the conventional sense to all orders in perturbation theory.

As it stands, it appears that the model suffers from two shortcomings : the appearance of a scalar σ and the presence of a direct S-wave coupling to the nucleon, both unsupported by experiments. We will see below, however, that these apparent shortcomings are naturally resolved by the relations between the parameters (5.10) following from chiral symmetry, giving the σ a large two-pion decay width and suppressing the S-wave scattering length in the pion-nucleon system.

5.2.2. *Nonlinear σ model*

To overcome the apparent shortcomings of the linear σ model, Weinberg[3] suggested to use solely the pion field $\vec{\pi}$, and express $\vec{\pi}$ in a nonlinear representation of $SU(2) \times SU(2)$. This representation is unique, modulo field redefinitions. Indeed, the nonlinear implementation of $SU(2) \times SU(2)$ on the pion field reads

$$\left[T^a, \pi^b\right] = i\epsilon^{abc}\pi^c, \qquad \left[X^a, \pi^b\right] = -if_{ab}(\vec{\pi}) \tag{5.11}$$

where $f_{ab}(\vec{\pi})$ is an arbitrary function of the pion field. The first relation comes from the requirement that the isospin is a symmetry. The global algebra (5.5) constrains the general form of f_{ab}. Indeed, the two Jacobi identities

$$\left[T^a, \left[X^b, \pi^c\right]\right] + \left[\pi^c, \left[T^a, X^b\right]\right] + \left[X^b, \left[\pi^c, T^a\right]\right] = 0,$$

$$\left[X^a, \left[X^b, \pi^c\right]\right] + \left[\pi^c, \left[X^a, X^b\right]\right] + \left[X^b, \left[\pi^c, X^a\right]\right] = 0 \quad (5.12)$$

imply respectively

$$\frac{\partial f_{bc}(\vec{\pi})}{\partial \pi_d} \epsilon^{ade} \pi^e = f_{bd}(\vec{\pi}) \epsilon^{acd} + f^{dc}(\vec{\pi}) \epsilon^{abd},$$

$$\frac{\partial f_{bc}(\vec{\pi})}{\partial \pi_d} f_{ad}(\vec{\pi}) - \frac{\partial f_{ac}(\vec{\pi})}{\partial \pi_d} f_{bd}(\vec{\pi}) = \delta_{ac}\vec{\pi}_b - \delta_{bc}\pi_a. \quad (5.13)$$

The general solution of (5.13) is given by

$$f_{bc}(\vec{\pi}) = \delta_{bc} f(\pi^2) + \pi_b \pi_c g(\pi^2) \quad (5.14)$$

where $f(\pi^2)$ is *arbitrary* function of π^2 and $g(\pi^2)$ is related to $f(\pi^2)$ through

$$g(\pi^2) = \frac{1 + 2f(\pi^2)f'(\pi^2)}{f(\pi^2) - 2\pi^2 f'(\pi^2)}. \quad (5.15)$$

The transformations (5.12) are form-invariant under a general local redefinition of the pion field : $\pi^a \to \pi^a \Phi$, $f \to f\Phi$ and $g \to (g\Phi + 2(f + g\pi^2)\Phi')/\Phi^2$, where Φ is short for $\Phi(\pi^2)$ and f for $f(\pi^2)$. This invariance is fundamental.

The nonlinear implementation of $SU(2) \times SU(2)$ on the nucleon field follows the same rationale as for the pion field. Indeed, the transformation will be sought in the form

$$\left[T^a, N\right] = -\frac{\tau^a}{2}N, \qquad \left[X^a, N\right] = v^{ab}(\pi^2)\frac{\tau^b}{2}N \quad (5.16)$$

with v^{ab} an arbitrary odd function of π. A rerun of the Jacobi identities for X, T and N yields

$$v^{ab} = \epsilon^{abc}\pi^c v(\pi^2) = \epsilon^{abc}\pi^c \left(f(\pi^2) + \sqrt{f^2(\pi^2) + \pi^2}\right)^{-1}. \quad (5.17)$$

It is straightforward to check the form-invariance of (5.16) under general reparametrization of the pion field.

The form-invariance of (5.11) and (5.16) suggests that the derivatives on π and N that are natural are covariant derivatives. In fact, there is a close analogy between nonlinear σ models and gravity[5]. This point will be elaborated on in Chapter 6. The covariant derivatives we seek transform homogeneously under axial isospin transformation. The covariant derivative for the pion field will be sought in the homogeneous form for the N field (5.16) as

$$\left[X^b, \nabla_\mu \pi^c\right] = -iv^{ab}\epsilon^{bcd}\nabla_\mu \pi^d, \qquad \left[T^a, \nabla_\mu \pi^c\right] = -i\epsilon^{acd}\nabla_\mu \pi^d. \quad (5.18)$$

The Jacobi identities yield the following result for the covariant derivative

$$\nabla_\mu \pi^a = \frac{1}{\sqrt{f^2(\pi^2)+\pi^2}}\partial_\mu \pi^a + \frac{f'(\pi^2)+v(\pi^2)/2}{f^2(\pi^2)+\pi^2}\pi^a\partial_\mu \pi^2. \quad (5.19)$$

Finally the covariant derivative of the nucleon field also transforms like N:

$$\left[T^a, \nabla_\mu N\right] = -\frac{\tau^a}{2}\nabla_\mu N, \qquad \left[X^a, \nabla_\mu N\right] = v^{ab}(\pi^2)\frac{\tau^b}{2}\nabla_\mu N. \quad (5.20)$$

The solution to (5.20) using the Jacobi identities is

$$\nabla_\mu N = \partial_\mu N + i\frac{v(\pi^2)}{\sqrt{f^2(\pi^2)+\pi^2}}\frac{\vec{\tau}}{2}\cdot(\vec{\pi}\times\partial_\mu\vec{\pi})N. \quad (5.21)$$

Again (5.18) and (5.20) are form-invariant under an arbitrary redefinition of the pion field.

The minimal chiral Lagrangian in the nonlinear representation of $SU(2)\times SU(2)$ is

$$\mathcal{L} = \overline{N}i\gamma\cdot\nabla N + \frac{1}{2}f_0^2\nabla_\mu\vec{\pi}\cdot\nabla^\mu\vec{\pi} + \left(\frac{g_{\pi NN}f_0}{2m_N}\right)\overline{N}i\gamma^\mu\gamma^5\frac{\vec{\tau}}{2}N\cdot\nabla_\mu\vec{\pi}. \quad (5.22)$$

This is a prototype of nonlinear σ model. A convenient parametrization for the induced nonlinear representation is

$$f(\pi^2) = \frac{F_0^2 - \pi^2}{2F_0} \quad (5.23)$$

where $F_0 = 2f_0$ is twice the bare pion decay constant. Other parametrizations are also possible. For instance, the standard nonlinear σ-model parametrization is $f(\pi^2) = -\sqrt{f_0^2 - \pi^2}$. In terms of (5.23), the covariant derivatives (5.19) and (5.21) simplify to

$$\nabla_\mu \pi^a = \frac{2F_0}{F_0^2 + \pi^2}\partial_\mu \pi^a,$$

$$\nabla_\mu N = \partial_\mu N + \frac{2i}{F_0^2 + \pi^2}\frac{\vec{\tau}}{2}\cdot(\vec{\pi}\times\partial_\mu\vec{\pi})N \quad (5.24)$$

respectively. Using (5.24), the minimal version of the nonlinear σ model (5.22) reads

$$
\begin{aligned}
\mathcal{L} =& \overline{N} i\gamma \cdot \partial N + \mathcal{L}_B \\
&+ \frac{1}{2} \left(\frac{F_0^2}{F_0^2 + \pi^2} \right)^2 \partial^\mu \vec{\pi} \cdot \partial_\mu \vec{\pi} - \frac{1}{2} \frac{F_0^2}{F_0^2 + \pi^2} m_0^2 \pi^2 \\
&- \frac{2}{F_0^2 + \pi^2} (\vec{\pi} \times \partial_\mu \vec{\pi}) \cdot \overline{N} \gamma^\mu \frac{1}{2} \vec{\tau} N \\
&+ \frac{g_{\pi NN}}{2 m_N} \frac{F_0^2}{F_0^2 + \pi^2} \partial_\mu \vec{\pi} \cdot i\overline{N} \gamma^\mu \gamma_5 \frac{1}{2} \vec{\tau} N.
\end{aligned} \tag{5.25}
$$

We have explicitly added a chiral-symmetry-breaking pion mass term[3] and put all other symmetry breaking terms in $\mathcal{L}_B$. The symmetry breaking is assumed to be a function of π^2 and transforming according to $(\frac{1}{2}, \frac{1}{2})$ representation of $SU(2) \times SU(2)$. Using the isomorphism of $SU(2) \times SU(2)$ with $SO(4)$ the symmetry breaking term is of form with $V_4 = -1/2 + \pi^2/(F_0^2 + \pi^2)$ transforming as the fourth component of a four-vector under $SO(4)$. This breaking is suggested from QCD as discussed in the preceding sections. It can be checked that the vector current associated to (5.25) is conserved, while the axial current is not,

$$
\partial^\mu A_\mu^a = -i \left[X^a, \mathcal{L}_B \right] \tag{5.26}
$$

where $\mathcal{L}_B$ is the symmetry breaking part of (5.25), in agreement with the PCAC hypothesis. As expected, both the vector and axial currents obey Gell-Mann algebra.

In the following chapters, we shall use the Goldstone field $U(\pi)$ defined as a representative of the coset space $SU(N_f) \times SU(N_f)/SU(N_f)$. The construction of the nonlinear σ model in this field is discussed in Chapter 6.

5.3. Applications

The chiral effective Lagrangians described above satisfy the two basic requirements: current algebra and PCAC. Thus, they ought to reproduce the soft-pion theorems at tree level, as proposed by Weinberg and proved by Dashen and Weinstein[6]. In this section, we will illustrate this fact by applying them to the pion-nucleon system at tree level.

5.3.1. *Goldberger-Treiman relation*

- *Linear σ model*

For tree-level processes, the nucleon is on-shell, $(i\not{\partial} - m_N)N = 0$. Thus, (5.26) is equivalent in the linear σ model (5.9) to

$$\mathcal{L}_{\pi NN} \sim -\frac{g_{\pi NN}}{2m_N}\partial^\mu\left(\overline{N}i\gamma_5\vec{\tau}\gamma_\mu N\right)\cdot\vec{\pi} \sim \frac{g_{\pi NN}}{2m_N}\overline{N}i\gamma_5\gamma_\mu\vec{\tau}N\cdot\partial^\mu\vec{\pi} \quad (5.27)$$

which follows from integration by parts. From (5.27), the axial coupling constant g_A can be simply read off:

$$g_A = \frac{g_{\pi NN}f_\pi}{m_N} = 1. \quad (5.28)$$

At tree level, the linear σ model reproduces the Goldberger-Treiman relation with $g_A = 1$.

- *Nonlinear σ model*

In this model, it is immediate that $g_A = f_0 g_{\pi NN}/m_N$, in agreement with the Goldberger-Treiman relation. This conclusion can be reached, by noting that from (5.25), we have $G_A = g_{\pi NN}/F_0 m_N$, while $G_V = 2/F_0^2$ for the normalized currents. Thus $g_A = G_A/G_V = f_0 g_{\pi NN}/m_N$.

5.3.2. *Weinberg-Tomozawa relation*

- *Linear σ model*

The pion-nucleon scattering amplitude in the linear σ model at tree level is given by the diagrams shown in Fig. 5.2. Their respective contribution to the on-shell T matrix reads

$$T^{ab}(\nu^2, t) = g_{\pi NN}^2\overline{N}(p_2)\tau^b\tau^a\gamma_5\frac{\not{p}_1 + \not{q}_1 + m_N}{m_\pi^2 + 2p_1\cdot q_1}\gamma_5 N(p_1)$$

$$+ g_{\pi NN}^2\overline{N}(p_2)\tau^a\tau^b\gamma_5\frac{\not{p}_1 - \not{q}_2 + m_N}{m_\pi^2 - 2p_1\cdot q_2}\gamma_5 N(p_1)$$

$$+ g_{\sigma NN}g_{\sigma\pi\pi}\overline{N}(p_2)\frac{-\delta^{ab}}{(q_1 - q_2)^2 - m_\sigma^2}N(p_1) \quad (5.29)$$

where

$$g_{\sigma NN} = g_{\pi NN}, \qquad g_{\sigma\pi\pi} = \frac{g_{\pi NN}}{2m_N}\left(m_\sigma^2 - m_\pi^2\right). \quad (5.30)$$

The amplitude (5.29) is a function of two of the three Mandelstam variables s, t, u, since $s + t + u = 2m_\pi^2 + 2m_N^2$. Here we have chosen to use $\nu = (s-u)/4m_N$ and t. We note that ν is crossing-odd and reduces to the pion energy in the Breit frame.

Fig. 5.2. Tree πN-scattering amplitude in the linear σ model.

Generically, the scattering amplitude (5.29) may be decomposed into

$$T^{ab} = \overline{N}(p_2)\left(\delta^{ab}T^+ + i\epsilon^{abc}T^-\tau^c\right)N(p_1) \tag{5.31}$$

where T^+ is the isoscalar and T^- the isovector part of the amplitude. Using (5.29) and (5.31), we obtain

$$\begin{aligned}
T^+(\nu^2,t) =& \frac{g_{\pi NN}^2}{2m_N}\frac{(\slashed{q}_1 + \slashed{q}_2)}{2}\left(\frac{1}{\nu_B - \nu} - \frac{1}{\nu_B + \nu - t/2m_N}\right) \\
&+ g_{\sigma NN}g_{\sigma \pi\pi}\frac{-1}{t - m_\sigma^2}
\end{aligned} \tag{5.32}$$

and

$$T^-(\nu^2,t) = \frac{g_{\pi NN}^2}{2m_N}\frac{(\slashed{q}_1 + \slashed{q}_2)}{2}\left(\frac{1}{\nu_B - \nu} + \frac{1}{\nu_B + \nu - t/2m_N}\right) \tag{5.33}$$

where $\nu_B = (t - 2m_\pi^2)/4m_N$. At threshold ($\nu = m_\pi$ and $t = 0$), the scattering amplitudes are related to the isoscalar and isovector scattering lengths through[*]

$$T^\pm(m_\pi^2, 0) = 4\pi\left(1 + \frac{m_\pi}{m_N}\right)a^\pm. \tag{5.34}$$

---

[*]We prefer here to use the normalization convention $\langle p|p'\rangle = (2\pi)^3\frac{E}{m}$, so (5.34) differs from (4.91).

Thus, for the isoscalar scattering length, we have

$$a^+ = \frac{1}{4\pi(1 + \frac{m_\pi}{m_N})} \left(\frac{g_{\sigma NN} g_{\sigma\pi\pi}}{m_\sigma^2} - \frac{g_{\pi NN}^2}{m_N} \frac{1}{1 - \frac{m_\pi^2}{4m_N^2}} \right) \tag{5.35}$$

while for the isovector scattering length, we have

$$a^- = \frac{1}{4\pi(1 + \frac{m_\pi}{m_N})} \frac{g_{\pi NN}^2}{2m_N} \left(\frac{\frac{m_\pi}{m_N}}{1 - \frac{m_\pi^2}{4m_N^2}} \right). \tag{5.36}$$

Using the relations (5.30) as dictated by symmetry in the linear σ model, the S-wave isoscalar scattering length simplifies to

$$a^+ = -\frac{g_{\pi NN}^2}{4\pi(m_N + m_\pi)} \left(\frac{m_\pi^2}{m_\sigma^2} + \frac{m_\pi^2}{4m_N^2} \right). \tag{5.37}$$

The isoscalar scattering length is proportional to m_π^2. Thus,

$$a^+ = \frac{1}{3} \left(a_{\frac{1}{2}} + 2a_{\frac{3}{2}} \right) = \mathcal{O}(m_\pi) \tag{5.38}$$

in agreement with the Tomozawa-Weinberg relation. We note that if it were not for (5.30) which induces the cancellation of the "pair term" in the nucleon Born diagrams and the σ-exchange term, the S-wave scattering length would in general be very large, in disagreement with the Tomozawa-Weinberg relation and also with the data. This problem is generic to the use of pseudoscalar couplings, and shows that the pion-nucleon interaction is intrinsically pseudovector in nature. In fact, derivative coupling is the key feature of the spontaneously broken chiral symmetry[4]. Such a coupling is *natural* in the nonlinear σ model.

• *Nonlinear σ model*

The pion-nucleon scattering amplitude in the nonlinear σ model at tree level is given by the diagrams shown in Fig.5.3. At threshold, the amplitude reads

$$\begin{aligned}
\mathcal{T}^{ab}(\nu^2, t) = &\frac{g_{\pi NN}^2}{4m_N^2} \overline{N}(p_2) \tau^b \tau^a \slashed{q}_2 \gamma_5 \frac{\slashed{p}_1 + \slashed{q}_1 + m_N}{m_\pi^2 + 2p_1 \cdot q_1} \slashed{q}_1 \gamma_5 N(p_1) \\
&+ \frac{g_{\pi NN}^2}{4m_N^2} \overline{N}(p_2) \tau^a \tau^b \slashed{q}_1 \gamma_5 \frac{\slashed{p}_1 - \slashed{q}_2 + m_N}{m_\pi^2 - 2p_1 \cdot q_2} \slashed{q}_2 \gamma_5 N(p_2) \\
&+ \frac{1}{F_0^2} \overline{N}(p_2) i\epsilon^{abc} \tau^c (\slashed{q}_1 + \slashed{q}_2) N(p_1).
\end{aligned} \tag{5.39}$$

The contribution to the isoscalar scattering length comes from the first two terms only. The result is

$$T^+(\nu^2, t) = \frac{g_{\pi NN}^2}{2m_N} \frac{2\nu_B^2}{\nu_B^2 - \nu^2}, \qquad (5.40)$$

with an isoscalar scattering amplitude

$$a^+ \propto -\frac{1}{4\pi(1 + \frac{m_\pi}{m_N})} \left(\frac{g_{\pi NN}}{2m_N}\right)^2 \frac{m_\pi^2}{1 - \frac{m_\pi^2}{4m_N^2}}, \qquad (5.41)$$

which is again consistent with the Weinberg-Tomozawa result. The leading contribution to the isovector scattering length stems solely from the last term in (5.39) and, modulo an isospin factor given in (4.91), is of the form

$$a^- \propto \frac{1}{4\pi(1 + \frac{m_\pi}{m_N})} \frac{2m_\pi}{F_0^2}. \qquad (5.42)$$

We note that $a^\pm$ in the nonlinear σ model follows from $a^\pm$ in the linear σ model by taking $m_\sigma \to \infty$. This procedure, while valid at tree order and for many other observables, does not apply to one or more loops since divergences are encountered in the nonlinear version. Finally, from (5.41-5.42), it follows that to order m_π,

$$a_{\frac{1}{2}} \approx \frac{1}{4\pi} \frac{m_\pi}{f_0^2} \frac{1}{1 + \frac{m_\pi}{m_N}}, \qquad a_{\frac{3}{2}} \approx -\frac{1}{8\pi} \frac{m_\pi}{f_0^2} \frac{1}{1 + \frac{m_\pi}{m_N}} \qquad (5.43)$$

where we have used $F_0 = 2f_0$. Many other results of current algebra may be recovered by the tree amplitudes. They will not be discussed here.

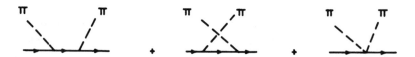

Fig. 5.3. Tree πN-scattering amplitude in the nonlinear σ model.

5.4. Chiral Perturbation Theory

Are Feynman diagrams in effective chiral theories a way to go beyond current algebra? The answer is not as straightforward as it may seem for two reasons : first, the need to organize the diagrammatic expansion for a strongly coupled theory, and second, the nonlinear effective formulations are in general not tractable in the ultraviolet in the conventional sense (*i.e.*, à la Dyson).

In 1971, Li and Pagels[7] suggested that corrections to the current algebra result could not follow from ordinary perturbation theory in the pion mass, about the chiral limit (massless pions), their main observation being that the S-matrix and the matrix elements of currents are in general non-analytic functions of m_π^2. Their effort, while not systematic, led to the advent of chiral perturbation theory.

5.4.1. *Chiral logarithms*

Consider the S-matrix following from an effective chiral Lagrangian for the process $\alpha \to \beta$. In general, the S-matrix is given by the Feynman diagrams, with one-pion insertion, two-pion insertion, etc. as shown in Fig.5.4. The contribution of the one-pion insertion term is

$$S^1_{\alpha\beta}(m_\pi^2) = \int \frac{d^4q}{q^2 - m_\pi^2} \sum_a T^{aa}_{\alpha\beta}(q) \tag{5.44}$$

where $T^{ab}_{\alpha\beta}(q)$ characterizes the off-shell process $\pi^a(q) + \alpha \to \pi^b(q) + \beta$. To lowest order, $T^{ab}_{\alpha\beta}(q)$ is independent of m_π^2. The dependence of (5.44) on the pion mass m_π^2 can be seen by taking a derivative,

$$\frac{\partial S^1_{\alpha\beta}(m_\pi^2)}{\partial m_\pi^2} = \int \frac{d^4q}{(q^2 - m_\pi^2)^2} \sum_a T^{aa}_{\alpha\beta}(q). \tag{5.45}$$

For a small pion mass, the integrand in (5.45) is infrared-sensitive. If we let $m_\pi^2 \to 0$, then (5.45) becomes

$$\frac{\partial S^1_{\alpha\beta}(m_\pi^2)}{\partial m_\pi^2} \sim \sum_a T^{aa}_{\alpha\beta}(m_\pi)\ln(m_\pi^2). \tag{5.46}$$

Thus, the contribution of the one-pion insertion to the S-matrix is not analytic around the chiral limit. The pion insertion gives a logarithmic contribution. The coefficient of the logarithmic contribution is given by the T-matrix element for the soft process $\pi + \alpha \to \pi + \beta$, thus calculable from current algebra and PCAC.

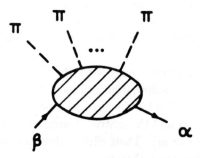

Fig. 5.4. Pion insertions in a generic S-matrix element.

So, on general grounds, one expects the two-pion, three-pion insertions to yield non-analytic contributions of the form $(\ln(m_\pi^2))^2$, $(\ln(m_\pi^2))^3$, etc. that may or may not be summable. Carruthers and Haymaker[8] have argued, using the linear σ model, that the expansion around the chiral limit has a very small radius of convergence. Their argument, however, applies to the minimum of the σ field and does not carry readily to S-matrix elements.

Assuming that the expansion has a nonzero radius of convergence, we will now illustrate its function by calculating the leading logarithmic corrections to the pion parameters away from the chiral limit, as well as to the nucleon mass, using the nonlinear σ model. These calculations will give us insights into the concept of chiral perturbation theory that will later be developed more systematically.

5.4.2. Chiral loops and the pion

Restricting (5.25) to the pion sector only, we have

$$\mathcal{L}_\pi = \frac{1}{2}\left(\frac{F_0^2}{F_0^2 + \pi^2}\right)^2 (\partial_\mu \vec{\pi}) \cdot (\partial^\mu \vec{\pi}) - \frac{1}{2}\left(\frac{F_0^2}{F_0^2 + \pi^2}\right) m_0^2 \pi^2. \quad (5.47)$$

Expanding (5.47) to leading order in $1/F_0^2$ gives

$$\mathcal{L}_\pi = +\frac{1}{2}(\partial_\mu \vec{\pi}) \cdot (\partial^\mu \vec{\pi}) - \frac{1}{2} m_0^2 \pi^2$$
$$-\frac{\pi^2}{F_0^2}(\partial_\mu \vec{\pi}) \cdot (\partial^\mu \vec{\pi}) + \frac{1}{2}\frac{m_0^2}{F_0^2}\pi^4 + \mathcal{O}(\frac{1}{F_0^4}). \quad (5.48)$$

To sum up the effects induced by m_0^2 to all orders, we will use the mean-field approximation. The idea is to resum all the tadpole diagrams shown in Fig.5.5. With this in mind, the mean-field version of (5.48) to order $1/F_0^4$ is

$$\overline{\mathcal{L}}_\pi \sim \frac{1}{2} \frac{(\partial_\mu \vec{\pi}) \cdot (\partial^\mu \vec{\pi})}{1 + 2\langle \pi^2 \rangle / F_0^2} - \frac{m_0^2}{2} \left(1 + \frac{2}{3} \frac{\langle \pi^2 \rangle}{F_0^2} \right) \frac{\pi^2}{1 + 2\langle \pi^2 \rangle / F_0^2} \quad (5.49)$$

where we have defined, using the $\overline{\text{MS}}$ scheme of dimensional regularization:

$$\langle \pi^2 \rangle = 3 \int \frac{d^4 p}{(2\pi)^4} \frac{1}{-p^2 + m_0^2} = \frac{3}{16\pi^2} m_0^2 \ln \frac{m_0^2}{\mu^2} \quad (5.50)$$

where μ is the induced scale. In the mean-field approximation, the pion field renormalizes through $\pi \to \pi/\sqrt{1 + 2\langle \pi^2 \rangle / F_0^2}$ and the pion mass is given by

$$m_\pi^2 = m_0^2 \left(1 + \frac{2}{3} \frac{\langle \pi^2 \rangle}{F_0^2} \right). \quad (5.51)$$

Note that the pion mass vanishes for $m_0^2 = 0$ as it should.

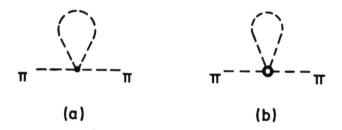

Fig. 5.5. Tadpole diagrams on the pion propagator in the nonlinear σ model. Here (a) and (b) correspond respectively to the first and second vertices of the quartic terms of eq.(5.47).

The divergence of the axial current operator follows from (5.26) using (5.25). To order $1/F_0^2$, the result is

$$\partial^\mu A_\mu^a \sim \frac{m_0^2 F_0}{2} \left(1 - \frac{\pi^2}{F_0^2} \right) \pi^a. \quad (5.52)$$

In the mean-field approximation, (5.52) becomes

$$\partial^\mu \overline{A}_\mu^a \sim \frac{m_0^2 F_0}{2} \left(1 - \frac{5}{3} \frac{\langle \pi^2 \rangle}{F_0^2} \right) \pi^a \quad (5.53)$$

or

$$\partial^\mu \overline{A}^a_\mu \sim \frac{F_0}{2} \left(1 - \frac{4}{3} \frac{\langle \pi^2 \rangle}{F_0^2} \right) m_\pi^2 \frac{\pi^a}{\sqrt{1 + 2\langle \pi^2 \rangle / F_0^2}}. \tag{5.54}$$

The pion decay constant to leading order in $1/F_0^2$ can be read from (5.54)

$$f_\pi = \frac{F_0}{2} \left(1 - \frac{4}{3} \frac{\langle \pi^2 \rangle}{F_0^2} \right). \tag{5.55}$$

Using the result (5.48) and the fact that $F_0 = 2f_0$ as mentioned earlier, we can rewrite (5.51) and (5.55) in the following form

$$m_\pi^2 = m_0^2 \left(1 + \frac{1}{32\pi^2 f_0^2} m_0^2 \ln \frac{m_0^2}{\mu^2} \right),$$

$$f_\pi = f_0 \left(1 - \frac{1}{16\pi^2 f_0^2} m_0^2 \ln \frac{m_0^2}{\mu^2} \right). \tag{5.56}$$

These results were first derived by Langacker and Pagels using different arguments[9]. They are also in agreement with the results of Gasser and Leutwyler[2] as derived in the context of one-loop chiral perturbation theory. These chiral logarithms can be calculated systematically in the context of chiral perturbation theory as pioneered by Weinberg and further developed by Gasser and Leutwyler. Some of these matters will be discussed in Chapter 6.

5.4.3. *Chiral loops and the nucleon*

The chiral corrections to the nucleon mass are in general complicated. We do not know of any systematic evaluation of these corrections beyond the leading non-analytic term that we will discuss below. This prompts us to ask whether such an expansion is meaningful, and if so, to what extent the coefficients of the expansion are controlled by soft-pion physics. While we do not have clear answers to these questions, we will point out that since the pion emission (absorption) is soft, what matters in the estimation of the leading chiral correction is the coefficient of the leading infrared singularity, *i.e.*, the Born term, much like in the pion mass correction. Since the pion mass is linear in the current quark mass $m \equiv (m_u + m_d)/2$, we will denote the corrections by either m_π^2 or m.

- *m-correction*

To estimate the effect of the current quark mass on the nucleon mass, consider the QCD Hamiltonian in a physical gauge (say time-like axial-gauge)

$$\mathbf{H} = \int d^3x \left(\Theta_*^{00}(0, \vec{x}) + \Theta_A^{00}(0, \vec{x}) \right) \quad (5.57)$$

where Θ^{00} is the time-component of the energy-momentum tensor as discussed in Chapter 1. The conventional part of the energy momentum tensor reads,

$$\Theta_*^{00} = \frac{1}{2}(E^2 + B^2) + q^\dagger(-i\alpha \cdot \nabla)q + \frac{3}{4}\sum_f m_f \bar{q}_f q_f \quad (5.58)$$

while the anomalous part reads

$$\Theta_A^{00} = \frac{1}{4}\sum_f (1 + \gamma_f)m_f(\bar{q}_f q_f) + \frac{1}{4}\left(\frac{\beta}{2g}G^2\right) \quad (5.59)$$

and follows from (1.46). If we were to use nucleon states that are normalized to one ($\langle N|N \rangle$) for simplicity, then the nucleon mass is just given by $m_N = \langle N|\mathbf{H}|N \rangle$. The quark-gluon content of (5.59) follows from the decomposition (5.57)[10]. We can write

$$m_N = m_0 + \int d^3x \sum_{f=u,d,s} m_f(1 + \frac{1}{4}\gamma_f)\langle N|\bar{q}_f q_f(0, \vec{x})|N \rangle_0 + \mathcal{O}(m^{3/2}) \quad (5.60)$$

where the subscript 0 refers to the condensate in the chiral limit, and m_0 is the bare nucleon mass also in the chiral limit,

$$m_0 = \int d^3x \langle N|\frac{1}{2}(E^2 + B^2) + \frac{\beta(g)}{4}(E^2 - B^2) + q^\dagger(-i\alpha \cdot \nabla)q|N \rangle_0. \quad (5.61)$$

Only the light flavors have been retained in the expansion (5.60). The heavy flavor masses are kept unexpanded. All matrix elements are understood in the connected sense (subtraction of the vacuum value). The non-analytic contribution to (5.60) (*i.e.*, the last term) comes from the leading pion loop (see below). Higher pion loops are expected to generate chiral logarithms as discussed by Langacker and Pagels.

If we denote by

$$B_f = \int d^3x (1 + \frac{1}{4}\gamma_f) \langle p | \bar{q}_f q_f(0, \vec{x}) | p \rangle \qquad (5.62)$$

the matrix elements of $\bar{u}u$, $\bar{d}d$ and $\bar{s}s$ in the proton state, then for the proton and neutron, we expect

$$m_p = m_0 + m_u B_u + m_d B_d + m_s B_s + \mathcal{O}(m^{3/2}),$$
$$m_n = m_0 + m_u B_d + m_d B_u + m_s B_s + \mathcal{O}(m^{3/2}). \qquad (5.63)$$

Similar relations may be derived for the octet baryons. To the chiral order retained, these relations are linear in the current quark masses. Assuming that the corrections (non-analytic plus chiral logarithms) are small, the linear mass relations can be used to derive the Gell-Mann-Okubo formulae and the Coleman-Glashow relations.

- The $m^{3/2}$ *correction*

The one-pion loop contribution can be readily written down using the pion-nucleon pseudovector coupling and the Goldberger-Treiman relation from the nonlinear σ model. The result is a shift in the nucleon mass of the form

$$\Delta m_N(p) \sim \frac{3i}{2} \frac{g_A^2}{f_\pi^2} \int \frac{d^4q}{(2\pi)^4} \frac{i}{q^2 - m_\pi^2} \overline{N}(p) \left(\slashed{q}\gamma_5 \frac{i}{\slashed{p} - \slashed{q} - m_N} \slashed{q}\gamma_5 \right) N(p) \quad (5.64)$$

where p is the nucleon momentum. To the order we are interested in, the value the nucleon mass on the right-hand side of (5.64) is irrelevant. Since we are interested in the leading shift in the nucleon mass, we can take the latter to infinity and use the non-relativistic reduction. Inserting

$$\frac{1}{\slashed{p} - \slashed{q} - m_N} \to -\frac{1 + \gamma^0}{2} \frac{1}{q^0} \qquad (5.65)$$

into (5.64) gives to leading order a momentum independent mass

$$\Delta m_N \sim 3 \left(\frac{g_A}{f_\pi} \right)^2 \frac{1}{2} \int \frac{d^4q}{(2\pi)^4} \frac{1}{2q^0} \frac{\vec{q}^2}{q^2 - m_\pi^2}. \qquad (5.66)$$

This is precisely the classical energy of a heavy source of charge $g_A \tau^a$, with pseudovector coupling to a massive pion field. The factor $3 = \tau^a \tau^a$

follows from isospin average. Corrections to the non-relativistic limit are suppressed by powers of the nucleon mass. The integral in (5.66) can be undone by using the Feynman parametrization and Wick-rotation to Euclidean space. The result is

$$\Delta m_N = -\frac{3}{32\pi}\frac{g_A^2 m_\pi^3}{f_\pi^2} + \mathcal{O}(m_\pi^4 \ln m_\pi^2). \qquad (5.67)$$

The shift is attractive and vanishes in the chiral limit. It is non-analytic in the current quark mass, that is, $\Delta m_N \sim \mathcal{O}(m^{3/2})$. [†]

This result warrants two comments. First, the appearance of such a term through one-pion loop suggests that the inclusion of the pseudoscalar nonet will cause large corrections[12] since for instance $(m_K/m_\pi)^3 \sim 46$, suggesting that the expansion is bad in the strange sector. Second, the fact that the correction is not analytic in the current quark mass suggests that the scalar quantities in the baryonic sector, such as masses, sigma terms etc., are not analytic either.

5.4.4. *The sigma term*

A good example of a scalar quantity in hadronic processes is the so-called pion-nucleon sigma term. To define it, let us reconsider the expression for the off-shell pion-nucleon scattering amplitude as discussed in Chapter 4. Using LSZ and the PCAC relation eqs.(4.83)-(4.89), we have

$$T^{ab}(k_1^2, k_2^2, \nu, \nu_B) = \frac{1}{f_\pi^2 m_\pi^4}(k_1^2 - m_\pi^2)(k_2^2 - m_\pi^2)$$
$$i \int d^4x e^{ik_2 \cdot x} \langle N(p_2)|T\left(\partial^\alpha A_\alpha^b(x)\partial^\beta A_\beta^a(0)\right)|N(p_1)\rangle. \qquad (5.68)$$

Off-shell, $\nu_B = (t - k_1^2 - k_2^2)/4m_N$, distinct from the Mandelstam variable t. Pulling out the derivatives in (5.68) through the T product and using the equal-time commutator (modulo Schwinger terms discussed in Chapter 4)

$$i\delta(x^0)\left[A_0^a(x), \partial^\beta A_\beta^b(0)\right] = \delta^{ab}\delta^4(x)\hat{\sigma}(0), \qquad (5.69)$$

[†]In the large N_c limit, the Δ resonance must be treated on the same footing as the nucleon in eq.(5.67). With the Δ included[11], eq.(5.67) must be multiplied by a factor of 3.

we obtain

$$
\begin{aligned}
T^{ab}(k_1^2, k_2^2, \nu, \nu_B) = {} & \\
& + \frac{1}{f_\pi^2 m_\pi^4}(k_1^2 - m_\pi^2)(k_2^2 - m_\pi^2)\langle N(p_2)| - \hat{\sigma}(0)\delta^{ab}|N(p_1)\rangle \\
& + \frac{1}{f_\pi^2 m_\pi^4}(k_1^2 - m_\pi^2)(k_2^2 - m_\pi^2)\langle N(p_2)|i\epsilon^{abc}k_2 \cdot V^c(0)|N(p_1)\rangle \\
& + \frac{1}{f_\pi^2 m_\pi^4}(k_1^2 - m_\pi^2)(k_2^2 - m_\pi^2) \\
& \quad i\int d^4x\, e^{ik_2 \cdot x}\langle N(p_2)|T\left(k_2 \cdot A^b(x)k_1 \cdot A^a(0)\right)|N(p_1)\rangle.
\end{aligned} \qquad (5.70)
$$

This is to be contrasted with eq.(4.89). In the present case, however, the axial-vector current is not one-pion reduced. At the soft point, we have $k_1^2 = k_2^2 = 0$, $\nu = 0$ and $\nu_B = 0$. The second term in (5.70) vanishes by current conservation, and so does the third term since the pole is at m_π^2. Thus, the amplitude (5.70) simplifies to

$$
T^{ab}(0,0,0,0) = -\frac{\delta^{ab}}{f_\pi^2}\langle N(p)|\hat{\sigma}(0)|N(p)\rangle \equiv -\frac{\delta^{ab}}{f_\pi^2}\sigma_{\pi N} \qquad (5.71)
$$

which defines the pion-nucleon sigma term $\sigma_{\pi N}$. It follows from (5.68), after a volume integration, that

$$
\begin{aligned}
\hat{\sigma}(0) &= \sum_{a=1}^{3}\left[X^a, \sum_{f=u,d,s} m_f \bar{q}_f i\gamma_5 T^a q_f\right] \\
&= \frac{1}{3}\sum_{a=1}^{3}\left[X^a, \left[X^a, \sum_{f=u,d,s} m_f \bar{q}_f q_f\right]\right]
\end{aligned} \qquad (5.72)
$$

where we have denoted by X^a the total axial charge, and used the non-conservation of the total axial flavor current in QCD form. In terms of (5.72), $\sigma_{\pi N}$ is just a measure of the amount of chiral symmetry breaking in the nucleon state.

Cheng and Dashen[13] have suggested a possible way to determine $\sigma_{\pi N}$ using the pion-nucleon scattering amplitude in the physical region. They first noticed that at the Cheng-Dashen point, $k_1^2 = k_2^2 = m_\pi^2$ and $\nu = \nu_B = 0$ ($t = 2m_\pi^2$), the scattering amplitude is given by

$$
T^{ab}(m_\pi^2, m_\pi^2, 0, 0) = \frac{\delta^{ab}}{f_\pi^2}\Sigma_{\pi N}(t = 2m_\pi^2). \qquad (5.73)
$$

Since the scattering amplitude obeys Adler's condition

$$T^{ab}(m_\pi^2, 0, 0, 0) = T^{ab}(0, m_\pi^2, 0, 0) = 0, \tag{5.74}$$

it follows through Taylor expansions that

$$\Sigma_{\pi N}(t = 2m_\pi^2) = \sigma_{\pi N} + \mathcal{O}(m^{3/2}). \tag{5.75}$$

In the chiral limit, $\Sigma_{\pi N}(0) = \sigma_{\pi N} \to 0$. We note that the leading current mass correction is non-analytic, much like in the nucleon mass case. Relying on the analyticity of the amplitude in the ν plane for fixed ν_B, and using dispersion relation, Hohler and others [14] were able to relate the scattering amplitude at the Cheng-Dashen point to the scattering amplitude in the physical region, thereby obtaining an empirical estimate for the $\sigma_{\pi N}$ term. Experimentally, it was found that $\Sigma_{\pi N} = 64 \pm 8$ MeV.

The leading non-analytical correction in (5.75) can be evaluated using the same arguments as those given above for the nucleon mass shift. This calculation was done by Pagels and Pardee[15] and the result is

$$\Sigma_{\pi N} = \sigma_{\pi N} + \frac{3g_A^2}{64\pi f_\pi^2} m_\pi^3 + \mathcal{O}(m_\pi^4 \ln m_\pi^2). \tag{5.76}$$

Assuming that the logarithmic corrections are small and using the central value at the Cheng-Dashen point for $\Sigma_{\pi N}$ with $g_A = 1.26$, one obtains $\sigma_{\pi N} \approx 58$ MeV.

The exact value of the pion-nucleon sigma term has been the object of long debates and controversies. Some of the reasons lie in the fact that the soft point is away from the Cheng-Dashen point, requiring the use of chiral perturbation theory and dispersion relations for an empirical fit, each of which causes an uncertainty. A recent analysis by Gasser, Leutwyler and Sainio[16] in the context of chiral perturbation theory plus several assumptions gives a lower value for the pion-nucleon sigma term, $\sigma_{\pi N} \approx 45$ MeV. An alternative approach to this quantity that relies only on on-shell information will be briefly discussed in Chapter 6.

5.5. Weinberg "Theorem"

In 1979, Weinberg[1] advanced the idea that effective chiral Lagrangians actually encompass the limited framework of current algebra and soft-pion physics. He put forward a proposition which he held as a self-evident –

though unproven – "theorem." Specifically, he postulated that the most general effective Lagrangian, including *all* terms consistent with analyticity, unitarity, cluster decomposition and symmetry, yields in a given order in perturbation theory the most general S-matrix elements consistent with perturbative unitarity, cluster decomposition and symmetry. This argument has been recently put on a more rigorous basis by Leutwyler[17].

Weinberg's strategy amounts to doing current algebra *without* current algebra. Clearly, this proposition would be useful only if a systematic method of calculation could be devised such that only a limited part of this general effective Lagrangian is to be used. Weinberg argued that for pion processes at low energy, a computational strategy could be devised based on chiral counting. This argument has more recently been extended to the case where matter fields, such as baryons, are introduced. (See Chapter 10.)

5.5.1. *Chiral counting*

To facilitate the rules for the chiral counting procedure, let us redefine the pion field in the nonlinear σ model to be dimensionless. This is achieved by redefining through $\pi/F_0 \to \pi$. In this form, the covariant derivative (5.24) in Weinberg's parametrization simplifies to

$$\nabla_\mu \pi^a = \frac{1}{1+\pi^2} \partial_\mu \pi^a. \qquad (5.77)$$

In terms of (5.77), Weinberg's master effective chiral Lagrangian for massless pions reads

$$\mathcal{L} = \frac{g_2}{2} \nabla_\mu \vec{\pi} \cdot \nabla_\mu \vec{\pi} + \frac{g_4^1}{4} \left(\nabla_\mu \vec{\pi} \cdot \nabla_\mu \vec{\pi}\right)^2$$
$$+ \frac{g_4^2}{4} \left(\nabla_\mu \vec{\pi} \cdot \nabla_\nu \vec{\pi}\right) \left(\nabla^\mu \vec{\pi} \cdot \nabla^\nu \vec{\pi}\right) + \frac{g_4^3}{4} \left(\nabla_\mu \nabla^\mu \vec{\pi}\right)^2 + \cdots. \qquad (5.78)$$

The couplings g_D^n carry dimensionality $4 - D$ with $g_2 = F_0^2$. The ellipsis stands for infinitely many terms. The ultraviolet divergences inherent to a perturbative treatment using (5.78) are to be absorbed in the infinitely many couplings at some renormalization scale μ to be specified. Because of the redefinition of the pion field, each pion propagator carries an additional $1/g_2$ factor and each external leg an additional factor of $1/\sqrt{g_2}$.

Now, consider a process with N_E external pions carrying energy of the order of E. Following the redefinition of the pion field the corresponding T

matrix carries dimension $D_1 = 4 - N_E$. The coupling constants multiplying the corresponding T-matrix carry dimensions

$$D_2 = \sum_D N_D \, (4 - D) - 2N_I - N_E$$

where N_D is the number of vertices of species g_D and N_I the number of internal pion legs, $-2N_I$ the contribution of $1/g_2$ brought by each propagator and $-N_E$ the contribution of $1/\sqrt{g_2}$ brought by each external line. It follows that the T-matrix element for this process should carry, after renormalization, an energy dependence E^{D_*},

$$T \sim g_2^{D_2} \, E^{D_*} \, f(\frac{E}{\mu}) \tag{5.79}$$

with

$$D_* = D_1 - D_2 = 4 + \sum_D N_D(D - 4) + 2N_I. \tag{5.80}$$

Since for a given topology with N_L loops,

$$N_L = N_I - \sum_D N_D + 1, \tag{5.81}$$

it follows that

$$D_* = 2 + \sum_D N_D(D - 2) + 2N_L. \tag{5.82}$$

At low energy (how low is the question that cannot always be answered), the dominant contribution to the S-matrix ($S - 1 \sim \delta^4 \, T$) is given by tree graphs ($N_L = 0$) and terms with the least number of derivatives ($D = 2$) so that $D_* = 2$. This is

$$\mathcal{L}_2 = \frac{g_2}{2} \nabla_\mu \vec{\pi} \cdot \nabla_\mu \vec{\pi} \tag{5.83}$$

which is just the minimal version of the nonlinear σ model for pions. This result says that the leading term of order E^2 in the T-matrix of a process with N_E external pions follows from (5.83) *only* at *tree* level. This is the current-algebra result. The power counting formula (5.82) suggests that the terms of order E^3, E^4, etc. could be systematically calculated by a proper combination of loops and tree diagrams using a restricted number of derivative terms of (5.78), thus going beyond current algebra. This observation is central in the effective chiral Lagrangian approach to pion physics. We proceed to illustrate its application to $\pi\pi$-scattering at low energy and in the chiral limit.

5.5.2. $\pi\pi$ *scattering*

Consider $\pi\pi$ scattering in the chiral limit. For definiteness, we choose the Mandelstam variables $s, t, u \sim E^2$ with $s + t + u = 0$ in the chiral limit. We normalize the S-matrix as

$$S - 1 = i(2\pi)^4 \delta^4(P_A + P_B - P_C - P_D)\frac{T_{ABCD}}{(2\pi)^6\sqrt{16E_A E_B E_C E_D}}. \quad (5.84)$$

The chiral counting given above suggests that the amplitude T should be organized in powers of a characteristic energy E. Let $E = \sqrt{s}$ be the center-of-mass energy for fixed angles and isospin indices, thus

$$T \sim A_2 E^2 + A_4 E^4 + A_6 E^6 + \cdots. \quad (5.85)$$

The contribution to A_2 comes solely from the tree diagrams associated with the term with $D = 2$ and/or two derivatives in (5.78) at tree level. The corresponding tree diagrams are shown in Fig.5.6, and their contribution is

$$T^2_{ABCD} = \frac{4}{g_2}\left(\delta_{AB}\delta_{CD}\ s + \delta_{AC}\delta_{BD}\ t + \delta_{AD}\delta_{BC}\ u\right). \quad (5.86)$$

This is the result derived by Weinberg[18] using current algebra.

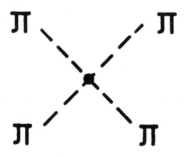

Fig. 5.6. Tree graphs in $\pi\pi$ scattering to order E^2.

The contribution to A_4 in (5.85) comes from graphs with $D = 2$ to *one loop* and the graphs with $D = 4$ at tree level, as shown in Fig.5.7. The one-loop graphs have a typical contribution of the form

$$\text{one loop} \sim \left(\frac{1}{\sqrt{g_2}}\right)^4 \cdot \frac{1}{g_2^2} \cdot g_2^2 \sim \frac{1}{g_2^2} \tag{5.87}$$

for external lines, propagators and vertices, respectively. The tree graphs have a typical contribution of the form

$$\text{tree} \sim \left(\frac{1}{\sqrt{g_2}}\right)^4 \cdot g_4^n \sim \frac{g_4^n}{g_2^2} \tag{5.88}$$

for external lines and vertices respectively. The contribution of the one-loop graphs is divergent. Using dimensional regularization, the result is

$$T^2_{ABCD} = \frac{\delta_{AB}\delta_{CD}}{g_2^2}\left(-\frac{1}{2\pi^2}s^2\ln(-s) - \frac{1}{12\pi^2}(u^2 - s^2 + 3t^2)\ln(-t)\right.$$

$$-\frac{1}{12\pi^2}(t^2 - s^2 + 3u^2)\ln(-u) + \frac{1}{3\pi^2}(s^2 + t^2 + u^2)(\frac{1}{\epsilon} + \ln(\mu^2))$$

$$\left. -\frac{1}{2}g_4^1 s^2 - \frac{1}{4}g_4^2(t^2 + u^2)\right) + \text{crossed terms.} \tag{5.89}$$

Since the pions are on-shell, the contribution of g_4^3 at tree level drops out.

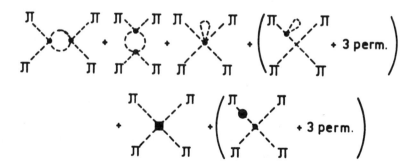

Fig. 5.7. One-loop contributions to $\pi\pi$ scattering to order E^4.

The divergences appearing in (5.89) may be reabsorbed into the $D = 4$ couplings, through

$$g_4^1(\mu) = g_4^1 - \frac{2}{3\pi^2}(\frac{1}{\epsilon} + \ln(\mu^2)),$$

$$g_4^2(\mu) = g_4^2 - \frac{4}{3\pi^2}(\frac{1}{\epsilon} + \ln(\mu^2)), \tag{5.90}$$

leading to a "renormalized" amplitude

$$T_{ABCD}^2 = \frac{\delta_{AB}\delta_{CD}}{g_2^2}\left(-\frac{1}{2\pi^2}s^2\ln(\frac{-s}{\mu^2}) - \frac{1}{12\pi^2}(u^2 - s^2 + 3t^2)\ln(\frac{-t}{\mu^2})\right.$$
$$\left. - \frac{1}{12\pi^2}(t^2 - s^2 + 3u^2)\ln(\frac{-u}{\mu^2}) - \frac{1}{2}g_4^1(\mu^2)s^2 - \frac{1}{4}g_4^2(\mu^2)(t^2 + u^2)\right)$$

+crossed terms. $\qquad(5.91)$

All logarithmic terms are fixed by g_2. This is natural since they reflect on the two-pion cuts in Fig.5.7. The polynomial terms are not fixed in the scattering amplitude. They also show up in the general context of chiral perturbation theory in other processes and provide constraints through relationships among low-energy amplitudes. This point will be discussed in Chapter 6.

5.5.3. *Renormalization of nonrenormalizable theories*

The master effective chiral Lagrangian (5.78) is nonrenormalizable in the conventional sense (that is, à la Dyson): the infinities triggered by the perturbative analysis would proliferate with the number of loops and not close.

For a truly fundamental theory, this would pose a serious obstacle since it would mean loss of genuine predictive power. However, as generally believed nowadays, all phenomena accessible by controlled laboratory experiments at present or in the foreseeable future are described by effective theories, with a possible fundamental theory being relegated to the Planck scale or higher. In fact, up to date, there is no such fundamental theory known.[‡] For the chiral effective Lagrangian (5.78) concerned with a very low-energy scale, this means that low-energy hadronic data are required order by order (in the chiral counting) from a set of given processes to fix the parameters of the effective theory. The information obtained from one set of experiments, however, can carry over to other processes through symmetry relations dictated by QCD. The organizational principle provided by the diagrammatic analysis and chiral counting yields relationships among

---

[‡]Even the notion of renormalizability has been undergoing a major revision. See Gomis and Weinberg[19] for a modern concept of renormalizability of gauge theories.

various processes, thus the possibility of making predictions for different observables.

To illustrate this strategy, consider an *arbitrary* soft-pion process with N_E external lines, as a function of $E = \sqrt{s}$, the center of mass energy for fixed angles and isospin. Generically,

$$T(E) \sim \frac{g_2}{(g_2)^{N_E/2}} \left(A_2 E^2 + \frac{1}{g_2} E^4 \left(F(\frac{E}{\mu}) + \sum_n A_4^n g_4^n(\frac{\mu}{\mu_0}) \right) + \cdots \right) . \quad (5.92)$$

The term of order E^2 is the tree contribution from $D = 2$. The term of order E^4 is the one-loop contribution from $D = 2$ (*i.e.*, the F term) and the tree contribution from $D = 4$. The term of order E^6 is the two-loop contribution from $D = 2$, the one-loop contribution from $D = 4$ and the tree contribution from $D = 6$ and mixed trees from $D = 4$ and so on. Note that (5.92) applies to all processes with N_E soft pions, not just $\pi\pi$. This is the generalized version of (5.85). The running couplings are *common* to all these processes. The fixed coefficients A_i^n and the functions F are process-dependent. Contrary to a conventional renormalizable theory, we see the proliferation of the running couplings.

The way in which the couplings g_i^n run is dictated by a renormalization group equation. Indeed, (5.92) should be independent of the choice of the arbitrary renormalization μ. In varying μ, the couplings and the functions vary so as to leave T invariant, *whatever* E. To order E^4,

$$\frac{E}{\mu} F'(\frac{E}{\mu}) = \sum_n A_4^n \, \mu g_4'^n(\frac{\mu}{\mu_0}). \quad (5.93)$$

This equation admits a solution if the two sides are μ-independent. Setting $\mu g_4'^n(\mu/\mu_0) = B_4^n$ yields

$$g_4^n(\frac{\mu}{\mu_0}) = B_4^n \ln(\frac{\mu}{\mu_0}) \quad (5.94)$$

and

$$F(\frac{E}{\mu}) - F(1) = \left(\sum_n A_4^n B_4^n \right) \ln(\frac{E}{\mu}). \quad (5.95)$$

From $\pi\pi$-scattering, it follows that $B_4^2 = 2B_4^1 = 4/3\pi^2$. Thus the parameters A_4^n calculable from tree amplitudes and the running couplings

$g_4^n(\mu/\mu_0)$ fixed by $\pi\pi$-scattering determine uniquely the E^4 dependence of the S-matrix in *all* pion processes, modulo an overall normalization $F(1)$. A rerun of this argument for E^6 yields a hierarchy of relationships among the other functions[1].

By being able to extract *finite relationships*, we have in a way tamed (renormalized) an *a priori* untamable (nonrenormalizable) formulation. Also, since the dependence $\ln(E/\mu)$ in F reflects the two-pion cut, as expected from perturbative unitarity, the expansion in E in $\pi\pi$-scattering – and thus *all* soft pion processes by induction – requires amendments beyond the ρ-meson threshold. In a way, this puts a limit on the range of applicability of *all* chiral expansions with only pions. But this is what characterizes effective theories.

5.5.4. *Chiral symmetry breaking*

Up to this point, we have ignored the pion mass to keep the counting rules and the overall discussion simple. What are the modifications brought about to (5.78) by a finite pion mass? To answer this question, recall that in QCD, chiral $SU(2) \times SU(2)$ is softly broken by current quark masses. Generically

$$
\begin{aligned}
\mathcal{L}_B &= m_u \bar{u}u + m_d \bar{d}d \\
&= \frac{1}{2}(m_u - m_d)(\bar{u}u - \bar{d}d) + \frac{1}{2}(m_u + m_d)(\bar{u}u + \bar{d}d).
\end{aligned} \tag{5.96}
$$

The first term transforms as the third component (V_3) of a four-vector under $SU(2) \times SU(2)$ and the second term as the fourth component (V_4) of a four-vector. The corresponding lowest-order chiral effective terms are

$$
(V_3, V_4) = (\frac{\pi_3}{1 + \pi^2}, -\frac{1}{2} + \frac{\pi^2}{1 + \pi^2}). \tag{5.97}
$$

Thus a possible chiral breaking term in Weinberg's master Lagrangian (5.78) which is isospin symmetric reads

$$
\mathcal{L}_B^{0,1} = -\frac{1}{2}g_2 m_0^2 \frac{\pi^2}{1 + \pi^2} \tag{5.98}
$$

which is the term quoted in the minimal version of the nonlinear σ model (5.25). Since $m_\pi^2 \sim (m_u + m_d)$, this is the lowest-order term in the general expression

$$
\mathcal{L}_B = \sum_{K,L} \mathcal{L}_B^{K,L} \sim \sum_{K,L} \mathcal{L}_{K,L}(\pi)(m_u - m_d)^K (m_u + m_d)^L. \tag{5.99}
$$

Near threshold, $E \sim m_\pi$. Thus, expanding in E also means expanding in m_π (modulo chiral logarithms). Adding (5.99) to (5.78) requires a change in the chiral counting rules. Thus, the leading term in E and m_π in the S-matrix carries a power

$$D_{**} = 2 + 2N_L + \sum_{D,K,L} N_{D,K,L}(D - 2 + 2K + 2L) \qquad (5.100)$$

for an arbitrary vertex with D derivatives, L powers of $(m_d + m_u)$ and K powers of $(m_d - m_u)$. This generalizes (5.82) away from the chiral limit. Again, the order E^2 term in the S-matrix corresponds to $(D, K, L) = (2, 0, 0)$ and $(0, 0, 1)$, in agreement with the minimal version of the nonlinear σ model and thus current algebra.

Notice that the term (5.98) is the unique term in all possible chiral breaking terms as summarized by (5.99) with a quadratic pion mass term, *i.e.*

$$\mathcal{L}_B^{0,1} = -\frac{m_0^2 g_2}{2}\pi^2 + \frac{m_0^2 g_2}{2}\pi^4 + \cdots. \qquad (5.101)$$

In a given process, the insertion of this term is not preempted by topology. It can be inserted in any diagram without changing the overall power behavior. Phrased more simply, this term should be summed to all orders by using massive pion propagators

$$\frac{1}{g_2}\frac{1}{q^2} \rightarrow \frac{1}{g_2}\frac{1}{q^2 - m_0^2}. \qquad (5.102)$$

Thus, in addition to powers of E and m_π, one expects branch points in E and m_0 as expected from perturbative unitarity. These are the chiral logarithms discussed earlier. They will considerably enhance the chiral effects at threshold and dominate over the power corrections.

References

1. S. Weinberg, Physica (Amsterdam) **96A**, 327 (1979).
2. J. Gasser and H. Leutwyler, Ann. Phys. **158**, 142 (1984).
3. S. Weinberg, Phys. Rev. **166**, 1568 (1968) and in *Lectures on Elementary Particles and Quantum Field Theory* Vol 1, eds, S. Deser, M. Grisaru and H. Pendleton (M.I.T. Press, Cambridge, MA, 1970) p.283.
4. H. Georgi, *Weak Interactions and Modern Particle Physics* (The Benjamin/Cummings Pub. Co., Menlo Park, CA., 1984).

5. J. Honerkamp, Nucl. Phys. **36**, 130 (1972).
6. R. Dashen and M. Weinstein, Phys. Rev. **183**, 1291 (1969).
7. L. Li and H. Pagels, Phys. Rev. Lett. **26**, 1204 (1971).
8. P. Carruthers and R.W. Haymaker, Phys. Rev. Lett. **27**, 455 (1971).
9. P. Langacker and H. Pagels, Phys. Rev. **D8**, 4595 (1973).
10. X. Ji, Phys. Rev. **D52**, 271 (1995).
11. E. Jenkins, Nucl. Phys. **B368**, 190 (1992).
12. E. Jenkins and A.V. Manohar, Phys. Lett. **B281**, 336 (1992).
13. T.P. Cheng and R. Dashen, Phys. Rev. Lett. **26**, 594 (1971).
14. G. Hohler, Nucl. Phys. **A508**, 525c (1990); R. Koch, Z. Phys. **C15**, 161 (1982).
15. H. Pagels and W.J. Pardee, Phys. Rev. **D4**, 3335 (1971).
16. J. Gasser, H. Leutwyler and M.E. Sainio, Phys. Lett. **B253**, 252 (1991).
17. H. Leutwyler, Ann. Phys. (N.Y.) **235**, 165 (1994); see also E. D'Hoker and S. Weinberg, Phys. Rev. **D50**, 6050 (1994).
18. S. Weinberg, Phys. Rev. Lett. **17**, 616 (1966).
19. J. Gomis and S. Weinberg, "Are nonrenormalizable gauge theories renormalizable?" hep-th/9510087.

CHAPTER 6

QCD EFFECTIVE ACTION

6.1. Introduction

In this chapter, we discuss the formal aspects of the material covered in the preceding chapters and consider in general terms the approach to low-energy nonperturbative QCD in effective field theories.

In 1984, Gasser and Leutwyler[1] proposed to extend Weinberg's program[2] for the S-matrix to the QCD correlation functions. Their argument relied on the fact that the low-energy effective Lagrangian could be sought as a solution of the chiral Ward identities around the chiral limit. The aim of the approach is to construct an effective action which is as close as possible to QCD at low energy.[3]

After reviewing the basic construction of the theory and some problems inherent to the definition of the effective action in field theories with an underlying geometrical structure, we proceed to discuss the basic symmetries of QCD required for an effective action formulation at low energy. The concept of anomalies will be discussed more precisely than in preceding chapters and the pertinent Ward identities will be derived. Their resolution will be sought using the chiral counting method devised by Weinberg[2]. We will explicitly construct the one-loop effective action and use it to derive the one-loop corrections to the pion mass, decay constant and scattering amplitudes. We will end this chapter with an alternative formulation, where the chiral Ward identities are exploited exactly and on-shell [4], with a novel result for $\pi\pi$ scattering. Although brief, this part shows that the constraints of chiral symmetry *can* go far beyond the threshold region.

6.2. Elements of Effective Action

In a field theory, perturbative Feynman techniques allow us to construct the effective action, which in turn contains all the information about the quantized theory. From the effective action follow the S-matrix, the quantum mean-field equations, as well as all the Green's functions. Let $S(\phi)$ be a classical non-gauge action of the field ϕ^a. The naive definition of the effective action $\Gamma(\Phi)$ for a given *quantum* field Φ is given by the Legendre transform of the connected generating functional $W(J)$,

$$\Gamma(\Phi) = W(J) - \Phi^a J_a \tag{6.1}$$

with $W(J)$ defined by

$$e^{\frac{i}{\hbar}W(J)} = \int [d\phi]\sqrt{\det g}\ e^{\frac{i}{\hbar}(S(\phi)+\phi^a J_a)} \tag{6.2}$$

for an external source J^a. The measure in (6.2) is the geodesic length in the manifold associated to the (classical) fields ϕ^a with metric g^{ab}. The latter can be read directly from the kinetic part in the action

$$S(\phi) = \int d^4x \left(\frac{1}{2}g_{ab}\partial_\mu\phi^a\partial^\mu\phi^b + \cdots\right). \tag{6.3}$$

It follows from (6.1) that

$$\frac{\partial\Gamma(\Phi)}{\partial\Phi^a} = -J_a. \tag{6.4}$$

For $J = 0$, this is the familiar quantum mean-field equation. In terms of (6.2) and (6.4) the expression for the effective action becomes

$$e^{\frac{i}{\hbar}\Gamma(\Phi)} = \int [d\phi]\sqrt{\det g}\ e^{\frac{i}{\hbar}(S(\phi)+(\Phi^a-\phi^a)\frac{\partial\Gamma(\Phi)}{\partial\Phi^a})}. \tag{6.5}$$

This equation can be solved by expanding in powers of $\hbar$. The result is

$$\Gamma(\Phi) = S(\Phi) + \frac{\hbar}{i}\ln g(\Phi) - \frac{\hbar}{2i}\operatorname{Tr}\ln\left(\frac{\partial^2 S(\Phi)}{\partial\Phi^a\partial\Phi^b}\right) + \mathcal{O}(\hbar^2). \tag{6.6}$$

The above discussion is standard and can be found in all textbooks. However there is a subtle point which could be a source of ambiguity associated with the coupling of the field ϕ^a in (6.2) to the external source.

Indeed, we could have used $f(\phi) \cdot J$, where $f^a(\phi)$ is an arbitrary function, instead of $\phi \cdot J$. Along the same arguments as above, one obtains

$$\begin{aligned}
\Gamma_*(\Phi) \;=\;\; & S(\Phi) + \frac{\hbar}{i}\ln g(\Phi) \\
& -\frac{\hbar}{2i}\,\mathrm{Tr}\,\ln\left(\frac{\partial^2 S(\Phi)}{\partial\Phi^a\partial\Phi^b} + \frac{\partial f^c}{\partial\Phi^a}\frac{\partial f^d}{\partial\Phi^b}\frac{\partial^2\Phi^e}{\partial f^c\partial f^d}\frac{\partial S}{\partial\Phi^e}\right) \\
& +\mathcal{O}(\hbar^2)
\end{aligned}$$

(6.7)

which differs from (6.6) in the third term. Thus the definition of the effective action *in the way formulated above* is *not unique*, as it depends on the way one chooses to parametrize the quantized fields. Note, however, that on-shell, the parametrization dependence drops out (at least) to one loop. Thus the S-matrix elements are unaffected by this ambiguity. The standard argument is that to one-loop, one can substitute $\partial S/\partial\Phi^a \to \partial\Gamma/\partial\Phi^a$, so that at the extremum there is no difference between Γ and Γ_*. The caveat to this argument, however, is that the definition of the mean field and the Green's functions do not involve Γ or Γ_*, but their *derivatives*. The derivatives are affected by a general redefinition of the source term even at the extremal point, so (6.2) is unsatisfactory.

A way out of this problem was suggested by Vilkovisky[5] who noted that the effective action should be constructed so that it is invariant under arbitrary (local) redefinitions of the classical fields. The idea is to recognize that $\phi \cdot J$ is not covariant under these reparametrizations and to replace it by $\sigma(\phi,\Phi) \cdot J$, where

$$-\sigma^a(\phi,\Phi) = (\phi - \Phi)^a + \frac{1}{2}\Gamma^a_{bc}(\phi-\Phi)^b(\phi-\Phi)^c + \cdots$$

(6.8)

with Γ^a_{bc} the connection in configuration space. When a proper metric is identified, say, g_{ab}, the connection is simply the Christoffel symbol,

$$\Gamma^a_{bc} = \frac{1}{2}g^{ad}\left(\partial_b g_{dc} + \partial_c g_{db} - \partial_d g_{bc}\right).$$

(6.9)

We note that $\sigma^a(\Phi,\phi)$ is the tangent to the geodesic going through Φ and ϕ from ϕ to Φ. It transforms as a vector with respect to the quantum field Φ and a scalar with respect to the classical field ϕ. In terms of (6.8) the effective action to one loop becomes

$$\Gamma(\Phi) = S(\Phi) + \frac{\hbar}{i}\ln g(\Phi) - \frac{\hbar}{2i}\,\mathrm{Tr}\ln\left(\nabla_a\nabla_b S(\Phi)\right) + \mathcal{O}(\hbar^2)$$

(6.10)

where we have converted the usual derivatives to covariant derivatives,

$$\nabla_a \nabla_b S(\Phi) = \frac{\partial^2 S(\Phi)}{\partial \Phi^a \partial \Phi^b} - \Gamma^c_{ab}(\Phi) \frac{\partial S(\Phi)}{\partial \Phi^c}. \tag{6.11}$$

Under local reparametrizations, the shift in the one-loop part is compensated by the shift in the measure. This argument can be generalized to all orders. The invariant effective action proposed by Vilkovisky reads

$$e^{\frac{i}{\hbar}\Gamma(\Phi)} = \int [d\phi] \sqrt{\det g} \; e^{\frac{i}{\hbar}\left(S(\phi) + \sigma^a(\Phi,\phi)\frac{\partial \Gamma(\Phi)}{\partial \Phi^a}\right)}. \tag{6.12}$$

The integrand in (6.12), including the measure, is a scalar under reparametrizations of both the classical and quantum fields. We note that for $\Gamma^a_{bc} = 0$, (6.12) reduces to (6.5). The modifications in the source are of higher order in the field and do not affect the S-matrix.

6.3. Role of the Effective Action

The QCD effective action we will seek below will not be in terms of underlying quarks and gluon degrees of freedom, but instead in terms of pseudoscalar bosons. In the framework of chiral perturbation theory, the latter is equivalent to the former, modulo some reservations to be discussed below. Green's functions can be seen to follow from $W(J)$, while S-matrix elements follow from $\Gamma(\phi)$ or equivalently from LSZ-reduced Green's functions. The only limitation in the substitution (quark-gluon fields for pseudoscalar fields) resides in the fact that the pseudoscalar effective action involves unknown coefficients that are fixed by experiments only. Thus, the above ambiguities do not affect S-matrix elements, but may affect Green's functions. They may also occur in the discussions of vacuum averages such as $\langle \bar{q}q \rangle$, and mean field treatments such as the definition of average pion fields in matter. Some issues related to the foundations of chiral perturbation theory have recently been discussed by Leutwyler[3].

In principle, one may think about the coefficients in the effective action as resulting from a field decomposition of the original QCD fields into soft but confined modes, and hard but unconfined modes. The effective action following from a suitable integration over the hard modes, will display coefficients that are all related to the QCD scale Λ. This procedure, which was first suggested by Wilson and recently discussed by Polchinski[6], may in practice be difficult to implement for a theory like QCD. However, it is not totally inappropriate to think about the effective action to be discussed below in this sense. What should be kept in mind is that QCD may induce dynamical constraints between the various coefficients of the effective action that cannot simply follow from symmetry arguments and general principles.

6.3.1. External gauge symmetries

Consider the QCD Lagrangian coupled with matrix-valued external vector $(\mathcal{V}_\mu)$, axial-vector $(\mathcal{A}_\mu)$, scalar $(\mathcal{S})$ and pseudoscalar $(\mathcal{P})$ fields,

$$\mathcal{L}[\mathcal{V}, \mathcal{A}, \mathcal{S}, \mathcal{P}] = \mathcal{L}_{QCD} + \bar{q}\gamma^\mu i(\mathcal{V}_\mu + \gamma_5 \mathcal{A}_\mu)q - \bar{q}(\mathcal{S} - i\gamma_5 \mathcal{P})q. \quad (6.13)$$

The quark mass term that explicitly breaks the chiral symmetry of QCD is included when the scalar field takes its mean-field value as $\mathcal{S} \sim \mathrm{diag}(m_u, m_d, m_s)$. The Lagrangian (6.13) displays a *local* $SU_L(3) \times SU_R(3)$ symmetry at the classical level

$$q \rightarrow g_R q_R + g_L q_L,$$
$$(\mathcal{S} + i\mathcal{P}) \rightarrow g_R(\mathcal{S} + i\mathcal{P})g_L^\dagger,$$
$$(\mathcal{V} + \mathcal{A}) \rightarrow g_R(\mathcal{V} + \mathcal{A})g_R^\dagger + g_R \partial g_R^\dagger,$$
$$(\mathcal{V} - \mathcal{A}) \rightarrow g_L(\mathcal{V} - \mathcal{A})g_L^\dagger + g_L \partial g_L^\dagger \quad (6.14)$$

with $g_{R,L}$ elements in local $SU_{R,L}(3)$. The QCD generating functional $W[\mathcal{V}, \mathcal{A}, \mathcal{S}, \mathcal{P}]$ can be expressed in the form[*]

$$e^{iW[\mathcal{V}, \mathcal{A}, \mathcal{S}, \mathcal{P}]} = \langle 0_+|0_-\rangle_{\mathcal{V}, \mathcal{A}, \mathcal{S}, \mathcal{P}} = \int d[QCD] \; e^{i\int d^4 x \mathcal{L}[\mathcal{V}, \mathcal{A}, \mathcal{S}, \mathcal{P}]} \quad (6.15)$$

where d[QCD] denotes the full QCD measure. The QCD effective action Γ follows from $W[\mathcal{V}, \mathcal{A}, \mathcal{S}, \mathcal{P}]$ by taking the appropriate Legendre transform. Even though $\mathcal{L}$ is invariant under local $SU_L(3) \times SU_R(3)$, W is not. The reason is that the fermion determinant following from (6.15) after integration over the quark fields cannot be regularized in a way that preserves *both* vector and axial-vector symmetries. This phenomenon is generic to chiral gauge theories, as discussed in Chapter 1 and reviewed extensively[7]. Here, it gives rise to the nonabelian chiral anomaly.

---

[*]Throughout we will work with zero vacuum angle $\theta = 0$ (see Chapter 1) and set $\hbar = 1$.

6.3.2. *Nonabelian chiral anomaly*

As discussed in Chapter 4, the concept of anomalies is central in quantum field theories with fermions. The most dramatic manifestation of the anomaly is the radiative decay $\pi^0 \rightarrow \gamma\gamma$ which was discussed in detail in Chapter 4. The physical mechanism behind the emergence of anomalies in field theory was explained in Chapter 1.

Here we would like to discuss chiral anomalies in more detail. We begin by recapitulating the distinction between the abelian anomaly and the nonabelian chiral anomaly.

We consider quantum action (generating functional)

$$e^{iW[A]} = \int [d\psi][d\bar\psi]\, e^{i \int d^4x \mathcal{L}}, \qquad (6.16)$$

where $\mathcal{L} = \bar\psi i\gamma^\mu(\partial_\mu + A_\mu)\psi$ is the Dirac Lagrangian for fermions in D dimensions in the *external* general vector field $A_\mu = \mathcal{V}_\mu + \gamma_{D+1}\mathcal{A}_\mu$. The γ_{D+1} in Minkowski space is defined as $\gamma_{D+1} = -i^{D/2} \prod_0^{D-1} \gamma_\mu$. It is the D-dimensional generalization of γ_5. The vector field is matrix-valued, *i.e.* $A_\mu = -iA_\mu^a T^a$, where T^a are generators of $SU(N_f)$.

The *abelian* (or $U_A(1)$) anomaly stems from the noninvariance of the fermionic path integral measure under the local chiral transformations[†]

$$\psi(x) \rightarrow e^{i\theta(x)\gamma_{D+1}}\psi(x)$$
$$\bar\psi(x) \rightarrow \bar\psi(x)e^{i\theta(x)\gamma_{D+1}}. \qquad (6.17)$$

Quantum action is invariant under this change of variables, *i.e.*

$$\left.\frac{\delta W[A]}{\delta\theta(x)}\right|_{\theta=0} = 0. \qquad (6.18)$$

The above equation generates an anomaly to the classically conserved axial current $J_\mu^A = \bar\psi\gamma_\mu\gamma_{D+1}\psi$

$$\partial^\mu J_\mu^A = -\frac{1}{16\pi^2}\epsilon^{\mu\nu\alpha\beta}\,\text{Tr}(F_{\mu\nu}F_{\alpha\beta})$$
$$= -\frac{1}{4\pi^2}\,\text{Tr}(\partial^\mu\epsilon_{\mu\nu\alpha\beta}(A^\nu\partial^\alpha A^\beta + \frac{2}{3}A^\nu A^\alpha A^\beta)). \qquad (6.19)$$

The *nonabelian* (or $SU_A(N)$ in our case) anomaly stems from the non-invariance of the quantum action W under nonabelian transformations of

---

[†]A formal presentation based on the Fujikawa method [9] is given in Appendix A_I.

the fermionic fields

$$\psi(x) \to e^{i\theta_a(x)\gamma_{D+1}T^a}\psi(x),$$
$$\bar{\psi}(x) \to \bar{\psi}(x)e^{i\theta_a(x)\gamma_{D+1}T^a} \tag{6.20}$$

and the simultaneous transformation of the background field

$$A_\mu \to A_\mu^\theta = e^{i\theta_a(x)\gamma_{D+1}T^a}(\partial_\mu + A_\mu)e^{-i\theta_a(x)\gamma_{D+1}T^a}, \tag{6.21}$$

so

$$W(A_\mu^\theta) - W(A_\mu) \neq 0. \tag{6.22}$$

The nonabelian anomaly G_i is defined as the infinitesimal change of the above difference

$$\delta_{\theta_i} W[A_\mu^\theta]\big|_{\theta_i=0} = G_i(x). \tag{6.23}$$

In terms of the axial nonabelian currents $J_{\mu,i}^A = \bar{\psi}\gamma_\mu\gamma_{D+1}T^i\psi$, the above equation reads

$$D^\mu J_{\mu i}^A = \frac{1}{4\pi^2}\,\text{Tr}[T^i(G^{(1)} + G^{(2)})], \tag{6.24}$$

where $D_\mu = \partial_\mu + A_\mu$,

$$G^{(1)} = \epsilon^{\mu\nu\alpha\beta}\left(\frac{1}{4}V_{\mu\nu}V_{\alpha\beta} + \frac{1}{12}A_{\mu\nu}A_{\alpha\beta}\right.$$
$$-\frac{2}{3}(A_\mu A_\nu V_{\alpha\beta} + A_\mu V_{\nu\alpha}A_\beta + V_{\mu\nu}A_\alpha A_\beta)$$
$$\left.+\frac{8}{3}A_\mu A_\nu A_\alpha A_\beta\right), \tag{6.25}$$

where

$$V_{\mu\nu} = \partial_\mu V_\nu - \partial_\nu V_\mu + [V_\mu, V_\nu] + [A_\mu, A_\nu]$$
$$A_{\mu\nu} = \partial_\mu A_\nu - \partial_\nu A_\mu + [V_\mu, A_\nu] + [A_\mu, V_\nu]. \tag{6.26}$$

Finally, the $G^{(2)}$ part is defined as

$$G^{(2)} = \frac{2}{3}[D_\mu^V A_\nu + D_\nu^V A_\mu, D_\nu^V A_\mu]_+ - \frac{1}{3}[D_\mu^V A_\mu, A_\nu^2]_+$$
$$+\frac{2}{3}[A_\mu, D_\alpha^V V_{\mu\alpha}] + \frac{1}{3}D_\nu^V D_\nu^V(D_\nu^V A_\nu)$$
$$+2A_\alpha(D_\mu^V A_\mu)A_\alpha, \tag{6.27}$$

and the covariant derivative acting on any X is defined as $D_\mu^V X = \partial_\mu X + [\mathcal{V}_\mu, X]$.

The $G^{(1)}$ term is usually referred to as the Bardeen anomaly[12].

The $G^{(2)}$ term can be understood as a consequence of local counterterms in four dimensions[‡] coming from the vector-gauge invariant Lagrangian[13]

$$\mathcal{L}_B = \mathrm{Tr}(\frac{4}{3}(D_\mu^V \mathcal{A}_\nu)^2 - 2(D_\mu^V \mathcal{A}_\mu)^2 - \frac{4}{3}[\mathcal{A}_\mu, \mathcal{A}_\nu]^2$$
$$+ \frac{4}{3}\mathcal{A}_\mu \mathcal{A}_\nu \mathcal{A}_\mu \mathcal{A}_\nu + \mathcal{V}_{\mu\nu}^2). \tag{6.28}$$

The role of these counterterms is to ensure that the vector-like nonabelian current, $J_{\mu i}^V = \bar{\psi}\gamma_\mu T^i \psi$, is conserved

$$D^\mu J_{\mu i}^V = 0. \tag{6.29}$$

It is sometimes more convenient to consider the left and right fields rather than the vector and axial ones. The relations are as follows:

$$\psi_{R,L} = \frac{1}{2}(1 \pm \gamma_5)\psi$$
$$A_{R,L} = \mathcal{V} \pm \mathcal{A}$$
$$D_{L,R}^\mu = \partial^\mu + A_{L,R}^\mu$$
$$J_{\mu i}^{L,R} = \bar{\psi}_{L,R}\gamma_\mu T^i \psi_{L,R}. \tag{6.30}$$

Nonabelian anomaly in this notation reads

$$D^\mu J_{\mu i}^L = -\frac{1}{24\pi^2}\,\mathrm{Tr}(T^i(\partial^\mu \epsilon_{\mu\nu\alpha\beta}(A_L^\nu \partial^\alpha A_L^\beta + \frac{1}{2}A_L^\nu A_L^\alpha A_L^\beta))),$$
$$D^\mu J_{\mu i}^R = +\frac{1}{24\pi^2}\,\mathrm{Tr}(T^i(\partial^\mu \epsilon_{\mu\nu\alpha\beta}(A_R^\nu \partial^\alpha A_R^\beta + \frac{1}{2}A_R^\nu A_R^\alpha A_R^\beta))). \tag{6.31}$$

We show in the Appendix, that the two forms of anomalies, Bardeen's (6.24), denoted hereafter as (B), and the symmetric one (6.31), denoted hereafter as (S), come from the different choice of the renormalization procedure.

Formal similarity between the divergence of the abelian current (6.19) and the *covariant* divergence of the left/right nonabelian currents (6.31) is accidental. There is, nevertheless, a deep relation between these two objects, requiring, however, going to a higher dimensions. The topological

---

[‡]In the $G^{(1)}$ case, this is impossible; however, one could write down the local action in five dimensions.

construction of the anomalies, relating the abelian anomaly in $D+2$ dimensions to the nonabelian one in D dimensions, is presented in the Appendix. The construction is preceded by a primer on differential forms. The explicit construction of the anomalies is possible due to the strong condition for nonabelian anomalies, known as the Wess-Zumino consistency condition[14], which we now formulate.

Let us define the current as the response of the quantum action under the chiral change $A_\mu \to A_\mu + \delta A_\mu$

$$J^\mu = \frac{\delta W}{\delta A_\mu}. \qquad (6.32)$$

Here A, J denote either right or left potential and current, respectively. Then, by definition, the nonabelian anomaly reads

$$G = \delta W(A) = W(A + \delta A) - W(A) = \int d^D x \, \text{Tr}(J^\mu(x)\delta A_\mu(x)). \qquad (6.33)$$

Parametrizing the infinitesimal chiral transformation by θ (where $g = 1-\theta$), we rewrite the infinitesimal version of gauge transformation (6.21) as

$$\delta A_\mu = \partial_\mu \theta + [A_\mu, \theta]. \qquad (6.34)$$

Then, the above equation reads

$$\begin{aligned} G(\theta, A) = \delta_\theta W(A) &= \int d^D x \, \text{Tr} J^\mu(x)(\partial_\mu \theta + [A_\mu, \theta]) \\ &= \int d^D x \, \text{Tr}((-\partial_\mu J^\mu)\theta + [J^\mu, A_\mu]\theta) \\ &= -\int d^D x \, \text{Tr}(\theta D_\mu J^\mu), \end{aligned} \qquad (6.35)$$

where in the second line, we have integrated the first term by parts and used the cyclic properties of the trace for the second term. Thus, the nonabelian anomaly expresses the covariance of the current

$$\frac{\delta W(A)}{\delta \theta(x)} = -D^\mu J_\mu(x) = G(x). \qquad (6.36)$$

Let us now perform the sequence of such transformations. In our compact notation, under the gauge transformation (1) (parametrized by θ_1), we pick the anomaly

$$G(\theta_1, A) = \int \text{Tr}(J \cdot \delta A_1) = \int \text{Tr}(J \cdot (\partial \theta_1 + [A, \theta_1])). \qquad (6.37)$$

Let us now consider how the anomaly itself changes under the infinitesimal gauge transformation (2) (parametrized by θ_2) of field A:

$$\delta_2 G(\theta_1, A) \equiv G(\theta_1, A + \delta A_2) - G(\theta_1, A)$$

$$= \int \text{Tr}(J \cdot (\partial \theta_1 + [A + \delta A_2, \theta_1])) - \int \text{Tr}(J \cdot (\partial \theta_1 + [A, \theta_1]))$$

$$= \int \text{Tr}(J \cdot [\delta A_2, \theta_1])$$

$$= \int \text{Tr}(J \cdot ([\partial \theta_2, \theta_1] + [[A, \theta_2], \theta_1])). \tag{6.38}$$

Similarly,

$$\delta_1 G(\theta_2, A) = \int \text{Tr}(J \cdot ([\partial \theta_1, \theta_2] + [[A, \theta_1], \theta_2])). \tag{6.39}$$

Subtracting (6.38) from (6.39), we get

$$\delta_1 G(\theta_2, A) - \delta_2 G(\theta_1, A) = \int \text{Tr}(J^\mu \partial_\mu (\theta_1 \theta_2 - \theta_2 \theta_1)$$

$$+ J^\mu ([[A_\mu, \theta_1], \theta_2] - [[A_\mu, \theta_2], \theta_1])). \tag{6.40}$$

Finally, using the Jacobi identity

$$[[A, \theta_1], \theta_2] - [[A, \theta_2], \theta_1] + [[\theta_1, \theta_2], A] = 0, \tag{6.41}$$

we immediately recognize that the right-hand side of equation (6.40) is the anomaly itself, corresponding to the local chiral transformation $\theta = [\theta_1, \theta_2]$:

$$\delta_1 G(\theta_2, A) - \delta_2 G(\theta_1, A) = \int \text{Tr}(J^\mu (\partial [\theta_1, \theta_2] + [A_\mu, [\theta_1, \theta_2]]))$$

$$= \int \text{Tr}(J \cdot \delta A_{[1,2]}) = G_{[1,2]}([\theta_1, \theta_2], A). \tag{6.42}$$

This is the Wess-Zumino consistency condition for nonabelian anomalies. As we have seen, it comes from the fundamental property of the Lie algebra (Jacobi identity). From (6.33), we can rewrite the Wess-Zumino condition in an equivalent form

$$\delta_{\theta_1} \delta_{\theta_2} W - \delta_{\theta_2} \delta_{\theta_1} W = \delta_{[\theta_1, \theta_2]} W, \tag{6.43}$$

where the Lie structure of the infinitesimal nonabelian chiral transformations is even more transparent.

It should be stressed that (6.43) is a highly nontrivial relation. It is a nonlinear equation. Therefore, knowing, for example, only one of the terms of $G^{(1)}$, we can reproduce the whole Bardeen anomaly by a successive application of the Wess-Zumino consistency condition.

The above arguments translate, verbatim, to the most general case including the presence of the scalar and pseudoscalar sources. As discussed in detail in Chapter 1, in the real world, global chiral symmetry is spontaneously broken. Since the vector flavor transformation is preserved in this process, we are looking for the quantum action preserving the vector transformation

$$\delta_V W[\mathcal{V}, \mathcal{A}, \mathcal{S}, \mathcal{P}] = 0,$$
$$\delta_A W[\mathcal{V}, \mathcal{A}, \mathcal{S}, \mathcal{P}] = G \tag{6.44}$$

leading to normal (vector) and anomalous (axial) Ward identities. Equations (6.44) are fundamental in QCD. They express that any effective mesonic theory obtained by integrating out quark degrees of freedom has to respond to chiral transformations in exactly the same way as the quark theory does.

Is there a way to solve them systematically without invoking explicit QCD degrees of freedom? While we are not aware of a general answer to this question, we will present a solution, making use of an effective action in terms of the degrees of freedom of the spontaneously broken chiral symmetry. To do this, we need to know more about the symmetry structure and properties of the Goldstone modes.

6.3.3. *Coordinates of the nonlinear σ model*

Weinberg's formulation of current algebra suggests that the pions as Goldstone modes are in the nonlinear representation of chiral $SU_L(2) \times SU_R(2)$. This point was discussed in Chapter 5. Here we would like to focus on the geometrical aspect of the space inhabited by these modes. More specifically, we shall try to seek a suitable coordinate representation for the Goldstone modes that exhibits the important property of reparametrization invariance of the chiral field, as first noted by Gürsey[15] and Nishijima,[16] and stressed by Weinberg (see Chapter 5).

Callan, Coleman, Wess and Zumino (CCWZ)[17] observed that the nonlinear realization of chiral symmetry as discussed by Weinberg is related to the geometry of the spontaneous breaking of group G into H, namely, the geometry of the coset $G/H \sim SU_L(3) \times SU_R(3)/SU_{L+R}(3)$ for chiral symmetry. The spontaneous breakdown of chiral symmetry is signaled in the form of light Goldstone bosons, *i.e.*, the light pions and kaons. In the notations of Chapter 5, we will refer by T^a to the generators of the conserved subgroup $H \sim SU_{L+R}(3)$ and by X^a to the generators of the coset (broken generators), *i.e.*, G/H. The ensuing Lie algebra structure reads

$$[T^a, T^b] = i f^{abc} T^c,$$
$$[T^a, X^b] = i f^{abc} X^c,$$
$$[X^a, X^b] = i f^{abc} T^c. \tag{6.45}$$

The first relation indicates that H has a group structure usually referred to as the "little group." Note that the T's and X's are not field operators but simply generators. Let $U(\phi)$ be an element on the coset G/H with the standard parametrization

$$U(\phi) = e^{i\phi^a X^a} = \exp\left(\frac{i}{f_0}\begin{pmatrix} \pi^0 + \frac{\eta_8}{\sqrt{3}} & \sqrt{2}\pi^+ & \sqrt{2}K^+ \\ \sqrt{2}\pi^- & -\pi^0 + \frac{\eta_8}{\sqrt{3}} & \sqrt{2}K^0 \\ \sqrt{2}K^- & \sqrt{2}\overline{K}^0 & -\frac{2\eta_8}{\sqrt{3}} \end{pmatrix}\right) \tag{6.46}$$

in terms of the pseudoscalar octet, where f_0 is the *bare* pion decay constant. The group elements g of G can act either on the left or on the right of the coset elements (6.46). The left action on (6.46) is given by

$$U^g(\phi) = g\, U(\phi) = U(\phi')\, h(\phi, g) \tag{6.47}$$

where the compensator $h(\phi, g)$ is an element of H. It says that under the action of any element of G, the parametrization (6.46) is not necessarily preserved. Indeed, if g is an element of H, then

$$U^g(\phi) = g\, U(\phi) = \left(e^{+i\alpha^a T^a} U(\phi) e^{-i\alpha^a T^a}\right) e^{+i\alpha^a T^a}$$
$$= U(\mathbf{R}_X(g)\phi)\, g \tag{6.48}$$

where

$$\mathbf{R}_X^{ab}(g) = \frac{1}{2} \mathrm{Tr}\left(X^a g X^b g^\dagger\right) \tag{6.49}$$

are the adjoint generators of X. The coordinate ϕ transforms *linearly* in this case: $\phi \to \mathbf{R}_X(g)\phi$ with ϕ-independent compensator $C = g$. This is not the case if g is not in H. The reader can easily verify that the transformation in this case is *nonlinear* in the field ϕ with field-dependent compensators.

The geometry of the coset space is best characterized by Maurer-Cartan left (L) and right (R) one-forms

$$-iL = U^\dagger dU = \Gamma_L + E_L = (\Gamma^a_{L,b}T^b + E^a_{L,b}X^b)d\phi^a,$$
$$-iR = UdU^\dagger = \Gamma_R + E_R = (\Gamma^a_{R,b}T^b + E^a_{R,b}X^b)d\phi^a \qquad (6.50)$$

where $\Gamma_{L,R}$ are the one-form L,R-connections on G/H and $E_{L,R}$ the one-form L,R-vielbeins on G/H. The latter form a local and oriented coordinate system on G/H, while the former define parallel transport on G/H. From (6.48), it follows that under the action of an element C of G,

$$E \to CEC^{-1}, \qquad \Gamma \to C\Gamma C^{-1} + CdC^{-1}. \qquad (6.51)$$

The vielbeins transform homogeneously while the connections transform inhomogeneously. The latter behave like nonabelian gauge fields. Thus, derivatives in the coset will be sought in the covariant form $\nabla = d + \Gamma$ and transform homogeneously under the action of an element of G.

Since $G \sim SU_R(3) \times SU_L(3)$, this means that the elements of G can be classified by parity, and that the elements of H correspond to $h = g_L = g_R$. A suitable representation for the U field in this case would be

$$U(\phi) = U_R(\phi)U_L^{-1}(\phi) \qquad (6.52)$$

as it transforms *linearly* under the action of G,

$$U^g(\phi) = g_R\, U(\phi)\, g_L^{-1}. \qquad (6.53)$$

The compensator C drops out in this case. The decomposition (6.52) is only dictated by parity. In "unitary gauge"

$$U_R(\phi) = U_L^{-1}(\phi) = \sqrt{U(\phi)}, \qquad (6.54)$$

we recover the general form (6.46). In this gauge, the connection Γ_L and the vielbein E_L are given by

$$\Gamma_L = \frac{1}{2}\left(\sqrt{U}d\sqrt{U^\dagger} + \sqrt{U^\dagger}d\sqrt{U}\right) = \Gamma_L^\mu(\phi)dx_\mu,$$
$$E_L = \frac{1}{2}\left(\sqrt{U}d\sqrt{U^\dagger} - \sqrt{U^\dagger}d\sqrt{U}\right) = \frac{i}{2}E_L^\mu(\phi)dx_\mu \qquad (6.55)$$

and similarly for Γ_R and E_R. These are the usual "induced" vector and axial-vector fields made up of Goldstone bosons familiar in chiral Lagrangians.

The parametrization (6.46) viewed as (6.53) in the unitary gauge constitutes the natural coordinate system for the pseudoscalar octet in the coset space, the space of Goldstone bosons. At low energy, the Goldstone bosons could be used as alternate (auxiliary) degrees of freedom to quarks and gluons, for reconstructing the QCD effective action using Weinberg's chiral counting rationale.

6.4. Construction of Effective Action: Chiral Perturbation Theory

Weinberg's counting rules[2] suggest that if we denote by p_μ some typical momentum scale, the analog of E for the S-matrix, then $U \sim p_\mu^0$, $\partial U, \mathcal{V}, \mathcal{A} \sim p_\mu$ and $\mathcal{S}, \mathcal{P} \sim p_\mu^2$ since $\mathcal{S} \sim m \sim m_\pi^2$. This means that the connected part of the generating functional $W[\mathcal{V}, \mathcal{A}, \mathcal{S}, \mathcal{P}]$, and equivalently the QCD effective action Γ, can be sought in powers of p_μ^2,

$$W = W_2 + W_4 + W_6 + \cdots \qquad (6.56)$$

much like the S-matrix. The W_{2n} obey the chiral Ward identities (6.44) order by order in p_μ^2. More importantly, the decomposition (6.56) can be reconstructed from a "master" effective Lagrangian

$$\mathcal{L}(U, \mathcal{V}, \mathcal{A}, \mathcal{S}, \mathcal{P}) = \mathcal{L}_2 + \mathcal{L}_4 + \mathcal{L}_6 + \ldots \qquad (6.57)$$

through

$$e^{iW[\mathcal{V}, \mathcal{A}, \mathcal{S}, \mathcal{P}]} = \int [dU] \, e^{i \int d^4 x \mathcal{L}(U, \mathcal{V}, \mathcal{A}, \mathcal{S}, \mathcal{P})} \qquad (6.58)$$

where $[dU]$ is the Haar measure in the coset G/H. The master effective Lagrangian (6.57) is the most general Lagrangian for U and the external sources, consistent with symmetries and the general precepts of field theory. This is chiral perturbation theory (ChPT) for the off-shell amplitudes as formulated by Gasser and Leutwyler[1].

In a way, (6.58) is the low-energy analog of (6.15). The important points are these: first, in (6.58), the pseudoscalars U are auxiliary fields, and therefore are integration variables that do not carry asymptotic weight in physical amplitudes. Second, the procedure follows from Weinberg's approach, and provides a systematic construction of the QCD effective action, given the freedom in writing (6.58). In principle, it is a theoretical edifice for reconstructing the effective action based on symmetries and perturbative unitarity that encapsulates all there is in QCD at low energy. Third, the QCD Ward identities (6.44) provide *new* constraints on the parameters of the master Lagrangian (6.57), which are absent in Weinberg's initial formulation discussed in Chapter 5. We now proceed to construct W in the context of chiral perturbation theory.

6.4.1. W_2

Since the extended QCD effective Lagrangian (6.13) exhibits local $G \sim SU_R(3) \times SU_L(3)$ invariance under (6.14), the natural variables in the master effective Lagrangian (6.57) are U, their covariant derivatives with respect to the external gauge fields $((A_R, A_L) = (\mathcal{V} + \mathcal{A}, \mathcal{V} - \mathcal{A}))$

$$\nabla_\mu U = \partial_\mu U + \left(A_\mu^R U - U A_\mu^L\right), \qquad (6.59)$$

the field strengths for the external gauge fields

$$\begin{aligned} F_R^{\mu\nu} &= \partial^\mu A_R^\nu - \partial^\nu A_R^\mu + [A_R^\mu, A_R^\nu], \\ F_L^{\mu\nu} &= \partial^\mu A_L^\nu - \partial^\nu A_L^\mu + [A_L^\mu, A_L^\nu] \end{aligned} \qquad (6.60)$$

and $\mathcal{S}$ and $\mathcal{P}$, all of which transform covariantly under G. According to Weinberg's power counting, the field strength $F \sim p_\mu^2$. Thus, the lowest-order $(O(p_\mu^2))$, locally chiral-invariant and parity-even effective Lagrangian reads

$$\mathcal{L}_2 = \frac{f_0^2}{4} \operatorname{Tr} \left(\nabla_\mu U \nabla^\mu U^\dagger + \chi U^\dagger + \chi^\dagger U\right) \qquad (6.61)$$

where

$$\chi = 2B_0 \left(\mathcal{S} + i\mathcal{P}\right). \qquad (6.62)$$

This is the nonlinear sigma model in the presence of external sources. It has two free parameters f_0 and B_0 to order p_μ^2. To this order, the generating functional is exactly the tree action at the extremum,

$$W_2[\mathcal{V}, \mathcal{A}, \mathcal{S}, \mathcal{P}] = \int d^4x \, \mathcal{L}_2(U_*, \mathcal{V}, \mathcal{A}, \mathcal{S}, \mathcal{P}) \qquad (6.63)$$

with $U_* = U_*[\mathcal{V}, \mathcal{A}, \mathcal{S}, \mathcal{P}]$ following from the classical equation of motion for $\mathcal{L}_2$,

$$(\nabla^2 U_*)U_*^\dagger - U_*(\nabla^2 U_*^\dagger) = \chi U_*^\dagger - U_* \chi^\dagger - \frac{1}{3} \text{Tr} \left(\chi U_*^\dagger - U_* \chi^\dagger \right) \, 1. (6.64)$$

We note that to order p_μ^2, $\delta_\theta W_2 = \delta_\omega W_2 = 0$. The anomalous terms are of order p_μ^4, hence do not affect W_2.

The parameters appearing in (6.61) have physical meaning. Indeed, to lowest order, the vector and axial-vector currents follow from (6.63) at the extremum. Equivalently, they can be extracted from (6.61) at tree order by identifying the auxiliary field ϕ in $U(\phi)$ with the "physical" mesonic field, using the vacuum conditions $\mathcal{V} = \mathcal{A} = \mathcal{P} = 0$ and $\mathcal{S} = M = \text{diag}(m_u, m_d, m_s)$. It should be stressed that this simplification does not carry to higher orders. Having said this, the Noether currents associated to (6.61) are

$$V^\mu = -\frac{if_0^2}{2\sqrt{2}} \left(U^\dagger \nabla^\mu U + U \nabla^\mu U^\dagger \right),$$

$$A^\mu = +\frac{if_0^2}{2\sqrt{2}} \left(U^\dagger \nabla^\mu U - U \nabla^\mu U^\dagger \right). \quad (6.65)$$

Taking the matrix element between the vacuum and the pseudoscalar modes

$$\langle 0 | A^{\mu,a}(x) | \pi^b(q) \rangle = i\delta^{ab} e^{-iq \cdot x} f_0 q^\mu \quad (6.66)$$

shows that $f_0 \sim f_\pi = 93$ MeV to lowest order in ChPT. The parameter B_0 in (6.61) relates to the quark condensate in the chiral limit, to leading order in ChPT,

$$\left(\frac{\delta W_2}{\delta S^{ab}} \right)_{\mathcal{V} = \mathcal{A} = \mathcal{S} = \mathcal{P} = 0} = -\langle 0 | \bar{q}^a q^b | 0 \rangle_0 = f_0^2 B_0 \, \delta^{ab}. \quad (6.67)$$

Thus $B_0 \sim -\langle 0 | \bar{u}u | 0 \rangle_0 / f_0^2$ is related to the quark condensate in the chiral limit. We note that while B_0 is not physical (that is, renormalization group noninvariant), the combination $B_0 \chi \sim B_0 m$ appearing in (6.61) is.

The spectrum of the pseudoscalar octet can be directly identified from (6.61) at tree level using the vacuum conditions. To this order, this is equivalent to using Green's functions. This simplification does not occur beyond the lowest order. Inserting (6.46) into (6.61) yields to leading order in ϕ

$$\mathcal{L}_2 = \frac{1}{4} \text{Tr} \left(\partial \phi \cdot \partial \phi\right) - \frac{1}{2} B_0 \, \text{Tr} \left(M \phi^2\right). \tag{6.68}$$

Thus to lowest order in ChPT, the pseudoscalar octet masses satisfy

$$
\begin{aligned}
m_\pi^2 &= (m_u + m_d) B_0, \\
m_{K^+}^2 &= (m_u + m_s) B_0, \\
m_{K^0}^2 &= (m_d + m_s) B_0, \\
m_{\eta_8}^2 &= \frac{1}{3} (m_u + m_d + 4 m_s) B_0
\end{aligned}
\tag{6.69}
$$

up to corrections of order $(m_u - m_d)$. These results imply a number of relations. Indeed, by eliminating B_0 from the above relations, we obtain, to lowest order in ChPT,

$$\frac{m_\pi^2}{m_u + m_d} = \frac{m_{K^+}^2}{m_u + m_s} = \frac{m_{K^0}^2}{m_d + m_s}. \tag{6.70}$$

These relations are also known as the current-algebra mass relations. The Gell-Mann-Okubo relation follows immediately from (6.69),

$$m_{\eta_8}^2 = \frac{4}{3} m_K^2 - \frac{1}{3} m_\pi^2. \tag{6.71}$$

The Gell-Mann-Oakes-Renner relation can be obtained from the relation for the pion mass after inserting the value of B_0 following from (6.67):

$$f_\pi^2 m_\pi^2 = -\frac{1}{2}(m_u + m_d)\langle 0 | \bar{u}u + \bar{d}d | 0 \rangle_0 \tag{6.72}$$

with $f_\pi \sim f_0$, to lowest order in ChPT.

6.4.2. W_4

The order p_μ^4 part W_4 in W follows from Weinberg's counting in the context of ChPT. W_4 through (6.58) receives contributions from : (1) the most general and symmetric $\mathcal{L}_4^N$ term at tree level; (2) the chiral anomaly $\mathcal{L}_4^A$; (3) the contribution of $\mathcal{L}_2$ to one loop. We now consider each of these contributions sequentially.

(i) Tree order

The most general form of $\mathcal{L}_4$ in the presence of the external sources, consistent with local $SU_L(3) \times SU_R(3)$, charge conjugation and parity, reads

$$
\begin{aligned}
\mathcal{L}_4^N =& G_1 \operatorname{Tr}\left(\nabla_\mu U^\dagger \nabla^\mu U\right)^2 + G_2 \operatorname{Tr}\left(\nabla_\mu U^\dagger \nabla_\nu U\right) \operatorname{Tr}\left(\nabla^\mu U^\dagger \nabla^\nu U\right) + \\
& G_3 \operatorname{Tr}\left(\nabla_\mu U^\dagger \nabla^\mu U \nabla_\nu U^\dagger \nabla^\nu U\right) + \\
& G_4 \operatorname{Tr}\left(\nabla_\mu U^\dagger \nabla^\mu U\right) \operatorname{Tr}\left(\chi^\dagger U + \chi U^\dagger\right) + \\
& G_5 \operatorname{Tr}\left(\nabla_\mu U^\dagger \nabla^\mu U \left(\chi^\dagger U + U^\dagger \chi\right)\right) + G_6 \operatorname{Tr}\left(\chi^\dagger U + U^\dagger \chi\right)^2 + \\
& G_7 \operatorname{Tr}\left(\chi^\dagger U - U^\dagger \chi\right)^2 + \\
& G_8 \operatorname{Tr}\left(\chi^\dagger U \chi^\dagger U + \chi U^\dagger \chi U^\dagger\right) + \\
& G_9 \operatorname{Tr}\left(\nabla_\mu U \nabla^\nu U^\dagger F_R^{\mu\nu} + \nabla_\mu U^\dagger \nabla^\nu U F_L^{\mu\nu}\right) - \\
& G_{10} \operatorname{Tr}\left(U^\dagger F_R^{\mu\nu} U F_{L\mu\nu}\right) - \\
& G_{11} \operatorname{Tr}\left(F_R^{\mu\nu} F_{R,\mu\nu} + F_L^{\mu\nu} F_{L,\mu\nu}\right) + G_{12} \operatorname{Tr}\left(\chi^\dagger \chi\right).
\end{aligned} \tag{6.73}
$$

It involves 12 arbitrary bare parameters, which are to be determined phenomenologically as originally suggested by Weinberg. Their detailed determination in the present context has been discussed by Gasser and Leutwyler[18]. The list of terms (6.73) is not exhaustive. Indeed, we have omitted total derivatives, as they do not affect the action in open or periodic spaces. We have also omitted terms of the type $\Box U$ as they differ from the terms quoted by a multiple of the classical equation of motion. These terms can be removed from the effective action by a suitable redefinition of the field variables U.

(ii) Wess-Zumino-Witten term

The anomalous variation of W as given by (6.44) is of order p_μ^4 in ChPT, and contributes to W_4. The master effective Lagrangian (6.57) reads (see (A6.66))

$$
S_{WZW}(A_L, A_R, g_L, g_R) = W(A_L, A_R) - W(A_L^{g_L}, A_R^{g_R}) \tag{6.74}
$$

through the identification $U \equiv g_L^{-1} g_R = \xi^2$. This is Witten's version[19] of the result first established laboriously by Wess and Zumino[14]. The construction presented in the Appendix for the case of the external electromagnetic current, extended now to the case of full nonabelian external currents, allows us to rewrite (6.74) in the following form

$$\int \mathcal{L}_4^A = -\frac{N_c}{240\pi^2} \int_{H^5} \text{Tr}\left(L^5\right)$$

$$-\frac{iN_c}{48\pi^2} \int_{S^4} \left(\mathcal{W}(U, A_R, A_L) - \mathcal{W}(1, A_R, A_L)\right) \qquad (6.75)$$

where L is the Maurer-Cartan one-form (6.50) and

$$\mathcal{W}(U, A_R, A_L) = \text{Tr}\Bigg((A_R dA_R + dA_R A_R + A_R^3 - (dU^{-1}U)^2 A_R)$$

$$(U^{-1}A_L U + U^{-1}dU) - (A_R dU^{-1}U A_R - dU^{-1}U dA_R)$$

$$U^{-1}A_L U + \frac{1}{2}A_L dUU^{-1}A_L dUU^{-1} - (L \leftrightarrow R, U \leftrightarrow U^{-1})$$

$$+\frac{1}{2}(A_L U A_R U^{-1})^2\Bigg). \qquad (6.76)$$

Note that the first term in (6.75) is local only in $H^5 \sim [0,1] \times S^4$, through a suitable extension of $U(\phi)$. However, by the previous arguments, it is a pure gauge and thus a boundary term. This term is important as it drives the anomalous processes of the QCD effective action in the pseudoscalar sector. The second term characterizes anomalous amplitudes with vector currents. This term can also be generated by trial-and-error gauging of the first term, using $A_{L,R} = (\mathcal{V} \pm \mathcal{A})$ as gauge fields[19].

(iii) One-loop order

Weinberg's chiral counting shows that to order p_μ^4, one has to include the one-loop effect from $\mathcal{L}_2$ for a description of the off-shell amplitudes that is consistent with chiral symmetry and perturbative unitarity. This is achieved by including the Gaussian fluctuations around the extremum. The procedure consists in expanding $\mathcal{L}_2(U, \mathcal{V}, \mathcal{A}, \mathcal{S}, \mathcal{P})$ around $U_*(\mathcal{V}, \mathcal{A}, \mathcal{S}, \mathcal{P})$, the solution to (6.64), using

$$U = \sqrt{U_*}\left(1 + i\xi - \frac{1}{2}\xi^2 + ...\right)\sqrt{U_*} \qquad (6.77)$$

with $\xi = \xi^a \lambda^a$ and $\frac{1}{2}\text{Tr}(\lambda^a \lambda^b) = \delta^{ab}$. To second order in ξ

$$\mathcal{L}_2 \sim \mathcal{L}_{2,*} - \int d^4x \frac{f_0^2}{2}\xi^a \mathcal{D}^{ab}\xi^b \qquad (6.78)$$

with

$$\mathcal{D}^{ab} = [\partial^\mu + \Gamma^\mu, \partial_\mu + \Gamma_\mu]^{ab} + \hat\sigma^{ab} \tag{6.79}$$

where Γ is the covariantized connection (6.55)

$$\Gamma_\mu = \frac{1}{2} \left(\sqrt{U_*} \nabla_\mu^R \sqrt{U_*} + \sqrt{U_*^\dagger} \nabla_\mu^L \sqrt{U_*^\dagger} \right) \tag{6.80}$$

with $\nabla^{R,L} = \partial + A_{R,L}$, and $\hat\sigma$ is

$$\hat\sigma^{ab} = +\frac{1}{2} \mathrm{Tr}\left([\lambda^a, \nabla_\mu][\lambda^b, \nabla^\mu] \right)$$
$$+ \frac{1}{8} \mathrm{Tr}\left([\lambda^a, \lambda^b]_+ \left(\sqrt{U_*} \xi^\dagger \sqrt{U_*} + \sqrt{U_*^\dagger} \xi \sqrt{U_*^\dagger} \right) \right). \tag{6.81}$$

The indices in the trace run over the adjoint representation. We note that with the present parametrization and to one-loop, the Haar measure linearizes,

$$[dU] \sim \prod_{a=1}^{3} d\xi^a. \tag{6.82}$$

(iv) Full action W_4

Putting together the above contributions to W_4 yields

$$W_4 = \int (\mathcal{L}_4^N + \mathcal{L}_4^A) + \frac{i}{2} \mathrm{Tr} \ln \mathcal{D} \tag{6.83}$$

where the last term follows from the Gaussian integration over the ξ field. The explicit evaluation of the determinant (trace-log term) is given in Appendix A_{II}. Generically, it has divergent parts and finite parts:

$$\frac{i}{2} \mathrm{Tr} \ln \mathcal{D} = \frac{i}{2} \mathrm{Tr} \ln \mathcal{D}_{\mathrm{div}} + \frac{i}{2} \mathrm{Tr} \ln \mathcal{D}_{\mathrm{fin}}. \tag{6.84}$$

The divergent contribution in (6.84) is composed of terms that have already appeared in (6.73) as expected from symmetry, with the exception of the terms multiplying G_{11} and G_{12}. The reason is that these terms do not contain internal pseudoscalars, thus *no* ξ. Their renormalization is prescription-dependent. Fortunately, they do not appear in physical observables at low energy. The effects of the loop integration is to turn the bare couplings of $\mathcal{L}_4^N$ into running couplings, plus finite terms. Generically $(i = 1, ..., 10)$,

$$G_i(\mu) = G_i(\mu_0) + \frac{\gamma_i}{32\pi^2}\ln\frac{\mu_0^2}{\mu^2} = \frac{\gamma_i}{32\pi^2}\left(\overline{G}_i + \ln\frac{\mu_0^2}{\mu^2}\right) \qquad (6.85)$$

as suggested by Weinberg, where the γ's are listed in Table 6.1. Up to a numerical factor, $\overline{G}_i$ is the value of the renormalized coupling constant at the scale $\mu \sim \mu_0$, with the identification $f_0^2\mu_0^2 = 2mB_0$ in the isospin symmetric case $m = m_u = m_d$. Thus, $\overline{G}_i \to -\ln\mu_0^2$ for $\mu_0 \to 0$ (chiral limit). With this in mind, we have

$$\begin{aligned}
S_4 &= \int \mathcal{L}_4^A + \left(\int \mathcal{L}_4^N + \frac{i}{2}\,\text{Tr}\,\ln\mathcal{D}_{\text{div}}\right) + \frac{i}{2}\,\text{Tr}\,\ln\mathcal{D}_{\text{fin}} \\
&= \int \left(\mathcal{L}_4^A + \mathcal{L}_4^{N,R}\right) + \frac{i}{2}\,\text{Tr}\,\ln\mathcal{D}_{\text{fin}}. \qquad (6.86)
\end{aligned}$$

The μ-dependence of the running couplings $G(\mu)$ in the renormalized part $\mathcal{L}_4^{N,R}$ is balanced by the μ-dependence in the threshold parameters of the finite part (last term). This is the pattern advocated by Weinberg for the S-matrix in $\pi\pi$-scattering. There, variations in the two-pion threshold were balanced by variations in the running couplings.

Table 6.1: Values of the coefficients γ.

i	γ_i	i	γ_i
1	3/32	6	11/144
2	3/16	7	0
3	0	8	5/48
4	1/8	9	1/4
5	3/8	10	−1/4

In the large N_c limit, the scale dependence drops out, since the one-loop contribution is down by $1/f_0^2 \sim 1/N_c$. To leading order in $1/N_c$, the coupling constants G_i's are pure numbers [§] We do not know of a systematic derivation of these couplings from large N_c QCD.

[§] Apart from G_7 which receives contribution of order N_c^2 from η' exchange[18].

An effective estimation of the various coupling constants entering the present formulation may be obtained by cutting the loop integrals originating from $\mathcal{L}_2$ at some fixed momentum P, and disregarding higher order terms in the derivative expansion. In this way, various threshold amplitudes will only involve a single cutoff P and will follow exclusively from the symmetries of $\mathcal{L}_2$. This procedure does not work for the isovector charge radius of the pion $\langle r^2 \rangle_V$ [20]. Indeed, $\langle r^2 \rangle_V$ in ChPT diverges logarithmically to one-loop, and for a reasonable cutoff $P \sim 1$ GeV, $\langle r^2 \rangle_V \sim 0.09$ fm^2 which is totally off the empirical mark of 0.44 fm^2. The latter value is reached only for very large cutoffs, $P \sim 50$ GeV. This procedure does not *seem* to work. A similar remark applies to the use of the renormalizable linear sigma model. Thus, it appears that the size of the higher order terms in the derivative expansion of the effective Lagrangian cannot be understood on the basis of leading term.

6.5. Some Applications

The external field approach discussed above does not involve Green's functions associated with a hermitian operator $\pi(x)$ or with a unitary operator $U(x)$. Thus the issue of the definition of the pion field and its renormalization does not arise. As noted above, the variables $\pi(x)$ and $U(x)$ are mere auxiliary variables devoid of physical meaning.

However, for the calculation of some quantities, such as the pion mass, it is more practical to use the effective action with the sources switched off and identify the pion field with $\pi(x)$. This procedure, while simpler, may be quite intricate in some cases and should be used with care. Having said this, we now proceed to evaluate certain low-energy parameters and scattering amplitudes using the effective action formulation described above.

6.5.1. *Pion parameters*

To evaluate the pion mass and decay constant up to one-loop, we expand $W \sim W_2 + W_4$ to second order in the external field $\mathcal{A}_\mu$. For simplicity, we will assume here that the mass of the strange quark is large compared to the up and down quark mass, and we neglect the kaon and eta meson contributions (two-flavor limit). In this case, the two-point function for the axial current $J_\mu^{(A),a}(x) = \bar{q}\gamma_\mu\gamma_5 T^a q(x)$ reads, to one-loop,

$$i \int d^4x e^{ip\cdot(x-y)}\langle 0|T\left(J_\mu^{(A),a}(x)J_\nu^{(A),b}(y)\right)|0\rangle \sim$$

$$\delta^{ab}\left(g_{\mu\nu}f_\pi^2 + p_\mu p_\nu \frac{f_\pi^2}{m_\pi^2 - p^2} + \left(p_\mu p_\nu - g_{\mu\nu}p^2\right)\frac{\overline{\Delta}}{48\pi^2}\right) \qquad (6.87)$$

where

$$m_\pi^2 \sim \mu_0^2\left(1 - \frac{\mu_0^2}{32\pi^2 f_0^2}\bar{l}_3\right),$$

$$f_\pi^2 \sim f_0^2\left(1 + \frac{\mu_0^2}{8\pi^2 f_0^2}\bar{l}_4\right). \qquad (6.88)$$

The parameters in (6.87-6.88) are related to the barred parameters of the effective action (6.85) through

$$\bar{l}_3 = 2\overline{G}_4 + 3\overline{G}_5 - \frac{22}{9}\overline{G}_6 - 15\overline{G}_8 - \frac{121}{9}\ln\frac{\mu_0^2}{\mu^2},$$

$$\bar{l}_4 = 2\overline{G}_4 + 3\overline{G}_5 + 4\ln\frac{\mu_0^2}{\mu^2}. \qquad (6.89)$$

The parameter $\overline{\Delta}$ will not be needed here. We note that since $\overline{G} \sim -\ln\mu_0^2$, both the pion mass and decay constant contain a chiral logarithm, in agreement with the relations (5.57) derived in Chapter 5.

6.5.2. *Fermion condensate*

The fermion condensate may be extracted in a similar way, by considering the one-point function $\langle 0|\bar{q}q|0\rangle$ in the chiral limit. In the isospin symmetric limit, the calculation is straightforward and the answer to one-loop is

$$\langle 0|\bar{q}q|0\rangle = -2f_0^2 B_0\left(1 + \frac{\mu_0^2}{32\pi^2 f_0^2}\left(\frac{22}{9}\overline{G}_6 + \frac{85}{6}\overline{G}_8 + \frac{5}{6}\overline{G}_{12} + \frac{493}{36}\ln\frac{\mu_0^2}{\mu^2}\right)\right) \qquad (6.90)$$

in agreement with the expression of Novikov *et al.*[21]

$$\langle 0|\bar{q}q|0\rangle = \langle 0|\bar{q}q|0\rangle_0\left(1 - \frac{3m_\pi^2}{32\pi^2 f_\pi^2}\ln\frac{m_\pi^2}{\mu^2}\right) \qquad (6.91)$$

in the chiral limit.

We remark that the results (6.88-6.91) show that the Gell-Mann-Oakes-Renner relation holds to one-loop. The validity of the Gell-Mann-Oakes-Renner relation has been challenged in the past by a few authors[22]. Recently, this has led some to suggest a generalized chiral perturbation theory where the chiral counting is modified so as to accommodate a modified Gell-Mann-Oakes-Renner relation[23]. These modifications have direct impact on physical observables such as the S-wave pion scattering lengths. This point will not be pursued in this book.

6.5.3. The $\pi\pi$ scattering

The $\pi\pi$ scattering amplitude is obtained by considering the four-point function and the reduction formula. The process

$$\pi^a(k_1) + \pi^b(k_2) \to \pi^c(p_1) + \pi^d(p_2)$$

is determined by a single scalar function $A(s,t,u)$ defined in terms of the isospin decomposition

$$\mathcal{T}^{ab,cd}(s,t,u) = \delta^{ab}\delta^{cd}\, A(s,t,u) + \delta^{ac}\delta^{bd}\, A(t,s,u) + \delta^{ad}\delta^{bc}\, A(u,t,s) \quad (6.92)$$

with $s = (k_1+k_2)^2$, $t = (k_1-p_1)^2$ and $u = (k_1-p_2)^2$ the three Mandelstam variables. To one-loop order, the result for the scattering amplitude is given by

$$A(s,t,u) = \frac{(s - m_\pi^2)}{f_\pi^2} + B(s,t,u) + C(s,t,u). \quad (6.93)$$

The first term is the tree contribution and was determined long ago by Weinberg as discussed in Chapter 5, the second term is the unitarity correction due to the loop contribution from $\mathcal{L}_2$, and the third term is the tree and tadpole contributions from $\mathcal{L}_4$ and $\mathcal{L}_2$. For more details, we refer to the paper of Gasser and Leutwyler[18] and Appendix A_{II}. Explicitly,

$$B(s,t,u) = \frac{1}{6f_\pi^4}\left(+3(s - m_\pi^2)\mathcal{J}(s)\right.$$
$$+\left(t(t-u) - 2m_\pi^2 t + 4m_\pi^2 u - 2m_\pi^4\right)\mathcal{J}(t)$$
$$\left.+(t \leftrightarrow u)\right) \quad (6.94)$$

and

$$C(s,t,u) = \frac{1}{96\pi^2 f_\pi^4}\left(2(\bar{l}_1 - \frac{4}{3})(s - 2m_\pi^2)^2 + (\bar{l}_2 - \frac{5}{6})(s^2 + (t-u)^2)\right.$$

$$\left. +12m_\pi^2 s(\bar{l}_4 - 1) - 3m_\pi^4(\bar{l}_3 + 4\bar{l}_4 - 5)\right). \qquad (6.95)$$

The function $\mathcal{J}(s)$ reads

$$\mathcal{J}(s) = -\frac{1}{16\pi^2}\int_0^1 dx \, \ln\left(1 - x(x-1)\left(\frac{s}{m_\pi^2}\right)\right) \qquad (6.96)$$

and the four low-energy parameters in (6.95) are related to the effective action couplings through

$$\bar{l}_1 = \frac{9}{8}\overline{G}_1 + \frac{1}{8}\ln\frac{\mu_0^2}{\mu^2},$$

$$\bar{l}_2 = \frac{9}{8}\overline{G}_2 + \frac{1}{8}\ln\frac{\mu_0^2}{\mu^2},$$

$$\bar{l}_3 = 2\overline{G}_1 + 3\overline{G}_5 - \frac{22}{9}\overline{G}_6 - 15\overline{G}_8 - \frac{121}{9}\ln\frac{\mu_0^2}{\mu^2},$$

$$\bar{l}_4 = 2\overline{G}_1 + 3\overline{G}_5 + 4\ln\frac{\mu_0^2}{\mu^2}. \qquad (6.97)$$

In terms of (6.93), the S-wave scattering length a_0^0 (in m_π^{-1} units) for the isosinglet channel in the $l = 0$ partial wave reads, to one-loop,

$$a_0^0 = \frac{7}{32\pi}\frac{m_\pi^2}{f_\pi^2}\left(1 + \frac{5}{84\pi^2}\frac{m_\pi^2}{f_\pi^2}\left(\bar{l}_1 + 2\bar{l}_2 - \frac{3}{8}\bar{l}_3 + \frac{21}{10}\bar{l}_4 + \frac{21}{8}\ln\frac{\mu_0^2}{\mu^2}\right)\right). \qquad (6.98)$$

In the chiral limit, all the low-energy coupling constants develop a chiral logarithm as discussed above. The quark mass expansion of the a_0^0 scattering length is therefore of the form

$$a_0^0 = \frac{7}{32\pi}\frac{m_\pi^2}{f_\pi^2}\left(1 - \frac{9}{32\pi^2}\frac{m_\pi^2}{f_\pi^2}\ln\frac{m_\pi^2}{\mu^2}\right). \qquad (6.99)$$

The correction to the Weinberg result (tree) is not of order m_π^2 in ChPT but of order $m_\pi^2\ln m_\pi^2$. At the scale $\mu \sim 1$ GeV, $a_0^0 = 0.20$. The correction is about 20% and positive. This correction narrows the discrepancy between the Weinberg-Tomozawa result

$$a_0^0(\text{tree}) = \frac{7}{32\pi}\frac{m_\pi^2}{f_\pi^2} = 0.16 \qquad (6.100)$$

and the observed value

$$a_0^0(\exp) = 0.26 \pm 0.05. \tag{6.101}$$

Similar relations can be worked out for the other scattering lengths and range parameters[1].

It is worth mentioning at this stage the recurrent appearance of the factor $1/16\pi^2$ in the one-loop correction to the pion parameters and scattering amplitude. This factor suggests that the expansion parameter at low energy is

$$\frac{q^2}{16\pi^2 f_\pi^2} \sim \frac{q^2}{1 \text{ GeV}^2} \tag{6.102}$$

rather than the naive $q^2/f_\pi^2 \sim q^2/0.01$ GeV^2. The two are seen to differ by a factor of 100. This type of low-energy expansion was first advocated by Georgi and Manohar in the context of the chiral constituent quark model [24].

6.5.4. *Other applications*

Since its inception more than a decade ago, the QCD effective action approach to low-energy phenomena has had a number of other applications. The formulation has been used to analyze the mesonic form factors[18], and η-decays[25]. Principally spurred by the physics program of the forthcoming DAFNE facility at Frascati, much activity has focused on the semi-leptonic K-decays[26]. A full two-loop calculation has also been carried out for the radiative pion transition $\gamma\gamma \to \pi^0\pi^0$ process[27], with some agreement with the data from the Crystal Ball collaboration[28].

A lot of work has also been invested in trying to understand the role of the omitted resonances in ChPT. The effective action formulation for spinless Goldstone modes coupled to mesons with spin larger or equal to one is not unique.¶ It was recently argued by Gasser and Leutwyler and their coworkers[30] that when the QCD short distance constraints are imposed on the effective action formulation, these ambiguities are then removed. The resulting effective action is found to be compatible with vector dominance, and most of the low-energy parameters were found to be saturated by vector mesons. The most general principles of field theory such as causality and unitarity, once combined with (broken) chiral symmetry,

---

¶It may also suffer from acausality problems as discussed by Velo and Zwanziger[29].

may not necessarily yield the QCD effective action uniquely. Indeed, QCD is likely to generate dynamical constraints that may restrict the effective action at low energy to a minimal version. While these constraints are theoretically unknown, they may be motivated by empirically established concepts, such as vector meson dominance.

Attempts to extend the effective action formulation to hadronic weak interactions have also been made. However, progress in these directions has been hampered by the large number of parameters involved to one-loop. Finally, the formulation has also been extended to include baryons as matter fields and has been successfully applied to πN interactions[31] and pion processes (such as nuclear forces and exchange currents) in nuclei[32]. A version of this extension and its application to the chiral nuclear problem will be discussed in Chapter 10.

6.6. Limitations of Effective Action

In ChPT, the chiral Ward identities are implemented systematically starting from the chiral point. In a way this is limited, and one would be more interested in an on-shell expansion scheme for which the low-energy constants are fixed to their physical values. Also, one would be interested in the general constraints imposed by chiral symmetry on various amplitudes, irrespective of what the pion momentum is.

As a method, ChPT raises some technical subtleties that we have so far ignored[33]. They are related to the use of the pion equations of motion in the nonlinear sigma model. If we were to specialize U_* in (6.64) to the two-flavor case, and use the parametrization $U_* = U_*^0 + i\vec{\tau} \cdot \vec{U}_*$, then to leading order the solution to (6.64) reads (see eq.(10.21) in Gasser and Leutwyler[1])

$$U_*^i = \frac{1}{\Box + m_0^2 - i\epsilon}\left(\partial^\mu \mathcal{A}_\mu^i + \chi^i\right) \qquad (6.103)$$

where Stückelberg-Feynman boundary conditions have been implemented. It follows that to leading order, the chiral field U_* evaluated at the stationary point is non-unitary, that is $U_* U_*^\dagger \neq 1$, which leads to some difficulties. The first is that we will obtain different answers for different Lagrangians:

$$+\frac{f_0^2}{4}\,\text{Tr}\left(\partial^\mu U\,\partial_\mu U^\dagger\right),\ +\frac{f_0^2}{4}\,\text{Tr}\left(\partial^\mu U\,\partial_\mu U^{-1}\right),\ -\frac{f_0^2}{4}\,\text{Tr}\left(U^\dagger\partial^\mu U\,U^\dagger\partial_\mu U\right).$$

The second is that the Haar measure $[dU]$ in (6.58) has to be redefined.

The third is that there is an infinite number of terms such as

$$\mathrm{Tr}\left(U_* U_*^\dagger\right)^n$$

which are invariant under rigid chiral transformations and may be added to the Lagrangian. All such difficulties disappear if we work in Euclidean space. However, the problem still remains as to whether we should continue U or U^{-1} or for instance, Green's functions, when coming to Minkowski space.

This problem is actually generic. In the process of constructing the effective action using Feynman rules in Minkowski space, one usually inherits complex mean-field equations and thus complex actions. The reason is that the Feynman rules define the mean field between two different vacua, $|\text{in}\rangle$ with negative frequencies, and $|\text{out}\rangle$ with positive frequencies. What is usually needed is the expectation value in the same vacuum, say $|\text{in}\rangle$ [34].

6.7. "Master Equation" Approach

The expansion around unphysical points such as soft point in current algebra and the chiral point in ChPT is, in some cases, unnecessary. Also some of the limitations described above can be avoided, by relying directly on an S-matrix formulation. In this section, we show that all low-energy theorems of current algebra can be obtained from a single formula – which we shall call "master equation," as recently suggested by Yamagishi and Zahed [4]. The idea is to start from the gauge-covariant version of the PCAC equation as discussed by Veltman and Bell [35], and convert it into an equation of motion for the pion field by a change of variable. Integration then gives a condition on the extended S matrix in the form of a master formula. The latter has the structure of a reduction formula. In this approach, the current-algebra program discussed in Chapter 5 can be exactly carried out on-shell.

The master-equation approach allows for an exact rewriting of scattering amplitudes in terms of correlation functions and form factors, some of which are directly measurable. This allows one to go beyond the threshold region in a way that is fully compatible with chiral symmetry. This point will be illustrated with $\pi\pi$ scattering. Furthermore this approach allows for certain processes to treat pions and nucleons on an equal footing [36].

For an illustration on how the program works, consider an action whose kinetic part is invariant under chiral $SU_L(2) \times SU_R(2)$ and a scalar-isoscalar mass term that transforms as a $(2,2)$ representation. Examples are two-flavor QCD or sigma models. The symmetry properties of the theory are conveniently expressed by gauging the kinetic part with c-number external fields V_μ^a and A_μ^a, and extending the mass term to include couplings with scalar and pseudoscalar fields S and $\mathcal{P}^a$. For two-flavor QCD, the fermionic action in the presence of c-number external fields reads

$$\mathbf{I} = \int d^4x \bar{q}\gamma^\mu \left(i\partial_\mu + G_\mu + V_\mu^a \frac{\tau^a}{2} + A_\mu^a \frac{\tau^a}{2}\gamma_5 \right) q -$$
$$\frac{m_q}{m_\pi^2} \int d^4x \bar{q} \left(m_\pi^2 + s - i\gamma_5 \tau^a p^a \right) q \qquad (6.104)$$

where m_π is the physical pion mass. Throughout, we will assume that $\phi = (V_\mu^a, A_\mu^a, s, p^a)$ are smooth functions that fall off rapidly at infinity.[||] Currents and densities $\mathbf{J} = (\mathbf{V}, \mathbf{A}, f_\pi\sigma, f_\pi\pi)$ may be introduced as

$$\mathbf{J}(x) = \frac{\delta \mathbf{I}}{\delta\phi(x)} \qquad (6.105)$$

which obey the Veltman-Bell equations[35]

$$\nabla^\mu \mathbf{V}_\mu + \underline{\mathbf{A}}^\mu \mathbf{A}_\mu + f_\pi \underline{p}\,\pi = 0, \qquad (6.106)$$

$$\nabla^\mu \mathbf{A}_\mu + \underline{\mathbf{A}}^\mu \mathbf{V}_\mu - f_\pi(m_\pi^2 + s)\pi + f_\pi p\sigma = 0 \qquad (6.107)$$

where $\nabla_\mu = \partial_\mu \mathbf{1} + \underline{\mathbf{V}}_\mu$ is the vector covariant derivative, $\underline{A}_\mu^{ac} = \epsilon^{abc}A_\mu^b$, $\underline{V}_\mu^{ac} = \epsilon^{abc}V_\mu^b$, and f_π is the physical pion decay constant. In the above, we have used the fact that the Bardeen anomaly[12] and the Wess-Zumino term[11] vanish for $SU_L(2) \times SU_R(2)$. Introducing the extended S-matrix $\mathcal{S}$, holding the incoming fields fixed and using the quantum action principle

$$\delta\langle\beta \text{ out}|\alpha \text{ in}\rangle = i\langle\beta \text{ out}|\delta\mathbf{I}|\alpha \text{ in}\rangle \qquad (6.108)$$

together with asymptotic completeness, yields the Peierls-Dyson formula[37]

$$\mathbf{J}(x) = -i\mathcal{S}^\dagger \frac{\delta\mathcal{S}}{\delta\phi(x)}. \qquad (6.109)$$

[||]Here for convenience we use different dimensions for the scalar and pseudoscalar fields (cf. the second term in (6.104)), therefore, in order to distinguish these new fields from the old ones, we denote them by (s, p) rather than by (S,P).

It follows from the Veltman-Bell equations (6.106-6.107) that

$$\left(\nabla^{ac}_\mu \frac{\delta}{\delta V^c_\mu(x)} + A^{ac}_\mu(x)\frac{\delta}{\delta A^c_\mu(x)} + p\frac{\delta}{\delta p}\right)S = \left(\mathbf{X}_V + p\frac{\delta}{\delta p}\right)S = 0, \quad (6.110)$$

$$\left(\nabla^{ac}_\mu \frac{\delta}{\delta A^c_\mu(x)} + A^{ac}_\mu(x)\frac{\delta}{\delta V^c_\mu(x)} - (m^2_\pi + s)\frac{\delta}{\delta p} + p\frac{\delta}{\delta s}\right)S$$

$$= \left(\mathbf{X}_A - (m^2_\pi + s)\frac{\delta}{\delta p} + p\frac{\delta}{\delta s}\right)S = 0. \quad (6.111)$$

It can be checked that $\mathbf{X}_V$ and $\mathbf{X}_A$ are the generators of local $SU_L(2) \times SU_R(2)$.

If we further require

$$\langle 0|\mathbf{A}^a_\mu(x)|\pi^b(p)\rangle = if_\pi\delta^{ab}p_\mu\, e^{-ip\cdot x}, \quad (6.112)$$

then in the absence of stable axial vector or pseudoscalar mesons, this is equivalent to the asymptotic conditions $(x^0 \to \mp\infty)$

$$\mathbf{A}^a_\mu(x) \to -f_\pi\partial_\mu\pi^a_{\text{in,out}}(x), \qquad \partial^\mu\mathbf{A}^a_\mu(x) \to f_\pi m^2_\pi\pi^a_{\text{in,out}}(x) \quad (6.113)$$

where π_{in} and π_{out} are free incoming and outgoing pion fields. Comparison of (6.113) with (6.107) shows that π is a normalized interpolating field. In the present formulation, there is no difficulty in defining such a field. In particular, the issues related to the nonmultiplicative character of the pion field in the effective formulations do not arise here.

To incorporate (6.113) into (6.110-6.111) we introduce a modified action

$$\hat{\mathbf{I}} = \mathbf{I} - f^2_\pi \int d^4x\left(s(x) + \frac{1}{2}A^\mu(x)\cdot A_\mu(x)\right), \quad (6.114)$$

the corresponding S-matrix

$$\hat{S} = S\exp\left(-if^2_\pi \int d^4x\left(s(x) + \frac{1}{2}A^\mu(x)\cdot A_\mu(x)\right)\right) \quad (6.115)$$

and a change of variables $p^a = J^a/f_\pi - \nabla^\mu A^a_\mu$. Taking $\hat{\phi} = (V^a_\mu, A^a_\mu, s, J^a)$ as independent variables, the modified current densities $\hat{\mathbf{J}} = (\mathbf{j}_V, \mathbf{j}_A, f_\pi\hat{\sigma}, f_\pi\hat{\pi})$ may be defined as

$$\hat{\mathbf{J}}(x) = \frac{\delta\hat{\mathbf{I}}}{\delta\hat{\phi}} = -i\hat{S}^\dagger\frac{\delta\hat{S}}{\delta\hat{\phi}}. \quad (6.116)$$

Application of the chain rule yields $(\mathbf{V}, \mathbf{A}, \sigma, \pi)$ in terms of $(\mathbf{j}_V, \mathbf{j}_A, \hat{\sigma}, \hat{\pi})$. Substitution into (6.106) gives

$$\nabla^\mu \mathbf{j}_{V\mu} + \underline{A}^\mu \mathbf{j}_{A\mu} + \underline{J}\pi = 0 \tag{6.117}$$

and therefore

$$\left(\mathbf{X}_V + \underline{J}\frac{\delta}{\delta J}\right)\hat{S} = 0. \tag{6.118}$$

On the other hand, substitution into (6.107) gives

$$\nabla^\mu \mathbf{j}_{A\mu} + \underline{A}^\mu \mathbf{j}_{V\mu} = -f_\pi^2 \nabla^\mu \mathcal{A}_\mu + f_\pi \nabla^\mu \nabla_\mu \pi - f_\pi \underline{A}^\mu \underline{A}_\mu \pi$$
$$+ f_\pi (m_\pi^2 + s)\pi - (J - f_\pi \nabla^\mu \mathcal{A}_\mu) f_\pi (\hat{\sigma} + f_\pi). \tag{6.119}$$

This equation may be integrated by introducing the retarded and advanced Green's functions

$$\left(-\nabla^\mu \nabla_\mu + \underline{A}^\mu \underline{A}_\mu - m_\pi^2 - s\right) G_{R,A}(x,y) = 1\delta^4(x-y) \tag{6.120}$$

and the kernel

$$\mathbf{K} = \nabla^\mu \nabla_\mu - \underline{A}^\mu \underline{A}_\mu + s - \square$$
$$= 2\underline{V}^\mu \partial_\mu + (\partial^\mu \underline{V}_\mu) + \underline{V}^\mu \underline{V}_\mu - \underline{A}^\mu \underline{A}_\mu + s \tag{6.121}$$

where we have adopted a condensed matrix notation. We then have the Yang-Feldman-Källen-like equations [38]

$$\pi = \left(1 + G_R \mathbf{K}\right)\pi_{\text{in}} - f_\pi G_R J + G_R \left(\nabla^\mu \mathcal{A}_\mu - J\right)\hat{\sigma} -$$
$$\frac{1}{f_\pi} G_R \left(\nabla^\mu \mathbf{j}_{A\mu} + \underline{A}^\mu \mathbf{j}_{V\mu}\right)$$
$$= \left(1 + G_A \mathbf{K}\right)\pi_{\text{out}} - f_\pi G_A J + G_A \left(\nabla^\mu \mathcal{A}_\mu - J\right)\hat{\sigma} -$$
$$\frac{1}{f_\pi} G_A \left(\nabla^\mu \mathbf{j}_{A\mu} + \underline{A}^\mu \mathbf{j}_{V\mu}\right). \tag{6.122}$$

When the external fields are switched off, (6.122) reduces to the PCAC equation (4.60). While the latter enforces chiral symmetry exactly at the soft point, the former does the same on-shell. Noting that $\pi_{\text{out}} = \hat{S}^\dagger \pi_{\text{in}} \hat{S}$, and using (6.116) we arrive at

$$
\begin{aligned}
\frac{\delta}{\delta J}\hat{S} &= -if_\pi^2 G_R J\hat{S} + if_\pi \hat{S}\left(1 + G_R\mathbf{K}\right)\pi_{in} \\
&\quad + G_R\left(\nabla^\mu A_\mu - J\right)\frac{\delta\hat{S}}{\delta s} - G_R\mathbf{X}_A\hat{S} \\
&= -if_\pi^2 G_A J\hat{S} + if_\pi\left(1 + G_A\mathbf{K}\right)\pi_{in}\hat{S} \\
&\quad + G_A\left(\nabla^\mu A_\mu - J\right)\frac{\delta\hat{S}}{\delta s} - G_A\mathbf{X}_A\hat{S}.
\end{aligned} \tag{6.123}
$$

Evidently, any result which is a consequence of (6.113) and symmetry (6.110-6.111) as well as general principles such as unitarity and causality must be contained in (6.118-6.123). Since (6.118) simply represents local isospin invariance, the nontrivial results of current algebra must be basically contained in (6.123). The latter is the master formula for chiral symmetry breaking[4].

To demonstrate this, we note that

$$
G_{R,A} = \Delta_{R,A} + \Delta_{R,A}\mathbf{K}G_{R,A} = \Delta_{R,A} + G_{R,A}\mathbf{K}\Delta_{R,A} \tag{6.124}
$$

where $\Delta_{R,A}$ are the Green's functions for free fields. Multiplying (6.123) by $(1 + G_A\mathbf{K})^{-1} = 1 - \Delta_A\mathbf{K}$ gives

$$
\begin{aligned}
\left[\pi_{\text{in}}, \hat{S}\right] &= \hat{S}\Delta\left(1 + \mathbf{K}G_R\right)\mathbf{K}\pi_{\text{in}} - f_\pi\Delta\left(1 + \mathbf{K}G_R\right)J\hat{S} - \\
&\quad \frac{i}{f_\pi}\Delta\left(1 + \mathbf{K}G_R\right)\left(\nabla^\mu A_\mu - J\right)\frac{\delta\hat{S}}{\delta s} + \frac{i}{f_\pi}\Delta\left(1 + \mathbf{K}G_R\right)\mathbf{X}_A\hat{S}
\end{aligned} \tag{6.125}
$$

where $\Delta = \Delta_R - \Delta_A$ is the Jordan-Pauli distribution. Fourier decomposing further yields

$$
\begin{aligned}
\left[a_{\text{in}}^a(k), \hat{S}\right] &= \int d^4y\, d^4z\, e^{ik\cdot y}\left(1 + \mathbf{K}G_R\right)^{ac}(y, z) \\
&\quad \left(-i\hat{S}(\mathbf{K}\pi_{\text{in}})^c(z) + if_\pi J^c(z)\hat{S} - \right.
\end{aligned}
$$

$$\frac{1}{f_\pi}\left(\nabla^\mu A_\mu - J\right)^c(z)\frac{\delta\hat{S}}{\delta s(z)} + \frac{1}{f_\pi}\mathbf{X}_A^c(z)\hat{S}\right)$$

(6.126)

where $a_{\text{in}}^a(k)$ is the annihilation operator of incoming pions with momentum k and isospin a. A similar relation for $a_{\text{in}}^a(k) \to a_{\text{in}}^{\dagger a}(k)$ may also be established, by using the hermitian part of (6.126) and the unitarity of the extended S-matrix. Iterations give the reduction formulas for two- and higher number of pions.

The Bogoliubov causality condition[39] implies that

$$T^*\left(\mathcal{O}(x_1)...\mathcal{O}(x_n)\right) = (-i)^n\hat{S}^\dagger\frac{\delta^n}{\delta\hat{\phi}(x_1)...\delta\hat{\phi}(x_n)}\hat{S}$$

(6.127)

where $\mathcal{O} = -i\hat{S}^\dagger\delta\hat{S}/\delta\hat{\phi}$. With this in mind, using the two-pion reduction formula, sandwiching between nucleon states and switching off the external fields, give the familiar πN scattering formula

$$\langle N(p_2)|\left[a_{\text{in}}^b(k_2), \left[\mathbf{S}, a_{\text{in}}^{a\dagger}(k_1)\right]\right]|N(p_1)\rangle =$$

$$-\frac{i}{f_\pi}m_\pi^2\delta^{ab}\int d^4y e^{-i(k_1-k_2)\cdot y}\langle N(p_2)|\hat{\sigma}(y)|N(p_1)\rangle$$

$$-\frac{1}{f_\pi^2}k_1^\alpha k_2^\beta\int d^4y_1 d^4y_2 e^{-ik_1\cdot y_1+ik_2\cdot y_2}\langle N(p_2)|T^*\left(\mathbf{j}_{A\alpha}^a(y_1)\mathbf{j}_{A\beta}^b(y_2)\right)|N(p_1)\rangle$$

$$+\frac{1}{f_\pi^2}k_1^\alpha\int d^4y e^{-i(k_1-k_2)\cdot y}\epsilon^{abe}\langle N(p_2)|\mathbf{V}_\alpha^e(y)|N(p_1)\rangle$$

(6.128)

where $\mathbf{S} = \hat{S}|_{\phi=0}$ is the on-shell S-matrix. At threshold, (6.128) yields the correct Tomozawa-Weinberg relation.

The extension to $\pi\pi$ scattering is straightforward in principle, although lengthy in practice. We find,

$$iT\left(k_1a, p_1c \to p_2d, k_2b\right) = iT_{\text{tree}} + iT_V + iT_S + iT_R$$

(6.129)

$$iT_{\text{tree}} = \frac{i}{f_\pi^2}\left(s - m_\pi^2\right)\delta^{ac}\delta^{bd} + 2\text{ perm.}$$

(6.130)

$$iT_V = \frac{i}{f_\pi^2}\epsilon^{abe}\epsilon^{cde}\left(\mathbf{F}_V(t) - 1 - \frac{t}{4f_\pi^2}\mathbf{\Pi}_V(t)\right) + 2\text{ perm.}$$

(6.131)

$$
iT_S = -\frac{2im_\pi^2}{f_\pi}\delta^{ab}\delta^{cd}\left(\mathbf{F}_S(t) + \frac{1}{f_\pi} - \frac{1}{2f_\pi^2}\langle 0|\hat{\sigma}(0)|0\rangle\right)
$$

$$
+\frac{m_\pi^4}{f_\pi^2}\delta^{ab}\delta^{cd}\int d^4y\, e^{-i(k_1-k_2)\cdot y}\langle 0|T^*\left(\hat{\sigma}(y)\hat{\sigma}(0)\right)|0\rangle_{\text{con.}}
$$

$$
+2\ \text{perm.} \tag{6.132}
$$

$$
iT_R = \frac{1}{f_\pi^4}k^{1\alpha}k^{2\beta}p^{1\gamma}p^{2\delta}
$$

$$
\int d^4y_1 d^4y_2 d^4y_3 \langle 0|T^*\left(\mathbf{j}_{A\alpha}^a(y_1)\mathbf{j}_{A\beta}^b(y_2)\mathbf{j}_\gamma^c(y_3)\mathbf{j}_{A\delta}^d(0)\right)|0\rangle_{\text{con.}}
$$

$$
\tag{6.133}
$$

where s, t, u are the canonical Mandelstam variables and

$$
\langle 0|a_{\text{in}}^d(p_2)\mathbf{V}_\alpha^e(y)a_{\text{in}}^{\dagger c}|0\rangle_{\text{con.}} = i\epsilon^{dec}(p_1 + p_2)_\alpha\,\mathbf{F}_V(t)e^{-i(p_1-p_2)\cdot y} \tag{6.134}
$$

is the pion electromagnetic form factor,

$$
i\int d^4x\, e^{iq\cdot x}\langle 0|T^*\left(\mathbf{V}_\alpha^a(x)\mathbf{V}_\beta^b(0)\right)|0\rangle = \delta^{ab}\left(-g_{\alpha\beta}q^2 + q_\alpha q_\beta\right)\mathbf{\Pi}_V(q^2)
$$

$$
\tag{6.135}
$$

is the isovector correlation function, and

$$
\langle 0|a_{\text{in}}^d(p_2)\hat{\sigma}(y)a_{\text{in}}^{\dagger c}|0\rangle_{\text{con}} = \delta^{cd}\,\mathbf{F}_S(t)e^{-i(p_1-p_2)\cdot y} \tag{6.136}
$$

is the scalar form factor. Experimentally, eqs.(6.134-6.135) are well described by ρ dominance (modulo subtraction constants). Their appearance is a direct manifestation of the chiral $SU_L(2) \times SU_R(2)$ structure. In the isospin-1 channel, the scalar contribution (6.132) has the wrong quantum numbers. The rest (6.133) is unknown, but since $\mathbf{j}_A$ is one-pion reduced, it is natural to assume that it has no strong resonant behavior. Hence, in the ρ region we expect $\pi\pi$ scattering to be dominated by (6.131-6.132).

This point is confirmed by the data. Ignoring (6.132-6.133), and using a one-resonance fit to $\mathbf{F}_V$ and $\mathbf{\Pi}_V$, we get the results shown in Fig. 6.1, for the real and imaginary parts of the amplitude, with isospin 1 and angular momentum 1 [4]. The results are in good agreement with the data. The bulk of the $\pi\pi$ data is indeed given by (6.131-6.132), as expected from chiral symmetry. This shows that the power of chiral symmetry can extend beyond the threshold region.

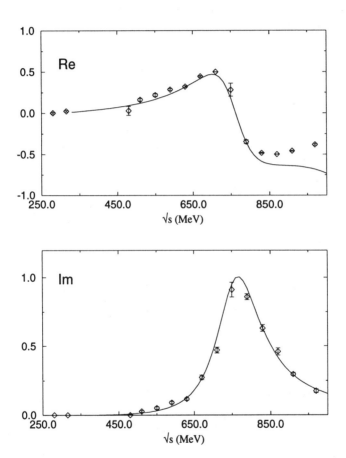

Fig. 6.1. Real (Re) and Imaginary (Im) parts of the P-wave $\pi\pi$ scattering amplitude versus $\sqrt{s}$. The solid line is the result (6.131-6.132) as discussed in the text.

Since f_π appears explicitly in the master equation (6.123), a systematic expansion in $1/f_\pi$ may be feasible. In the baryon-free sector and to leading order, this expansion is analogous to one-loop ChPT. The only distinction is that it is done directly on-shell, with no need to expand quantities such as f_π and m_π. Also, there can be advantages in this formulation for elementary processes in the baryon sector [8]. Their discussion, however, goes beyond the scope of this book.

Appendix

(A$_I$) Anomalies

- ### 1. Fujikawa construction[9]

Here we briefly present the path-integral construction of the abelian anomaly as the consequence of noninvariance of the fermionic measure $[d\bar\psi][d\psi]$ under the chiral transformation (6.17). We start with the Euclidean version of the partition function (6.16)

$$e^{-W(A)} = \int [d\psi][d\bar\psi] e^{-\int d^D x \ \bar\psi i \not{D}\psi}. \qquad (A6.1)$$

Now, under the infinitesimal gauge transformation,

$$\begin{aligned}
\psi &\to \psi + \theta\gamma_{D+1}\psi, \\
\bar\psi &\to \bar\psi + \bar\psi\gamma_{D+1}\theta,
\end{aligned} \qquad (A6.2)$$

the classical action undergoes a shift

$$\int d^D x \bar\psi i \not{D}\psi \to \int d^D x \bar\psi i \not{D}\psi - i \int d^D x \theta(x)\partial_\mu J_A^\mu. \qquad (A6.3)$$

We expand then the quark fields in the basis of the solutions of the Dirac equation $i\not{D}\phi_n = \lambda\phi_n$ (note that the coefficients are Grassmann variables),

$$\begin{aligned}
\psi &= \sum_n a_n \phi_n, \\
\bar\psi &= \sum_n \bar a_n \phi_n^\dagger,
\end{aligned} \qquad (A6.4)$$

so, under chiral transformation (A6.2),

$$a_m \to \sum_n (\delta_{nm} + \int \theta(x)\phi_m^\dagger \gamma_{D+1}\phi_n)a_n. \qquad (A6.5)$$

The Jacobian for this infinitesimal transformation of the Grassmann variables reads

$$\det(\delta_{nm} + \sum_n \int \theta(x)\phi_m^\dagger \gamma_{D+1}\phi_n) = e^{2\sum_n \int \theta(x)\phi_n^\dagger \gamma_{D+1}\phi_n}, \qquad (A6.6)$$

where factor 2 comes from taking the $\bar a$ into account. The nontriviality of this Jacobian manifests itself as the anomaly. Obviously, the sum in the exponent requires regularization,

$$\begin{aligned}
\mathcal{M} &= \int d^D x\, \theta(x) \left(\sum \phi_n(x)^\dagger \gamma_{D+1}\phi_n(x)\right)_{\text{reg}} \\
&= \lim_{\Lambda\to\infty} \int d^D x\, \theta(x) \sum \phi_n(x)^\dagger \gamma_{D+1}\phi_n(x) e^{-\lambda_n^2/\Lambda^2} \\
&= \lim_{\Lambda\to\infty} \int d^D x\, \theta(x)\, \text{Tr}\left(\int \frac{d^D p}{(2\pi)^D}\gamma_{D+1}e^{ipx}e^{-i(\slashed{D}/\Lambda)^2}\right)e^{-ipx} \\
&= \lim_{\Lambda\to\infty} \int d^D x\, \theta(x)\, \text{Tr}\left(\int \frac{d^D p}{(2\pi)^D}\gamma_{D+1}e^{-B}\right) \qquad (A6.7)
\end{aligned}$$

and

$$B = \frac{i}{2\Lambda^2}\left((-p_\mu + iA_\mu)^2 + \frac{1}{2}[\gamma_\mu,\gamma_\nu][D^\mu,D^\nu]\right), \qquad (A6.8)$$

where the regularized sum has been evaluated in the plane wave basis. The calculation of the anomaly is now straightforward. Doing the algebra of the gamma matrices leads to

$$\text{Tr}(\gamma_{\mu_1}\dots\gamma_{\mu_{2k}}) = \begin{cases} 0 & \text{if } k < D/2 \\ \frac{1}{2}(-2i)^{D/2+1}\epsilon_{\mu_1\dots\mu_D} & \text{if } k = D/2. \end{cases} \qquad (A6.9)$$

All terms with $k < D/2$ vanish. Terms with $k \geq D/2$ vanish by taking the cutoff Λ to infinity. Thus the anomaly is given by the only remaining term ($n = D/2$)

$$\mathcal{M} = \frac{1}{n!}\left(\frac{i}{2\pi}\right)^n \int d^D x\theta(x)\, \text{Tr}(\epsilon^{\mu_1\nu_1\dots\mu_n\nu_n}F_{\mu_1\nu_1}\dots F_{\mu_n\nu_n}). \, (A6.10)$$

Equations (6.18), (A6.3) and (A6.10) give immediately the abelian anomaly (in D dimensions). Continuing the result to the Minkowski spacetime, we finally get

$$\partial_\mu J_A^\mu = \frac{1}{2n!}\left(\frac{i}{4\pi}\right)^n \mathrm{Tr}(\epsilon^{\mu_1\nu_1\ldots\mu_n\nu_n}F_{\mu_1\nu_1}\ldots F_{\mu_n\nu_n}). \qquad \text{(A6.11)}$$

- **2. Primer on differential forms.**

We encourage the reader unfamiliar with differential forms to read this section. Here we introduce basic concepts and techniques of the differential forms calculus[41]. The reader familiar with forms may skip this section. Forms provide an extremely simple and compact notation for operations on multi-index tensors.** We define matrix-valued p-forms that result from the contraction of an antisymmetric matrix-valued tensor field $\omega_{\mu_1\mu_2\ldots\mu_p}$ of rank p with antisymmetric product $dx^{\mu_1} \wedge dx^{\mu_2} \wedge \ldots \wedge dx^{\mu_p}$

$$\omega_p = \frac{1}{p!}\omega_{\mu_1\mu_2\ldots\mu_p}dx^{\mu_1} \wedge dx^{\mu_2} \wedge \ldots \wedge dx^{\mu_p} \qquad \text{(A6.12)}$$

where we have denoted the antisymmetric products by wedge, *i.e.*

$$dx^\mu \wedge dx^\nu = -dx^\nu \wedge dx^\mu. \qquad \text{(A6.13)}$$

The numerical factor $1/p!$ in (A6.12) is conventional. Generally, a tensor field $\omega_{\mu_1\ldots\mu_p}$ is a matrix-valued object, *i.e.* $\omega_{\mu_1\ldots\mu_p} = \omega_{\mu_1\ldots\mu_p}^a G^a$, where G^a are matrices (*e.g.*, in our case $G^a = -iT^a$) where T^a are generators of some Lie algebra.

The scalar function $f(x)$ is an example of the 0-form.

Vector gauge fields are the 1-forms: $A = A_\mu dx^\mu$.

The external derivative is also a 1-form: $d = dx^\mu\frac{\partial}{\partial x^\mu}$; therefore $df = \frac{\partial f(x)}{\partial x^\mu}dx^\mu$ is a 1-form too.

The product of p-form ω and a q-form v is a (p+q)-form

$$\omega \wedge v = \frac{1}{p!q!}\omega_{\mu_1\ldots\mu_p}v_{\nu_1\ldots\nu_q}dx^{\mu_1} \wedge \ldots \wedge dx^{\mu_p} \wedge dx^{\nu_1} \wedge \ldots \wedge dx^{\nu_q}$$
$$= (-1)^{pq}v \wedge \omega. \qquad \text{(A6.14)}$$

The differentiation chain rule reads

$$d(\omega \wedge v) = (d\omega) \wedge v + (-1)^p\omega \wedge dv, \qquad \text{(A6.15)}$$

**Another advantage of forms is obvious when one is typing multi-index tensors in TEX.

where we define

$$d\omega = \partial_\nu \omega_{\mu_1 \dots \mu_p} \frac{1}{p!} dx^\nu \wedge dx^{\mu_1} \wedge \dots \wedge dx^{\mu_p}. \tag{A6.16}$$

In the above formulae (A6.14, A6.15), the sign factors are the consequences of the anticommutation properties of the wedge product. Finally, we define the commutator of two forms

$$[\omega, v] = \omega \wedge v \pm v \wedge \omega \tag{A6.17}$$

where the sign (+) appears if both forms are odd. For example, $[A, A] = 2A \wedge A$. Note that A is matrix-valued, $A_\mu = -iA_\mu^a T^a$.
The tensor field strength comes from a 2-form $F = dA + A \wedge A$. Indeed,

$$\begin{aligned} F &= dA + A \wedge A = \partial_\mu A_\nu dx^\mu \wedge dx^\nu + A_\mu A_\nu dx^\mu \wedge dx^\nu \\ &= (\partial_\mu A_\nu - \partial_\nu A_\mu + A_\mu A_\nu - A_\nu A_\mu) \frac{1}{2} dx^\mu \wedge dx^\nu \\ &= \frac{1}{2} F_{\mu\nu} dx^\mu \wedge dx^\nu. \end{aligned} \tag{A6.18}$$

Note that d^2 is always zero, e.g., $d^2 f = 0$ expresses the fact that the rotation of the gradient of any scalar function f vanishes.
Defining the covariant derivative acting on any object X as $DX = dX + [A, X]$, we immediately prove the Bianchi identity

$$\begin{aligned} DF &= dF + [A, F] = d(dA + A \wedge A) + [A, dA + A \wedge A] \\ &= dA \wedge A - A \wedge dA + A \wedge dA - dA \wedge A = 0. \end{aligned} \tag{A6.19}$$

Note also, that

$$\begin{aligned} d\,\mathrm{Tr}F \wedge F &= \mathrm{Tr}(dF \wedge F + F \wedge dF) = 2\,\mathrm{Tr}dF \wedge F = -2\,\mathrm{Tr}[A, F] \wedge F \\ &= -2\,\mathrm{Tr}(A \wedge F \wedge F) + 2\,\mathrm{Tr}(F \wedge A \wedge F) = 0, \end{aligned} \tag{A6.20}$$

where we have used properties of the trace and the Bianchi identity. The above result (A6.20) is true in *any* number of dimensions. In 4 dimensions, it is trivial, since the $d\,\mathrm{Tr}F \wedge F$ is the 5-form, which has to vanish in 4 dimensions by definition (A6.12). It is useful to define a dual (or Hodge) operation $*$ for a p-form in D dimensions

$$*(dx^{\mu_1} \wedge \dots \wedge dx^{\mu_p}) = \frac{1}{(D-p)!} \epsilon_{\mu_1 \dots \mu_D} dx^{\mu_{p+1}} \wedge \dots \wedge dx^{\mu_D}. \tag{A6.21}$$

The Hodge operation "transmutes" the p-form into (D-p)-form. The double Hodge operation reproduces the form, modulo a sign

$$** \omega_p = (-1)^{p(D-p)} \omega_p. \tag{A6.22}$$

As an example, we can calculate the divergence of the 1-form (current) J in $D = 4$ dimensions as

$$d * J = d(\epsilon_{\mu_1 \ldots \mu_4} J^{\mu_1} \frac{1}{3!} dx^{\mu_2} \wedge dx^{\mu_3} \wedge dx^{\mu_4})$$
$$= \partial_\alpha J^\alpha \frac{1}{4!} \epsilon_{\nu_1 \nu_2 \nu_3 \nu_4} dx^{\nu_1} \wedge dx^{\nu_2} \wedge dx^{\nu_3} \wedge dx^{\nu_4}. \tag{A6.23}$$

Integration of the p-form ω is based on Stokes' theorem, which in our notation reads

$$\int_{\mathcal{M}} d\omega = \int_{\partial \mathcal{M}} \omega \tag{A6.24}$$

where $\mathcal{M}$ is a $(p+1)$-dimensional manifold (surface) and $\partial \mathcal{M}$ is a boundary of this manifold. This compact formula summarizes the Gauss-Ostrogradski and Stokes formulae known from the traditional vector analysis. We relate the integration over forms to usual multidimensional integration by the replacement:

$$dx^{\mu_1} \wedge \ldots \wedge dx^{\mu_p} \rightarrow \epsilon^{\mu_1 \cdots \mu_p} dV \tag{A6.25}$$

where ϵ is the Levi-Civita tensor and dV is an infinitesimal element of the volume in p dimensions.

We introduce now two fundamental definitions for the forms. We say that a form ω is *closed*, if $d\omega = 0$. We say that a form ω is *exact*, if $\omega = d\alpha$. Since $d^2 = 0$, an *exact* form is always *closed*. However, the converse statement is generally not true. Poincaré lemma states that a *closed* form is only *locally exact*.

We visualize the above statement on the basis of an important example. Let a 0-form $g(x)$ be an element on the Lie group G, and the point x belongs to D-dimensional space. Then 1-form $l = g^{-1} dg$ is an element of the corresponding Lie algebra. Let us consider the D-form

$$\omega = \text{Tr}(\underbrace{l \wedge l \wedge \ldots \wedge l}_{D}). \tag{A6.26}$$

This form is closed, *i.e.*, $dw = 0$, as a direct consequence of the identity $dl + l \wedge l = 0$ (known as the Cartan-Maurer equation), which we prove immediately:

$$
\begin{aligned}
dl + l \wedge l &= d(g^{-1}dg) + l \wedge l = dg^{-1} \wedge dg + l \wedge l \\
&= -g^{-1}(dg)g^{-1} \wedge dg + l \wedge l \\
&= -l \wedge l + l \wedge l = 0
\end{aligned}
\tag{A6.27}
$$

where in the second line we have used $gg^{-1} = 1$. Let us consider the following integral Q over the D-dimensional sphere S_D,

$$
Q = \int_{S_D} \omega.
\tag{A6.28}
$$

If our closed form ω (A6.26) is *globally* exact, the integral has to vanish. If the closed form is only *locally* exact (Poincaré lemma), the integral Q is different from zero.

Properly normalized integral Q is defined as the topological charge, characterizing the mapping of each point x of sphere S_D onto the group element $g(x) \in G$. Types of topological (global) obstructions (singularities) of this mapping are denoted by $\Pi_D(G)$ and form the D-th homotopy group. For example, let us consider $U(1)$, parametrized as $g = e^{i\alpha(x)}$. Then $-ig^{-1}dg = d\alpha$ is a 1-form. This form is obviously closed, $d^2\alpha = 0$. Parametrizing $\alpha \in [0, 2\pi]$ we see that

$$
Q \equiv \frac{1}{2\pi} \int_{S_1} d\alpha = \frac{1}{2\pi} \int_0^{2\pi} \left(\frac{d\alpha}{dx}\right) dx = 1.
\tag{A6.29}
$$

Other parametrizations ($\theta \in [2\pi n, 2\pi(n+1)]$) give $Q = n$, where $n \in \mathbf{Z}$, set of integers. Therefore $\Pi_1(U(1)) = \mathbf{Z}$.

An important example of a higher homotopy group is $\Pi_3(SU(2)) = \mathbf{Z}$. Instantons and skyrmions are the direct consequences of the nontrivial $\Pi_3(SU(2))$.

- **3. Topological construction of the anomalies**[42].

In this section we construct the anomalies using the topological methods. The compact notation of the forms allows us to perform the construction in a transparent way.

We start from the case of an abelian anomaly. In $D = 2n$ dimensions, the anomaly is given by a $2n$-form (Chern character)

$$\Omega_{2n}(A) = \text{Tr}\underbrace{(F \wedge \cdots \wedge F)}_{n} \qquad (A6.30)$$

where F is a two-form corresponding to the field tensor. For brevity, we will use the simpler notation for the strings of the same forms, denoting, e.g.,

$$\underbrace{F \wedge \ldots \wedge F}_{n} = F^n. \qquad (A6.31)$$

For the case $n = 2$ ($D = 4$), we reproduce, up to normalization, the abelian anomaly (6.19). The (A6.30) is gauge-invariant. Since the field tensor transforms covariantly under gauge transformations, for an infinitesimal chiral gauge transformation, $\delta_v F = [F, v]$, and

$$\delta_v \Omega_{2n}(A) = n \, \text{Tr}[F, v] \wedge F^{n-1} = 0 \qquad (A6.32)$$

as a consequence of the property of the trace. Ω is also a closed form:

$$d\Omega_{2n} = n \, \text{Tr}(dF \wedge F^{n-1}) = -n \, \text{Tr}([A, F] \wedge F^{n-1}) \qquad (A6.33)$$

where, along with the properties of trace, the Bianchi identity has been used. Therefore, the Poincaré lemma says that Ω_{2n} could be *locally* represented by a $(2n - 1)$ form ω_{2n-1}^0 (called Chern-Simons secondary form)

$$\Omega_{2n}(A) = d\omega_{2n-1}^0(A). \qquad (A6.34)$$

We could find the explicit form of ω_{2n-1}^0 using the following trick. For any space-time point x, let γ be a curve in a field space, connecting 0 and the field A. We parametrize this curve by t, *i.e.* we define the mapping: $\gamma : t \in [t_a, t_b] \to A(t)$, where $A(t_a) = 0$ and $A(t_b) = A$. In this way, instead of one field A, we have the whole family of fields $A(t)$. Then we have $F(t) = dA(t) + A(t) \wedge A(t)$. Let us now rewrite (A6.30) in the following trivial way:

$$\Omega_{2n} = \text{Tr} F^n = \int_{t_a=0}^{t_b} dt \left(\frac{d}{dt} \text{Tr} F(t)^n \right). \qquad (A6.35)$$

Note that the integration over the real parameter t is the "usual integration", *i.e.*, dt is not a form, and commutes with the forms dx. Then, denoting differentiation over t by dot, we continue with the string of equalities (A6.35)

$$
\int_{t_a=0}^{t_b} dt \frac{d}{dt} \mathrm{Tr} F(t)^n = n \int_{t_a}^{t_b} dt \, \mathrm{Tr}(\dot{F}(t) \wedge F(t)^{n-1})
$$

$$
= n \int_{t_a}^{t_b} dt \, \mathrm{Tr}([d\dot{A} + \dot{A} \wedge A(t) + A(t) \wedge \dot{A}] \wedge F(t)^{n-1})
$$

$$
= n \int_{t_a}^{t_b} dt \, \mathrm{Tr}(d\dot{A} \wedge F^{n-1} + \dot{A} \wedge [A, F^{n-1}])
$$

$$
= n \int_{t_a}^{t_b} dt \, \mathrm{Tr}(d\dot{A} \wedge F^{n-1} - \dot{A} \wedge dF^{n-1})
$$

$$
= d \left(n \int_{t_a}^{t_b} dt \, \mathrm{Tr}(\dot{A} \wedge F^{n-1}) \right) \tag{A6.36}
$$

where we have used the Bianchi identity, properties of trace and rules for differentiation of the product of forms. From the last line, we read (up to the total derivative $d\rho_{2n-2}$, since $d^2 = 0$) the solution for ω_{2n-1}

$$
\omega_{2n-1}^0(A) = n \int dt \, \mathrm{Tr} \dot{A}(t) \wedge F^{n-1} + d\rho_{2n-2}. \tag{A6.37}
$$

As an example, we calculate the anomaly in $D = 4$ dimensions ($n = 2$). We take t belonging to interval $[0, 1]$, and we choose the mapping $A(t) = tA$. Obviously $A(1) = A$ and $A(0) = 0$. Then $\dot{A}(t) = A$, $F(t) = tdA + t^2 A^2$, and the trivial integration over t yields

$$
\omega_3^0(A) = 2 \int_0^1 dt \, \mathrm{Tr}(A \wedge F(t)) = 2 \int_0^1 dt \, \mathrm{Tr} A \wedge (tdA + t^2 A^2)
$$

$$
= \mathrm{Tr}(A \wedge dA + \frac{2}{3} A^3). \tag{A6.38}
$$

This is the anomaly (6.19). How can we fix the unknown coefficient in front of the anomaly? We have noticed in Chapter 2 that the integrated anomaly equation is the Atiyah-Singer index theorem. The change of the chirality, being the integer, has to match another integer, the topological charge, *i.e.*, the properly normalized integral over the Chern character. This requirement gives the proportionality constant equal to $-(16\pi^2)^{-1}$, reproducing exactly the abelian anomaly (6.19).

The same argument works in higher spacetime dimensions, (only the normalization of the higher topological charges changes) yielding the unambiguous prediction for an abelian anomaly in $D = 2n$ dimensions

$$\partial_\mu J_A^\mu = c_n \epsilon_{\mu_1 \nu_1 \ldots \mu_n \nu_n} \text{Tr}(F^{\mu_1 \nu_1} \ldots F^{\mu_n \nu_n}) = c_n 2^n \text{Tr} * F^n \quad \text{(A6.39)}$$

with

$$c_n = \frac{i^n}{2^{2n-1} \pi^n n!}. \quad \text{(A6.40)}$$

- **4. Topological construction of the nonabelian anomaly[42].**

We will now demonstrate how to obtain the Bardeen anomaly from topological considerations, without calculating a single triangle Feynman diagram. The crucial element for our construction is the consistency condition obtained by Wess and Zumino. By definition, any object $Y(A)$ depending solely on the gauge fields *and* fulfilling the Wess-Zumino consistency condition gives the nonabelian anomaly. Zumino, Wu and Zee have guessed[42] that the abelian anomaly in $D + 2$ dimensions determines the nonabelian anomaly in D dimensions. Topological justification of this relation came later[43].

Let us recapitulate their reasoning. We consider the Chern form

$$\Omega_{D+2}(A_L, A_R) = \text{Tr} F_R^{D/2+1} - \text{Tr} F_L^{D/2+1}. \quad \text{(A6.41)}$$

Note that Ω has the form of the difference of two abelian anomalies in $D + 2$ for right and left fermions. The relative negative sign comes from the fact that the nonabelian anomaly is odd under parity (see G^1). In the previous section, we have noticed that any Chern character is a closed form, therefore locally

$$\Omega_{D+2} = d\omega_{D+1}. \quad \text{(A6.42)}$$

From gauge invariance of the Chern character, we read

$$0 = \delta_\theta \Omega_{D+2} = \delta_\theta(d\omega_{D+1}) = d(\delta_\theta \omega_{D+1}) \quad \text{(A6.43)}$$

therefore, again applying Poincaré lemma, we get that *locally*

$$\delta_\theta \omega_{D+1} = d\omega_D^1(\theta, A) \quad \text{(A6.44)}$$

where ω_D^1 is the D-form and the superscript 1 reminds us that this form is *linear* in θ. If we define gauge-variant functional $Y(A) = \int_{H_{D+1}} \omega_{D+1}(A)$, then $Y(A)$ automatically fulfills the Wess-Zumino consistency condition, because the Wess-Zumino condition is just the consequence of the Lie structure of the generators,

$$\delta_\theta \delta_\xi - \delta_\xi \delta_\theta = \delta_{[\theta,\xi]} \tag{A6.45}$$

acting on any gauge-variant functional $Y(A)$. H_{D+1} is a $(D+1)$-dimensional hemisphere. The boundary of this hemisphere is a D-dimensional "equator". In the case of $D = 4$, this "equator" is a sphere in four dimensions, S_4, which, by stereographic projection, can be mapped onto the Euclidean space-time volume. Therefore the nonabelian anomaly comes as:

$$\delta_v Y(A) = \int_{H_5} \delta_v \omega_5 = \int_{S_4} \omega_4^1(\theta, A) \tag{A6.46}$$

and we may identify our function $Y(A)$ with the effective action $W(A)$, modulo terms that are completely chirally-invariant. We can now construct the explicit form of the nonabelian anomaly in $D = 4$ dimensions. Repeating our trick (A6.36),

$$\Omega_6 = \text{Tr}(F_R^3 - F_L^3) = \int_{t_a}^{t_b} dt \left(\frac{d}{dt} \text{Tr} F(t)^3 \right)$$
$$= d \left(3 \int_{t_a}^{t_b} dt\, \text{Tr} \dot{A}(t) \wedge F(t)^2 \right), \tag{A6.47}$$

we immediately read ω_5 from (A6.42)

$$\omega_5(A_L, A_R, \gamma) = 3 \int_{t_a}^{t_b} dt\, \text{Tr}(\dot{A}(t) \wedge F(t)^2) + d\rho_4 \tag{A6.48}$$

where we have explicitly denoted by γ the dependence on the integration path. Since $d^2 = 0$, we know ω_5 up to $d\rho_4$. We will now show that different paths in the space of A correspond to different renormalization schemes. Hereafter we follow the elegant discussion by Petersen[44].

In the case of the symmetric scheme (S), we choose the path γ from A_L to 0, and then from 0 to A_R, as depicted in Fig. 6.2a. We have

$$\omega_5(A_L, A_R, (S)) = \omega_5(A_L, 0) + \omega_5(0, A_R)$$
$$= \omega_5(0, A_R) - \omega_5(0, A_L) \tag{A6.49}$$

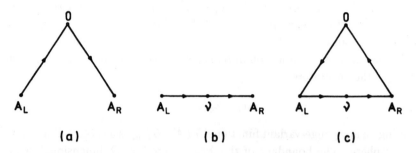

Fig. 6.2. Integration trajectories for anomalies: (a) symmetric scheme, (b) Bardeen scheme, (c) renormalization term $G^{(2)}$.

where the last equality comes from an elementary property of the integral. Let us now choose the simple parametrization: $A_{L,R}(t) = tA_{L,R}$, $t \in [0,1]$, so that $\dot{A}_{L,R}(t) = A_{L,R}$ and

$$F_{L,R}(t) = dA_{L,R}(t) + A^2_{L,R}(t) = tdA_{L,R} + t^2 A^2_{L,R}. \qquad (A6.50)$$

Thus we get

$$\omega_5 = 3 \int_{H_5} \int_0^1 dt \, \mathrm{Tr} A_R \wedge (tdA_R + t^2 A_R^2)^2 - (R \to L) \qquad (A6.51)$$

and integrating over monomials in t, we get

$$\omega_5 = \int_{H_5} \mathrm{Tr}(A_R \wedge F_R^2 - \frac{1}{2} A_R^3 \wedge F_R + \frac{1}{10} A_R^5) - (R \to L). \qquad (A6.52)$$

The next step is to calculate the change of ω_5 under an infinitesimal chiral gauge transformation,

$$\delta A_{L,R} = d\theta_{L,R} + [A_{L,R}, \theta_{L,R}],$$
$$\delta F_{L,R} = [F_{L,R}, \theta_{L,R}]. \qquad (A6.53)$$

An elementary algebra leads to

$$\delta \omega_5 = d \int (10\theta_R(F_R^2 - \frac{1}{2}(A_R^2 F_R + A_R F_R A_R + F_R A_R^2) + \frac{1}{2} A_R^4))$$
$$- d \int (10\theta_L(F_L^2 - \frac{1}{2}(A_L^2 F_L + A_L F_L A_L + F_L A_L^2) + \frac{1}{2} A_L^4))$$

$$(A6.54)$$

from which, with the $F = dA + A^2$, we reproduce the nonabelian anomaly (6.31),

$$\text{Tr}(\theta d(A_R dA_R + \frac{1}{2} A_R^3)) - \text{Tr}(\theta d(A_L dA_L + \frac{1}{2} A_L^3)). \quad \text{(A6.55)}$$

The Bardeen anomaly corresponds to the choice of the path depicted in Fig. 6.2b. Here the obvious parametrization is $A(t) = V + tA$, where $V = \frac{1}{2}(A_R + A_L)$ and $A = \frac{1}{2}(A_R - A_L)$ and $t \in [-1, 1]$. The infinitesimal vector transformations are

$$\delta_V V = d\theta + [V, \theta]$$
$$\delta_V A = [A, \theta]$$
$$\delta_V F(t) = [F(t), \theta] \quad \text{(A6.56)}$$

and the corresponding axial transformations are

$$\delta_A V = [A, \theta],$$
$$\delta_A A = d\theta + [V, \theta]. \quad \text{(A6.57)}$$

We can simply check that this choice of the trajectory leads automatically to the conserved vector current:

$$\delta_V \omega_5 = 3\delta_V \int_{H_5} \int_{-1}^{1} dt\, \text{Tr}\dot{A}(t) \wedge F^2$$
$$= 3 \int_{H_5} \int_{-1}^{1} dt\, \text{Tr}(\delta_V A F^2 + A\delta_V F^2)$$
$$= 3 \int_{H_5} \int_{-1}^{1} dt\, \text{Tr}([A, \theta]F^2 + A[F, \theta]F + AF[F, \theta]) = 0.$$
$$\text{(A6.58)}$$

By definition, the anomaly ω_4 in the Bardeen scheme is given by

$$\delta_A \omega_5 = d\omega_4. \quad \text{(A6.59)}$$

A rerun of the above step plus a little additional algebra leads to the form for the axial anomaly

$$\omega_4 = 3 \int_{H_5} \int_{-1}^{1} dt\, \text{Tr}(\theta(F^2(t) + 3(1 - t^2)A^2 F(t))) \quad \text{(A6.60)}$$

from which the integration reproduces

$$6\,\text{Tr}\theta[F_V^2 + \frac{1}{3}F_A^2 + \frac{8}{3}A^4 - \frac{4}{3}(F_V A^2 + A F_V A + A^2 F_V)] \quad \text{(A6.61)}$$

[here F_V, F_A is the form notation for (6.26)], which is precisely the $G^{(1)}$ term rewritten in the form notation.

To trace back to the origin of the $G^{(2)}$ term, let us look at Fig. 6.2. The Bardeen anomaly corresponds to the integration along the base of the triangle. The symmetric (left-right) anomaly corresponds to the integral along the two remaining arms of the triangle. Therefore a circulation around the triangle (Fig. 6.2c) gives

$$\left(\int_{\text{site}A_V A_R} + \int_{\text{site}A_R 0} + \int_{\text{site}0 A_L}\right)\omega_5 = \oint_{\text{triangle}A_L A_R 0} \omega_5 \quad \text{(A6.62)}$$

or, generically,

$$\text{Anomaly(B)} - \text{Anomaly(S)} = \oint_{\text{triangle}A_L A_R 0} \omega_5. \quad \text{(A6.63)}$$

We can calculate this integral using Stokes' theorem. Let us parametrize every point of the surface of the triangle in the A-space by $A(s,t) = sA_L + tA_R$, where $s, t \in [0,1]$, so $F(s,t) = dA(s,t) + A(s,t)^2$. Circulating along the contour of the "vector field"

$$\begin{pmatrix} \text{Tr}(A_L F^2(s,t)) \\ \text{Tr}(A_R F^2(s,t)) \end{pmatrix} \quad \text{(A6.64)}$$

is equal to the flux of the rotation of the vector field through the surface of the triangle

$$\oint \omega_5 = 3 \int ds\,dt\,\text{Tr}\left(A_R \frac{\partial F^2(s,t)}{\partial s} - A_L \frac{\partial F^2(s,t)}{\partial t}\right)$$

$$= 3 \int_0^1 ds \int_0^{1-s} dt\,\text{Tr}(A_R A_L F(s,t) - A_R F(s,t) A_L). \quad \text{(A6.65)}$$

which leads to the $G^{(2)}$ term. It is also clear now why this term is local in 4 dimensions, contrary to the anomalous term $G^{(1)}$, which, as observed by Witten[19], could be rewritten in the local form only after increasing the number of dimensions to five.

- **5. Wess-Zumino-Witten effective action.**

The infinitesimal change of the quantum action under chiral transformation is given by the anomaly. Wess and Zumino have noticed that it is possible to calculate the finite change under the chiral transformation by combining the sequence of infinitesimal transformations. Witten[19] has discovered how to do it in a more effective way, by increasing the number of dimensions by one. The Wess-Zumino-Witten action reads

$$S_{WZW}(A_L, A_R, g_L, g_R) = W(A_L, A_R) - W(A_L^{g_L}, A_R^{g_R}). \quad \text{(A6.66)}$$

Let us calculate the WZW action for the simplest case when both fields A_L and A_R vanish. We perform the integration in the Bardeen scheme, *i.e.* we impose the conserved vector current. Then the axial field comes only from pure-gauge terms

$$\mathcal{A} = \frac{1}{2}(A_R^{g_R} - A_L^{g_L}) = \frac{1}{2}(g_R dg_R^{-1} - g_L dg_L^{-1}). \quad \text{(A6.67)}$$

The Wess-Zumino-Witten action reads

$$\begin{aligned} S_{WZW}(0, 0, g_L, g_R) &= -W(A_L^{g_L}, A_R^{g_R}) \\ &= -4b_2 \int_{H_5} \int_{-1}^{1} dt \, \text{Tr}(\dot{A}(t) F^2(t)). \end{aligned} \quad \text{(A6.68)}$$

In our case, $A(t) = t\mathcal{A}$, $F(t) = dA(t) + A(t)^2 = (t^2 - t)\mathcal{A}^2$, where we have used the fact that $\mathcal{A}$ is a pure gauge, *i.e.* $d\mathcal{A} + \mathcal{A}^2 = 0$, and the integration over t leads to

$$S_{WZW}(0, 0, g_L, g_R) = b_2 \int_{H_5} \text{Tr}(U^{-1} dU)^5 \quad \text{(A6.69)}$$

where in the last line we have changed variables by introducing $U = g_L^{-1} g_R$. This is the only combination of g_L and g_R that the effective action could depend on in the vector scheme. Finally, we can fix the unknown coefficient b_2.

Note that there is an ambiguity in the definition of the hemisphere H_5. We have two possible choices, "northern" hemisphere H_5^N, or "southern" hemisphere H_5^S. Both have the same boundary, the "equator" S_4. Physics should not depend on the choice of the hemisphere, so

$$e^{i \int_{H_5^N} \omega_5} = e^{i \int_{H_5^S} \omega_5} \quad \text{(A6.70)}$$

or, equivalently, the difference

$$\int_{H_5^N} \omega_5 - \int_{H_5^S} \omega_5 = 2\pi N \tag{A6.71}$$

must be a multiple of 2π. On the other hand,

$$\int_{H_5^N} \omega_5 - \int_{H_5^S} \omega_5 = \int_{H_5^N \cup H_5^S} \omega_5$$

$$= \int_{S_5} \omega_5 = \int_{S_5} \text{Tr}(U^{-1}dU)^5 \tag{A6.72}$$

where we have used: (1) that the fixed orientation on equator S_4 induces relative opposite orientations on both hemispheres; and (2) that gluing together two hemispheres along the equator S_4 produces a compact "hypersphere" S_5. The last equation should give the topological charge corresponding to $\Pi_5(SU(N_f))$ (*cf.* (A6.28)), and this fixes the normalization. Generally,

$$b_n = \frac{i^{n-1}N}{(4\pi)^n(2n+1)!!} \tag{A6.73}$$

so, finally, in the case of $D = 2n = 4$, the anomalous action is

$$S_{WZW}(0,0,g_L,g_R) = \frac{iN}{240\pi^2}\int_{H_5} \text{Tr}(U^{-1}dU)^5. \tag{A6.74}$$

In an analogous way, we can construct the Wess-Zumino-Witten action for the non-zero external currents. Let us quote here the result for the abelian (electromagnetic) current. In this case, $A_L = A_R = -ieQA_{em}$, $F_{em} = dA_{em}$, where A_{em} is the electromagnetic potential and Q is the charge matrix for the quarks. Then the anomalous action for Goldstone bosons interacting with photons reads

$$S_{WZW}(A_{em},U) = \frac{iN}{240\pi^2}\int_{H_5} \text{Tr}(U^{-1}dU)^5$$

$$+ \int_{S_4} \text{Tr}\left(\frac{eN_c}{48\pi^2}A_{em} \wedge ((dU\,U^{-1})^3 - (dU^{-1}U)^3)Q\right.$$

$$+ i\frac{N_c e^2}{48\pi^2}F_{em} \wedge A_{em} \wedge (2(dU^{-1}UQ^2 - dUU^{-1}Q^2)$$

$$\left. + UQU^{-1}QdUU^{-1} - U^{-1}QUQdU^{-1}U)\right). \tag{A6.75}$$

Expanding U in pion fields, we see that (A6.75) relates the amplitude for the $\gamma \to 3\pi$ (second line) to the amplitude for $\pi^0 \to 2\gamma$ (third line) and the amplitude for $2\gamma \to 3\pi$ (last line). Each of these amplitudes (*e.g.* (1.53)) fixes the arbitrary integer N to the number of colors, *i.e.*, $N = N_c$.

(A$_{II}$) Determinant

In this Appendix, we detail some of the calculations that are needed to evaluate the divergent and finite parts of the one-loop contribution in (6.84). The calculation involves the determinant of the differential operator $\mathcal{D}$ in a series expansion of the external fields. For that, we define

$$\mathcal{D} = \overset{0}{\mathcal{D}} + \tilde{\mathcal{D}} \tag{A6.76}$$

where the free propagator

$$\overset{0}{\mathcal{D}}_{PQ} = \mathbf{1}_{PQ}\left(\square + \overset{0}{m}_P{}^2\right) \tag{A6.77}$$

with pseudoscalar mass $\overset{0}{m}_P$ given at tree level is defined in the physical basis $\{\lambda_P\}$,

$$\lambda_{\pi^\pm} = \mp\frac{1}{\sqrt{2}}\left(\lambda^1 \pm i\lambda^2\right),$$

$$\lambda_{K^\pm} = \mp\frac{1}{\sqrt{2}}\left(\lambda^4 \pm i\lambda^5\right),$$

$$\lambda_{K^0, \overline{K}^0} = -\frac{1}{\sqrt{2}}\left(\lambda^6 \pm i\lambda^7\right),$$

$$\lambda_{\pi^0} = +\lambda^3\cos\epsilon + \lambda^8\sin\epsilon,$$

$$\lambda_\eta = -\lambda^3\sin\epsilon + \lambda^8\cos\epsilon \tag{A6.78}$$

with ϵ the $\pi^0\eta$-mixing angle. In the decomposition (A6.76), $\tilde{\mathcal{D}}$ is given by

$$\tilde{\mathcal{D}} = [\hat{\gamma}^\mu, \partial_\mu]_+ + \hat{\Gamma}^\mu\hat{\Gamma}_\mu + \overline{\sigma} - \overset{0}{m}{}^2 \mathbf{1} \tag{A6.79}$$

with

$$(\hat{\Gamma}_\mu)_{PQ} = -\frac{1}{2}\operatorname{Tr}\left([\lambda_P, \lambda_Q^\dagger]\left(\frac{1}{2}\left(\sqrt{U_*}\nabla_\mu^R\sqrt{U_*} + \sqrt{U_*^\dagger}\nabla_\mu^L\sqrt{U_*^\dagger}\right)\right)\right) \tag{A6.80}$$

and

$$\bar{\sigma}_{PQ} = +\frac{1}{2} \text{Tr}\left([\lambda_P, \nabla_\mu][\lambda_Q^\dagger, \nabla^\mu] \right)$$
$$+ \frac{1}{8} \text{Tr}\left([\lambda_P, \lambda_Q^\dagger]_+ \left(\sqrt{U_*}\xi^\dagger\sqrt{U_*} + \sqrt{U_*^\dagger}\xi\sqrt{U_*^\dagger} \right) \right) - \mathbf{1}_{PQ}\, \overset{0}{m}_P^2 .$$

$$(A6.81)$$

Since $\tilde{\mathcal{D}}$ vanishes if the external fields are switched off, we may expand the determinant in powers of $\tilde{\mathcal{D}}$

$$\frac{i}{2}\ln \det\mathcal{D} = \frac{i}{2}\ln \det\overset{0}{\mathcal{D}} + \frac{i}{4}\ln \det\left(\overset{0}{\mathcal{D}}^{-1}\tilde{\mathcal{D}} \right) - \frac{i}{4}\ln \det\left(\overset{0}{\mathcal{D}}^{-1}\tilde{\mathcal{D}}\overset{0}{\mathcal{D}}^{-1}\tilde{\mathcal{D}} \right)$$
$$+ \cdots .$$

$$(A6.82)$$

The second term in (A6.82) is the sum of all tadpoles. The third term in (A6.82) collects all graphs with two vertices in the loop. To the order quoted, (A6.82) is the full one-loop contribution. Using dimensional regularization, we explicitly have

$$\frac{i}{2}\ln \det\mathcal{D} = \frac{i}{2}\sum_P \Delta_P(0) \int d^4x\, \bar{\sigma}_{PP}(x)$$
$$+ \sum_{PQ} \int d^x d^4y \left(\mathbf{M}_{\mu\nu}(x-y)\hat{\Gamma}^\mu_{PQ}(x)\hat{\Gamma}^\nu_{QP}(y) \right.$$
$$\left. + \mathbf{K}_\mu(x-y)\hat{\Gamma}^\mu_{PQ}(x)\bar{\sigma}_{QP}(y) + \frac{1}{4}\mathbf{J}(x-y)\bar{\sigma}_{PQ}(x)\bar{\sigma}_{QP}(y) \right)$$

$$(A6.83)$$

with

$$\mathbf{M}_{\mu\nu}(z) = \frac{i}{4}\left(\partial_\mu\Delta_P\partial_\nu\Delta_Q + \partial_\nu\Delta_P\partial_\mu\Delta_Q \right.$$
$$\left. - \partial_{\mu\nu}\Delta_P\,\Delta_Q - \Delta_P\partial_{\mu\nu}\Delta_Q + g_{\mu\nu}\delta(z)\left(\Delta_P(0) + \Delta_Q(0) \right) \right)$$
$$= (\partial_{\mu\nu} - g_{\mu\nu}\Box)\mathcal{M}(z) - g_{\mu\nu}\mathcal{L}(z) - \frac{\tilde{\gamma}}{6}(\partial_{\mu\nu} - g_{\mu\nu}\Box)\delta(z), \quad (A6.84)$$

$$\mathbf{K}_\mu(z) = \frac{i}{2}\left(\partial_\mu\Delta_P\,\Delta_Q - \Delta_P\partial_\mu\Delta_Q \right) = -\partial_\mu\mathcal{K}(z), \quad\quad (A6.85)$$

$$\mathbf{J}(z) = -i\Delta_P\,\Delta_Q = \tilde{\mathcal{J}}(z) - 2\tilde{\gamma}\delta(z) \quad\quad\quad (A6.86)$$

where the renormalized quantities in (A6.86) can be found in the original work by Gasser and Leutwyler[18]. The divergences are contained in the parameter $\tilde{\gamma}$ as defined through

$$\tilde{\gamma} = \frac{\mu^{d-4}}{16\pi^2} \left(\frac{1}{d-4} - \frac{1}{2} \ln 4\pi + \Gamma'(1) + 1 \right) \qquad \text{(A6.87)}$$

in d dimensions. Here μ is the renormalization scale brought in by dimensional regularization. More details on this calculation can be found in the original work by Gasser and Leutwyler[1,18].

References

1. J. Gasser and H. Leutwyler, Ann. Phys. **158**, 142 (1984).
2. S. Weinberg, Physica (Amsterdam) **96A**, 327 (1979).
3. H. Leutwyler, Ann. Phys. (N.Y.) **235**, 165 (1994).
4. H. Yamagishi and I. Zahed, SUNY-NTG-94-56.
5. G. A. Vilkovisky, Nucl. Phys. **234**, 125 (1984).
6. J. Polchinski, in *Recent Directions in Particle Physics*, eds. J. Harvey and J. Polchinski (World Scientific, Singapore, 1993) p.235.
7. R.D. Ball, Phys. Repts. **182**, 1 (1989).
8. J. Steele, H. Yamagishi and I. Zahed, SUNY-preprint, SUNY-NTG-95.
9. K. Fujikawa, Phys. Rev. **D21**, 2848 (1980).
10. A.P. Balachandran, in *High-Energy Physics 1985*, Vol. 1, eds. M.J. Bowick and F. Gürsey (World Scientific, Singapore, 1986).
11. B. Zumino, in *Relativity, Groups and Topology II*, eds. B. DeWitt and R. Stora (North-Holland, Amsterdam, 1984).
12. W.A. Bardeen, Phys. Rev. **184**, 1848 (1969).
13. A.P. Balachandran, G. Marmo, V.P. Nair and C.G. Trahern, Phys. Rev. **D25**, 2713 (1982).
14. J. Wess and B. Zumino, Phys. Lett. **B37**, 95 (1971).
15. F. Gürsey, Nuov. Cimento (Ser. X) **50A**, 129 (1967).
16. K. Nishijima, Nuov. Cimento (Ser. X) **11**, 698 (1959).
17. C. Callan, S. Coleman, J. Wess and B. Zumino, Phys. Rev. **177**, 2247 (1969).
18. J. Gasser and H. Leutwyler, Nucl. Phys. **B250**, 465 (1985).
19. E. Witten, Nucl. Phys. **B233**, 422 (1983).
20. H. Leutwyler, Nucl. Phys. B (Proc. Suppl.) **7A**, 42 (1989).
21. V.A. Novikov, M.A. Shifman, A.I. Vainshtein, and V.I. Zakharov, Nucl. Phys. **B191**, 301 (1981).
22. M.D. Scadron and H.F. Jones, Phys. Rev. **D10**, 967 (1974); H. Sazdjian and J. Stern, Nucl. Phys. **94**, 163 (1975); R.J. Crewther, Phys. Lett. **176**, 172 (1986).
23. J. Stern, H. Sazdjian and N.H. Fuchs, Phys. Rev. **D47**, 3814 (1993).
24. H. Georgi and A. Manohar, Nucl. Phys. **B234**, 189 (1984).

25. J. Gasser and H. Leutwyler, Nucl. Phys. **250**, 539 (1985).
26. J. Bijnens, G. Colangelo, G. Ecker and J. Gasser, "Semileptonic Kaon Decays", to be published.
27. S. Belluci, J. Gasser and M.E. Sainio, Nucl. Phys. **B423**, 80 (1994); erratum **B439**, 413 (1994).
28. H. Marsiske *et al.*, Phys. Rev. **D41**. 3324 (1990).
29. G. Velo and D. Zwanziger, Phys. Rev. **186**, 1337 (1969).
30. G. Ecker, J. Gasser, H. Leutwyler, A. Pich and E. de Rafael, Phys. Lett. **B233**, 425 (1989).
31. E. Jenkins and A.V. Manohar, in *Proc. of the Workshop on Effective Field Theories of the Standard Model*, Dobogókö, Hungary, Aug. 1991, ed. U.-G. Meissner (World Scientific, Singapore, 1992); U.G. Meissner, Rep. Prog. Phys. **56**, 903 (1993)
32. S. Weinberg, Phys. Lett. **B251**, 288 (1990); Nucl. Phys. **B363**, 3 (1991); T.-S. Park, D.-P. Min and M. Rho, Phys. Repts. **233**, 341 (1993); C. Ordóñez, L. Ray and U. van Kolck, Phys. Rev. Lett. **72**, 1982 (1994).
33. H. Yamagishi and I. Zahed, SUNY-preprint 1988, unpublished.
34. V.P. Frolov and G.A. Vilkovisky, in *Quantum Gravity*, ed. M.A. Markov (Plenum, London, 1983)
35. M. Veltman, Phys. Rev. Lett. **17**, 553 (1966); J.S. Bell, Nuov. Cimento (Ser. X) **50A**, 129 (1967).
36. H. Yamagishi and I. Zahed, SUNY preprint NTG-94-57.
37. R.E. Peierls, Proc. Roy. Soc. **A214**, 143 (1952).
38. C.N. Yang and D. Feldman, Phys. Rev. **79** (1950) 972; G. Källen, Ark. Fys. **2** (1950) 187, 371.
39. N.N. Bogoliubov and D.V. Shirkov, *Introduction to the Theory of Quantized Fields* (Interscience Publishers, Inc., New York, 1959).
40. T. Eguchi, P.B. Gilkey and A.J. Hanson, Phys. Repts. **66**, 213 (1980).
41. H. Flanders, *Differential Forms with Applications to Physical Sciences* (Academic Press, 1963).
42. B. Zumino, Wu Yong-Shi and A. Zee, Nucl. Phys. **B239**, 477 (1984).
43. L. Alvarez-Gaumé and P. Ginsparg, Nucl. Phys. **B243**, 449 (1984).
44. J.L. Petersen, Acta Phys. Pol. **B16**, 271 (1985).

CHAPTER 7

CHIRAL SOLITONS

7.1. Introduction

In Chapter 3, the notion of large number of colors (N_c) was used to derive effective Lagrangians of QCD applicable at low energy in terms of finite number of long wavelength meson excitations. If such a Lagrangian is the *entire* content of QCD at low energy, then where are the ground-state baryons? In what follows, we discuss in some detail how in large-N_c QCD baryons can emerge as solitons in mesonic effective theories. First we describe the ground-state systems, the nucleon and the Δ. Then we discuss the possible extensions of the model to the case of the hyperons. In particular, we show the various ways of incorporating explicitly the strangeness into the model.

That baryons must appear as solitons in meson field theory[1,2] is plausible but not fully established. The reason is that even in the limit $N_c = \infty$, the exact solution of QCD is not available. In any event, as we learned in Chapter 3, in the large N_c limit, QCD is a weakly interacting meson field theory, with the meson mass going as $O(1)$ in $1/N_c$ and the n-Goldstone boson function going as $\sim N_c^\lambda$, $\lambda = (2 - n)/2$. The meson scattering amplitude is described by the tree graphs of this effective Lagrangian with the coupling constant g proportional to $1/\sqrt{N_c}$, so goes to zero as N_c tends to infinity. Now this theory has no fermions and hence no *explicit* baryons. In quark picture, one needs N_c quarks to make up a baryon; thus the mass of a baryon must go like

$$m_B \sim N_c \sim \frac{1}{g^2}. \tag{7.1}$$

Such an object can occur from mesons only if there is a soliton in the meson field configuration. Such a configuration is known to exist and well understood in less than four dimensions. The soliton mass in weakly coupled meson field theory does indeed go like $1/\alpha$ where α is the coupling constant of the meson fields. Thus it must be that a baryon in QCD is a soliton with a large mass $\sim N_c$ with a size of $O(1)$. In the same vein, a nucleus made of A nucleons must be a soliton of mass $\sim AN_c$ of the same mesonic Lagrangian. In treating baryons as solitons, the basic assumption is that the bulk of baryon physics is dictated by the leading N_c contributions and next $1/N_c$ corrections are in some systematic fashion calculable. It is not obvious that such corrections are necessarily small. For instance, nuclear binding is of $O(1/N_c)$ and hence nuclei exist in the subleading order. So how useful is the concept of nucleons and nuclei as solitons? This is the question addressed below and in Chapter 10.

7.2. Soliton

7.2.1. *Stability of the soliton*

The large N_c consideration *suggests* that baryons can only arise from mesonic effective theories as solitons. Let us first see whether a stable soliton can indeed be obtained from simple meson Lagrangians.

The longest wavelength chiral Lagrangian is the leading nonlinear σ-model Lagrangian

$$\mathcal{L}_2 = \frac{f^2}{4} \operatorname{Tr}(\partial_\mu U \partial^\mu U^\dagger). \tag{7.2}$$

Does this Lagrangian by itself give a stable soliton? One can answer this question quite simply.

Before proceeding to answer this question, we have to introduce a basic notion of a soliton in the way that we are using it here[3]. (The notion of a soliton as used in particle and nuclear physics is different from other fields where the term soliton is applied to localized solutions which maintain their form even in scattering processes.) We will be considering only topological solitons, and not the non-topological variety one finds in the literature. The topology involves the space of fields of U. A sufficient condition for the existence of a soliton is that

$$U(\vec{x}, t) \to 1 \quad \text{as} \quad |\vec{x}| = r \to \infty \tag{7.3}$$

with the rate of approach of U to 1 sufficiently fast so that the energy is finite, $E < \infty$. Consider the usual hedgehog ansatz made by Skyrme[4]

$$U_0(\vec{x}) = e^{i\hat{x}\cdot\vec{\tau}F(r)}. \tag{7.4}$$

The pion field $\vec{\pi}$ in this ansatz depends only on the radial variable, and is aligned in isospin along the normal spatial vector $\hat{x} \equiv \vec{x}/|\vec{x}|$, *i.e.* $\pi_i = f_\pi F(r)\hat{x}_i$. Then the static energy functional is

$$\begin{aligned}
E &= 2\pi f^2 \int_\epsilon^\infty r^2 dr \left[(dF/dr)^2 + \frac{2}{r^2}\sin^2 F\right] \\
&= 2\epsilon\pi f^2 \int_1^\infty y^2 dy \left[(dF/dy)^2 + \frac{2}{y^2}\sin^2 F\right]
\end{aligned} \tag{7.5}$$

where the second equality is obtained with the change of variable $y \equiv r/\epsilon$. We have set ϵ as the lower limit for a reason that will be apparent soon. If we let $\epsilon = 0$, then the energy of the system is zero. Since the integral is positive definite, this is the lowest energy. The constant f has a dimension of mass and since the only other scale parameter possible is the size of the system (R), the energy must be proportional to f^2R, and hence the lowest-energy solution corresponds to the zero size, namely, a collapsed system. This is the well-known result of the "Hobart-Derrick theorem"[5]. Since there is no known massless point-like hadron with fermion quantum number, this must be an artifact of the defects of the theory.

There is one simple way[6] to avoid the Hobart-Derrick collapse without introducing any further terms in the Lagrangian or any other degrees of freedom, which is, in a way, physically relevant. It is to simulate short-distance physics by "drilling" a hole into the skyrmion or equivalently by taking a nonzero ϵ in the energy functional (7.5). This is not as academic an issue as it seems since a hole of that type arises naturally in the chiral bag model discussed in the next chapter. Let us for the moment ignore what is inside that hole and see instead what the hole does to the system. Since the detail of the analysis that establishes the existence of a stable soliton for $\epsilon \neq 0$ is not very relevant to us as there is a natural way of assuring a stable system using a realistic quark bag, it suffices to merely summarize what transpires from the analysis: For *any* nonzero value of ϵ, there is always a solution which is classically stable against the Hobart-Derrick collapse and is stable against a small fluctuation η of the solution F:

$$F(y) \to F(y) + \eta(y). \tag{7.6}$$

Since the system collapses for $\epsilon = 0$, it is clearly a singular point. This singular behavior is probably related to the presence or absence of a topological quantum number in the chiral bag[7]. Indeed a chiral bag can be shrunk to a point without losing its topological structure as long as a proper chiral boundary condition is imposed. Without the boundary condition, the topological quantum number is not conserved and the bag decays into a vacuum. Thus a nontrivial topological structure requires a finite ϵ to allow the boundary condition to be imposed at the bag boundary and if the bag is shrunk, then at the center. A relevant physical question is how to determine ϵ. One practical approach suggested by Balakrishna *et al*[6] is to elevate it to a quantum variable, the expectation value of which is related to the "bag radius." From this point of view, a bag is a necessary physical ingredient. On the other hand, although this is highly suggestive, the discussion on the Cheshire Cat phenomenon given in the next chapter clearly indicates that the procedure of replacing a quark bag by a quantized cutoff variable can only be an academic one.

Another way of stabilizing the system for $\epsilon = 0$ might be to introduce quantum fluctuations before solving for the minimum configuration. It is plausible that quantum effects could stabilize the system by themselves just as the collapse of the S-wave state of the classical hydrogen atom can be prevented by quantum fluctuations. It turns out that merely quantizing a few collective degrees of freedom such as scale variable does not work without a certain "confinement" artifact of the sort mentioned above. So far, nobody has been able to stabilize the skyrmion of (7.5) with $\epsilon = 0$ in a fully convincing way. This of course does not mean that such a scheme is ruled out. There are many other collective variables than the scale variable to quantize and it is possible that by a suitable quantization of appropriate variables, the system could be stabilized. We want to stress here however that these are all academic issues. We are working with effective Lagrangians that are supposed to model QCD. Hence there are many other degrees of freedom in nature, be they in quark variables as in the chiral bag or in meson variables such as the strong vector mesons. We saw above that a quark-bag does trivially avoid the Hobart-Derrick collapse. As one eliminates the quark bag, higher derivative terms in the chiral field arise through various induced gauge fields, among which we can have the quartic term

$$\frac{1}{32g^2} \operatorname{Tr}[U^\dagger \partial_\mu U, U^\dagger \partial_\nu U]^2 \tag{7.7}$$

the so-called "Skyrme quartic term" which Skyrme introduced (in a somewhat *ad hoc* way) to stabilize the soliton. We know now that such a term can arise in some particular way from the ρ meson when the latter is integrated out. We can readily verify that this term is enough to stabilize the soliton. To see this, let us scale $U(x) \to U(\lambda x)$. Then the energy of the system changes

$$E_\lambda = \lambda^{2-d} E_{(2)} + \lambda^{4-d} E_{(4)} \tag{7.8}$$

where the first term is the contribution from the quadratic term and the second from the quartic term, with subscripts labeling them appropriately and d is the space dimension (equal to 3 in nature). The minimum condition implies

$$dE_\lambda/d\lambda = 0 \to E_{(2)}/E_{(4)} = -(d-4)/(d-2) \tag{7.9}$$

and the stability implies

$$d^2 E_\lambda/d\lambda^2 > 0 \to 2(d-2)E_{(2)} > 0. \tag{7.10}$$

Thus for stability, we need $d > 2$. For $d = 3$, we have $E_{(2)} = E_{(4)}$, the virial theorem. What this implies is that once one has more massive strong interaction degrees of freedom, the stability issue no longer arises. Indeed there has been considerable controversy as to which vector mesons stabilize the soliton, in particular, the question as to whether the ρ meson can do so by itself. It is now understood that while the Skyrme quartic term which can be traced to the ρ meson in origin stabilizes the system, the finite-range ρ meson cannot and it is instead the ω meson which plays the role of stabilization. There is no lack of candidates for the role if some do not do the job. Our attitude will be that the skyrmion is safe from the Hobart-Derrick collapse and so we can focus on the physics of the skyrmion.

7.2.2. *Structure of the soliton*

The original Skyrme model captures the essence of the soliton structure, so we will focus on the Lagrangian

$$\mathcal{L}_{sk} = \frac{f^2}{4} \operatorname{Tr}(\partial_\mu U \partial^\mu U^\dagger) + \frac{1}{32g^2} \operatorname{Tr}[U^\dagger \partial_\mu U, U^\dagger \partial_\nu U]^2. \tag{7.11}$$

A quantitatively more precise prediction can be made with a Lagrangian that contains the strong vector mesons (*e.g.*, in consistency with hidden gauge symmetry) but the physics is not qualitatively modified by it.[*]

The static energy of the skyrmion given by (7.11) reads

$$E = \int d^3x \left\{ -\frac{f^2}{4} \operatorname{Tr} L_i^2 - \frac{1}{32g^2} \operatorname{Tr}[L_i, L_j]^2 \right\} \qquad (7.12)$$

with $L_i = U^\dagger \partial_i U$. It is convenient to scale the length by $1/gf$ by setting $x \to (1/gf)x$ and give the energy in units of $f/2g$. In these units, the energy reads

$$E = \int d^3x \left\{ -\frac{1}{2} \operatorname{Tr} L_i^2 - \frac{1}{16} \operatorname{Tr}[L_i, L_j]^2 \right\} . \qquad (7.13)$$

One can rewrite this as

$$\begin{aligned} E &= \int d^3x \left(-\frac{1}{2} \right) \operatorname{Tr} \left\{ L_i \pm \frac{1}{4}\epsilon_{ijk}[L_j, L_k] \right\}^2 \\ &\pm 12\pi^2 \int d^3x \frac{1}{24\pi^2}\epsilon_{ijk} \operatorname{Tr}(L_i L_j L_k). \end{aligned} \qquad (7.14)$$

The first term is non-negative and the second integral measures the winding number which can be identified with the baryon number B

$$B = \frac{1}{24\pi^2} \int d^3x \epsilon_{ijk} \operatorname{Tr}(L_i L_j L_k) \qquad (7.15)$$

and thus we have the topological bound on the energy

$$E \geq 12\pi^2 |B|. \qquad (7.16)$$

This bound is known as the Faddeev bound[9] analogous to Bogomol'nyi bound in magnetic monopoles. This bound would be saturated if we had the equality

$$L_i + \frac{1}{4}\epsilon_{ijk}[L_j, L_k] = 0. \qquad (7.17)$$

---

[*]Although we shall not discuss in detail hidden gauge symmetry (HGS) in this volume, we should note that there is a definite advantage in adhering to the hidden gauge symmetry strategy. As emphasized by Georgi[8], it enables us to make a systematic chiral counting, particularly when the vector-meson masses can be treated as "light" as we do in Chapter 10. If the vector mesons are introduced disregarding HGS, one has to make sure that the chiral counting comes out correctly, which is not always transparent.

But it turns out that the only solution to this equation is the trivial one[10], namely, $U = constant$ and so $L_i = 0$. On the other hand, if the space is curved, this bound *can* actually be satisfied and plays an interesting role in connection with chiral symmetry restoration.

To proceed further, we shall now take an ansatz for the chiral field U. In particular, we assume the hedgehog form

$$U(\vec{x}) = e^{iF(r)\hat{x}\cdot\tau}. \tag{7.18}$$

This hedgehog configuration is believed to give the absolute classical minimum of the Skyrme Lagrangian[11]. With this, the energy (7.13) is

$$\begin{aligned}
E &= 4\pi \int_0^\infty dr \left\{ r^2[(dF/dr)^2 + 2r^{-2}\sin^2 F] \right. \\
&\quad \left. + \sin^2 F[r^{-2}\sin^2 F + 2(dF/dr)^2] \right\} \\
&= 4\pi \int_0^\infty dr[r^2(dF/dr + r^{-2}\sin^2 F)^2 + 2\sin^2 F(dF/dr + 1)^2] \\
&\quad - 24\pi \int_0^\infty dr \sin^2 F \frac{dF}{dr}. \tag{7.19}
\end{aligned}$$

For the energy to be finite, U, (7.18), must approach a constant at infinity which we will choose to be the unit matrix. This means that the space is compactified to three-sphere (S^3) and U represents a map from S^3 to $SU(2)$ which is topologically S^3, the winding number of which was given by the right-hand side of (7.15). The implication of this on eq. (7.19) is that since the "profile function" F at the origin and at infinity must be an integral multiple of π, the second term of (7.19) must correspond to $12\pi[F(0) - F(\infty)] = 12\pi^2 B$, the Faddeev bound. As mentioned above, the first integral of (7.19) cannot vanish for a nontrivial U in flat space, so the energy is always greater than $12\pi^2 B$.

Minimizing the energy with respect to the profile function F gives us the classical equation of motion. The exact solution with appropriate boundary conditions at the origin and at infinity is known numerically [12,13]. An exact analytical solution may also exist, but for the present qualitative discussion, we do not need a precise form for F. Indeed one can infer a simple analytic form by the following argument[10]. First of all, it turns out numerically that the first integral in (7.19) is small so that the soliton energy is only slightly greater than the Faddeev bound. The equation of motion shows that $F(r)$ goes like $B\pi - \alpha(r/r_0)$ as $r \to 0$ and like $\beta(r_0/r)^2$ as $r \to \infty$ with some constant r_0. Let us therefore take the form for $B = 1$

$$
\begin{aligned}
F &= \pi - \alpha(r/r_0), &\quad r \leq r_0, \\
&= \beta(r_0/r)^2, &\quad r \geq r_0.
\end{aligned}
\tag{7.20}
$$

Demanding the continuity of F and of its first derivative with respect to r at $r = r_0$, one determines

$$
\alpha = 2\pi/3, \quad \beta = \pi/3.
\tag{7.21}
$$

Substituting (7.20) with (7.21) into (7.19), one finds

$$
E \approx 4\pi(4.76 r_0 + 7.55/r_0)
\tag{7.22}
$$

where the coefficients of r_0 and $1/r_0$ are estimated numerically. Minimizing with respect to r_0, one gets $r_0 = 1.26$ (in units of $(gf)^{-1}$) and $E = 151$ (in units of $f/2g$) which is 1.03 times the exact energy evaluated numerically. The ansatz (7.20) is compared in Fig.7.1 with the exact numerical solution. It is possible that there are more accurate analytic (and simple) forms available in the literature but our aim of showing the explicit form of (7.20) was to illustrate the simplicity of the function F.

It is mentioned above that the true energy of the soliton is only slightly greater than the Faddeev bound $12\pi^2$. Numerically it turns out to be about 23 % above. An interesting question is: When can the Faddeev bound be saturated? As stated, this can never happen in flat space, which is the case in nature since the space we are dealing with is R^3 or $S^3(\infty)$, the three-sphere with an infinite radius. Consider instead the space $S^3(L)$ with L finite. Let its (inverse) metric be g^{ij} and denote the dreibein by e^j_m so that

$$
g^{ij} = e^i_m e^j_m.
\tag{7.23}
$$

If we define

$$
X_m \equiv e^i_m \partial_i U U^\dagger
\tag{7.24}
$$

then the static energy of the system is (for $n_B = 1$)

$$
E = \int d^3x \sqrt{g} [-\frac{1}{2}(X_m + \frac{1}{4}\epsilon_{mnp}[X_n, X_p])^2] + 12\pi^2.
\tag{7.25}
$$

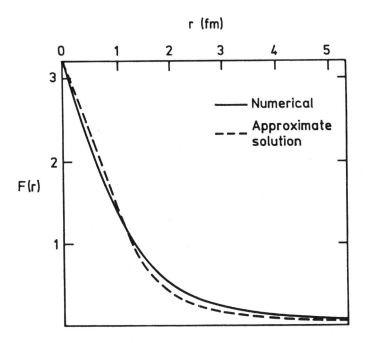

Fig. 7.1. The approximate profile function (7.20) compared with the exact (numerical) profile function.

Note that the topological term, $12\pi^2$, is independent of the metric as expected. Now the bound is saturated provided

$$X_m + \frac{1}{4}\epsilon_{mnp}[X_i, X_j] = 0. \qquad (7.26)$$

When the space is flat (*i.e.*, $L = \infty$), this equation reduces to eq. (7.17) which has no nontrivial solution as mentioned above. In curved space, however, it has a solution when $L = 1$. (A simple way of showing this is to use a geometrical technique patterned after nonlinear elasticity theory. See Manton[14].) This is an identity map, inducing no distortion and corresponds to isometry. One can perhaps associate this particular solution with a state of symmetry different from the spontaneously broken chiral symmetry for which the bound can never be satisfied.

7.2.3. *Quantization of the soliton*

To give a physical meaning, the soliton has to be quantized. As it is, the hedgehog solution is not invariant under separate rotation in configuration space or in isospace. (It is localized in space also, so breaks translational invariance. We will ignore this problem as it is not essential for our purpose.) It is invariant, however, under the simultaneous rotation of the two, with the conserved quantum number being the "grand spin" which is the vector sum of the spin and the isospin

$$\vec{K} = \vec{J} + \vec{I}. \tag{7.27}$$

The hedgehog configuration breaks the rotational symmetry as mentioned, as a consequence of which there are three zero modes corresponding to the triplet of the isospin direction. For $S \in SU(2)$, the equation of motion of the hedgehog is invariant under the transformation

$$U(\vec{x}) \rightarrow SU(\vec{x})S^\dagger \tag{7.28}$$

so the states obtained by rotating the hedgehog are degenerate with the unrotated hedgehog. To restore the symmetry, one has then to elevate the constant matrix S to a quantum variable by endowing it a time dependence and quantize the zero modes. This is essentially equivalent to rotating the hedgehog. It is also equivalent to projecting out good angular momentum and isospin states. Since this rotation leads to next order in $1/N_c$ in the N_c counting in physical quantities, we are implicitly making the adiabatic approximation.

Substituting $U(\vec{x}, t) = S^\dagger(t)U(\vec{x})S(t)$ into the Skyrme Lagrangian (7.11), we get

$$L_{sk} = -\mathcal{M} + \mathcal{I}\,\mathrm{Tr}(\dot{S}\dot{S}^\dagger) \tag{7.29}$$

with

$$\mathcal{I} = \frac{2\pi}{3}\frac{1}{g^3 f}\int_0^\infty s^2 ds \sin^2 F[1 + 4F'^2 + \frac{4\sin^2 F}{s^2}] \tag{7.30}$$

where $s = fgr$ is the dimensionless radial coordinate and $\mathcal{M} = E$, the soliton energy. Note that in terms of N_c, both $\mathcal{M}$ and $\mathcal{I}$ are of order N_c while $\dot{S}$ is of order $1/N_c$. The theory can now be quantized canonically in a standard way. To do so, introduce the collective coordinates $(a_0, \vec{a})$ by

$$S = a_0 + i\vec{\tau} \cdot \vec{a} \tag{7.31}$$

with

$$S^\dagger S = 1 = a_0^2 + \vec{a}^2. \tag{7.32}$$

The momentum conjugate to a_k is

$$\pi_k = \frac{\partial L_k}{\partial \dot{a}_k} = 4\mathcal{I}\dot{a}_k. \tag{7.33}$$

The Hamiltonian is

$$H = \sum_{k=0}^{3} \pi_k \dot{a}_k - L_{sk} = \mathcal{M} + \frac{1}{8\mathcal{I}} \sum_{k=0}^{3} \pi_k^2. \tag{7.34}$$

Quantization is effected by taking $\pi_k = -i(\partial/\partial a_k)$; thus

$$H = \mathcal{M} + \frac{1}{8\mathcal{I}} \sum_{k=0}^{3} \left(-i \frac{\partial}{\partial a_k} \right)^2. \tag{7.35}$$

The second term is just the Laplacian on S^3 whose eigenstates are the Jacobi polynomials in a's. Using the Noether method, the spin J and isospin I can be calculated:

$$J_k = \frac{i}{2} \left(a_k \frac{\partial}{\partial a_0} - a_0 \frac{\partial}{\partial a_k} - \epsilon_{klm} a_l \frac{\partial}{\partial a_m} \right), \tag{7.36}$$

$$I_k = \frac{i}{2} \left(a_0 \frac{\partial}{\partial a_k} - a_k \frac{\partial}{\partial a_0} - \epsilon_{klm} a_l \frac{\partial}{\partial a_m} \right) \tag{7.37}$$

with

$$\vec{J}^2 = \vec{I}^2 = \frac{1}{4} \sum_{k=0}^{3} \left(-\frac{\partial^2}{\partial a_k^2} \right). \tag{7.38}$$

Therefore the Hamiltonian describes a spherical top:

$$H = \mathcal{M} + \frac{1}{2\mathcal{I}} \vec{J}^2. \tag{7.39}$$

The energy eigenstates of this Hamiltonian can be explicitly constructed as polynomials in a's. Whether they are even or odd polynomials depends upon whether the object satisfies bosonic or fermionic statistics. A detailed discussion on this matter will be given in the next subsection. If one considers fermions, then for the ground state $J = I = 1/2$ and its wave function will then be a polynomial in a's. Specifically,

$$|p \uparrow\rangle = \frac{1}{\pi}(a_1 + ia_2), \qquad |p \downarrow\rangle = -\frac{i}{\pi}(a_0 - ia_3), \qquad (7.40)$$

$$|n \uparrow\rangle = \frac{1}{\pi}(a_0 + ia_3), \qquad |n \downarrow\rangle = -\frac{1}{\pi}(a_1 - ia_2). \qquad (7.41)$$

These can be identified with the physical proton with spin up or down and neutron with spin up or down. A similar construction can be made for the excited states with $J = I = 3/2$ etc. A more general construction will be given later when we include the strangeness in the scheme.

7.3. Baryons as skyrmions

In what follows, we supply evidence that we are indeed dealing with objects (solitons) which correspond to baryons expected in QCD. We shall start with phenomenological aspects with $SU(2)$ skyrmions and later give more mathematical aspects associated with the Wess-Zumino term.

Some people have difficulty accepting the skyrmion as a bona-fide baryon consistent with QCD. They ask where the quark-gluon imprints are in the skyrmion description. They even attempt to find in skyrmions what should be there in QCD. In other words, their suspicion is that the skyrmion "misses" something basic of QCD. Throughout what follows, we will argue that this attitude is unfounded.

7.3.1. *SU(2) skyrmions*

Masses

Let us assume that for N_c odd, we have a fermion and when quantized as sketched above, it is a baryon. For $N_c = 3$, we have two states $J = I = 1/2$ and $J = I = 3/2$. The first is the nucleon, the ground state and the second the rotationally excited state Δ. In terms of N_c, their masses are of the form

$$M_J = M_1 + M_0 + M^J_{-1} + O(N_c^{-2}) \qquad (7.42)$$

where the subscript stands for n in N_c^n. The $O(N_c)$ term corresponds to the soliton mass $\mathcal{M}$ of eq. (7.39) and the $O(N_c^{-1})$ to the "hyperfine structure" term, *i.e.*, the second term of (7.39). The "fine structure" term M_0 has not yet been encountered. It is the first ($O(\hbar)$) quantum correction to the mass known as Casimir energy which describes the shift in energy of the vacuum produced by the presence of the soliton. Note that the two terms of $O(N_c)$ and $O(1)$ define the ground state of the baryons and the dependence on specific quantum numbers appears at $O(1/N_c)$. When heavy flavors are introduced, there will be an *additional* overall shift at $O(1)$ for each flavor. An important point is that the first two terms of (7.42) are common to all baryons, be they light-quark or heavy-quark baryons.

We will later present an argument that the term M_0 is attractive and can be of order of ~ -0.5 GeV. For the moment, let us treat it as a parameter independent of quantum numbers and estimate how much it could be from phenomenology. Let us consider the nucleon and Δ. For definiteness, we will consider the Skyrme Lagrangian $\mathcal{L}_{sk}$ (7.11). There are two constants in the model, f and g. We fix the constant f to the pion decay constant $f_\pi = 93$ MeV. The constant g can be obtained from a more general Lagrangian containing the HGS vector mesons but here we shall fix it from imposing that the g_A of the neutron decay come out to be the experimental value 1.25 (see below how g_A is calculated)[12]. The resulting value is $g \approx 4.76$. Restoring the energy unit $f/2g$, the soliton mass is

$$M_1 = \mathcal{M} = 1.23 \times 12\pi^2 \left(\frac{f}{2g} \right) \approx 1423 \text{ MeV}. \tag{7.43}$$

Similarly from (7.30), we get numerically

$$\mathcal{I}^{-1} \approx 193 \text{ MeV}. \tag{7.44}$$

We can now *predict* the mass difference between Δ and N

$$\Delta M = \frac{3}{2} \mathcal{I}^{-1} \approx 290 \text{ MeV} \tag{7.45}$$

to be compared with the experiment 296 MeV. The Casimir energy is a difficult quantity to calculate accurately and since our knowledge on it is quite uncertain, we cannot make a prediction for the absolute value of the ground state baryon (*i.e.*, nucleon) mass. We can turn the procedure around and get some idea on the magnitude of the Casimir energy *required* for consistency with nature. Taking the nucleon mass formula with $J = 1/2$, we have $M_1 + \frac{3}{8} \mathcal{I}^{-1} \approx 1495$ MeV and hence

$$M_0 \approx (940 - 1495) \text{ MeV} = -555 \text{ MeV}. \tag{7.46}$$

This may look large but it is in the range expected theoretically as discussed below [15] and also consistent with the expectation based on the N_c counting that

$$M_1 > |M_0| > M_{-1}. \tag{7.47}$$

There is another caveat to this discussion which would modify the numerics somewhat. We have not taken into account the translational mode; the center-of-mass correction, which is of $O(N_c^{-1})$, should increase the centroid energy. Therefore the value of M_0 may even be higher in magnitude than (7.46).

7.3.2. *Static properties*

As an example of static quantities, consider the axial-vector coupling constant g_A. The nucleon matrix element of the axial current is

$$\langle N(p_l)|A_\mu^a|N(p)\rangle = \bar{U}(p')\frac{\tau^a}{2}\left[g_A(q^2)\gamma_\mu\gamma_5 + h_A(q^2)q_\mu\gamma_5\right]U(p) \tag{7.48}$$

with $q_\mu = (p - p')_\mu$ and $U(p)$ the Dirac spinor for the nucleon. We will continue working with the chiral limit, so the pseudoscalar form factor h has the zero-mass pion pole. Thus in the limit $q_\mu \to 0$, we have

$$
\begin{aligned}
\lim_{q\to 0}\langle N(p')|A_i^a|N(p)\rangle &= \lim_{q\to 0} g_A(0)(\delta_{ij} - \hat{q}_i\hat{q}_j)\langle N|\frac{\tau^a}{2}\sigma_j|N\rangle \\
&= \frac{2}{3}g_A(0)\langle N|\frac{\tau^a}{2}\sigma_i|N\rangle
\end{aligned} \tag{7.49}
$$

where we have expressed the matrix element in the nucleon rest frame. We now obtain the relevant matrix element in the present model. Using the axial current suitably "rotated", we have

$$\int d\vec{x}A_i^a(\vec{x}) = \frac{1}{2}\bar{g}\,\mathrm{Tr}(\tau_i S^\dagger \tau_a S), \tag{7.50}$$

with

$$
\begin{aligned}
\bar{g} &= \frac{2\pi}{3g^2}\int_0^\infty s^2 ds \left(F' + \frac{\sin 2F}{s}(1 + 4F'^2)\right. \\
&\quad \left. + 8F'\frac{\sin^2 F}{s^2} + \frac{4\sin^2 F}{s^3}\sin 2F\right)
\end{aligned} \tag{7.51}
$$

where we have used the dimensionless radial coordinate $s = (fg)r$. Note that given the profile function, $\bar{g}$ depends only on the constant g that figures in the Skyrme quartic term. The relevant matrix element is

$$\lim_{q \to 0} \int d\vec{x} e^{i\vec{q}\cdot\vec{x}} \langle N|A_i^a(\vec{x})|N\rangle = \frac{1}{2}\bar{g}\langle N| \operatorname{Tr}(\tau_i S^\dagger \tau_a S)|N\rangle$$

$$= \frac{2}{3}\bar{g}\langle N|\frac{\tau^a}{2}\sigma_i|N\rangle. \tag{7.52}$$

We are therefore led to identify

$$g_A(0) = \bar{g}. \tag{7.53}$$

Given g_A from experiment, one can then determine the constant g from this relation. The value $g \approx 4.76$ used in the mass formula was determined in this way.

Other static quantities like magnetic moments and various radii are calculated in a similar way and we will not go into details here, referring to standard review articles[16].

7.3.3. *The Casimir energy*

The calculation of the $O(N_c^0)$ Casimir energy is in practice a very difficult problem although in principle it is well-defined, given an effective Lagrangian. The reason is that there are ultraviolet divergences that have to be canceled by counterterms and since the effective theory is non-renormalizable in the conventional sense, increasing number of counterterms intervene as one makes higher-order computations. This problem was discussed in Chapters 5 and 6. As noted there, in principle, given that the finite counterterms can be completely determined from experiments, the non-renormalizability is not a serious obstacle to computing the Casimir contribution. The main problem, however, is that chiral perturbation theory applicable in the Goldstone boson sector cannot be naively applied because of the baryon mass which is not small compared with the chiral scale $\Lambda_\chi \sim 4\pi f_\pi$. We will describe in Chapter 10 how this problem is avoided in chiral expansion when baryon fields are explicitly introduced as heavy matter fields but it is not clear how to do so in the skyrmion structure.

The basic idea of computing the Casimir energy of $O(N_c^0)$ is as follows: Let us consider specifically the Skyrme Lagrangian (7.11). The argument applies to any effective Lagrangian with more or less complications. Let

$L_\mu \equiv i\tau \cdot \mathbf{L}_\mu$. Then (7.11) takes the form

$$\mathcal{L}_{sk} = \frac{f^2}{2} \left(\mathbf{L}_\mu \cdot \mathbf{L}^\mu + \frac{1}{2\kappa^2}[(\mathbf{L}_\mu \cdot \mathbf{L}^\mu)^2 - (\mathbf{L}_\mu \cdot \mathbf{L}_\nu)^2] \right) \qquad (7.54)$$

with $\kappa = gf$. We denote by $\tilde{\mathbf{L}}_\mu$ the static solution of the Euler-Lagrange equation of motion

$$\partial^\mu \left(\tilde{L}_\mu^a + \frac{1}{\kappa^2}(\tilde{L}_\mu^a \tilde{\mathbf{L}}_\nu \cdot \tilde{\mathbf{L}}_\nu - \tilde{\mathbf{L}}_\mu \cdot \tilde{\mathbf{L}}_\nu \tilde{L}_\nu) \right) = 0. \qquad (7.55)$$

In terms of the chiral field U, the solution is of course the hedgehog we have seen above which we will denote by $U_0(\vec{x})$. We introduce fluctuations around U_0 by

$$U(x) = U_0 e^{i\tau \cdot \phi} \qquad (7.56)$$

and L_μ as

$$\mathbf{L}_\mu = \tilde{\mathbf{L}}_\mu + \partial_\mu \phi + \partial_\mu \phi \wedge \phi \cdots \qquad (7.57)$$

Substitution of this expansion into (7.11) gives rise to a term zeroth order in ϕ which is what we studied above for the nucleon and Δ, a term quadratic in ϕ and higher orders. There is no term linear in ϕ because its coefficient is simply the equation of motion (7.55). To the next-to-leading order, the relevant Lagrangian is the quadratic term which takes the form

$$\Delta \mathcal{L} = \phi^a \left(H^{ab} + \cdots \right) \phi^b, \qquad (7.58)$$

$$H^{ab} = -\nabla_\mu^2 \delta^{ab} - 2\epsilon^{abc} \tilde{L}_\mu^c \nabla_\mu \qquad (7.59)$$

where the ellipsis stands for the term coming from fluctuations around the Skyrme quartic term and higher derivative terms (and/or other matter fields in a generalized Lagrangian). Ignoring terms higher order in ϕ's, the effective potential V^{eff} we wish to calculate is

$$e^{-i \int d^4x V^{eff}(\tilde{L})} = \int [d\phi] e^{i \int d^4x \Delta \mathcal{L}}. \qquad (7.60)$$

The fluctuation contains the zero modes corresponding to three translational modes and three (isospin) rotational modes in addition to other "vibrational modes". We have already quantized above the three rotational modes to obtain the baryons with correct quantum numbers. Quantizing the translational zero modes is to restore the translational invariance to the system. We are here interested in the other modes which are orthogonal to these zero modes. This requires constraints or "gauge fixing." Now doing the integral over the modes minus the six zero modes, we get formally

$$\int d^4 x V^{eff}(\tilde{L}) = -\frac{1}{2} \text{Tr}' \ln(H + \cdots) \qquad (7.61)$$

where the prime on trace means that zero energy modes of H be omitted in the sum. This generates one-loop corrections to the effective potential. The $O(N_c^0)$ Casimir energy, to one-loop order, is then given by

$$M_0 = \frac{1}{T} \int d^4 x \left(V^{eff}(\tilde{\mathbf{L}}) - V^{eff}(0) \right). \qquad (7.62)$$

This can be evaluated by chiral perturbation theory. In doing so, one can ignore the quartic and higher derivative terms in (7.61) in the one-loop graphs since it leads to $O(Q^2)$ correction relative to the one-loop terms evaluated with the quadratic term H. In accordance with general strategy of chiral perturbation theory, the ultraviolet divergences are to be eliminated by counterterms involving four derivatives, with finite constants fixed by experiments. Those constants are available from analyses on $\pi\pi$ scattering. Using this strategy, several people have estimated this Casimir contribution to the nucleon mass. Because of the somewhat dubious validity in using the derivative expansion for skyrmions mentioned above, it is difficult to pin down the magnitude: the magnitude depends on the dynamical details of the Lagrangian and chiral loop corrections. The calculations, however, agree generally on its sign which is negative. Numerically it is found to range[15] from ~ -200 MeV to ~ -1 GeV. The recent more reliable calculations [17] give $\sim -(500 - 600)$ MeV, quite consistent with what seems to be needed for the ground-state baryon mass.

7.3.4. *SU*(3) *skyrmions*

Here we study the extensions of the $SU(2)$ flavor picture to the case of the strangeness. We shall see that this extension is highly non-trivial, from both mathematical and physical reasons. We start with the simplest embedding of the $SU(2)$ ansatz into $SU(3)$. Then we shall show that treating the mass of the strange quark in the solitonic sector as a perturbation leads to wrong phenomenology. Two possible scenarios for improving the agreement with the data will be discussed: the Yabu-Ando and Callan-Klebanov pictures.

(i) Eight-fold way

Consider the system of three flavors u, d and s in the unrealistic situation where all three quarks are taken massless. The idea that one can perhaps treat the three flavors on an equal footing, with the mass of the strange quark treated as a "small" perturbation turns out to be not good at all, but it provides a nice theoretical framework to study the role of the anomalous (Wess-Zumino-Witten) term which remains important in realistic treatments given below. Putting the chiral field in $SU(3)$ space

$$U = e^{i\lambda^a \pi^a / f} \tag{7.63}$$

where a runs over octet Goldstone bosons, we can generalize (7.11) to apply to $SU(3)$

$$\mathcal{L}_n = -\frac{f^2}{4} \text{Tr} L_\mu L^\mu + \frac{1}{32g^2} \text{Tr}[L_\mu, L_\nu]^2 \tag{7.64}$$

and

$$S_n = \int d^4x \mathcal{L}_n \tag{7.65}$$

with L_μ defined with $U \in SU(3)$. Here the subscript n stands for "normal part." While this Lagrangian was sufficient in $SU(2)$ space, it is not in $SU(3)$ space. This can be seen in the following way[18].

Consider two discrete symmetries P_1 and P_2

$$P_1 : \qquad U(\vec{x}, t) \rightarrow U(-\vec{x}, t), \tag{7.66}$$

$$P_2 : \qquad U(\vec{x}, t) \rightarrow U^\dagger(\vec{x}, t). \tag{7.67}$$

It is easy to see that (7.64) is invariant under both P_1 and P_2 and of course under parity

$$P = P_1 P_2 \tag{7.68}$$

which transforms

$$U(\vec{x}, t) \rightarrow U^\dagger(-\vec{x}, t). \tag{7.69}$$

QCD is of course invariant under parity but does not possess the symmetries P_1 and P_2 separately. For instance the well-known five pseudoscalar process $K^+K^- \to \pi^+\pi^-\pi^0$ is allowed by QCD but is not by the Lagrangian (7.64). Thus there must be an additional term that breaks P_1 and P_2 separately while preserving P. This information is encoded in the Wess-Zumino-Witten term which we have encountered before in lower dimensions. In (3+1) dimensions, it is given by (see the Appendix, Chapter 6)

$$S_{WZ} = -N_c \frac{i}{240\pi^2} \int_D d^5x \epsilon^{ijklm} \, \text{Tr}(L_i L_j L_k L_l L_m) \qquad (7.70)$$

with $U(\vec{x}, t, s = 0) = 1$ and $U(\vec{x}, t, s = 1) = U(\vec{x}, t)$, with D a five-dimensional disk whose boundary is the space time. It is easy to see that this has the right symmetry structure.

The action we will study is then the sum

$$S_{sk} = S_n + S_{WZ}. \qquad (7.71)$$

As in the nonstrange hadrons, a marked improvement in quantitative agreement with experiments can be gained by introducing vector and axial-vector fields but the qualitative structure is not modified by those fields, so we will continue our discussion with the action (7.71) which we will call Skyrme action although the Wess-Zumino-Witten term is a new addition.

The $SU(3)$ skyrmion is constructed by first taking the ansatz that embeds the hedgehog in $SU(3)$ as

$$U_c = \begin{pmatrix} U_0 & 0 \\ 0 & 1 \end{pmatrix},$$
$$U_0 \equiv e^{i\vec{\tau}\cdot\hat{r}F(r)}. \qquad (7.72)$$

This is one of two possible spherically symmetric ansätze that one can make. Instead of embedding the $SU(2)$ group, one could embed the $SO(3)$ group. However the latter turns out to give baryon-number even objects. While it may be related to dibaryons, we shall not pursue this latter embedding since it is not clear that such an embedding is physically relevant for the time being.

Substitution of (7.72) into (7.71) gives us the equations for an $SU(2)$ soliton with a vanishing Wess-Zumino-Witten term as the hedgehog is sitting only in that space. To generate excitations in the isospin as well as in the strangeness direction, we rotate the hedgehog with an $SU(3)$ matrix S

$$U = S(t)U_c S^\dagger(t). \tag{7.73}$$

The resulting Lagrangian from (7.71), after some standard – though tedious – algebra[19], is

$$L_{sk} = \int d^3x \mathcal{L}_{sk} = -\mathcal{M} + \frac{a(U_c)}{2}[\text{Tri}\lambda_a S^\dagger \dot{S}]^2 + \frac{b(U_c)}{2}[\text{Tri}\lambda_A S^\dagger \dot{S}]^2$$
$$+ L_{\text{wz}} \tag{7.74}$$

where $\mathcal{M}$ is the static soliton energy and

$$a = \frac{2\pi}{3fg^3}\int_0^\infty ds \, \sin^2 F \left(s^2 + 4[s^2 F'^2 + \sin^2 F]\right),$$

$$b = \frac{\pi}{fg^3}\int_0^\infty ds \, \sin^2(F/2)\left(s^2 + [s^2 F'^2 + 2\sin F]\right)$$

where λ_α are the Gell-Mann matrices for $a = 1, 2, 3$ and $A = 4, 5, 6, 7$ and

$$L_{\text{wz}} = -i\frac{N_c}{2}B(U_c)\,\text{Tr}[YS^\dagger\dot{S}] \tag{7.75}$$

is the Wess-Zumino-Witten term. (Putting the Wess-Zumino-Witten term in this form requires some work. If the reader is diligent, this is an exercise. Otherwise, consult A.P. Balachandran's lecture note[3].) Here $B(U_c)$ is the baryon number corresponding to the configuration U_c and Y is the hypercharge $Y = \frac{1}{\sqrt{3}}\lambda_8$.

- *The Wess-Zumino-Witten term and $U(1)$ gauge symmetry*

The Wess-Zumino-Witten term can now be shown to have some remarkable roles in making the skyrmion correspond precisely to the physical baryon. The Lagrangian (7.74) has invariance under right $U(1)$ local (time-dependent) transformation

$$S(t) \rightarrow S(t)e^{iY\alpha(t)}. \tag{7.76}$$

This is simply because under this transformation $U = SU_c S^\dagger$ is invariant as U_c commutes with $e^{iY\alpha(t)}$. This can be seen explicitly on L_n in (7.74) as

$$\text{Tr}[\lambda_\alpha S^\dagger \dot{S}] \rightarrow \text{Tr}[e^{iY\alpha(t)}\lambda_\alpha e^{-iY\alpha(t)}S^\dagger\dot{S}] + i\dot\alpha\,\text{Tr}[\lambda_\alpha Y] \tag{7.77}$$

and $\text{Tr}[\lambda_\alpha Y] = \frac{2}{\sqrt{3}}\delta_{\alpha 8}$. The Wess-Zumino-Witten term is also invariant, modulo a total derivative term:

$$L_{\text{WZ}} \rightarrow L_{\text{WZ}} + \frac{N_c B}{3}\dot{\alpha}. \qquad (7.78)$$

However the last term, being a total derivative, does not affect the equation of motion. Normally a group of symmetries implies conservation laws but this is not the case with a *time-dependent* gauge symmetry. It imposes instead a constraint. To see this, let $\hat{Y}$ be the operator that generates this $U(1)$ transformation which we will call right hypercharge generator. By Noether's theorem,

$$\hat{Y}_{\text{R}} = \frac{N_c B}{3} \qquad (7.79)$$

and hence the allowed quantum state must satisfy the eigenvalue equation

$$\hat{Y}_{\text{R}}\psi = \frac{N_c B}{3}\psi. \qquad (7.80)$$

We will see later that this is the condition to be imposed on the wave function that will be constructed.

- *A proof that the skyrmion is a fermion*

We can now show explicitly that the skyrmion is a fermion for three number of colors (*i.e.*, $N_c = 3$). To do so, we note that the flavor $SU(3)_f$ and the spin $SU(2)_s$ are related to the left and right transformations, respectively

$$\begin{aligned} S &\rightarrow BS & B \in SU_L(3), \\ S &\rightarrow SC^\dagger & C \in SU(2) \subset SU_R(3). \end{aligned} \qquad (7.81)$$

The proof is simple. Under $SU(3)_f$ transformation, we have

$$SU_c S^\dagger \rightarrow B(SU_c S^\dagger)B^\dagger = (BS)U_c(BS)^\dagger \qquad (7.82)$$

so it is effectuated by a left multiplication on S, while under the space rotation $e^{i\alpha_i \hat{L}_i}$ with $\hat{L}_i$ ($i = 1, 2, 3$) the angular momentum operator, we have

$$e^{i\alpha_i \hat{L}_i}(SU_c S^\dagger)e^{-i\alpha_i \hat{L}_i} = (Se^{-i\alpha_i \lambda_i/2})U_c(Se^{-i\alpha_i \lambda_i/2})^\dagger \qquad (7.83)$$

where we have used the fact that S commutes with $\hat{L}_i$ and $L_i + \lambda_i/2 = 0$ with the hedgehog U_c. This shows that the spatial rotation is a right multiplication.

Now let us see what happens to the wave function of the system when we make a space rotation by angle 2π about the z axis for which we have

$$C = \begin{pmatrix} -1 & 0 & 0 \\ 0 & -1 & 0 \\ 0 & 0 & 1 \end{pmatrix}. \tag{7.84}$$

This induces the transformation on S

$$S \to SC^\dagger = Se^{3\pi i Y}. \tag{7.85}$$

Therefore the wave function is rotated to

$$\psi \to \psi e^{i\pi N_c B} \tag{7.86}$$

and hence for $N_c = 3$ and $B = 1$, the object is a fermion.

- *SU(3) skyrmion as a top in the magnetic monopole field*

We present here two quantum mechanical analogies that could clarify the structure of the $SU(3)$ skyrmion, which enable us to obtain the eigenvalues and eigenfunctions of the Hamiltonian. We rewrite

$$S^\dagger \partial_0 S = -i\lambda_\alpha \Omega_\alpha = -i\lambda_\alpha e_\beta^\alpha \dot{\theta}^\beta \tag{7.87}$$

where Ω^α are the angular velocities, θ^α are the eight Euler-like angles parametrizing the $SU(3)$ matrix S, and $e_\beta^\alpha(\theta)$ are the local frames ("vielbeins"). Fortunately their explicit forms are not needed for our discussion.

With (7.87), the Lagrangian (7.74) can be rewritten as

$$L_{sk} = -\mathcal{M} + \frac{1}{2}g_{mn}\dot{\theta}^m\dot{\theta}^n + \frac{N_c}{2\sqrt{3}}w_m\dot{\theta}^m \tag{7.88}$$

where $\mathcal{M}$ is the classical mass of a soliton, and the metric g_{mn} and the "gauge potential" w_m are given by

$$\begin{aligned} g_{mn} &= \frac{1}{2}\text{diag}(a, a, a, b, b, b, b, 0)_{ab}e_m^a e_n^b, \\ w_m &= \frac{3}{2}\text{Tr}(Y\lambda_a)e_m^a \end{aligned} \tag{7.89}$$

with all the indices running from 1 to 8. The fact that the metric (7.89) is not invertible, *i.e.*, $\det g = 0$, is a consequence of the gauge symmetry (7.76). Thus the space of independent variables is a seven-dimensional coset space $SU(3)/U(1)_Y$ instead of the original $SU(3)$ manifold. Note the close analogy[19,20] between the $SU(3)$ skyrmion and magnetic monopole mentioned above, the configuration space of which is spanned by the coset $SU(2)/U(1)$.

We illustrate the analogy by the following comparison: After the di-

Table 7.1. Analogies between the magnetic monopole and the $SU(3)$ skyrmion

Magnetic monopole	$SU(3)$ skyrmion
Structure of the local Lagrangian	
on the coset $X = G/H$	
$\mathcal{L} = \frac{1}{2}m\dot{x}_i^2 + egA_i\dot{x}_i$	$\mathcal{L} = \frac{1}{2}g_{mn}\dot{\theta}_m\dot{\theta}_n + \lambda w_m\dot{\theta}^m$
Nontrivial topology	
A_i - singular Dirac potential	ω_i - $U(1)_Y$ gauge potential
Homotopy class:$\Pi_1(S_1) = \mathbf{Z}$	Homotopy class $\Pi_2(X) = \Pi_1(H) = \mathbf{Z}$
Constraint of Hilbert space	
$n_i J_i|\psi\rangle = \frac{1}{2}|\psi\rangle$	$Y_R|\psi\rangle = \frac{N_c}{3}B|\psi\rangle$

mensional reduction of the Wess-Zumino-Witten term (7.75), the gauge potential w has the same topology as the magnetic monopole. We therefore expect that the "charge" λ be topologically quantized, and indeed, this is the case. The canonical construction of the Hamiltonian H gives

$$H = p_m\dot{\theta}^m - L, \quad p_m = \frac{\partial L}{\partial\dot{\theta}_m}. \tag{7.90}$$

This is the first analogy. The second analogy is based on the observation that the Hamiltonian so constructed is identical to the symmetric top in seven-dimensional configuration space[21]. We therefore compare the quantum mechanical properties of the symmetric top with the moments of inertia I_a, I_b to the $SU(3)$ skyrmion. In the case of the symmetric top, the eigenvectors are given by the D functions. The fact that we know simultaneously *two* components of the angular momentum, K and M, is not in

Table 7.2. Analogies between the symmetric top and the $SU(3)$ skyrmion

Symmetric top	**$SU(3)$ skyrmion**
Hamiltonian	
$H = \frac{1}{2I_a}(J_\xi^2 + J_\eta^2) + \frac{1}{2I_b}J_\rho^2$	$H = \frac{1}{2a}\sum_i^3 2R_i^2 + \frac{1}{2b}\sum_4^7 2R_i^2$
J_i - generators of rotations	R_i - generators of $SU(3)_R$
Symmetries	
$SU(2)_L$ - symmetry of space	$SU(3)_L$ - flavor group
$U(1)_R$ - symmetry of the top	$SU(2)_R$ - spin group
	$U(1)_R$ - "right hypercharge" group
Eigenfunctions	
$\psi = D^J_{KM}$	$\psi = D^{p,q}_{AB}$
J labels IR of $SU(2)$	p, q label IR of $SU(3)$
M - value of third	$A \equiv (I, I_3, Y)$ - set of isospin
component of $\vec{J}$	and hypercharge
K - value of the scalar $\vec{J} \cdot \vec{n}$,	$B \equiv (J, -J_3, Y_R = N_c B/3)$
where $\vec{n}$ picks the axis	constrained set of spin components
of the top	

contradiction with quantum mechanics, since they come from two different frames[22]. The $SU(3)$ skyrmion is a symmetric top, with the moments of inertia a in three non-strange directions and the moments of inertia b in remaining four strange directions. The eigenfunctions are again the D functions. However since the flavor group is $SU(3)$, they are labeled by the irreducible representations (IR) and multiple indices, *i.e.*, sets of quantum numbers. In addition, the $SU(3)$ skyrmion is not rotating freely, but is subjected to the influence of the non-trivial magnetic field. This field imposes a constraint on the eighth component of the "right" hypercharge (unrelated to the physical hypercharge, which is the "left" hypercharge), which picks from the set of all irreducible representations only those that fulfill, for $B = 1$, the condition (7.80),

$$\hat{Y}_R D^{p,q}_{AB} = \frac{N_c}{3} D^{p,q}_{AB}. \tag{7.91}$$

This condition is met only by the representations for which the triality – defined as the difference $(p - q)$ modulo three – is zero. The lowest representations fulfilling the triality-zero condition are the octet and the decuplet, so the constraint (7.91) picks only those representations of the baryons that agree with the "eightfold way" of the $SU(3)$ quark model.

We can now formally construct the eigenfunctions and verify that our intuitive identification of the quantum numbers is indeed correct. The irreducible representations (IR) of the $SU(3)$ group are characterized by two integers (p, q). The element $S \in SU(3)$ is represented by the operator $D^{(p,q)}(S)$, with the basis vector on which it acts denoted by $|I, I_3, Y\rangle$. If we denote the generators of the $SU(2)$ subgroup of $SU(3)$ by $\mathcal{I}_a$ ($a=1,2,3$), with the commutation relations $[\mathcal{I}_a, \mathcal{I}_b] = i\epsilon_{abc}\mathcal{I}_c$, and the hypercharge operator by $\hat{Y}$, we have

$$\begin{aligned}
\mathcal{I}_a^2|I, I_3, Y\rangle &= I(I+1)|I, I_3, Y\rangle, \\
\mathcal{I}_3|I, I_3, Y\rangle &= I_3|I, I_3, Y\rangle, \\
\hat{Y}|I, I_3, Y\rangle &= Y|I, I_3, Y\rangle.
\end{aligned} \tag{7.92}$$

The basis for the corresponding Hilbert space is

$$D^{(p,q)}_{(I,I_3,Y),(I',I'_3,Y')} \equiv \langle I, I_3, Y|D^{(p,q)}(S)|I', I'_3, Y'\rangle. \tag{7.93}$$

The right hypercharge constraint then requires that $Y' = N_c B/3$ which is 1 for $N_c = 3$ and baryon number 1. We will write the wave function as

$$\psi^{(p,q)}_{ab} = \sqrt{\dim(p,q)}\langle a|D^{(p,q)}(S)|b\rangle. \tag{7.94}$$

To determine the quantum numbers involved, let us see how the wave functions transform under relevant operations. Under flavor $SU(3)$, S gets transformed by left multiplication, namely, $S \to BS$. Under this transformation,

$$\begin{aligned}
D^{(p,q)}_{(I,I_3,Y),(I',I'_3,Y')}(S) \to\; & D^{(p,q)}_{(I,I_3,Y),(I'',I''_3,Y'')}(B) \\
& \times D^{(p,q)}_{(I'',I''_3,Y''),(I',I'_3,Y')}(S)
\end{aligned} \tag{7.95}$$

which is just the consequence of group property. From this one sees (once again) that the flavor is carried by the left group (I, I_3, Y). Thus I and I_3 are the isopin and its projection respectively and Y is the hypercharge. Under space rotation, S is multiplied on the right $S \to SC^\dagger$ with $C^\dagger$ a $2I' + 1$ dimensional IR of $SU(2)$. Therefore we have

$$D^{(p,q)}_{(I,I_3,Y),(I',I'_3,Y')}(S) \quad \to \quad D^{(p,q)}_{(I,I_3,Y),(I'',I''_3,Y')}(S)$$
$$\times D^{(p,q)}_{(I'',I''_3,Y'),(I',I'_3,Y')}(C^\dagger). \tag{7.96}$$

This means that the space rotation is characterized by the right group (I', I'_3). Thus I'_a is the angular momentum J_a, I'_3 its projection $-J_3$ and Y' the right hypercharge.

The system we are interested in has $N_c = 3$, $B = 1$, and $Y' = 1$ for which the lowest allowed IR's are $(p = 1, q = 1)$ corresponding to the octet with $J = 1/2$ and $(p = 3, q = 0)$ corresponding to the decuplet with $J = 3/2$. In this wave function representation, the Casimir operators have the eigenvalues:

$$\hat{J}^2 \psi^{(p,q)}_{..J..} = J(J+1)\psi^{(p,q)}_{..J..},$$
$$\hat{C}^2_2 \psi^{(p,q)}_{ab} = \frac{1}{3}[p^2 + q^2 + 3(p+q) + pq]\psi^{(p,q)}_{ab}.$$

The diagonalization of the Hamiltonian (see Table 7.2)

$$H = \frac{1}{2a}\sum_i^3 2R_i^2 + \frac{1}{2b}\sum_4^7 2R_i^2 \tag{7.97}$$

is now straightforward. Adding and subtracting the squares of generators in 1,2,3 and 8 directions, imposing the constraint and using the fact that the Casimir operators built from left generators have the same action as the ones built from right ones, we arrive at

$$E = \mathcal{M} + 2(a^{-1} - b^{-1})J(J+1)$$
$$+2b^{-1}\{\frac{1}{3}[p^2 + q^2 + 3(p+q) + pq] - \frac{3}{4}\}. \tag{7.98}$$

Calculation of the matrix elements relevant to static properties and other observables involves standard (generalized) angular momentum algebra and is straightforward, so we will not go into details here.

- *Elegant disaster*

The mass formula (7.98) is valid for the chiral limit at which all three quarks have zero mass. The u and d quarks are practically massless but in reality, the s-quark mass is not. Given the scheme sketched above, it is tempting to retain the elegance of the $SU(3)$ collective rotation by assuming that the mass difference can be treated as perturbation. This does not sound so bad in view of the fact that the Gell-Mann-Okubo mass formula which is proven to be so successful is obtained by taking the quark mass terms as a perturbation and that current algebra works fairly well even with s-quark hadrons. The procedure to implement the strange quark mass then is that one takes the $SU(3)$ wave functions constructed above and takes into account the symmetry breaking to first order with

$$\mathcal{L}_{sb} = a\,\text{Tr}\,[U(M-1) + \text{h.c.}] \qquad (7.99)$$

where M is the mass matrix which we can choose to be diagonal (since we are focusing on the strong interaction sector only), $M = \text{diag}\,(m_u, m_d, m_s)$. This is the standard form of symmetry breaking transforming as $(3, \bar{3}) + (\bar{3}, 3)$. The spectrum then will be given by the mass formula (7.98) plus the lowest-order perturbative contribution from (7.99). The matrix elements of physical operators will, however, be given by the unperturbed wave functions constructed above. Elegant though it may be, this scheme *does not work!* The phenomenology is a simple disaster[23] not only for the spectrum but also for other observable matrix elements, showing that considering the s-quark mass to be light in the present context is incorrect. From this point of view, it seems that the s-quark cannot be classed in the chiral family.

(ii) Yabu-Ando method

That the eight-fold way of treating collective variables fails completely seems at odds with the relative success one has with current algebras with strangeness. Why is it that current algebras work with kaons to $\sim 20\,\%$ accuracy if the strange quark mass is too heavy to make chiral symmetry meaningless?

A way out of this contradiction was suggested by the method of Yabu and Ando[24]. The idea is to preserve all eight zero modes as relevant dynamical variables but treat the symmetry breaking due to the s quark nonperturbatively. The result indicates that in consistency with three-flavor current algebras, the s quark *can still be* classed in the chiral family. This together with what we find next (the "heavy-quark" model of Callan and Klebanov) presents a puzzling duality that the s quark seems at the same time *light* and *heavy*. Testing experimentally which description is closer to nature will be discussed after presenting the Callan-Klebanov scenario.

We start with the eight-fold way Hamiltonian (7.97) which we will call H_0 plus a symmetry breaking term

$$H_{\text{SB}} = \frac{1}{2}\gamma(1 - D_{88}) \tag{7.100}$$

with adjoint representation

$$D_{ab}(S) \equiv \frac{1}{2}\text{Tr}(\lambda_a S \lambda_b S^\dagger). \tag{7.101}$$

The coefficient γ measures the strength of the symmetry breaking that gives the mass difference between the pion and the kaon. If we adopt the explicit parametrization of $SU(3)$ used by Yabu and Ando,

$$S = R(\alpha,\beta,\gamma)e^{-i\nu\lambda_4}R(\alpha',\beta',\gamma')e^{-i\frac{\rho}{\sqrt{3}}\lambda_8}, \tag{7.102}$$

where $R(\alpha,\beta,\gamma)$ is the $SU(2)$ isospin rotation matrix with the Euler angles α,β,γ and the eighth right generator R_8 and ρ are the conjugate variables, then

$$H_{\text{SB}} = \frac{3}{4}\gamma\sin^2\nu. \tag{7.103}$$

The idea then is to diagonalize exactly the total Hamiltonian ($H_0 + H_{\text{SB}}$) in the basis of the eight-fold way wave function. Specifically, write the wave function as

$$\begin{aligned}
\Psi_{YII_3,JJ_3} &= (-1)^{J-J_3} \sum_{M_{\text{L}},M_{\text{R}}} D^{(I)*}_{I_3,M_{\text{L}}}(\alpha,\beta,\gamma) f_{M_{\text{L}}M_{\text{R}}}(\nu)e^{i\rho} \\
&\quad \times D^{(J)*}_{M_{\text{R}},-J_3}(\alpha',\beta',\gamma').
\end{aligned} \tag{7.104}$$

Then the eigenvalue equation

$$[\hat{C}_2^2 + \omega^2\sin^2\nu]\Psi = \epsilon_{\text{SB}}\Psi \tag{7.105}$$

leads to a set of coupled differential equations for the functions $f_{M_{\text{L}}M_{\text{R}}}(\nu)$. Here $\omega^2 = \frac{3}{2}\gamma b$ with b the moment of inertia in the strange direction appearing in (7.97). When the parameter γ or ω^2 is zero, the wave function is just the pure $SU(3)$ symmetric one corresponding to the *eightfold way* given above, namely the matrix element of an irreducible representation of $SU(3)$ containing states that satisfy $Y = 1$ and $I = J$.

The mass spectrum is still very simple. It is given by the $SU(3)$ symmetric formula (7.98) supplemented by the eigenvalue ϵ_{SB} of the H_{SB}

$$
\begin{aligned}
M &= \mathcal{M} + 2(a^{-1} - b^{-1})J(J+1) \\
&\quad + 2b^{-1}\{\frac{1}{3}[p^2 + q^2 + 3(p+q) + pq] - \frac{3}{4}\} + 2b^{-1}\epsilon_{\mathrm{SB}}. \quad (7.106)
\end{aligned}
$$

The wave functions are also quite simple. A simple way of understanding the structure of the resulting wave function is to write it in perturbation theory. For instance, since the symmetry-breaking Hamiltonian mixes the **8** with $\overline{\mathbf{10}}$ and **27** for spin 1/2 baryons, the proton wave function would look like [25]

$$
|p\rangle = |p; \mathbf{8}\rangle + \frac{\omega^2}{9\sqrt{5}}|p; \overline{\mathbf{10}}\rangle + \frac{\sqrt{6}\omega^2}{75}|p; \mathbf{27}\rangle + \cdots \quad (7.107)
$$

- *Phenomenology with Yabu-Ando method*

Treating the s-quark mass term exactly improves considerably the baryon spectrum as well as their static properties. Once the ground state is suitably shifted, the excitation spectra are satisfactory for physical values of the coupling constants f, g etc. Other static observables such as magnetic moments, radii, weak matrix elements etc. are also considerably improved[26]. The problem of the too big ground-state energy gets worse, however, for $SU(3)$ if one takes physical values of the coupling constants but as mentioned above, this is expected as the Casimir energy is proportional to the number of zero modes and the Casimir attraction will be 7/3 times greater than the two-flavor case. Again a careful Casimir calculation will be needed to confront nature in a quantitative way.

A surprising thing is that the results obtained in the Yabu-Ando $SU(3)$ description are rather similar to the Callan-Klebanov method described next, though the basic assumption is quite different. In the latter, while the vacuum is assumed to be $SU(3)$ symmetric, the excitation treats the strangeness direction completely asymmetrically with respect to the chiral direction.

An important caveat with the exact treatment of the symmetry breaking is that it is not consistent with chiral perturbation theory. The point is that the symmetry-breaking Hamiltonian H_{SB} is a linear term in quark-mass matrix M and in principle one can have in the Lagrangian terms higher order in M in combination with derivatives. Treating the linear mass term exactly while ignoring higher order mass term is certainly not consistent. On the other hand, as mentioned before, the naive chiral expansion does not make sense in the baryon sector. Since the Yabu-Ando procedure seems to work rather satisfactorily, this may indicate that while the higher order chiral expansion is meaningful in the meson sector, it may not be in the baryon sector. This is somewhat like the six-derivative term in the skyrmion structure which is formally higher order in derivative but is more essential for the baryon than higher orders in quark mass term. This suggests that chiral perturbation theory powerful in certain aspect (see Chapter 10) is not very useful in *baryon structure*.

(iii) Callan-Klebanov scheme

Since the s-quark is not "light," let us imagine that it is heavy. In a strict sense, this does not sound right. After all the s-quark mass in the range of 130 to 190 MeV is comparable to the QCD scale parameter Λ_{QCD}, not much larger as would be required to be in the heavy-quark category. As we will see later, heavy-quark hadrons exhibit a heavy-quark symmetry which is not explicitly visible but known to be present in the QCD Lagrangian that applies to the quarks whose mass satisfies $m_H \gg \Lambda_{QCD}$. The b quark belongs to this category and to some extent so does the charm (c) quark. In this subsection, we will simply assume that the s quark can be treated as "heavy." This leads to the Callan-Klebanov model[27] of hyperon skyrmion which will turn out to be remarkably successful.

Let us consider a simplified model based on the following intuitive idea. Imagine that the s quark is very heavy. Instead of "rotating" into the s-flavor direction as we did above, we now treat it as an independent degree of freedom. The effective field carrying the s-quark flavor is the kaon which can be put into isospin doublet corresponding to the quark configurations $s\bar{u}$ and $s\bar{d}$. Let K represent the doublet $K^T = (K^-\ \bar{K}^0)$. (In anticipation of what is to come, let us mention that other "heavy" quark flavors Q will also be put into independent doublets, the Q replacing the s quark.) Suppose one puts a kaon into the $SU(2)$ soliton field. One can then ask what happens to the soliton-kaon system. The Lagrangian density takes the form:

$$\mathcal{L}_{toy} = \mathcal{L}_{\text{SU}(2)} + \mathcal{L}_{\text{K-SU}(2)}. \tag{7.108}$$

Here the first term describes the chiral (u- and d–quark) sector:

$$\mathcal{L}_{\text{SU}(2)} = \frac{f^2}{4}\,\text{Tr}(\partial_\mu U \partial^\mu U^\dagger) + \cdots \tag{7.109}$$

where $U = e^{i\vec{\tau}\cdot\vec{\pi}/f}$ is the $SU(2)$ chiral field and the ellipsis stands for terms of higher derivative, mass term etc. For simplicity, we will consider a Lagrangian density constructed entirely of the pion field since more realistic Lagrangians do not modify the discussion qualitatively. To write the second term of (7.108), introduce as before the "induced" vector field V_μ and axial field A_μ:

$$V_\mu = \frac{1}{2}(\partial_\mu \xi \xi^\dagger + \partial_\mu \xi^\dagger \xi), \tag{7.110}$$

$$A_\mu = \frac{1}{2i}(\partial_\mu \xi \xi^\dagger - \partial_\mu \xi^\dagger \xi) \tag{7.111}$$

with $\xi = \sqrt{U}$. We recall that under chiral transformation $\xi \to L\xi g^\dagger = g\xi R^\dagger$, V_μ and A_μ transform as

$$V_\mu \to g(\partial_\mu + V_\mu)g^\dagger, \tag{7.112}$$

$$A_\mu \to gA_\mu g^\dagger \tag{7.113}$$

with g a complicated local function of L, R and π. We assume that under chiral transformation, K transforms

$$K \to gK. \tag{7.114}$$

(Later on we will implement this with heavy-quark spin transformation when the quark mass becomes heavy compared with the chiral symmetry scale, as for instance in the case of D and B mesons containing an s and b quarks respectively. Since K^* is still much higher in mass, we need not worry about this transformation here.) Therefore in $\mathcal{L}_{\text{K-SU}(2)}$, we must have the "covariant" kinetic energy term $(D_\mu K)^\dagger D^\mu K$ with $D_\mu K = (\partial_\mu + V_\mu)K$ and mass term $m_K^2 K^\dagger K$. We must also have a potential term $K^\dagger \mathcal{V} K$ with $\mathcal{V}$ transforming $\mathcal{V} \to g\mathcal{V}g^\dagger$ in which we could have terms like A_μ^2 etc. If the kaon is very massive, we can ignore terms higher order than quadratic in the kaon field. In accordance with the spirit of chiral perturbation theory, we ignore higher derivative terms involving pion fields. There is one term of higher derivatives, however, that we cannot ignore, namely, the term involving the baryon current B_μ which contains three derivatives but while higher order in the chiral counting in the meson sector, is of $O(1)$ in the baryon sector. Therefore we also have a term of the form $(K^\dagger D^\mu K B_\mu + \text{h.c.})$. Putting these together, the minimal Lagrangian density we will consider is

$$
\begin{aligned}
\mathcal{L}_{toy} = \; & \frac{f^2}{4} \text{Tr}(\partial_\mu U \partial^\mu U^\dagger) + \cdots \\
& + (D_\mu K)^\dagger D^\mu K - K^\dagger(m_K^2 + V)K + \kappa(K^\dagger D^\mu K B_\mu + \text{h.c.}) \\
& + \cdots
\end{aligned}
\tag{7.115}
$$

with κ a dimensionful constant to be fixed later.

- *Isospin-spin transmutation*

We will now show that in the field of the soliton the isospin of the kaon in (7.108) transmutes to a spin, the latter behaving effectively as an s quark. This is in analogy to the transmutation of a scalar doublet into a fermion in the presence of a magnetic monopole[28] except that in the present case, the scalar is "wrapped" by the soliton while in the other case, the scalar is "pierced" by the monopole. To see the isospin-spin transmutation in our soliton-kaon complex, we first note that the Lagrangian density (7.108) with (7.109) and (7.115) (and with $U = U_c = e^{i\vec{\tau}\cdot\hat{r}F(r)}$) is invariant under the (global) rotation $S \in SU(2)$:

$$
\begin{aligned}
U &\rightarrow SUS^\dagger, \\
K &\rightarrow SK
\end{aligned}
\tag{7.116}
$$

which implies $D_\mu K \rightarrow S(D_\mu K)$. Thus we again have three zero modes. These zero modes have to be quantized to get the quantum numbers. Thus we are invited to write

$$
K(\vec{x}, t) = S(t)\tilde{K}(\vec{x}, t). \tag{7.117}
$$

Here $\tilde{K}$ is the kaon field defined in the soliton rotating frame. Under an isospin rotation B which is a subgroup of the left transformation $L \in SU(3)$,

$$
S \rightarrow BS, \qquad \tilde{K} \rightarrow \tilde{K}. \tag{7.118}
$$

Table 7.3. Even-parity $J = \frac{1}{2}$ and $\frac{3}{2}$ baryons containing strange and charm quarks. S and C stand for the strangeness and charm numbers, respectively. R is the rotor angular momentum which is equal to the isospin of the rotor I.

particle	I	J	S	C	R	J_1	J_2	J_m
Λ	0	$\frac{1}{2}$	-1	0	0	$\frac{1}{2}$	0	$\frac{1}{2}$
Σ	1	$\frac{1}{2}$	-1	0	1	$\frac{1}{2}$	0	$\frac{1}{2}$
Σ^*	1	$\frac{3}{2}$	-1	0	1	$\frac{1}{2}$	0	$\frac{1}{2}$
Ξ	$\frac{1}{2}$	$\frac{1}{2}$	-2	0	$\frac{1}{2}$	1	0	1
Ξ^*	$\frac{1}{2}$	$\frac{3}{2}$	-2	0	$\frac{1}{2}$	1	0	1
Ω	0	$\frac{3}{2}$	-3	0	0	$\frac{3}{2}$	0	$\frac{3}{2}$
Λ_c	0	$\frac{1}{2}$	0	1	0	0	$\frac{1}{2}$	$\frac{1}{2}$
Σ_c	1	$\frac{1}{2}$	0	1	1	0	$\frac{1}{2}$	$\frac{1}{2}$
Σ_c^*	1	$\frac{3}{2}$	0	1	1	0	$\frac{1}{2}$	$\frac{1}{2}$
Ξ_{cc}	$\frac{1}{2}$	$\frac{1}{2}$	0	2	$\frac{1}{2}$	0	1	1
Ξ_{cc}^*	$\frac{1}{2}$	$\frac{3}{2}$	0	2	$\frac{1}{2}$	0	1	1
Ω_{ccc}	0	$\frac{3}{2}$	0	3	0	0	$\frac{3}{2}$	$\frac{3}{2}$
Ξ_c	$\frac{1}{2}$	$\frac{1}{2}$	-1	1	$\frac{1}{2}$	$\frac{1}{2}$	$\frac{1}{2}$	1,0
Ξ_c'	$\frac{1}{2}$	$\frac{1}{2}$	-1	1	$\frac{1}{2}$	$\frac{1}{2}$	$\frac{1}{2}$	1,0
Ξ_c^*	$\frac{1}{2}$	$\frac{3}{2}$	-1	1	$\frac{1}{2}$	$\frac{1}{2}$	$\frac{1}{2}$	1
Ω_c	0	$\frac{1}{2}$	-2	1	0	1	$\frac{1}{2}$	$\frac{1}{2}$
Ω_c^*	0	$\frac{3}{2}$	-2	1	0	1	$\frac{1}{2}$	$\frac{3}{2}$
Ω_{cc}	0	$\frac{1}{2}$	-1	2	0	$\frac{1}{2}$	1	$\frac{1}{2}$
Ω_{cc}^*	0	$\frac{3}{2}$	-1	2	0	$\frac{1}{2}$	1	$\frac{3}{2}$

This shows that the kaon field in the rotating frame carries no isospin. Under a spatial rotation,

$$\tilde{K} \rightarrow e^{i\vec{\alpha}\cdot\vec{L}}\tilde{K} = e^{-i\vec{\alpha}\cdot\vec{I}}e^{i\vec{\alpha}\cdot\vec{T}}\tilde{K},$$
$$U_c \rightarrow e^{i\vec{\alpha}\cdot\vec{L}}U_c e^{-i\vec{\alpha}\cdot\vec{L}}$$
$$= e^{-i\vec{\alpha}\cdot\vec{I}}U_c e^{i\vec{\alpha}\cdot\vec{I}} \tag{7.119}$$

where $\vec{T}$ is the grand spin $\vec{J} + \vec{I}$ under which the hedgehog U_c is invariant. Thus we have

$$S \rightarrow S e^{-i\vec{\alpha}\cdot\vec{I}}, \tag{7.120}$$

and

$$\tilde{K} \rightarrow e^{i\vec{\alpha}\cdot\vec{T}}\tilde{K}. \tag{7.121}$$

This means that the grand spin operator $\vec{T}$ is the *angular momentum operator* on $\tilde{K}$. For the kaon, the grand spin $\vec{T}$ is the vector sum $\vec{L} + \vec{I}$ with $I = 1/2$ and since, as we will see later, the lowest kaon mode has $L = 0$, the kaon seen from the soliton rotating frame carries spin $1/2$. This shows that in the soliton rotating frame, the isospin lost from the kaon field is transmuted to a spin. This is the analog to the transmutation of the isospin of the scalar doublet to a spin in the presence of a magnetic monopole. There is one difference however. While the kaon acquires spin in this way, it retains its bosonic statistics while in the case of the monopole-scalar, the boson changes into a fermion. Since the soliton-kaon complex must be a baryon (fermion), the soliton which will be quantized with *integer* spin will carry *fermion* statistics. With one kaon in the system, the soliton carries $J = I = 0, 1, \cdots$

Suppose we have two kaons in the soliton rotating frame. They now have integer spins $T = 0, 1$, so the soliton carries $J = I = 1/2, 3/2, \cdots$. For three kaons in the system, the kaons carry $T = 1/2, 3/2, \cdots$ with the soliton carrying $J = I = 0, 1, \cdots$ and so on. One can show that this pattern of quantization follows naturally from the CK theory[29]. The resulting quantum numbers and particle identification are given in Table 7.3. It is remarkable that they are identical to what one expects from quark models. Thus *effectively*, the kaon behaves like an s quark in all respect except for its statistics. By field redefinition one can indeed make it behave *precisely* like an s quark, thereby making the picture identical to the quark-model picture.

- *Heavy baryons from chiral symmetry*

In discussing the dynamical structure of the baryons in the CK approach, we have to specify the quantities that enter into the Lagrangian (7.115). Suppose that we want to arrive at (7.115) starting from a chiral Lagrangian as Callan and Klebanov did. Then the basic assumption to be made is that the vacuum is $SU(3)$-symmetric while the quark "particle-hole" excitation may not be. This means that we can start with a Lagrangian that contains a piece that is $SU(3)$ symmetric with the U field in $SU(3)$ plus a symmetry-breaking term. Thus the f that appears in (7.115) is directly related to the pion decay constant f_π and the $\mathcal{V}$ and κ will be specified. In particular the coefficient κ will be fixed by the topological

Wess-Zumino term to

$$\kappa = N_c/4f^2. \tag{7.122}$$

The structure of the resulting baryons can be simply described as follows. In the large N_c limit, the $SU(2)$ soliton contributes a mass $\mathcal{M}$ which is of $O(N_c)$. The kaon field which will be bound by the Wess-Zumino term of (7.115) (with some additional contribution from the potential term $\mathcal{V}$) contributes a fine-structure splitting $n\omega$ of $O(N_c^0)$ where n is the number of bound kaons and ω is the eigenenergy of the bound kaon (so $< m_K$). This is the $O(1)$ term dependent on "heavy flavors," mentioned above. The quantization of the collective rotation of the soliton brings in the rotational energy of the soliton as well as the hyperfine-structure splitting of $O(1/N_c)$.

Let us now show how this structure arises in detail. We write the energy of the baryons as before:

$$E_{JIs} = \mathcal{M} + M_0 + M_{-1} \tag{7.123}$$

where the subscript labels n in N_c^n expansion. We know how $\mathcal{M}$ – which is of $O(N_c)$ – is obtained. To obtain the term M_0 which depends on flavor, we may write from (7.115) the equation of motion satisfied by the kaon field and solve it to $O(N_c)$. Alternatively we calculate the Lagrangian (7.115) to $O(N_c)$ by taking the hedgehog $U = U_c = e^{iF(r)\tau\cdot\hat{r}}$. To do this, we note that the hedgehog is symmetric under the "grand spin" rotation $\vec{T} = \vec{L} + \vec{I}$ (we are changing the notation for the "grand spin" to T), so we may write the kaon field as

$$K = k(r)Y_{TLT_z}. \tag{7.124}$$

The Lagrangian is then (using the length unit $1/gf$ as before, so that $s = (gf)r$ is dimensionless)

$$
\begin{aligned}
L_K = \ & 4\pi \int ds s^2 \left[\dot{k}^\dagger \dot{k} + i\lambda(s)(k^\dagger \dot{k} - \dot{k}^\dagger k) \right. \\
& \left. - \frac{d}{ds}k^\dagger \frac{d}{ds}k - k^\dagger k(m_K^2 + \mathcal{V}(s; T, L)) \right]
\end{aligned}
\tag{7.125}
$$

where

$$\lambda(s) = -\frac{N_c g^2}{2\pi^2 s^2} F' \sin^2 F \tag{7.126}$$

is the contribution from the Wess-Zumino term. This expression differs from what one sees in the literature[30] in that the kinetic energy terms are not multiplied by nonlinear functions of F which come from the Skyrme term involving kaon fields. They are not essential, so we will not include them. The Euler-Lagrange equation of motion obtained from (7.125) is

$$\frac{d^2}{dt^2}k - 2i\lambda(s)\frac{d}{dt}k + \mathcal{O}k = 0 \qquad (7.127)$$

where

$$\mathcal{O} = -\frac{1}{s^2}\frac{d}{ds}s^2\frac{d}{ds} + m_\kappa^2 + \mathcal{V}(s;T,L). \qquad (7.128)$$

Expanded in its eigenmodes, the field k takes the form

$$k(s,t) = \sum_{n>0}\left(\tilde{k}(s)e^{i\tilde{\omega}_n t}b_n^\dagger + k_n(s)e^{-i\omega_n t}a_n\right) \qquad (7.129)$$

with $\omega_n \geq 0$ and $\tilde{\omega}_n \geq 0$. The Euler-Lagrange equation gives the eigenmode equations:

$$\left(\omega_n^2 + 2\lambda(s)\omega_n + \mathcal{O}\right)k_n = 0, \qquad (7.130)$$
$$\left(\tilde{\omega}_n^2 - 2\lambda(s)\tilde{\omega}_n + \mathcal{O}\right)\tilde{k}_n = 0. \qquad (7.131)$$

An important point to note here is the sign difference in the term linear in frequency associated with the Wess-Zumino term, which is of course due to the linear time derivative. We shall discuss its consequence shortly. Now recognizing that $\mathcal{O}$ is hermitian, one can derive the orthonormality conditions

$$4\pi \int ds s^2 k_n^* k_m \left((\omega_n + \omega_m) + 2\lambda(s)\right) = \delta_{nm}, \qquad (7.132)$$

$$4\pi \int ds s^2 \tilde{k}_n^* \tilde{k}_m \left((\tilde{\omega}_n + \tilde{\omega}_m) - 2\lambda(s)\right) = \delta_{nm}, \qquad (7.133)$$

$$4\pi \int ds s^2 k_n^* \tilde{k}_m \left((\omega_n - \tilde{\omega}_m) + 2\lambda(s)\right) = 0. \qquad (7.134)$$

The creation and annihilation operators of the modes satisfy the usual bosonic commutation rules

$$[a_n, a_m^\dagger] = \delta_{nm}, \qquad [b_n, b_m^\dagger] = \delta_{nm} \qquad (7.135)$$

with the rest of the commutators vanishing. The diagonalized Hamiltonian so quantized takes the form

$$H = \sum_{n>0} \left(\omega_n a_n^\dagger a_n + \tilde{\omega}_n b_n^\dagger b_n \right) \tag{7.136}$$

and the strangeness charge is given by

$$S = \sum_{n>0} \left(b_n^\dagger b_n - a_n^\dagger a_n \right). \tag{7.137}$$

It follows that the mode k_n carrying $S = -1$ receives an attractive contribution from the Wess-Zumino term while the mode $\tilde{k}_n$ carrying $S = +1$ gets a repulsive contribution. The potential $\mathcal{V}$ is numerically small compared with the strength of the Wess-Zumino term, so the $S = -1$ state is bound whereas the $S = +1$ state is not. The former has the strangeness quantum number of the hyperons seen experimentally and the latter the quantum number of what is called "exotics" in quark models.

To $O(N_c^0)$, at which the kaon binding takes place, no quantum numbers of the bound system other than the strangeness are defined. It is at $O(N_c^{-1})$ that the proper quantum numbers are recovered. What we have at this point is a bound object in which the kaon has $T = 1/2$ and $L = 1$ with $S = -1$. We shall see shortly that the bound object, when collective-quantized, has the right quantum numbers to describe Λ, Σ and Σ^*. The corresponding frequency ω_n constitutes the "fine-structure" contribution M_0 to eq. (7.123).

To get the "hyperfine-structure" term M_{-1} and the quantum numbers, we have to quantize the three zero modes encountered above. Let

$$U(\vec{r},t) = S(t)U_0(\vec{r})S^\dagger(t), \qquad \tilde{K}(\vec{r},t) = S(t)K(\vec{r},t). \tag{7.138}$$

As defined, K is the kaon field "seen" in the soliton rotating frame and $\tilde{K}$ is the kaon field defined in its rest frame. As before, we let

$$S^\dagger \partial_0 S = -i\lambda_\alpha \dot{q}_\alpha \equiv \lambda_\alpha \Omega_\alpha \tag{7.139}$$

defining the rotational velocities Ω_α where α here runs over 1 to 3. When (7.138) is substituted into (7.115) and integrated over space, one obtains the Hamiltonian at $O(N_c^{-1})$ of the following form

$$H_{-1} = 2\mathcal{I}\vec{\Omega}^2 - 2\vec{\Omega} \cdot \vec{Q}. \tag{7.140}$$

Here $\mathcal{I}$ is the moment of inertia of the rotating $SU(2)$ soliton which can be determined by the ΔN mass difference as discussed before. Note that (7.140) is a generic form of the Hamiltonian with a linear time derivative encountered before. The quadratic velocity-dependent term arises from the kinetic energy term of the soliton rotation. The linear velocity-dependent term was seen above as arising from integrating out certain fast degrees of freedom. After a straightforward and tedious algebra, one can get an explicit form for $\vec{Q}$ from (7.115). It is given by spatial integral of a model-dependent function χ_i, the explicit form of which is not needed for the discussion:

$$Q_i = \int d^3r\chi_i(K, U_0).\tag{7.141}$$

Canonically quantizing (7.140), we obtain

$$H_{-1} = \frac{1}{2\mathcal{I}}\left(R_i - \int d^3r\chi_i\right)^2\tag{7.142}$$

where R_i is the right generator we encountered above which in the present case is the rotor angular momentum which we will denote by $\vec{J}_l$ (the subscript l stands for "light" quark making up the soliton). Explicitly evaluating with the known solutions for the soliton and the bound kaon, one finds that

$$\int d^3r\chi_i = cJ_{\mathrm{K}}^i\tag{7.143}$$

where J_{K}^i is the angular momentum stored in the bound kaon, identified with the grand spin T^i. An explicit form of the coefficient c can be written down for a given dynamical model. Both the functions χ_i and the coefficient have been worked out for the Skyrme-type Lagrangian[31]. The Hamiltonian (7.142) which can be rewritten in a form familiar from the previous chapter

$$H_1 = \frac{1}{2\mathcal{I}}\left(\vec{J}_l + c\vec{J}_K\right)^2\tag{7.144}$$

gives the hyperfine spectrum

$$E_{-1} = \frac{1}{2\mathcal{I}}\left(cJ(J+1) + (1-c)J_l(J_l+1) + c(c-1)J_{\mathrm{K}}(J_{\mathrm{K}}+1)\right)\tag{7.145}$$

where $\vec{J} = \vec{J}_l + \vec{J}_{\mathrm{K}}$ is the total angular momentum of the bound system.

It should be stressed that although the numerical evaluation of the coefficient c and the moment of inertia $\mathcal{I}$ requires (in the absence of direct information from QCD) dynamical models, the form of the spectrum (7.145) *is completely general*. This is because the generic form of the Hamiltonian (7.142) could be derived on the basis of general assumptions made in identifying the so-called "fast" and "slow" degrees of freedom in the Born-Oppenheimer approximation. The integration over the fast degrees of freedom gives rise to the presence of a geometric (Berry) phase. We shall study this problem in detail in the next chapter. The quantity c, as we will see shortly, can be interpreted in terms of a Berry charge quite analogous to the charge $(1 - \kappa)$ that figures in the spectrum of the diatomic molecule discussed as a toy example in Chapter 8. The hyperfine spectrum can be used to relate the c coefficient to the masses of the hyperons. It takes the form (using the particle symbol for the masses)

$$c = \frac{2(\Sigma^* - \Sigma)}{2\Sigma^* + \Sigma - 3\Lambda} \approx 0.62 \tag{7.146}$$

where the last equality comes from experiments. Note that this is close to unity. Were it precisely equal to 1, then the hyperfine spectrum would have been that of a symmetric top depending only on the total angular momentum J. Deviation from 1 will become accentuated as the mass of the heavy meson increases. As the heavy meson mass becomes infinite, the charge c will go to zero. This particular result will be proven in Chapter 9, where the consequences of the heavy quark symmetry (Isgur-Wise symmetry[32]) will be studied in detail.

References

1. E. Witten, Nucl. Phys. **B160**, 57 (1979);
2. V. Kazakov and A.A. Migdal, Phys. Lett. **B236**, 214 (1981)
3. For a more precise discussion, A.P. Balachandran, in *High-Energy Physics 1985*, vol. 1, ed. by M.J. Bowick and F. Gürsey (World Scientific, Singapore, 1986).
4. T.H.R. Skyrme, Nucl. Phys. **31**, 556 (1962).
5. R.H. Hobart, Proc. R. Soc. London **82**, 201 (1963); **85**, 610 (1964); G.H. Derrick, J. Math. Phys. **5**, 1252 (1964).
6. B.S. Balakrishna, V. Sanyuk, J. Schechter and A. Subbaraman, Phys. Rev. **D45**, 344 (1992).
7. J. Baacke and A. Schenke, Z. Phys. **C41**, 259 (1988).

8. H. Georgi, Nucl. Phys. **B331**, 311 (1990).
9. L.D. Faddeev, Lett. Math. Phys. **1**, 289 (1976).
10. N.S. Manton and P.J. Ruback, Phys. Lett. **B181**, 137 (1986).
11. See V.G. Makhanov, Y.P. Rybakov and V.I. Sanyuk, *The Skyrme Model*, Springer Series in Nuclear and Particle Physics (Springer-Verlag, Berlin 1993).
12. A.D. Jackson and M. Rho, Phys. Rev. Lett. **51**, 751 (1983).
13. G.S. Atkins, C.R. Nappi and E. Witten, Nucl. Phys. **B228**, 352 (1983).
14. N.S. Manton, Commun. Math. Phys. **111**, 469 (1987).
15. I. Zahed, A. Wirzba and U.-G. Meissner, Phys. Rev. **D33**, 830 (1986);
16. I. Zahed and G.E. Brown, Phys. Repts. **142**, 1 (1986).
17. G. Holzwarth, Phys. Lett. **B291**, 218 (1992).
18. E. Witten, Nucl. Phys. **B223**, 422, 433 (1983).
19. P.O. Mazur, M.A. Nowak and M. Praszałowicz, Phys. Lett. **147B**, 137 (1984).
20. E. Guadagnini, Nucl. Phys. **B236**, 35 (1984).
21. S. Jain and S. W. Wadia, Nucl. Phys. **B258**, 713 (1984).
22. L.D. Landau and E.M. Lifshitz, *Quantum Mechanics* (Pergamon Press, 1976).
23. See M. Praszalowicz, Phys. Lett. **B158**, 264 (1985); M. Chemtob, Nucl. Phys. **B256**, 600 (1985).
24. H. Yabu and Ando, Nucl. Phys. **B301**, 601 (1988).
25. J.H. Kim, C.H. Lee and H.K. Lee, Nucl. Phys. **A501**, 835 (1989); N.W. Park, J. Schechter and H. Weigel, Phys. Lett. **B224**, 171 (1989).
26. N.W. Park and H. Weigel, Nucl. Phys. **A541**, 453 (1992).
27. C.G. Callan and I. Klebanov, Nucl. Phys. **B262**, 365 (1985); N.N. Scoccola, H. Nadeau, M.A. Nowak and M. Rho, Phys. Lett. **B201**, 425 (1986);C.G. Callan, H. Hornbostel and I. Klebanov, Phys. Lett. **B202**, 425 (1986).
28. R. Jackiw and C. Rebbi, Phys. Rev. Lett. **36**, 1116 (1976); P. Hasenfratz and G. 't Hooft, Phys. Rev. Lett. **36**, 1119 (1976).
29. C.C. Callan and I. Klebanov[27] ; N.N. Scoccola and A. Wirzba, Phys. Lett. **B66**, 1130 (1991).
30. *E.g.*, I. Klebanov, Lectures in *Proc. of the Nato Advanced Summer Institute on Nonperturbative quantum Field Theory*, ed. by D. Vautherin , F. Lenz and J.W. Negele (Plenum Press, NY, 1990).
31. M. Rho, Phys. Rep. **240**, 1 (1994).
32. N. Isgur and M.B. Wise, Phys. Rev. Lett. **66**, 1130 (1991) and references given therein.

CHAPTER 8

CHIRAL BAGS

8.1. Introduction

The general discussion presented in the preceding chapters strongly suggests that at sufficiently low energies or long distances, QCD in the limit of large N_c can be accurately approximated by an effective field theory written in terms of weakly interacting meson fields. The price to pay to obtain the desired accuracy is that one is limited to only very long-wavelength processes or else to increase the number of meson fields and/or the number of derivatives, requiring an increasing number of experimental inputs. This point was emphasized in Chapters 5 and 6. In Chapter 7, we discussed how to describe baryons with the same effective theory of meson fields. How the baryons seen in experiments emerge from a theory consisting of meson fields only is the most fascinating new development in hadronic physics. The important point of this new development is not just that it provides an approach to nonperturbative aspect of QCD below the QCD scale Λ. Its main virtue is that it provides a description of hadrons that encompasses in a unified way from elementary particle properties to strongly interacting many-body systems or complex nuclei. (Chapter 10 deals with some aspects of the latter.) At present there is no other approach of comparable versatility, be that "derived" from QCD or purely phenomenological.

Since we do not yet know how to "bosonize" exactly the (3+1)-dimensional gauge theory, QCD, we can only approach the problem with the help of a model that captures the generic structure of the underlying features described in the preceding chapters. As such, there is a great deal of guess work involved here. Nonetheless there has been an impressive amount of progress in unraveling the complex structure of hadronic systems based on modeling of QCD in a partially bosonized form. The key concept that plays a crucial role here is the so-called "Cheshire Cat Mechanism" and the notion of a hierarchy of induced gauge structure associated with the Cheshire Cat mechanism. We will thus start with the notion of Cheshire Cat principle (referred to as CCP in short)[1].

It turns out that one can formulate the CCP as a gauge symmetry as will be discussed later and hence consider it as a simple and profitable means of "deriving" an effective model of QCD at low energy. It might be considered as a "derivation" from QCD of the skyrmion model first proposed by Skyrme for baryons[2], later given a QCD interpretation by Witten[3], as discussed in detail in the preceding chapter. In this chapter, we will confine ourselves to chiral-quark systems, in particular chiral baryons consisting of the up (u), down (d) and strange (s) quarks with zero mass. The problem of mass will be discussed in Chapter 9 where we will also consider the heavy flavors, charm (c) and bottom (b).

8.2. The Cheshire Cat Mechanism

8.2.1. *(1+1) dimensions*

(i) Fermion charge leakage

In (1+1) dimensions the Cheshire Cat theory can be formulated exactly; we shall therefore develop the idea in these low dimensions first. We shall use for definiteness the chiral bag picture[4]. The chiral bag is a simple and useful model for capturing the essence of the physics involved which has the virtue of being "derivable" from QCD. In fact this model emerges in the large N_c limit, as shown by several authors[5,6]. For the present purpose, what matters is that the result is generic and should not sensitively depend upon details of the models. Thus other models of similar structure containing the same symmetries could be equally used. We will leave the construction to the aficionados of their own favorite models.

In the spirit of a chiral bag, consider a massless free single-flavored fermion ψ confined to a region of "volume" V ("inside") coupled on the surface ∂V to a massless free boson ϕ living in a region of "volume" $\tilde{V}$ ("outside"). Of course in one-space dimension, the "volume" is just a segment of a line but we will use this symbol in analogy to higher dimensions. We will assume that the action is invariant under global chiral rotations and parity. Interactions invariant under these symmetries can be introduced without changing the physics, so the simple system that we consider captures all the essential points of the construction. The action contains three terms[*]

$$S = S_V + S_{\tilde{V}} + S_{\partial V} \tag{8.1}$$

where

$$S_V = \int_V d^2x \, \bar{\psi} i \gamma^\mu \partial_\mu \psi + \cdots, \tag{8.2}$$

$$S_{\tilde{V}} = \int_{\tilde{V}} d^2x \, \frac{1}{2} (\partial_\mu \phi)^2 + \cdots \tag{8.3}$$

and $S_{\partial V}$ is the boundary term which we will specify shortly. Here the ellipsis stands for other terms such as interactions, masses etc. consistent with the assumed symmetries of the system on which we will comment later. For instance, there can be a coupling to a $U(1)$ gauge field in (8.2) and a boson mass term in (8.3) as would be needed in Schwinger model. Without loss of generality, we will simply ignore what is in the ellipsis unless we absolutely need it. Now the boundary term is essential in making the connection between the two regions. Its structure depends upon the physics ingredients we choose to incorporate. We will assume that chiral symmetry holds on the boundary even if it is as in nature broken both inside and outside by mass terms. As long as the symmetry breaking is gentle, this should be a good approximation since the surface term is a δ function. We should also assume that the boundary term does not break the discrete symmetries $\mathcal{P}$, $\mathcal{C}$ and $\mathcal{T}$. Finally we demand that it give no decoupled boundary conditions, that is to say, boundary conditions that involve only ψ or ϕ field. These three conditions are sufficient in (1+1) dimensions to give a unique term[7]

$$S_{\partial V} = \int_{\partial V} d\Sigma^\mu \left\{ \frac{1}{2} n_\mu \bar{\psi} e^{i\gamma_5 \phi/f} \psi \right\} \tag{8.4}$$

---

[*]For convenience, our convention is specified once more: The metric is $g_{\mu\nu} = \mathrm{diag}(1, -1)$ with Lorentz indices $\mu, \nu = 0, 1$ and the γ matrices are in Weyl representation, $\gamma_0 = \gamma^0 = \sigma_1$, $\gamma_1 = -\gamma^1 = -i\sigma_2$, $\gamma_5 = \gamma^5 = \sigma_3$ with the usual Pauli matrices σ_i.

with the ϕ "decay constant" $f = 1/\sqrt{4\pi}$ where $d\Sigma^\mu$ is an area element with n^μ the normal vector, *i.e.*, $n^2 = -1$ and picked outward-normal. As we will mention later, we cannot establish the same unique relation in (3+1) dimensions but we will continue using this simple structure even when there is no rigorous justification in higher dimensions.

At classical level, the action (8.1) gives rise to the bag "confinement," namely that inside the bag the fermion (which we shall call "quark" from now on) obeys

$$i\gamma^\mu \partial_\mu \psi = 0 \qquad (8.5)$$

while the boson (which we will call "pion") [†]satisfies

$$\partial^2 \phi = 0 \qquad (8.6)$$

subject to the boundary conditions on ∂V

$$in^\mu \gamma_\mu \psi = -e^{i\gamma_5 \phi/f} \psi, \qquad (8.7)$$

$$n^\mu \partial_\mu \phi = f^{-1}\bar{\psi}(\frac{1}{2}n^\mu \gamma_\mu \gamma_5)\psi. \qquad (8.8)$$

Equation (8.7) is the familiar "MIT confinement condition" which is simply a statement of the classical conservation of the vector current $\partial_\mu j^\mu = 0$ or $\bar{\psi}\frac{1}{2}n^\mu \gamma_\mu \psi = 0$ at the surface while eq. (8.8) is just the statement of the conserved axial vector current $\partial_\mu j_5^\mu = 0$ (ignoring the possible explicit mass of the quark and the pion at the surface)[‡] The crucial point of our argument is that these classical observations are invalidated by quantum mechanical effects. In particular while the axial current continues to be conserved, the vector current fails to do so due to quantum anomaly. Note that this is the reverse of what happens in open space where anomaly is in the axial current.

There are several ways of seeing that something is amiss with the classical conservation law, with slightly different interpretations. The easiest way is as follows. For definiteness, let us imagine that the quark is "confined" to the space $-\infty \leq r \leq R$ with a boundary at $r = R$. Now the vector current $j_\mu = \frac{1}{2}\bar{\psi}\gamma_\mu \psi$ is conserved inside the "bag"

$$\partial^\mu j_\mu = 0, \qquad r < R. \qquad (8.9)$$

[†]We recall from Chapter 3 that in two dimensions, the "pion" is not strictly a Goldstone boson, but should be understood as a BKT (Berezinski-Kosterlitz-Thouless) boson.
[‡]Here the currents are defined with a factor 1/2: $j^\mu = \bar{\psi}\frac{1}{2}\gamma^\mu \psi$, $j_5^\mu = \bar{\psi}\frac{1}{2}\gamma^\mu \gamma_5 \psi$.

If one integrates this from $-\infty$ to R in r, one gets the time-rate change of the fermion (*i.e.*, quark) charge

$$\dot{Q} \equiv \frac{d}{dt}Q = 2\int_{-\infty}^{R} dr\partial_0 j_0 = 2\int_{-\infty}^{R} dr\partial_1 j_1 = 2j_1(R) \qquad (8.10)$$

which is just

$$\bar{\psi}n^\mu\gamma_\mu\psi, \quad r = R. \qquad (8.11)$$

This vanishes classically as we mentioned above. But this is not correct quantum mechanically because it is not well-defined locally in time. In other words, $\psi^\dagger(t)\psi(t+\epsilon)$ goes like ϵ^{-1} and so is singular as $\epsilon \to 0$. To circumvent this difficulty which is related to vacuum fluctuation, we regulate the bilinear product by point-splitting in time

$$j_1(t) = \bar{\psi}(t)\frac{1}{2}\gamma_1\psi(t+\epsilon), \quad \epsilon \to 0. \qquad (8.12)$$

Now using the boundary condition (at $r = R$)

$$\begin{aligned} i\gamma_1\psi(t+\epsilon) &= e^{i\gamma_5\phi(t+\epsilon)/f}\psi(t+\epsilon) \\ &\approx e^{i\gamma_5\phi(t)/f}[1 + i\epsilon\gamma_5\dot{\phi}(t)/f]e^{\frac{1}{2}[\phi(t),\phi(t+\epsilon)]}\psi(t+\epsilon)(8.13) \end{aligned}$$

and the commutation relation

$$[\phi(t), \phi(t+\epsilon)] = i \,\text{sign}\,\epsilon, \qquad (8.14)$$

we obtain

$$j_1(t) = -\frac{i}{4f}\epsilon\dot{\phi}(t)\psi^\dagger(t)\psi(t+\epsilon) = \frac{1}{4\pi f}\dot{\phi}(t) + O(\epsilon), \quad r = R \qquad (8.15)$$

where we have used $\psi^\dagger(t)\psi(t+\epsilon) = \frac{i}{\pi\epsilon} + [\text{regular}]$. Thus quarks can flow out or in if the pion field varies in time. But by fiat, we declared that there be no quarks outside, so the question is what happens to the quarks when they leak out? They cannot simply disappear into nowhere if we impose that the fermion (quark) number is conserved. To understand what happens, rewrite (8.15) using the surface tangent

$$t^\mu = \epsilon^{\mu\nu}n_\nu. \qquad (8.16)$$

We have

$$t \cdot \partial\phi = -\frac{1}{2f}\bar{\psi}n \cdot \gamma\psi = \frac{1}{2f}\bar{\psi}t \cdot \gamma\gamma_5\psi, \quad r = R \tag{8.17}$$

where we have used the relation $\bar{\psi}\gamma_\mu\gamma_5\psi = \epsilon_{\mu\nu}\bar{\psi}\gamma^\nu\psi$ valid in two dimensions. Equation (8.17) together with (8.8) is nothing but the bosonization relation

$$\partial_\mu\phi = f^{-1}\bar{\psi}(\frac{1}{2}\gamma_\mu\gamma_5)\psi \tag{8.18}$$

at the point $r = R$ and time t. As is well known, this is a unique feature of (1+1) dimensional fields that makes one-to-one correspondence between fermions and bosons[8]. In fact, one can even write the fermion field in terms of the boson field as

$$\psi(x) = \exp\left\{-\frac{i}{2f}\int_{x_0}^{x} d\xi[\pi(\xi) + \gamma_5\phi'(\xi)]\right\}\psi(x_0) \tag{8.19}$$

where $\pi(\xi)$ is the conjugate field to $\phi(\xi)$ and $\phi'(\xi) = \frac{d}{d\xi}\phi(\xi)$.

Equation (8.10) with (8.15) is the vector anomaly, *i.e.*, quantum anomaly in the vector current: The vector current is not conserved due to quantum effects. What it says is that the vector charge which in this case is equivalent to the fermion (quark) number inside the bag is not conserved. Physically what happens is that the amount of fermion number ΔQ corresponding to $\Delta t\dot{\phi}/\pi f$ is pushed into the Dirac sea at the bag boundary and so is lost from inside[§] This accumulated baryon charge must be carried by something residing in the meson sector. Since there is nothing but "pions" outside, it must be the pion field that carries the leaked charge. This can happen only if the pion field supports a soliton. This is rather simple to verify in (1+1) dimensions using arguments given in the work of Coleman and Mandelstam, referred to above. This can also be shown to be the case in (3+1) dimensions as we shall see below. In the present model, we find that one unit of fermion charge $Q = 1$ is partitioned as

$$Q = 1 = Q_V + Q_{\bar{V}}, \tag{8.20}$$

[§] One can see this more clearly by rewriting the change in the quark charge in the bag as

$$\Delta Q_V = \frac{1}{2}\int_V \langle 0|[\psi^\dagger(x), \psi(x)]|0\rangle - 1 = -\frac{1}{2}\left\{\sum_{E_n>0} 1 - \sum_{E_n<0} 1\right\}$$

$$Q_{\bar{V}} = \theta/\pi,$$
$$Q_V = 1 - \theta/\pi$$

with

$$\theta = \phi(R)/f. \tag{8.21}$$

We thus learn that the quark charge is partitioned into the bag and outside of the bag, without however any dependence of the total on the size or location of the bag boundary. In the (1+1)-dimensional case, one can calculate other physical quantities such as the energy, response functions and more generally, partition functions and show that the physics does not depend upon the presence of the bag[9]. We could work with quarks alone, or pions alone or any mixture of the two. If one works with the quarks alone, we have to do a proper quantum treatment to obtain something which one can obtain in mean-field order with the pions alone. In some situations, the hybrid description is more economical than the pure ones. *The complete independence of the physics on the bag – size or location– is called the "Cheshire Cat Principle" (CCP) with the corresponding mechanism referred to as "Cheshire Cat Mechanism" (CCM).* Of course the CCP holds exactly in (1+1) dimensions because of the exact bosonization rule. There is no exact CCP in higher dimensions since fermion theories cannot be bosonized exactly in higher dimensions but we will see later that a fairly strong case of CCP can be made for (3+1)-dimensional models. Topological quantities like the fermion (quark or baryon) charge satisfy an exact CCP in (3+1) dimensions while nontopological observables such as masses, static properties and also some nonstatic properties satisfy it approximately but rather well.

(ii) "Color" anomaly

So far we have been treating the quark as "colorless," in other words, in the absence of a gauge field. Let us consider that the quark carries a $U(1)$ charge e, coupling to a $U(1)$ gauge field A^μ. We will still continue working with a single-flavored quark, treating multi-flavored cases later. Then inside the bag, we have essentially the Schwinger model, namely, (1+1)-dimensional QED[10]. It is well established that the charge is confined in the Schwinger model, so there are no charged particles in the spectrum. If now our "leaking" quark carries the color (electric) charge, this will at first sight pose a problem since the anomaly obtained above says that there will be a leakage of the charge by the rate

$$\dot{Q}^c = \frac{e}{2\pi}\dot{\phi}/f. \tag{8.22}$$

This means that the charge accumulated on the surface will be

$$\Delta Q^c = \frac{e}{2\pi}\phi(t,R)/f. \tag{8.23}$$

Unless this is compensated, we will have a violation of charge conservation, or breaking of gauge invariance. This is unacceptable. Therefore we are invited to introduce a boundary condition that compensates the induced charge, *i.e.*, by adding a boundary term

$$\delta S_{\partial V} = -\int_{\partial V} d\Sigma \frac{e}{2\pi}\epsilon^{\mu\nu}n_\nu A_\mu \frac{\phi}{f} \tag{8.24}$$

with $n^\mu = g^{\mu\nu}n_\nu$. The action is now of the form[11]

$$S = S_V + S_{\tilde{V}} + S_{\partial V} \tag{8.25}$$

with

$$S_V = \int_V d^2x \left\{ \bar{\psi}(x)\left[i\partial_\mu - eA_\mu\right]\gamma^\mu\psi(x) - \frac{1}{4}F_{\mu\nu}F^{\mu\nu} \right\}, \tag{8.26}$$

$$S_{\tilde{V}} = \int_{\tilde{V}} d^2x \left\{ \frac{1}{2}(\partial_\mu\phi(x))^2 - \frac{1}{2}m_\phi^2\phi(x)^2 \right\}, \tag{8.27}$$

$$S_{\partial V} = \int_{\partial V} d\Sigma \left\{ \frac{1}{2}\bar{\psi}e^{i\gamma^5\phi/f}\psi - \frac{e}{2\pi}\epsilon^{\mu\nu}n_\mu A_\nu \frac{\phi}{f} \right\}. \tag{8.28}$$

We have included the mass of the ϕ field, the reason of which will become clear shortly.

In this example, we are having the "pion" field (more precisely the soliton component of it) carrying the color charge. In general, though, the field that carries the color could be different from the field that carries the soliton. In fact, in (3+1) dimensions, it is the η' field that will be coupled to the gauge field (although η' is colorless) while it is the pion that supports a soliton. But in (1+1) dimensions, the soliton is lodged in the $U(1)$ flavor sector (whereas in (3+1) dimensions, it is in $SU(n_f)$ with $n_f \geq 2$). For simplicity, we will continue our discussion with this Lagrangian.

We now illustrate how one can calculate the mass of the ϕ field using the action (8.25) and the CCM. This exercise will help in understanding a similar calculation of the mass of the η' in (3+1) dimensions. The additional ingredient needed for this exercise is the $U_A(1)$ (Adler-Bell-Jackiw) anomaly[12] discussed in detail in Chapters 1 and 6 which in (3+1) dimensions reads

$$\partial_\mu j_5^\mu = -\frac{e^2}{32\pi^2}\epsilon_{\mu\nu\rho\sigma}G^{\mu\nu}G^{\rho\sigma} \tag{8.29}$$

and in (1+1) dimensions takes the form

$$\partial_\mu j_5^\mu = -\frac{e}{4\pi}\epsilon_{\mu\nu}F^{\mu\nu} = -\frac{1}{2\pi}eE. \tag{8.30}$$

Instead of the configuration with the quarks confined in $r < R$, consider a small bag "inserted" into the static field configuration of the scalar field ϕ which we will call η' in anticipation of the (3+1)-dimensional case we will consider later. Let the bag be put in $-\frac{R}{2} \leq r \leq \frac{R}{2}$ as in Figure 8.1. The CCP states that the physics should not depend on the size R, hence we can take it to be as small as we wish. For a small enough size, the charge accumulated at the two boundaries can be reduced to $\pm\frac{e}{2\pi}\phi/f$. This means that the electric field generated inside the bag is

$$E = F_{01} = \frac{e}{2\pi}(\phi/f). \tag{8.31}$$

Therefore from the $U_A(1)$ anomaly (8.30), the axial charge created or

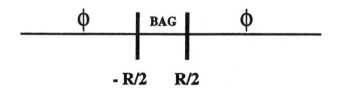

Fig. 8.1. A small bag $-\frac{R}{2} \leq r \leq \frac{R}{2}$ inserted into the static field ϕ to calculate the mass m_ϕ using the CCP.

destroyed (depending on the sign of the scalar field) in the static bag is

$$2\int_V dr\,\partial_\mu j_5^\mu = 2(-j_1^5|_{\frac{R}{2}} + j_1^5|_{-\frac{R}{2}}) = -\frac{e^2}{2\pi^2}(\phi/f)R. \tag{8.32}$$

From the bosonization condition (8.18), we have

$$2f\left(\partial_1\phi|_{\frac{R}{2}} - \partial_1\phi|_{-\frac{R}{2}}\right) = \frac{e^2}{2\pi^2}(\phi/f)R. \tag{8.33}$$

Now taking R to be infinitesimal by the CCP, we have, on canceling R from both sides,

$$\partial^2\phi - \frac{e^2}{4\pi^2 f^2}\phi = 0 \tag{8.34}$$

which then gives the mass, with $f^{-2} = 4\pi$,

$$m_\phi^2 = \frac{e^2}{\pi}, \tag{8.35}$$

the well-known scalar mass in the Schwinger model which can also be obtained by bosonizing the $(QED)_2$. A completely parallel reasoning will be used later to calculate the η' mass in (3+1) dimensions.

(iii) The Wess-Zumino-Witten term

Let us consider the case of more than one flavor. For simplicity, we take two flavors. In the (3+1) dimensions, we have an extra topological term called Wess-Zumino term if we have more than two flavors. As shown in Chapter 3, in (1+1) dimensions, such an extra term arises already for two flavors[13], for which the relevant flavor symmetry is then $U(2) \times U(2)$. The $U_V(1)$ symmetry associated with the fermion number does not concern us, so we may drop it, leaving the symmetry $U_A(1) \times SU(2) \times SU(2)$. The scalar field discussed above corresponds to the $U_A(1)$ field, so we will not consider it anymore and instead focus on scalars living in $SU(2)_V$ to which $SU(2) \times SU(2)$ is supposed to break down. The scalars will be called π^a with $a = 1, 2, 3$. As stated before, the soliton lives in the $U(1)$ sector, so the $SU(2)$ scalars have nothing to do with the "baryon" structure but its nonabelian nature brings in an extra ingredient associated with the Wess-Zumino-Witten term which we will now show emerges from the boundary condition. Let the scalar multiplets be summarized in the form

$$U = e^{i\vec{\tau}\cdot\vec{\pi}/f} \tag{8.36}$$

with $f = 1/\sqrt{4\pi}$.

As before the action consists of three terms

$$S = S_V + S_{\tilde{V}} + S_{\partial V}. \tag{8.37}$$

The bag action is given by

$$S_V = \int_V d^2x \bar{\psi} i \gamma^\mu \partial_\mu \psi + \cdots, \tag{8.38}$$

where the fermion field is a doublet $\psi^T = (\psi_1, \psi_2)$. A straightforward generalization of the abelian boundary condition consistent with chiral invariance suggests the boundary term of the form

$$S_{\partial V} = \int_{\partial V} d\Sigma^\mu \left\{ \frac{1}{2} n_\mu \bar{\psi} U^{\gamma_5} \psi \right\} \tag{8.39}$$

with

$$U^{\gamma_5} = e^{i\gamma_5 \vec{\tau} \cdot \vec{\pi}/f}. \tag{8.40}$$

The meson sector contains an extra term which plays an important role in multi-flavor systems even for noninteracting bosons. Thus the outside (meson sector) contribution to the action takes the form

$$S_{\tilde{V}} = \int_{\tilde{V}} d^2x \frac{f^2}{4} \operatorname{Tr}(\partial_\mu U \partial^\mu U^\dagger) + \cdots + S_{WZW}. \tag{8.41}$$

As seen, the main new ingredient in this action is the presence of the extra term S_{WZW} in the boson sector. This is the Wess-Zumino term we discussed in Chapter 3 which in the (1+1)-dimensional system we are interested in will be written as

$$S^{\tilde{V}}_{WZW} = \frac{2\pi n}{24\pi^2} \int_{\tilde{V} \times [0,1]} d^3x \epsilon^{ijk} \operatorname{Tr}(L_i L_j L_k) \tag{8.42}$$

with an integer n corresponding to the number of "colors", where $L_\mu = g^\dagger \partial_\mu g$ with g "homotopically" extended as $g(s = 0, x) = 1$, $g(s = 1, x) = U(x)$. The action (8.42) is defined in the partial space $\tilde{V}$. As we have seen before, the WZW term plays a crucial role in the skyrmion physics in (3+1) dimensions, so the question we wish to address here is: How does the Cheshire Cat manifest itself for the WZW term? To answer this question, we imagine shrinking the bag to a point and ask what remains from what corresponds to the WZW term coming from the space V in which quarks live. The obvious answer would be that the bag gives rise to the portion of the WZW term complement to (8.42) so as to yield the total defined in the full space. This is of course the right answer but it is not at all obvious that the chiral bag theory as formulated does this properly. For instance, consider the variation of the action of the meson sector (8.42)

$$\delta S_{WZW}^{\tilde{V}} = 3\frac{n}{12\pi}\int_{\tilde{V}} d^2x\epsilon^{\mu\nu}\,\text{Tr}((g^\dagger\delta g)L_\mu L_\nu)$$

$$+ 3\frac{n}{12\pi}\int_0^1 ds\int_{\partial\tilde{V}} d\Sigma_\mu\epsilon^{\mu\nu\alpha}\,\text{Tr}((g^\dagger\delta g)L_\nu L_\alpha). \quad (8.43)$$

The second term depends on both the homotopy extension and the surface. By the CCP, such a term is unphysical and must be canceled by a complementary contribution from the bag. Thus the consistency of the model requires that the cancellation occur through the boundary condition. We will now show this is exactly what one obtains within the model so defined[14].

Consider the action for the quark bag with the quark satisfying the Dirac equation

$$i\,\partial\!\!\!/\,\psi = 0 \quad (8.44)$$

subject to the boundary condition

$$(e^{i\theta(\beta,t)\gamma_5} - \hat{n}\!\!\!/)\psi(\beta,t) = 0 \quad (8.45)$$

where the bag wall is put at $x = \beta$ and $\theta \equiv \vec{\tau}\cdot\vec{\pi}/f$. We will move the bag wall such that the bag will shrink to zero radius. The general solution for (8.44) with (8.45) is

$$\psi(x,t) = \exp\left\{-i[\frac{1}{2}(1+\gamma_5)\theta_- + \frac{1}{2}(1-\gamma_5)\theta_+]\right\}\psi_0(x,t) \quad (8.46)$$

where $\theta_\pm = \theta(x \pm t)$ and ψ_0 is the solution for $\theta = 0$ (for the MIT bag). The boundary condition is satisfied if

$$\theta_-(\beta,t) - \theta_+(\beta,t) = \theta(\beta,t). \quad (8.47)$$

Since we are dealing with a time-dependent boundary condition,¶ it is more convenient to use the path-integral method instead of a canonical formal-

¶An alternative way is to make a local chiral rotation on the quark field so as to make the boundary condition time-independent and generate an induced gauge potential (known as Berry potential or connection). This elegant procedure is discussed below.

ism. Let $\partial\!\!\!/_\theta$ denote the Euclidean Dirac operator associated with the boundary condition (8.45). Then integrating the quark field in the bag action, we have

$$e^{-S_F} = \det\left(\partial\!\!\!/_\theta / \partial\!\!\!/_0\right) \tag{8.48}$$

or

$$S_F = -\text{tr}\,\ln(\partial\!\!\!/_\theta\,\partial\!\!\!/_0^{-1}) = -\int_0^1 ds\,\text{tr}(\partial_s\,\partial\!\!\!/_\theta\,\partial\!\!\!/_\theta^{-1}) \tag{8.49}$$

where we denote by S_F the effective Euclidean action for the bag. On the right-hand side of eq.(8.49), θ is homotopically extended as $\theta(x,t) \rightarrow \theta(x,t,s)$ such that $\theta(x,t,s=1) = \theta(x,t)$ and $\theta(x,t,s=0) = 0$. The trace tr goes over the "color", the space-time, γ matrices and the flavor (τ) matrices. Since the θ dependence appears only in the boundary condition, we can use the latter to rewrite

$$S_F = -\int_0^1 ds\,\text{tr}\,\left(\Delta_B\partial_s[e^{i\theta\gamma_5} - \not{n}]\,\partial\!\!\!/_\theta\right) \tag{8.50}$$

where Δ_B is the surface delta function. This expression makes sense only when it is suitably regularized. We use the point-splitting regularization

$$S_F = -\lim_{\epsilon\to 0+}\int_{V\times[0,1]} ds\,d^2x\,\Delta_B\text{tr}\,\left(\partial_s e^{i\gamma_5\theta}G_\theta(x,t+\epsilon;x,t-\epsilon)\right) \tag{8.51}$$

with G the Dirac propagator inside the chiral bag which can be written explicitly using the time-dependent solution (8.46)

$$G_\theta(x,t;x',t') = \begin{pmatrix} e^{-i\theta_-(x-t)} & 0 \\ 0 & e^{-i\theta_+(x+t)} \end{pmatrix}$$
$$G_0(x,t;x',t')\begin{pmatrix} e^{i\theta_+(x'+t')} & 0 \\ 0 & e^{i\theta_-(x'-t')} \end{pmatrix} \tag{8.52}$$

where G_0 is the bagged-quark MIT propagator which has been computed by multiple-reflection method[15]

$$\lim_{x,x'\to\beta} G_0(x,t+\epsilon;x',t-\epsilon) = \frac{1}{4}(1+\gamma_\beta^1)(\gamma^0/4\pi\epsilon)(3-\gamma_\beta^1) + O(1) \tag{8.53}$$

with $\gamma_\beta^1 = \gamma^1\cdot n_\beta$ and $n_\beta = n_\pm = \pm 1$. The WZW term arises from the imaginary part of the action, so calculating Im S_F to leading order in the derivative expansion, we get

$$\operatorname{Im} S_F = i\frac{2\pi n}{24\pi^2} \int_0^1 ds \int_V dx dt \operatorname{Tr}\left(\partial_s \theta \partial_t \theta \partial_x \theta\right) + \cdots \tag{8.54}$$

where the trace Tr now runs only over the flavor indices and the ellipsis stands for higher derivative terms. Invoking chiral symmetry and going to the Minkowski space, we finally obtain

$$S_{WZW}^V = \frac{2\pi n}{24\pi^2} \int_{V \times [0,1]} d^3 x \epsilon^{\mu\nu\alpha} \operatorname{Tr}\left(L_\mu L_\nu L_\alpha\right) \tag{8.55}$$

which is exactly the complement to the outside contribution (8.42). Notice that there is no surface-dependent term whose presence would have signaled the breakdown of the CCP.

So far our discussion has been somewhat cavalier. A lot more rigorous derivation was given by Falomir *et al.*[14] who computed the path integral

$$Z = \int [d\bar{\psi}][d\psi] e^{-S_E} \tag{8.56}$$

where

$$S_E = \int_V d^2 x \bar{\psi} e^{\gamma_5 \theta} i \not{\partial} e^{\gamma_5 \theta} \psi + \frac{i}{2} \int_{\partial V} d^1 x \bar{\psi} (1 - \not{n}) \psi. \tag{8.57}$$

A careful analysis showed indeed that there is no surface term in the effective action, corroborating the above derivation: the model is fully consistent with the CCP. Note that in (8.57), the field θ is defined in the interior of the volume V whereas in the above consideration all the *action* occurred on the surface. The physics is equivalent, however, since our argument was based on "shrinking" the bag whereas in the formulation of these authors, the quarks are being plainly integrated out from the bag. (The real part of the effective action gives $S_V = \frac{f^2}{4} \int_V d^2 x \operatorname{Tr}\left(\partial_\mu U \partial^\mu U^\dagger\right) +$ higher derivative terms, just to complement the meson sector.)

(iv) A recapitulation on the WZW term

The WZW term was described in various parts of the preceding chapters. Here we summarize, using Balachandran's lecture[16], some of the elements which will prove to be useful for understanding the skyrmion structure. We will treat the (1+1)-dimensional case but the argument can be easily extended to higher dimensions.

Consider a field g valued in the group $SU(N)$ with $N \geq 2$ (in the case of two dimensions) which we can represent by $N \times N$ unitary matrices with unit determinant. The configuration space Q is given by fields in one spatial dimension. If we impose the boundary condition that $g \to 1$ at $x = \pm\infty$, so that at a fixed time, the space is a circle S^1, then Q is just the set of maps

$$Q : S^1 \quad \to \quad SU(N)$$
$$x^1 \quad \to \quad g(x^1). \tag{8.58}$$

We will consider the WZW term in the full space $\Omega = V \cup \tilde{V}$. Now let D be a disc in Q parametrized by x^0 corresponding to time and an s with $0 \leq s \leq 1$ such that the boundary D is just the space-time Ω. The field is then $g(x^0, x^1, s)$ chosen so that $g(x^0, x^1, 0) = 1$ and $g(x^0, x^1, 1) = g(x^1) = U$. Then the WZW term is

$$S_{WZW} = \frac{2\pi n}{24\pi^2} \int_{D[\partial D = \Omega \times [0,1]]} \epsilon_{ijk} \; \mathrm{Tr}(g^\dagger \partial_i g g^\dagger \partial_j g g^\dagger \partial_k g). \tag{8.59}$$

This action has the properties that the quantity n is quantized to an integer value and that the action is invariant under the infinitesimal variation in the interior of the disc D. One can see the latter in various ways. One way is to note that $S_{WZW}/(2\pi n)$ is the charge of the winding number current

$$J_\mu = \frac{1}{24\pi^2} \epsilon_{\mu\nu\alpha\beta} \, \mathrm{Tr}[g^\dagger \partial^\nu g g^\dagger \partial^\alpha g g^\dagger \partial^\beta g], \tag{8.60}$$

namely

$$S_{WZW}/(2\pi n) = \int d^3x \, J_0(x) \tag{8.61}$$

which is precisely the conserved winding number. Another way of seeing is to note that S_{WZW} is just an integral of a closed three form. Now to show that n is quantized, it suffices to notice that $D \cup \bar{D} = S^2 \times S^1$ (where $\bar{D}$ is the disc complement to D) and that

$$\frac{2\pi n}{24\pi^2} \int_{S^2 \times S^1} \mathrm{Tr}[g^\dagger dg]^3 = 2\pi \times \text{integer}. \tag{8.62}$$

This proves the assertion. Similar quantization of the coefficient of the WZW term in four dimensions turned out to be crucial for skyrmion physics.

8.2.2. *(3+1) dimensions*

Since we cannot even bosonize completely a free Dirac theory in four dimensions (not to mention highly nonlinear QCD), we will essentially follow step-by-step the $(1+1)$-dimensional reasoning and construct in complete parallel a simple, yet hopefully sufficiently realistic chiral bag model which will then be *partially* bosonized. The roles of CCP and bosonization will then be reversed here: we will invoke CCP in bosonizing the chiral bag. Here again the boundary conditions play a key role. Our attitude here will be to consider CCP as a means to bosonize (*albeit* partially) QCD. Ultimately we will motivate a modeling of QCD at low energies in terms of effective chiral Lagrangian theory which we will use to describe low-energy hadronic/nuclear processes.

(i) The model

For generality, we consider the flavor group $SU(3) \times SU(3) \times U_A(1)$ appropriate to up, down and strange quark systems. The chiral symmetry $SU(3) \times SU(3)$ is spontaneously broken down to $SU(3)$, with the octet Goldstone bosons denoted by $\pi = \frac{\lambda^a}{2}\pi^a$ with λ the Gell-Mann matrices and the $U_A(1)$ symmetry, we know, is broken by anomaly, with the associated massive scalar denoted by η'. Vector mesons can be suitably introduced through hidden gauge symmetry (HGS) but we will not use them here for simplicity. We do not yet have a fully satisfactory theory of Cheshire Cat mechanism in the presence of vector-meson degrees of freedom. We will develop our strategy using a Lagrangian that contains only the octet π fields and limit ourselves to the lowest-order derivative order. We will for the moment consider the chiral limit with all quark masses equal to zero. The symmetry breaking due to quark masses will be discussed later.

We consider quarks "confined" within a volume V which we take spherical for definiteness, surrounded by the octet Goldstone bosons π and the singlet η' populating the outside space of volume $\tilde{V}$. This is pictorially represented in Figure 8.2.

The action is again given by the three terms

$$S = S_V + S_{\tilde{V}} + S_{\partial V}. \tag{8.63}$$

Including gluons, the "bag" action is given by

$$S_V = \int_V d^4x \left(\bar\psi i \not{D} \psi - \frac{1}{2}\mathrm{tr}\, G_{\mu\nu}G^{\mu\nu} \right) + \cdots \tag{8.64}$$

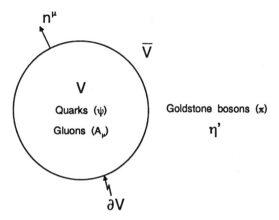

Fig. 8.2. A spherical chiral bag for deriving a "realistic" Cheshire Cat mechanism.

where the trace (tr) goes over the color index. The action for the meson sector must include the usual Goldstone bosons as well as the massive η' taking the form

$$
S_{\tilde{V}} = \frac{f^2}{4} \int_{\tilde{V}} d^4x \left(\text{Tr } \partial_\mu U^\dagger \partial^\mu U + \frac{1}{4N_f} m_{\eta'}^2 \left[\text{Tr} \left(\ln U - \ln U^\dagger \right) \right]^2 \right) + \cdots
$$
$$
+ S_{WZW} \tag{8.65}
$$

where $N_f = 3$ is the number of flavors and

$$
U = e^{i\eta'/f_0} e^{2i\pi/f}, \tag{8.66}
$$
$$
f_0 \equiv \sqrt{N_f/2} f.
$$

The WZW term is the five-dimensional analog of the three-dimensional one given above,

$$
S_{WZW} = -N_c \frac{i}{240\pi^2} \int_{D_5[\partial D_5 = V \times [0,1]]} \epsilon_{\mu\nu\lambda\rho\sigma} \text{Tr}(L^\mu L^\nu L^\lambda L^\rho L^\sigma) \tag{8.67}
$$

with $L = g^\dagger(x, s) dg(x, s)$. We will come back to this WZW term later. Now the surface action is nontrivial. It has the usual term

$$
\frac{1}{2} \int_{\partial V} d\Sigma^\mu (n_\mu \bar{\psi} U^{\gamma_5} \psi) \tag{8.68}
$$

with

$$U^{\gamma_5} = e^{i\eta'\gamma_5/f_0}e^{2i\pi\gamma_5/f}, \tag{8.69}$$

but it has more because of what is called "color anomaly," analogous to what we had in two dimensions, which we need to analyze[17]. The trouble with the boundary condition (8.68) is that it is not consistent with quantum anomaly effect and hence breaks color gauge invariance at quantum level. To see what happens, first we clarify what the color confinement is at classical level. With (8.68), the gluons inside the bag are subject to the boundary conditions

$$\hat{n} \cdot \vec{E}^a = 0, \quad \hat{n} \times \vec{B}^a = 0 \tag{8.70}$$

with $\hat{n}$ the outward normal to the bag surface and a the color index. These are the usual MIT bag boundary conditions. What these conditions mean is as follows. A positive helicity fermion near the surface with its momentum along the direction of the color magnetic field moves freely in the lowest Landau orbit, because the energy $-|gB|/\omega$ (where ω is its total energy and g is the gluon coupling constant) resulting from its magnetic moment cancels exactly its zero point energy $|gB|/\omega$. Since the color electric field points along the surface, it cannot force color charge to cross the bag wall and hence the color charge will be strictly confined within.

We will now show that quantum mechanical anomaly induces the leakage of color. The main cause is the boundary condition associated with the η' field

$$\left\{i\gamma \cdot \hat{n} + e^{i\gamma_5\eta(\beta)}\right\}\psi(\beta) = 0 \tag{8.71}$$

where as before β is a point on the bag surface $\Sigma = \partial V$ and we define

$$\eta \equiv \eta'/f_0. \tag{8.72}$$

Consider now the color charge inside the bag

$$Q^a = \int_V d^3x g(\psi^\dagger \frac{\lambda^a}{2}\psi + f^{abc}\vec{A}^b \cdot \vec{E}^c). \tag{8.73}$$

The color current is conserved inside the bag, so integrating over the spatial volume of the divergence of the current (which is zero), we have

$$\dot{Q}^a = -\oint_{\Sigma=\partial V} d\beta \vec{j}^a \cdot \hat{n} \tag{8.74}$$

where using temporal gauge, $A_0^a = 0$,

$$\vec{j}^a = g\bar\psi\vec\gamma\frac{\lambda^a}{2}\psi + gf^{abc}(\vec{A^b} \times \vec{B^c}). \tag{8.75}$$

As mentioned above, classically there is no color flux leaking out, so $\dot{Q}^a = 0$. However as in the case of $(1+1)$ dimensions encountered above, the integrand on the right-hand side of eq.(8.74) is not well-defined without regularization. Let us regularize it by time-splitting it as before. To simplify further our discussion, we will make the quasi-abelian approximation. After the calculation is completed, we can extract the nonabelian structure by inspection. In this case, we can write

$$\vec{j}^a(\beta) = g\,\mathrm{Tr}[\frac{\lambda^a}{2}\vec\gamma S^+(\beta_-;\beta_+)] \tag{8.76}$$

where the fermion propagator S^+ is defined as

$$S^+(\beta,\beta') = \lim_{(x,x')\to(\beta,\beta')} S(x,x') \tag{8.77}$$

and

$$\beta_\pm \equiv \beta \pm \frac{\epsilon}{2}. \tag{8.78}$$

Using (8.71) for β replaced by $\beta - \epsilon/2$ and the hermitian conjugate of it for $\beta + \epsilon/2$, we have

$$\dot{Q}^a = \frac{i}{2}g \lim_{\epsilon\to 0_+} \epsilon \oint_\Sigma \mathrm{Tr}[\frac{\lambda^a}{2}\gamma_5\vec\gamma\cdot\hat{n}S^+(\beta_-;\beta_+)]\dot\eta. \tag{8.79}$$

Therefore if we take, for the moment, η to depend on time only (this assumption will be lifted later), we can write

$$\frac{dQ^a}{d\eta} = \frac{i}{2}g \lim_{\epsilon\to 0_+} \epsilon \oint_\Sigma \mathrm{Tr}[\frac{\lambda^a}{2}\gamma_5\vec\gamma\cdot\hat{n}S^+(\beta_-;\beta_+)]. \tag{8.80}$$

Using the multiple reflection method in four dimensions[18], in Euclidean space,

$$S^+(\beta_-;\beta_+) = \frac{1}{4}(1 + \gamma_E\cdot\hat{n}U^5)S(\beta_-;\beta_+)(3 - U^5\gamma_E\cdot\hat{n}) \tag{8.81}$$

with $U^5 = e^{i\gamma_5\eta}$, $\gamma_E = -i\gamma$, $A_{Ei}^a = -A_i^a$ for $i = 1,2,3$, $A_{E0}^a = -iA_0^a$ and

$$S(\beta_-;\beta_+) = \int_{R\times V} d^4x\, S_0(\beta_- - x)(-ig\gamma_E^\alpha A_E^{ba}(x)\cdot\frac{1}{2}\lambda^b)S_0(x - \beta_+) \tag{8.82}$$

where $R \times V$ is the Euclidean space-time and S_0 is the free field Euclidean propagator

$$S_0(x - x') = \frac{1}{2\pi^2} \frac{\gamma_E^\alpha (x - x')^\alpha}{(x - x')^4}.$$ (8.83)

Substituting (8.81) and (8.82) into (8.80) and returning to Minkowski space, we arrive at

$$\frac{dQ^a}{d\eta} = -\frac{g^2}{4\pi^2} \epsilon^{0\mu\alpha\sigma} \oint_\Sigma d\beta \int_{R\times V} d^4 x \delta^4 (\beta - x) \partial_\sigma A_\alpha^a(x) n_\mu.$$ (8.84)

In the quasi-abelian case that we are considering, we can write

$$\frac{dQ^a}{d\eta} = \frac{g^2}{8\pi^2} \int_\Sigma d\beta \vec{B}^a(\beta) \cdot \hat{n}.$$ (8.85)

Since Q^a is covariant under small gauge transformations that are constant on the bag surface, eq.(8.85) equally holds for a nonabelian color magnetic field. Furthermore, one can generalize (8.85) to a variation with a space-time dependent $\eta(\vec{x}, t)$ by replacing the normal derivative by a functional derivative with it.

Roughly speaking, what (8.85) means is that color charge "disappears" from the bag following the time variation of the η field. Coming from the regularization, this is a quantum effect which signals a disaster to the model *unless* it is canceled by something else. A simple remedy to this is to impose, in a complete analogy to the (1+1)-dimensional Schwinger model considered above, a boundary condition which can be effected in the Lagrangian by a surface term of the type

$$\mathcal{L}_{CT} \sim -\frac{g^2 N_F}{8\pi^2} \oint_\Sigma d\beta A_0^a \vec{B}^a \cdot \hat{n} \eta(\beta)$$ (8.86)

where we have taken into account the number of light flavors involved. Heavy-quark flavors are irrelevant to the issue ‖ Chiral invariance, covariance and general gauge transformation properties allow us to rewrite (8.86)

‖Why this is so is easy to understand. When the fermions have a mass m, there is a gap between the positive and negative energy Landau levels. Now if the time variation in the η field is large on the scale of $1/m$, then there is no net flow out of the Dirac sea. In terms of anomaly, this is seen as follows. In the presence of heavy fermion masses, the anomaly is formally nonzero. However this is balanced by the term $2m\langle\bar{\psi} i\gamma_5 \psi\rangle$ coming from the explicit chiral symmetry breaking term in the Lagrangian.

in a general form

$$\mathcal{L}_{CT} = i\frac{g^2}{32\pi^2} \oint_\Sigma K_5^\mu n_\mu (\text{Tr}\,\ln U^\dagger - \text{Tr}\,\ln U) \qquad (8.87)$$

where K_5^μ is the (properly regularized) Chern-Simons current

$$K_5^\mu = \epsilon^{\mu\nu\alpha\beta}(A_\nu^a F_{\alpha\beta}^a - \frac{2}{3}g f^{abc} A_\nu^a A_\alpha^b A_\beta^c). \qquad (8.88)$$

Note that (8.87) is invariant under neither large gauge transformation (because of the Chern-Simons current) nor small gauge transformation (because of the surface). Thus at classical level, the Lagrangian is not gauge invariant. However at quantum level, it is because of the cancellation between the anomaly term and the surface term (8.87). A concise way to see all this is to note that the surface term (8.68), when regularized in a gauge-invariant way, takes the form

$$-\frac{1}{2}\int d^4x \Delta_{\partial V} \bar{\psi}(x+\epsilon/2) e^{i\gamma_5 \eta} \mathcal{P} e^{-ig\int_{x-\epsilon/2}^{x+\epsilon/2} A_\mu d\xi^\mu} \psi(x-\epsilon/2) \qquad (8.89)$$

with Δ the surface δ function. Taking the limit of ϵ going to zero, one then recovers (8.68) suitably normal-ordered and the Chern-Simons term (8.87). It may strike somewhat bizarre that we have here a theory which violates gauge invariance classically and becomes gauge invariant upon proper quantization. But the point is that when anomalies are involved as in this hybrid description, only quantum theory makes sense.

In summary, the consistent Cheshire bag model is given by

$$
\begin{aligned}
S &= S_V + S_{\tilde{V}} + S_{\partial V}, \qquad (8.90)\\
S_V &= \int_V d^4x \left(\bar{\psi}i\,\slashed{D}\psi - \frac{1}{2}\text{tr}\,G_{\mu\nu}G^{\mu\nu}\right) + \cdots\\
S_{\tilde{V}} &= \frac{f^2}{4}\int_{\tilde{V}} d^4x \left(\text{Tr}\,\partial_\mu U^\dagger \partial^\mu U + \frac{1}{4N_f}m_{\eta'}^2\left[\text{Tr}\,(\ln U - \ln U^\dagger)\right]^2\right) + \cdots\\
&\quad + S_{WZW},\\
S_{\partial V} &= \frac{1}{2}\int_{\partial V} d\Sigma^\mu \left\{(n_\mu \bar{\psi} U^{\gamma_5}\psi) + i\frac{g^2}{16\pi^2}K_{5\mu}(\text{Tr}\,\ln U^\dagger - \text{Tr}\,\ln U)\right\}.
\end{aligned}
$$

(ii) An application: η' mass

The presence of the Chern-Simons term in the surface action affects at classical level the boundary conditions satisfied by the gluon fields. In place of the MIT conditions (8.70), we have instead

$$\hat{n} \cdot \vec{E}^a = \frac{N_F g^2}{8\pi^2 f_0} \hat{n} \cdot \vec{B}^a \, \eta', \tag{8.91}$$

$$\hat{n} \times \vec{B}^a = -\frac{N_F g^2}{8\pi^2 f_0} \hat{n} \times \vec{E}^a \, \eta'. \tag{8.92}$$

The fact that radial color electric fields can exist contrary to the MIT conditions must have an important ramification on the so-called "proton spin" problem. No work has been done so far on this interesting issue.

Here we will discuss how to do a calculation analogous to that of the scalar mass in the (1+1)-dimensional Schwinger model to obtain the mass for the η' [19]. As there, we shall make use of the $U_A(1)$ anomaly.

Consider the situation where the η' excitation is a long wavelength oscillation of zero frequency and let V be a small volume in space. The CCP states that it should not matter whether we describe the η' in V in terms of QCD variables (quarks and gluons) or in terms of effective variables (*i.e.*, mesonic degrees of freedom). Since we are looking at the static situation, $\dot{Q}_5 = 0$, where Q_5 is the gauge-invariant singlet axial charge $Q_5 = \int_V d^3x j_5^0(x)$. The $U_A(1)$ anomaly translates into the statement that the flux of the singlet axial current through the bag wall is just the axial anomaly

$$\oint_{\partial V} \vec{j}_5 \cdot \hat{n} \approx \frac{N_F g^2}{4\pi^2} \int_V d^3x \vec{E}^a \cdot \vec{B}^a(x). \tag{8.93}$$

In terms of the bosonized variables, the flux is given by

$$\oint_{\partial V} \vec{j}_5 \cdot \hat{n} = -2 \oint_{\partial V} f_0 \vec{\nabla} \eta' \cdot \hat{n} = -2 \int_V f_0 \nabla^2 \eta' \tag{8.94}$$

where $f_0 \equiv \sqrt{\frac{N_f}{2}} f$. Note that unlike in the above consideration, the bosonization is done *inside* the volume V. Combining (8.93) and (8.94), we obtain

$$f_0^2 \nabla^2 \eta'(0, \vec{x}) \approx -f_0 \frac{N_F g^2}{8\pi^2} \vec{E}^a \cdot \vec{B}^a(0, \vec{x}). \tag{8.95}$$

This is an approximate bosonization condition. The left-hand side of (8.95) is

$$\nabla^2 \eta' \sim m_{\eta'}^2 \eta' \tag{8.96}$$

while the right-hand side can be evaluated using (8.91) and (8.92)

$$\vec{E}^a \cdot \vec{B}^a \approx -\frac{N_F g^2}{8\pi^2} \frac{\eta'}{f_0} \left(1 - \left\{ \frac{N_F g^2}{8\pi^2} \right\}^2 \frac{\eta'^2}{f_0^2} \right)^{-1} \frac{1}{2} G^2. \tag{8.97}$$

Expanding to linear order in η', we obtain from (8.95)

$$m_{\eta'}^2 f_0^2 \approx 2 \left(\frac{N_F g^2}{16\pi^2} \right)^2 \times \langle G^2 \rangle_V. \tag{8.98}$$

This is the four-dimensional analog of (8.35), the scalar mass in the Schwinger model. In this expression, the gluon condensate density is averaged over the bag volume V. A similar result was obtained by Novikov, Shifman, Vainshtein and Zakharov [20] by saturating the functional integral with only self-dual gauge configurations discussed in Chapter 2. In obtaining (8.98), we have been a bit cavalier about the definition of the gluon coupling constant. In principle, it should be a running coupling constant and it makes a difference whether it is defined on the surface as in eqs.(8.91) and (8.92) or in the volume as in eq.(8.93). We are using both in eq.(8.98) and one should distinguish them. If one puts in rough numerical factors for the gluon condensate and appropriate running coupling constants into (8.98), we find that the η' mass comes out to be in the range 315–1650 MeV to be compared with the experimental value 958 MeV. There is a large uncertainty in this prediction for the reason that it is difficult to pin down the relevant scales involved in the running of the coupling constant.

8.2.3. *Cheshire Cat as a gauge symmetry*

To summarize, the Cheshire Cat phenomenon means that the "bag" in the sense of the bag model is an artifact that has no strict physical meaning. An extremely attractive way to view this is to interpret the "bag" as a gauge artifact in the sense of gauge symmetry. Thus a description in terms of a particular "bag" radius can be interpreted as a particular gauge fixing. Therefore physics is strictly equivalent for *all* gauge choices ("bag" radii) one makes. This is a drastically novel way of looking at the classical confinement embodied by the MIT bag. Let us briefly discuss this fascinating and elegant approach[21]. The idea can be best described in (1+1) dimensions where the Cheshire Cat is exact.

Consider the case of massless fermions coupled to external vector $\mathcal{V}_\mu$ and axial-vector $\mathcal{A}_\mu$ fields, the generating functional of which (in Minkowski space) is

$$Z[\mathcal{V}, \mathcal{A}] = \int [d\psi][d\bar{\psi}] e^{iS}, \tag{8.99}$$

$$S = \int d^2 x \bar{\psi} \gamma^\mu \left(i\partial_\mu + \mathcal{V}_\mu + \mathcal{A}_\mu \gamma_5 \right) \psi.$$

One can get from this functional the massless Thirring model if one adds a term $\sim \mathcal{V}^\mu \mathcal{V}_\mu$ and integrate over the vector field. One can get the massive Thirring model if one adds scalar and pseudoscalar sources in addition. In this sense, the model is quite general. Now do the field redefinition while *enlarging the space*

$$\psi(x) = e^{i\theta(x)\gamma_5} \chi(x) , \quad \bar{\psi}(x) = \bar{\chi}(x) e^{i\theta(x)\gamma_5}. \tag{8.100}$$

Since this is just a change of symbols, the generating functional is unmodified:

$$Z[\mathcal{V}, \mathcal{A}] = \int [d\chi][d\bar{\chi}] \, J \, e^{iS'}, \tag{8.101}$$

$$S' = \int d^2 x \bar{\chi} \gamma^\mu \left(i\partial_\mu + \mathcal{V}_\mu + \mathcal{A}_\mu \gamma_5 - \partial_\mu \theta(x) \gamma_5 \right) \chi$$

where J is the Jacobian of the transformation which can be readily calculated:

$$J = \exp \left\{ i \int d^2 x \left(\frac{1}{2\pi} \partial_\mu \theta \partial^\mu \theta + \frac{1}{\pi} \epsilon^{\mu\nu} \mathcal{V}_\mu \partial_\nu \theta - \frac{1}{\pi} \mathcal{A}_\mu \partial^\mu \theta \right) \right\}. \tag{8.102}$$

This nontrivial Jacobian arises due to the fact that the measure is not invariant under local chiral transformation (8.100)[22]. The important thing to note is that since *by definition* (8.99) and (8.101) are the same, the physics should not depend upon $\theta(x)$. The latter is redundant. Therefore modulo an infinite constant which requires gauge fixing as described below, we can just do a functional integral over $\theta(x)$ in (8.101) without changing physics. Doing the integral in the path integral amounts to elevating the redundant $\theta(x)$ to a dynamical variable. Now if we define a Lagrangian $\mathcal{L}'$ by

$$Z[\mathcal{V}, \mathcal{A}] = \int [d\chi][d\bar{\chi}] e^{i \int d^2 x \mathcal{L}'}, \tag{8.103}$$

$$\begin{aligned}\mathcal{L}' &= \bar{\chi}\gamma^\mu \left(i\partial_\mu + \mathcal{V}_\mu + \mathcal{A}_\mu \gamma_5 - \partial_\mu \theta \gamma_5\right)\chi \\ &\quad + \frac{1}{2\pi}\partial_\mu \theta \partial^\mu \theta + \frac{1}{\pi}\epsilon^{\mu\nu}\mathcal{V}_\mu \partial_\nu - \frac{1}{\pi}\mathcal{A}_\mu \partial^\mu \theta, \end{aligned} \tag{8.104}$$

we note that there is a local gauge symmetry, namely, that (8.104) is invariant under the transformation

$$\chi(x) \to e^{i\alpha(x)\gamma_5}\chi(x), \quad \bar{\chi}(x) \to \bar{\chi}(x)e^{i\alpha(x)\gamma_5},$$

$$\theta(x) \to \theta(x) - \alpha(x) \tag{8.105}$$

provided of course the Jacobian is again taken into account. Thus by introducing a new dynamical field θ, we have gained a gauge symmetry at the expense of enlarging the space.

In order to quantize the theory, we have to fix the gauge. This means that we pick a θ, say, by the following condition

$$\Phi(\theta) = 0. \tag{8.106}$$

Then following the standard Faddeev-Popov method [23], we write

$$Z[V, A] = \int [d\chi][d\bar{\chi}][d\theta]\delta(\Phi[\theta])|\det(\frac{\delta\Phi}{\delta\theta})|e^{i \int d^2 x \mathcal{L}'}. \tag{8.107}$$

Note that if we choose $\Phi = \theta$, we recover the original generating functional (8.99) which describes everything in terms of fermions. In choosing the gauge condition relevant to the physics of the chiral bag, we recall that the axial current plays an essential role. Damgaard et al.[24] choose the following gauge fixing condition

$$\Phi[\theta, \chi, \bar{\chi}] = \Delta \int_{x_0}^{x} d\eta^\nu \bar{\chi}(\eta)\gamma_\nu\gamma_5\chi(\eta) + (1 - \Delta)\frac{1}{\pi}\theta(x) = 0 \tag{8.108}$$

which corresponds to saying that the total axial current consists of the fraction Δ of the pionic piece and the fraction $(1 - \Delta)$ of the fermionic piece. This is essentially a Cheshire Cat gauge fixing. We will not dwell on the proof which is rather technical but just mention that the functional integral does not depend upon the path of this line integral. The Faddeev-Popov determinant corresponding to (8.108) is calculated to be

$$\det\left(\frac{\delta\Phi}{\delta\alpha}\right) = \det\left(-\frac{1}{\pi}\right). \tag{8.109}$$

The Cheshire Cat structure is manifested in the choice of the Δ. If one takes $\Delta = 0$, we have the pure fermion theory as one can see trivially. If one takes instead $\Delta = 1$, one obtains after some calculation (the detail of which is given in the paper by Damgaard et al.[24])

$$Z[\mathcal{V}, \mathcal{A}] = \text{const.} \times \int [d\theta] e^{i \int d^2x \mathcal{L}^\theta} \tag{8.110}$$

with

$$\mathcal{L}^\theta = \frac{1}{2}\partial_\mu \theta \partial^\mu \theta - \frac{1}{\sqrt{\pi}}\epsilon^{\mu\nu}\mathcal{V}_\mu \partial_\nu \theta + \frac{1}{\sqrt{\pi}}A_\mu \partial^\mu \theta. \tag{8.111}$$

This is just the bosonized form of the fermion theory coupled to the vector fields. The Cheshire Cat statement is that the theory is identical for *any* arbitrary value of Δ.

One can go further and take Δ to be a local function. In this case, one can continuously change representation from one region of space-time to another, choosing different gauge fixing conditions in different space-time regions. This leads to a "smooth Cheshire bag." The standard chiral bag model[4] corresponds to the gauge choice

$$\Delta(x) = \Theta(\mathbf{x} - \mathbf{z}) \tag{8.112}$$

with $\mathbf{z} = \hat{r}R$. The usual bag boundary conditions employed in phenomenological studies arise after certain suitable averaging over part of the space and hence are a specific realization of the Cheshire Cat scheme. Possible discontinuities or anomalies caused by sharp boundary conditions used in actual calculations should not be considered to be defects of the effective model. An interesting conclusion of this point of view is that one can in principle construct an *exact* Cheshire Cat model without knowing the exact bosonized version of the theory considered. An intriguing possibility that emerges from a recent conjecture[5] that deserves further work is that the CCP is a line of fixed points in the large N_c QCD, with the bag radius R corresponding to such a line.

8.3. Induced (Berry) Gauge Fields in Hadrons

In this subsection, we discuss a slightly different aspect of the Cheshire Cat mechanism, with the objective to generalize the concept to excitations. So far, we have been focusing on ground-state properties. That topological properties satisfy the CCP exactly even in (3+1) dimensions (such as the baryon charge discussed above) is not surprising. This is a ground-state property associated with symmetry. It is a different matter when one wants to establish an approximate CCP for dynamical properties such as excitations, responses to external fields etc. To establish that certain response functions can be formulated in terms of effective variables, it proves to be highly fruitful to introduce and exploit the notion of induced gauge structure familiar in other areas of physics[25]. In "hidden gauge-symmetry" approach to effective chiral Lagrangians[26], gauge fields in internal (flavor) space emerge as in CP^1 model and the gauge fields of suitable flavor group can be identified as the strong vector mesons ρ, ω etc. This phenomenon suggests that there could be a hierarchy of hidden gauge structure that can be generated by symmetries at different energy scales. In this subsection, we discuss how such a hierarchy of "vector-field" degrees of freedom can be *induced* in strong interactions and how they can lead to a natural setting for describing *excited states and their dynamic properties*[27]. The intriguing question as to whether or not the induced gauge fields can be related to the hidden gauge fields of QCD such as the ρ, ω etc. remains unanswered.

8.3.1. *"Magnetic monopoles" in flavor space*

A useful concept in understanding baryon excitations is the concept of magnetic monopoles and instantons, *viz*, topological objects, in order-parameter or in our case flavor space. First we consider a toy model, the quantum mechanical spin-solenoid system à la Stone[28] which illustrates clearly the emergence of Berry phase.

(i) A toy model: (0+1) dimensional field theory

Consider a system of slowly rotating solenoid coupled to a fast spinning object (call it "electron") described by the (Euclidean) action

$$S_E = \int dt \left(\frac{\mathcal{I}}{2} \dot{\vec{n}}^2 + \psi^\dagger (\partial_t - \mu \hat{n} \cdot \vec{\sigma}) \psi \right) \tag{8.113}$$

where $n^a(t)$, $a=1,2,3$, is the rotator with $\vec{n}^2 = 1$, $\mathcal{I}$ its moment of inertia,

ψ the spinning object ("electron") and μ a constant. We will assume that μ is large so that we can make an adiabatic approximation in treating the slow-fast degrees of freedom. We wish to calculate the partition function

$$Z = \int [d\vec{n}][d\psi][d\psi^\dagger]\delta(\vec{n}^2 - 1)e^{-S_E} \tag{8.114}$$

by integrating out the fast degree of freedom ψ and $\psi^\dagger$. Formally this yields the familiar fermion determinant, the evaluation of which is the physics of the system. In the adiabatic approximation, this can be done as follows which brings out the essence of the method useful for handling complicated situations which will interest us later.

Imagine that $\vec{n}(t)$ rotates slowly. At each instant $t = \tau$, we have an instantaneous Hamiltonian $H(\tau)$ which in our case is just $-\mu\vec{\sigma} \cdot \hat{n}(\tau)$ and the "snap-shot" electron state $|\psi_0(\tau)\rangle$ satisfying

$$H(\tau)|\psi^0(\tau)\rangle = \epsilon(\tau)|\psi^0(\tau)\rangle. \tag{8.115}$$

In terms of these "snap-shot" wave functions, the solution of the time-dependent Schrödinger equation

$$i\partial_t|\psi(t)\rangle = H(t)|\psi(t)\rangle \tag{8.116}$$

is

$$|\psi(t)\rangle = e^{i\gamma(t)-i\int_0^t \epsilon(t')dt'}|\psi^0(t)\rangle. \tag{8.117}$$

Note that this has, in addition to the usual dynamical phase involving the energy $\epsilon(t)$, a nontrivial phase $\gamma(t)$ – known as Berry phase – which substituted into (8.116) is seen to satisfy

$$i\frac{d\gamma}{dt} + \langle\psi^0|\frac{d}{dt}\psi^0\rangle = 0. \tag{8.118}$$

This allows us to do the fermion path integrals to the leading order in adiabaticity and to obtain (dropping the trivial dynamical phase involving ϵ)

$$Z = const \int [d\vec{n}]\delta(\vec{n}^2 - 1)e^{-S^{eff}}, \tag{8.119}$$

$$S^{eff}(\vec{n}) = \int \mathcal{L}^{eff} = \int [\frac{\mathcal{I}}{2}\dot{\vec{n}}^2 - i\vec{A}(\vec{n}) \cdot \dot{\vec{n}}]dt$$

where

$$i\vec{\mathcal{A}}(\vec{n}) = -\langle \psi^0(\vec{n})| \frac{\partial}{\partial \vec{n}} \psi^0(\vec{n})\rangle \tag{8.120}$$

in terms of which γ is

$$\gamma = \int \vec{\mathcal{A}} \cdot d\vec{n}. \tag{8.121}$$

$\mathcal{A}$ so defined is known as Berry potential or connection and γ is known as Berry phase[29]. That $\mathcal{A}$ is a gauge field with coordinates defined by $\vec{n}$ can be seen as follows. Under the transformation

$$\psi^0 \rightarrow e^{i\alpha(\vec{n})}\psi^0, \tag{8.122}$$

$\vec{\mathcal{A}}$ transforms as

$$\vec{\mathcal{A}} \rightarrow \vec{\mathcal{A}} - \frac{\partial}{\partial \vec{n}}\alpha(\vec{n}) \tag{8.123}$$

which is just a $U(1)$ gauge transformation. The theory is gauge-invariant in the sense that under the transformation (8.123), the theory (8.119) remains unchanged. (We are assuming that the surface term can be dropped.)

One should note that the gauge field we have here is first of all defined in "order parameter space", not in real space like electroweak field or gluon field and secondly it is induced when fast degrees of freedom are integrated out. This is a highly generic feature we will encounter time and again. Later on, we will see that the space on which the gauge structure emerges is usually the flavor space like isospin or hypercharge space in various dimensions.

We shall now calculate the explicit form of the potential $\mathcal{A}$. For this, let us use the polar coordinate and parametrize the solenoid as

$$\vec{n} = (\sin\theta\cos\phi, \sin\theta\sin\phi, \cos\theta) \tag{8.124}$$

with the Euler angles $\theta(t)$ and $\phi(t)$ assumed to be slowly changing (slow compared with the scale defined by the fermion mass μ) as a function of time. Then the relevant Hamiltonian can be written as

$$\delta H = -\mu\vec{\sigma} \cdot \hat{n}(t) = S(t)\delta H_0 S^{-1}(t), \tag{8.125}$$
$$\delta H_0 \equiv -\mu\sigma_3$$

with

$$S(\vec{n}(t)) = \begin{pmatrix} \cos\frac{\theta}{2} & -\sin\frac{\theta}{2}e^{-i\phi} \\ \sin\frac{\theta}{2}e^{i\phi} & \cos\frac{\theta}{2} \end{pmatrix}. \tag{8.126}$$

Since the eigenstates of δH_0 are $\begin{pmatrix} 1 \\ 0 \end{pmatrix}$ with eigenvalue $-\mu$ and $\begin{pmatrix} 0 \\ 1 \end{pmatrix}$ with eigenvalue $+\mu$, we can write the "snap-shot" eigenstate of $H(t)$ as

$$|\psi^0_{+\uparrow}\rangle = S\begin{pmatrix} 1 \\ 0 \end{pmatrix} = \begin{pmatrix} \cos\frac{\theta}{2} \\ \sin\frac{\theta}{2}e^{i\phi} \end{pmatrix} \tag{8.127}$$

where the arrow in the subscript denotes the "spin-up" eigenstate of δH_0 and $+$ denotes the upper hemisphere to be specified below. The eigenstate $|\psi^0_{+\downarrow}\rangle$ is similarly defined with the "down spin". Now note that for $\theta = \pi$, (8.127) depends on ϕ which is undefined. This means that (8.127) is ill-defined in the lower hemisphere with string singularity along $\theta = \pi$. On the other hand, (8.127) is well-defined for $\theta = 0$ and hence in the upper hemisphere. The meaning of the $+$ in (8.127) is that it has meaning only in the upper hemisphere, thus the name "wave section" rather than wave function.

Given (8.127), we can use the definition (8.120) for the Berry potential to obtain

$$-i\vec{A}_+(\vec{n}) \cdot d\vec{n} = \langle \uparrow |S^{-1}dS| \uparrow \rangle = \frac{i}{2}(1 - \cos\theta)d\phi \tag{8.128}$$

given here in one-form. The explicit form of the potential is **

$$\vec{A}_+ = -\frac{1/2}{(1 + \cos\theta)}(-\sin\theta\sin\phi, \sin\theta\cos\phi, 0) \tag{8.129}$$

which is singular at $\theta = \pi$ as mentioned above. This is the well-known Dirac string singularity. Since we have the gauge freedom, we are allowed to do a gauge transformation

$$\psi^0_+ \to e^{-\phi}\psi^0_+ \equiv \psi^0_- \tag{8.130}$$

which corresponds to defining a gauge potential regular in the lower hemisphere (denoted with the subscript $-$)

$$\vec{A}_- \cdot d\vec{n} = \vec{A}_+ \cdot d\vec{n} + d\phi \tag{8.131}$$

---

*If the Hamiltonian commutes for different times, then the gauge field can be made to vanish. We are considering the case where the Hamiltonians do not commute.

giving

$$\vec{A}_- \cdot d\vec{n} = \frac{1}{2}(1 + \cos\theta)d\phi. \tag{8.132}$$

This potential has a singularity at $\theta = 0$. Thus we have gauge-transformed the Dirac string from the lower hemisphere to the upper hemisphere. This clearly shows that the string is an artifact and is unphysical. In other words, physics should not be dependent on the string. Indeed the field strength tensor, given in terms of the wedge symbol and forms,

$$\mathcal{F} = d\mathcal{A} = \frac{1}{2}d\theta \wedge d\phi = \frac{1}{2}d(Area) \tag{8.133}$$

is perfectly well-defined in both hemispheres and unique. A remarkable fact here is that the gauge potential or more properly the field tensor is completely independent of the fermion "mass" μ. This means that the potential does not depend upon how fast the fast object is once it is decoupled adiabatically. This indicates that the result may be valid even if the fast-slow distinction is not clear-cut. We will come back later to this matter in connection with applications to excitation spectra of light-quark baryons.

We now explain the quantization rule. For this consider a cyclic path. We will imagine that the solenoid is rotated from $t = 0$ to $t = T$ with large T such that the parameter $\vec{n}$ satisfies $\vec{n}(0) = \vec{n}(T)$. We are thus dealing with an evolution, with the trajectory of $\vec{n}$ defining a circle C. The parameter space manifold is two-sphere S^2 since $\vec{n}^2 = 1$. Call the upper hemisphere D and the lower hemisphere $\bar{D}$ whose boundary is the circle C, i.e., $\partial D = C$. Then using Stokes' theorem, we have from (8.121) for cyclic evolution Γ

$$\gamma(\Gamma) = \int_{C=\partial D} \vec{A} \cdot d\vec{n} \equiv \int_{\partial D=C} \mathcal{A} = -\int_D d\mathcal{A} = -\int_D \mathcal{F}. \tag{8.134}$$

Since the gauge field in $\bar{D}$ is related to that in D by a gauge transform, we could equally well write γ in terms of the former. Thus we deduce that

$$e^{i\int_C \mathcal{A}} = e^{i\int_D \mathcal{F}} = e^{-i\int_{\bar{D}} \mathcal{F}} \tag{8.135}$$

which implies

$$e^{i\int_{D+\bar{D}=S^2} \mathcal{F}} = 1. \tag{8.136}$$

Thus we get the quantization condition

$$\int_{S^2} \mathcal{F} = 2\pi n \tag{8.137}$$

with n an integer. This just means that the total magnetic flux going through the surface is quantized. Since in our case the field strength is given by (8.133), our system corresponds to $n = 1$ corresponding to a "monopole charge" $g = 1/2$ located at the center of the sphere. In general, as we will see later in real systems in (3+1) dimensions, the "monopole charge" need not be precisely $1/2$; it could be some quantity, say, g. What we learn from the above exercise is that consistency with quantum mechanics demands that it be a multiple of $1/2$. Otherwise, the theory makes no sense. It will turn out later that real systems in the strong interactions involve nonabelian gauge fields which do not require such "charge" quantization, making the consideration somewhat more delicate.

We should point out another way of looking at the action (8.113) which is useful for understanding the emergence of Berry potentials in complex systems. Since

$$\hat{n}(t) \cdot \vec{\sigma} = S(t)\sigma_3 S^{-1}(t) \tag{8.138}$$

we make the field redefinition

$$\psi \to \psi' = S^{-1}(t)\psi. \tag{8.139}$$

Then (8.113) can be rewritten as

$$S_E = S_E^0 + \int dt \psi^\dagger (S^{-1}\partial_t S)\psi. \tag{8.140}$$

Here the first term contains no coupling between the fast and slow degrees of freedom. The second term linear in time derivative generates the Berry structure analyzed above. We will encounter this structure in complex systems.

(ii) Connection to the Wess-Zumino-Witten term

The key point which will be found useful later is this: when the fast degree of freedom (the "electron") is integrated out, we wind up with a gauge field as a relic of the fast degree of freedom integrated out. The effective Lagrangian that results has the form, in Minkowski space,

$$L^{eff} = \frac{1}{2}I\dot{\vec{n}}^2 + \vec{A}(\vec{n}) \cdot \dot{\vec{n}}. \tag{8.141}$$

We now show that the Berry structure is closely related to what is called the Wess-Zumino-Witten term defined in two dimensions. For this, we note that the gauge field in (8.141) is an induced one, coming out of the solenoid $\vec{n}$. Therefore one should be able to rewrite the second term of (8.141) in terms of $\vec{n}$ alone. However this cannot be done *locally* because of the Dirac singularity but the corresponding action can be written locally in terms of $\vec{n}$ by extending (homotopically) to one dimension higher. This is the (one-dimensional) Wess-Zumino-Witten term. When written in this way, the gauge structure will be "hidden" in some sense.

Let us look at the action

$$S_{WZ} \equiv \int \vec{A}(\vec{n}) \cdot \dot{\vec{n}}dt. \tag{8.142}$$

There is a standard way of expressing this in a local form. The procedure is quite general and goes as follows. First extend the space from the physical dimension d which is 1 in our case to $d + 1$ dimensions. This extension is possible ("no obstruction") if

$$\pi_d(\mathcal{M}) = 0, \quad \pi_{d+1}(\mathcal{M}) \neq 0 \tag{8.143}$$

where π is the homotopy group and $\mathcal{M}$ is the parameter space manifold. In our case

$$\pi_1(S^2) = 0, \quad \pi_2(S^2) = Z. \tag{8.144}$$

We extend the space to $\tilde{n}(s)$, $0 \leq s \leq 1$ such that

$$\tilde{n}(s = 0) = 1, \quad \tilde{n}(s = 1) = \vec{n}. \tag{8.145}$$

Now the next step is to construct a winding number density $\tilde{Q}$ for π_{d+1} which is then to be integrated over a region of $\mathcal{M}$ ($\approx S^2$ here) bounded by d-dimensional fields n^i. The winding number density $\tilde{Q}(x)$ (say $x_1 = t$ and $x_2 = s$) is

$$\tilde{Q} = \frac{1}{8\pi}\epsilon^{ij}\epsilon^{abc}\tilde{n}^a\partial_i\tilde{n}^b\partial_j\tilde{n}^c \tag{8.146}$$

and the winding number n (which is 1 for $\pi \leq \theta \leq 0$, $2\pi \leq \phi \leq 0$)[††]

$$n = \int_{S^2} d^2x \tilde{Q}(x). \qquad (8.147)$$

Comparing with (8.137), we deduce

$$S_{WZ} = 4\pi g \int_{\mathcal{D}=t\times[0,1]} d^2x \tilde{Q}(x). \qquad (8.148)$$

This is the familiar form of the Wess-Zumino term defined in two dimensions. As we saw before, this has a "monopole charge" $g = 1/2$. Below we will carry g as an integral multiple of $1/2$. This way of understanding the Wess-Zumino term complements the approach given above and brings out the universal characteristic of Berry phases.

(iii) Quantizing the model

There are numerous ways of quantizing the effective action (hereupon we will work in Minkowski space)

$$S^{eff} = S_0 + S_{WZ} \qquad (8.149)$$

where the Wess-Zumino action is given by (8.148) and

$$S_0 = \oint dt \frac{\mathcal{I}}{2} \dot{\vec{n}}^2. \qquad (8.150)$$

Here we will consider the time compactified as defined above, so the time integral is written as a loop integral. We will choose one way[30] which illustrates other interesting properties and go through the reasoning in some detail. We shall not repeat the argument for the more complicated four-dimensional cases considered later.

[††]This can be calculated as follows. The surface element is

$$d\Sigma^i = \left(\frac{\partial \vec{n}}{\partial s} ds \times \frac{\partial \vec{n}}{\partial t} dt \right)^i = \frac{1}{2} \epsilon^{ab} \epsilon^{ijk} \frac{\partial n^j}{\partial x^a} \frac{\partial n^k}{\partial x^b} d^2x$$

and hence

$$\int_{S^2} \vec{n} \cdot d\vec{\Sigma} = 4\pi.$$

The manifold has $SO(3)$ invariance corresponding to $\sum_{i=1}^{3} n_i^2 = 1$. Consider now the complex doublet $z \in SU(2)$

$$z(t) = \begin{pmatrix} z_1 \\ z_2 \end{pmatrix}, \tag{8.151}$$

$$z^\dagger z = |z_1|^2 + |z_2|^2 = 1. \tag{8.152}$$

Then we can write

$$n_i = z^\dagger \sigma_i z \tag{8.153}$$

with σ_i the Pauli matrices. There is a redundant (gauge) degree of freedom since under the $U(1)$ transformation $z \rightarrow e^{i\alpha} z$, n_i remains invariant. This is as it should be since the manifold is topologically S^2 and hence corresponds to the coset $SU(2)/U(1)$. We are going to exploit this $U(1)$ gauge symmetry to quantize our effective theory.

Define a 2-by-2 matrix h

$$h = \begin{pmatrix} z_1 & -z_2^\star \\ z_2 & z_1^\star \end{pmatrix} \tag{8.154}$$

and

$$a(t) = \sum_{k=1}^{2} \frac{i}{2} (z_k^\star \overleftrightarrow{\partial}_k z_k). \tag{8.155}$$

Then it is easy to obtain (setting $\mathcal{I} = 1$) that

$$S_0 = \frac{1}{2} \oint dt \, \mathrm{Tr} \left[\overrightarrow{D}_t^\dagger h^\dagger h \overleftarrow{D}_t \right] \tag{8.156}$$

where

$$\overrightarrow{D}_t = \overrightarrow{\partial}_t - ia\sigma_3. \tag{8.157}$$

Let

$$\tilde{a}_\mu = \frac{i}{2} (\tilde{z}_k^\star \overleftrightarrow{\partial}_\mu \tilde{z}_k) \tag{8.158}$$

where the index μ runs over the extended coordinate (s, t). As noted before, there is no topological obstruction to this extension. This is defined in such a way that $\tilde{z}(s=0) = \begin{pmatrix} 1 \\ 1 \end{pmatrix}$ and $\tilde{z}(s=1) = z$. Then the Wess-Zumino action can be written in a "Chern-Simons" form

$$S_{WZ} = 2g \int_{\mathcal{D}} d^2x \epsilon^{\mu\nu} \partial_\mu \tilde{a}_\nu = 2g \oint dt a(t). \qquad (8.159)$$

Introducing an auxiliary function A, we can write the partition function

$$Z = \int [dz][dz^\dagger][dA] exp\, i \oint dt \left(\frac{1}{2} \text{Tr}(\vec{\partial}_t + iA\sigma_3)h^\dagger h(\overleftarrow{\partial}_t - iA\sigma_3) \right.$$
$$\left. + 2gA + \frac{1}{4}(2g)^2 \right). \qquad (8.160)$$

The $U(1)$ gauge invariance is manifest in this action. Indeed if we make the (local) transformation $h \to exp\, i\alpha(t)\frac{\sigma_3}{2}h$ and $A \to A - \frac{1}{2}\partial_t\alpha(t)$ with the boundary condition $\alpha(T) - \alpha(0) = 4\pi N$ where N is an integer, then the action remains invariant. This means that we have to gauge-fix the "gauge field" A in the path integral. The natural gauge choice is the "temporal gauge" $A = 0$. The resulting gauge-fixed action is $\oint \mathcal{L}_{gf}$ with

$$\mathcal{L}_{gf} = \text{Tr}(\partial_t h^\dagger \partial_t h) + g^2. \qquad (8.161)$$

Since there is no time derivative of A in (8.160), there is a Gauss' law constraint which is obtained by taking $\frac{\delta S}{\delta A}|_{A=0}$ from the action (8.160) *before* gauge fixing:

$$\frac{i}{2} \text{Tr}\left(\sigma_3 h^\dagger \partial_t h - \partial_t h^\dagger h\sigma_3 \right) + 2g = 0$$

which is

$$i\, \text{Tr}\left(\partial_t h^\dagger h \frac{\sigma_3}{2} \right) = g. \qquad (8.162)$$

The left-hand side is identified as the right rotation around third axis J_3^R, so the constraint is that

$$J_3^R = g. \qquad (8.163)$$

Since (8.161) is invariant under $SU(2)_{L,R}$ multiplication, we have that

$$\vec{J}^2 = \vec{J}_L^2 = \vec{J}_R^2. \qquad (8.164)$$

The Hamiltonian is (restoring the moment of inertia $\mathcal{I}$)

$$H = \frac{1}{2\mathcal{I}}\left(\vec{J}^2 - g^2\right) \tag{8.165}$$

which has the spectrum of a tilted symmetric top. Now adding the energy of the "electron," the total energy is

$$E = \epsilon + \frac{1}{2\mathcal{I}}\left(J(J+1) - g^2\right) \tag{8.166}$$

with the allowed values for J

$$J = |g|, |g| + 1, \cdots. \tag{8.167}$$

The rotational spectrum is the well-known Dirac monopole spectrum. The corresponding wave function is given by monopole harmonics.

- *Summary*

When a fast spinning object coupled to slowly rotating object is integrated out, a Berry potential arises gauge-coupled to the rotor. The effect of this gauge coupling is to "tilt" the angular momentum of the rotor in the spectrum of a symmetric top. It supplies an extra component to the angular momentum, along the third direction. The gauge field is abelian and has an abelian (Dirac) monopole structure. The abelian nature is inherited from one nondegenerate level crossing another nondegenerate level. When degenerate levels cross, the gauge field can be nonabelian and this is the generic feature we shall later encounter in strong interaction physics.

We went into some details above with the simple toy model since the manipulations that go into more realistic models that we will study later are quite similar although somewhat more complicated.

8.3.2. *Induced nonabelian gauge field*

(i) Diatomic molecules in Born-Oppenheimer approximation

In hadronic systems we will study below, we typically encounter nonabelian induced gauge potentials. This is because degeneracy is present. In order to understand this situation, we first discuss here a case in which a nonabelian gauge structure arises in a relatively simple quantum mechanical system. To do so, we will study a simple toy-model example of the induced nonabelian gauge fields and Berry phases in the Born-Oppenheimer approximation[31]. When suitably implemented, the treatment can be applied to a realistic description of the spectrum of a diatomic molecule, wherein this approximation is usually described as a separation of slow (nuclear) and fast (electronic) degrees of freedom. This separation is motivated by the fact that the rotation of the nuclei does not cause the transitions between the electronic levels. In other words, the splittings between the fast variables are much larger than the splittings between the slow ones. We will demonstrate how the integrating-out of the fast degrees of freedom generates in the slow-variable space an induced vector potential of the Dirac monopole type in certain special situations.

Before we develop the main argument, we should mention one point which may not be clear at each stage of development but should be kept in mind. In the strong interactions at low energy, it is not always easy to delineate the "fast" and "slow" degrees of freedom. In particular in chiral (light-quark) systems that we consider in this section, it is not even clear whether it makes sense to make the distinction. Nonetheless we will see that once the delineation is made, whether it is sharp or not, the result does not depend on how good the distinction is. This was already clear from the quantum mechanics of the solenoid-spin system we discussed above where the final result had no memory of how strong the coupling of the spin to the solenoid – namely, the coefficient μ – was. In Chapter 9 where we deal with heavy-quark systems, we will see that the concept developed here applies much better than in light-quark systems.

Let us define a generic Hamiltonian describing a system consisting of the slow ("nuclear") variables $\vec{R}(t)$ (with $\vec{P}$ as conjugate momenta) and the fast ("electronic") variables $\vec{r}$ (with $\vec{p}$ as conjugate momenta) coupled through a potential $V(\vec{R}, \vec{r})$

$$H = \frac{\vec{P}^2}{2M} + \frac{\vec{p}^2}{2m} + V(\vec{R}, \vec{r}) \qquad (8.168)$$

where we have reserved the capitals for the slow variables and lower-case letters for the fast variables. We expect the electronic levels to be stationary under the adiabatic (slow) rotation of the nuclei. We split therefore the Hamiltonian into the slow and fast parts,

$$H = \frac{\vec{P}^2}{2M} + h$$

$$h(\vec{R}) \;=\; \frac{\vec{p}^2}{2m} + V(\vec{r}, \vec{R}) \tag{8.169}$$

where the fast Hamiltonian h depends *parametrically* on the slow variable $\vec{R}$. The snapshot Hamiltonian (for fixed $\vec{R}$) leads to the Schrödinger equation:

$$h\phi_n(\vec{r}, \vec{R}) = \epsilon_n(\vec{R})\phi_n(\vec{r}, \vec{R}) \tag{8.170}$$

with electronic states labeled by the set of quantum numbers n. The wave function for the whole system is

$$\Psi(\vec{r}, \vec{R}) = \sum_n \Phi_n(\vec{R})\phi_n(\vec{r}, \vec{R}). \tag{8.171}$$

Substituting the wave function into the full Hamiltonian and using the equation for the fast variables we get

$$\sum_n \left[\frac{\vec{P}^2}{2M} + \epsilon_n(\vec{R}) \right] \Phi_n(\vec{R})\phi_n(\vec{r}, \vec{R}) = E \sum_n \Phi_n(\vec{R})\phi_n(\vec{r}, \vec{R}) \tag{8.172}$$

where E is the energy of the whole system. Note that the operator of the kinetic energy of the slow variables acts on *both* slow and fast parts of the wave function. We can now "integrate out" the fast degrees of freedom. A bit of algebra leads to the following effective Schrödinger equation

$$\sum_m H_{nm}^{eff} \Phi_m = E\Phi_n \tag{8.173}$$

where the explicit form of the matrix-valued Hamiltonian (with respect to the fast eigenvectors) is

$$H_{mn}^{eff} = \frac{1}{2M} \sum_k \vec{\Pi}_{nk}\vec{\Pi}_{km} + \epsilon_n \delta_{nm} \tag{8.174}$$

where

$$\vec{\Pi}_{nm} = \delta_{nm}\vec{P} - i\langle \phi_n(\vec{r}, \vec{R})|\vec{\nabla}_R|\phi_m(\vec{r}, \vec{R})\rangle \equiv \delta_{mn}\vec{P} - \vec{A}_{nm}. \tag{8.175}$$

The above equation is exact. We see that the effect of the fast variables is summarized by an effective gauge field $\vec{A}$. The vector part couples minimally to the momenta with the fast eigenvalue acting like a scalar potential. The vector field is in general nonabelian and corresponds to a nonabelian Berry potential first discussed by Wilczek and Zee[32].

In case one can neglect the off-diagonal transition terms in the induced gauge potentials (*i.e.*, if the adiabatic approximation is valid), then the Hamiltonian simplifies to

$$H_n^{eff} = \frac{1}{2M}(\vec{P} - \vec{A}_n)^2 + \epsilon_N \qquad (8.176)$$

where the diagonal component of the Berry phase A_{nn} is denoted by A_n. Suppose that the electronic eigenvalues are degenerate so that there are G_n eigenvectors with a degenerate eigenvalue ϵ_n. Then instead of a single Berry phase, we have a whole set of the $G_n \times G_n$ Berry phases, forming the matrix

$$A_n^{k,k'} = i\langle n, k|\nabla|n, k'\rangle \quad k, k' = 1, 2, ..., G_n. \qquad (8.177)$$

The gauge field generated in such a case is nonabelian valued in the gauge group $U(G_n)$. In practical calculations, one truncates the infinite sum in (8.171) to a few finite terms. Usually the sum is taken over the degenerate subspace corresponding to the particular eigenvalue ϵ_n. This is the so-called Born-Huang approximation, which we will use in what follows. (A Berry potential built of the whole space would be a pure gauge type and would have a vanishing stress tensor, so it would be trivial.)

As stated, the fast variable will be taken to be the motion of the electron around the internuclear axis. The slow variables are the vibrations and rotations of the internuclear axis. This case corresponds to the situation where the energy of the spin-axis interaction is large compared with the energy splittings between the rotational levels, so that the adiabaticity condition holds. This is called "Hund case a". Following the standard notation[33], we introduce the unit vector $\vec{N}$ along the internuclear axis and define the following quantum numbers

$$
\begin{aligned}
\Lambda &= \text{eigenvalue of } \vec{N} \cdot \vec{L} \\
\Sigma &= \text{eigenvalue of } \vec{N} \cdot \vec{S} \\
\Omega &= \text{eigenvalue of } \vec{N} \cdot \vec{J} = |\Lambda + \Sigma|, \qquad (8.178)
\end{aligned}
$$

so Λ, Σ, Ω are the projections of the orbital momentum, spin and total angular momentum of the electron on the molecular axis, respectively. For simplicity we focus on the simple case of $\Sigma = 0$, $\Lambda = 0, \pm 1$. The $\Lambda = 0$ state is referred to as Σ state and the $\Lambda = \pm 1$ states are called π, a degenerate doublet. We are interested in the property of these triplet states, in particular in the symmetry associated with their energy splittings.

To analyze the model, let us write the Lagrangian corresponding to the Hamiltonian (8.176)

$$L_{nm}^{eff} = \frac{1}{2}M\ddot{\vec{R}}(t)^2\delta_{mn} - i\vec{A}_{mn}[\vec{R}(t)] \cdot \dot{\vec{R}}(t) - \epsilon_m\delta_{mn}. \tag{8.179}$$

This can be rewritten in non-matrix form (dropping the trivial electronic energy ϵ)

$$\mathcal{L} = \frac{1}{2}M\dot{\vec{R}}^2 + i\theta_a^\dagger(\delta_{ab}\frac{\partial}{\partial t} - i\vec{A}^\alpha T_{ab}^\alpha \cdot \dot{\vec{R}})\theta_b \tag{8.180}$$

where we have introduced a Grassmannian variable θ_a as a trick to avoid using the matrix form of (8.179)[34] and $\mathbf{T}^\alpha$ is a matrix representation in the vector space in which the Berry potential lives satisfying the commutation rule

$$[\mathbf{T}^\alpha, \mathbf{T}^\beta] = if^{\alpha\beta\gamma}\mathbf{T}^\gamma. \tag{8.181}$$

To get to (8.176) from this form of Lagrangian, we use the standard quantization procedure to obtain the following commutation relations,

$$[R_i, p_j] = i\delta_{ij}, \qquad \{\theta_a, \theta_b^\dagger\} = i\delta_{ab}. \tag{8.182}$$

The Hamiltonian

$$H = \frac{1}{2M}(\vec{P} - \mathbf{A})^2 \tag{8.183}$$

follows with

$$\begin{aligned} \mathbf{A} &= \vec{A}^\alpha I^\alpha, \\ I^\alpha &= \theta_a^\dagger T_{ab}^\alpha \theta_b. \end{aligned} \tag{8.184}$$

Using the commutation relations, it can be verified that

$$[I^\alpha, I^\beta] = if^{\alpha\beta\gamma}I^\gamma. \tag{8.185}$$

The Lagrangian, eq.(8.180), is invariant under the gauge transformation

$$\begin{aligned} \vec{A}^\alpha &\rightarrow \vec{A}^\alpha + f^{\alpha\beta\gamma}\Lambda^\beta\vec{A}^\gamma - \vec{\nabla}\Lambda^\alpha, \tag{8.186} \\ \theta_a &\rightarrow \theta_a - i\Lambda^\alpha T_{ab}^\alpha\theta_b. \tag{8.187} \end{aligned}$$

It should be noted that eq.(8.187) corresponds to the gauge transformation on $|a\rangle$.

The procedure of constructing conserved angular momentum in a diatomic molecule in which a Berry potential couples to the dynamics of slow degrees of freedom, corresponding to the nuclear coordinate $\vec{R}^{31}$ is a complete parallel to the case of a particle coupled to an external gauge field of 't Hooft-Polyakov monopole[35]. A discussion on the 't Hooft-Polyakov monopole is given in the Appendix.

As mentioned before, the Berry potential is defined on the space spanned by the electronic states $\pi(|\Lambda| = 1)$ and $\Sigma(\Lambda = 0)$, where Λ's are eigenvalues of the third component of the orbital angular momentum of the electronic states. The electronic states responding to slow rotation of $\vec{R}$, $U(\vec{R})$ defined by

$$U(\vec{R}) = \exp(-i\phi L_z)\exp(i\theta L_y)\exp(i\phi L_z), \qquad (8.188)$$

induce a Berry potential of the form

$$\vec{A} = i\langle \Lambda_a | U(\vec{R})\vec{\nabla}U(\vec{R})^\dagger | \Lambda_b \rangle \qquad (8.189)$$

$$= \frac{\mathbf{A}_\theta}{R}\hat{\theta} + \frac{\mathbf{A}_\phi}{R\sin\theta}\hat{\phi}, \qquad (8.190)$$

where

$$\mathbf{A}_\theta = \kappa(R)(\mathbf{T}_y \cos\phi - \mathbf{T}_x \sin\phi),$$

$$\mathbf{A}_\phi = \mathbf{T}_z(\cos\theta - 1) - \kappa(R)\sin\theta(\mathbf{T}_x \cos\phi + \mathbf{T}_y \sin\phi). \qquad (8.191)$$

$\vec{\mathbf{T}}$'s are spin-1 representations of the orbital angular momentum $\vec{L}$ and κ measures the transition amplitude between the Σ and π states

$$\kappa(R) = \frac{1}{\sqrt{2}}|\langle \Sigma | L_x - iL_y | \pi \rangle|. \qquad (8.192)$$

The nonvanishing field strength tensor is given by

$$\vec{B} = \frac{F_{\theta\phi}}{R^2 \sin\theta} = -\frac{(1 - \kappa^2)}{R^2}T_z\hat{R}. \qquad (8.193)$$

Following the procedure of introducing a Grassmannian variable for each electronic state and replacing $\mathbf{T}$ by $\mathbf{I}$ defined in eq.(8.184) and quantizing the corresponding Lagrangian, we obtain the Hamiltonian

$$H = \frac{1}{2M}(\vec{P} - \vec{A})^2, \qquad (8.194)$$

where $\vec{A} = \vec{A}^\alpha I^\alpha$. It should be noted that the presence of the constant κ – which is not quantized – that appears in the Berry potential is a generic feature of nontrivial nonabelian Berry potentials as can be seen in many examples[25].

To proceed, we need a solution of eq.(A8.8) and eq.(A8.12). For this purpose, we can exploit the gauge freedom and rewrite the Hamiltonian in the most symmetric form. This can be done by a gauge transform of the form

$$\vec{A}' = V^\dagger \vec{A} V + iV^\dagger \vec{\nabla} V \qquad (8.195)$$

$$\mathbf{F}' = V^\dagger \mathbf{F} V \qquad (8.196)$$

where V is an inverse operation of U in eq.(8.188), *i.e.*, $V = U^\dagger$. Then

$$\mathbf{A}'_\theta = (1 - \kappa)(I_x \sin\phi - I_y \cos\phi),$$
$$\mathbf{A}'_\phi = (1 - \kappa)\{-I_z \sin^2\theta + \cos\theta \sin\theta(I_x \cos\phi + I_y \sin\phi)\}, \quad (8.197)$$

or more compactly

$$\vec{A}' = (1 - \kappa)\frac{\hat{R} \times \vec{I}}{R^2},$$

and

$$\vec{B}' = -(1 - \kappa^2)\frac{\hat{R}(\hat{R} \cdot \mathbf{I})}{R^2}. \qquad (8.198)$$

It is remarkable that the above Berry potential has the same structure as the 't Hooft-Polyakov monopole, eq.(A8.1) and eq.(A8.2), except for the different constant factors $(1 - \kappa)$ for vector potential and $(1 - \kappa^2)$ for magnetic field. Because of these two different factors, however, one cannot simply take eq.(A8.13) as a solution of (A8.12) for the case of nonabelian Berry potentials. Using the following identities derived from eq. (8.197),

$$\vec{R} \cdot \vec{A}' = 0, \qquad (8.199)$$

$$\vec{R} \times \vec{A}' = -(1 - \kappa)\{\vec{I} - (\vec{I} \cdot \hat{R})\hat{R}\}, \qquad (8.200)$$

the Hamiltonian, eq.(8.194), can be written as

$$H = -\frac{1}{2MR^2}\frac{\partial}{\partial R}R^2\frac{\partial}{\partial R} + \frac{1}{2MR^2}(\vec{L}_o + (1 - \kappa)\vec{I})^2$$
$$- \frac{1}{2MR^2}(1 - \kappa)^2(\vec{I} \cdot \hat{R})^2. \qquad (8.201)$$

Now one can show that the conserved angular momentum $\vec{L}$ is

$$\vec{L} = \vec{L}_o + \vec{I}, \tag{8.202}$$
$$= M\vec{R} \times \dot{\vec{R}} + \vec{Q}, \tag{8.203}$$

with

$$\vec{Q} = \kappa\vec{I} + (1 - \kappa)\hat{R}(\hat{R} \cdot \vec{I}). \tag{8.204}$$

Hence, in terms of the conserved angular momentum $\vec{L}$, the Hamiltonian becomes

$$H = -\frac{1}{2MR^2}\frac{\partial}{\partial R}R^2\frac{\partial}{\partial R} + \frac{1}{2MR^2}(\vec{L} - \kappa\vec{I})^2 - \frac{1}{2MR^2}(1 - \kappa)^2 \tag{8.205}$$

where $(\vec{I} \cdot \hat{R})^2 = 1$ has been used.

It is interesting to see what happens in the two extreme cases of $\kappa = 0$ and 1. For $\kappa = 1$, the degenerate Σ and π states form a representation of the rotation group and hence the Berry potential (and its field tensor) vanishes or becomes a pure gauge. Then $\vec{Q} = \vec{I}$ and $\vec{L} = M\vec{R} \times \dot{\vec{R}} + \vec{I}$. Now $\vec{I}$ can be understood as the angular momentum of the electronic system which is decoupled from the spectrum. One can also understand this as the restoration of rotational symmetry in the electronic system. Physically $\kappa \to 1$ as $R \to \infty$.

For $\kappa = 0$, the Σ and π states are completely decoupled and only the $U(1)$ monopole field can be developed on the π states. It becomes identical to eq.(A8.17) as κ goes to zero and the Hamiltonian can be written as

$$H = -\frac{1}{2MR^2}\frac{\partial}{\partial R}R^2\frac{\partial}{\partial R} + \frac{1}{2MR^2}(\vec{L} \cdot \vec{L} - 1) \tag{8.206}$$

which is a generic form for a system coupled to a $U(1)$ monopole field. Physically this corresponds to small internuclear distance at which the Σ and π states decouple. Therefore, a truly nonabelian Berry potential can be obtained only for κ which is not equal to zero or one. We will encounter in Chapter 9 an analogous situation in heavy-quark baryons.

8.3.3. *Berry phases in the chiral bag model*

(i) Induced gauge field

We now turn to hadronic systems in (3+1) dimensions and use the intuition we gained in quantum mechanical cases, exploiting largely the generic structure of the resulting expressions. Consider the generating functional for the chiral bag in $SU(2)$ flavor space

$$Z = \int [d\pi] [d\psi][\bar{\psi}]e^{iS} \qquad (8.207)$$

with the (Minkowski) action S for the chiral bag given by

$$S = \int_V \bar{\psi}i\gamma^\mu \partial_\mu \psi - \frac{1}{2}\int \Delta_s \,\bar{\psi}\,S\,e^{i\gamma_5 \vec{\tau}\cdot\hat{r}F(r)}\,S^\dagger\,\psi + S_M(SU_0S^\dagger) \quad (8.208)$$

where F is the chiral angle appearing in $U_0 = e^{i\vec{\tau}\cdot\hat{r}F(r)}$, Δ_s is a surface delta function and the space rotation has been traded in for an isospin rotation $(S(t))$ due to the hedgehog symmetry considered here. The quarks are assumed to be confined to the spatial volume V and the purely mesonic terms outside the bag are described by S_M.

Suppose that we adiabatically rotate the bag in space. As discussed in the preceding chapters, the "rotation" is required to quantize the zero modes to obtain the quantum states of baryons. In this model, the rotation implies transforming the hedgehog configuration living outside the bag as

$$U_0 \rightarrow S(t)U_0 S^\dagger(t) \qquad (8.209)$$

to which the quarks in the bag respond in a self-consistent way. This leads to computing the path integral in (8.207) with (8.208). Because of the degeneracy of the Dirac spectrum, mixing between quark levels is expected no matter how small the rotation is. This mixing takes place in each quark band and leads to a nonabelian Berry or gauge field. Instead of the boundary term rotating as a function of time, it is more convenient to work with a static boundary condition by "unwinding" the boundary with the redefinition $\psi \rightarrow S\psi$, leading to

$$S = S_{S=1} + \int \psi^\dagger \,S^\dagger i\partial_t S\,\psi\,. \qquad (8.210)$$

The effect of the rotation on the fermions inside the bag is the same as a time-dependent gauge potential. This is the origin of the induced Berry potential analogous to the solenoid-electron system described above. Now as the meson field outside rotates, the quarks inside the bag could either remain in their ground state which is in the $K = 0^+$ orbit (where $\vec{K} = \vec{J} + \vec{I}$ is the grand spin, introduced in Chapter 7) or excited to higher orbits. In the former case, we have a ground-state baryon and no Berry phases remain active: There is no Wess-Zumino term in the $SU(2)$ case. In the latter case, as sketched in Fig.8.3, there are, broadly, two classes of quark excitations that can be distinguished.

- Class A: One or more quarks get excited to $K \neq 0^+$ levels, changing no flavors;

- Class B: One or more quarks get excited to flavor Q levels through flavor change where Q could be strange, charm or bottom quark.

In what follows, we will address the problem using Class A but the discussion applies equally well to Class B as we will detail later.

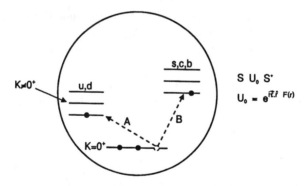

Fig. 8.3. Two classes of excitation spectrum described by Berry potentials: Class A for excitations within up and down quark space; Class B for excitations involving heavy quark $Q = s, c, b \cdots$

To expose the physics behind the second term of (8.210) which is the principal element of our argument, we expand the fermionic fields in the complete set of states ψ_{KM} with energies ϵ_K in the unrotating bag corresponding to the action $S_{S=1}$ in (8.210), and M labels $2K + 1$ projections of the grand spin K. Generically,

$$\psi(t, x) = \sum_{K,M} c_{KM}(t)\psi_{KM}(x) \tag{8.211}$$

where the c's are Grassmannians, so that

$$S = \sum_{KM} \int dt \, c_{KM}^\dagger (i\partial_t - \epsilon_K) c_{KM} + \sum_{KMK'N} \int dt \, c_{KM}^\dagger A_{MN}^{KK'} c_{K'N} \tag{8.212}$$

where

$$A_{MN}^{KK'} = \int_V d^3x \, \psi_{KM}^\dagger S^\dagger i\partial_t S \psi_{K'N} . \tag{8.213}$$

No approximation has been made up to this point. If the A of (8.213) were defined in the whole K space, then A would take the form of a pure gauge and the field strength tensor would be identically zero. As we saw in the case of the diatomic molecule, however, the vanishing of the field tensor does not imply that there is no effect. It describes the restoration of certain symmetry. See later a similar phenomenon in heavy-quark baryons. However we are forced to truncate the space. As in the precedent chapter, we can use now the adiabatic approximation and neglect the off-diagonal terms in K, *i.e.*, ignore the effect of adiabatic rotations, which can cause the jumps between the energy levels of the fast quarks. Still, for every $K \neq 0$ the adiabatic rotation mixes $2K+1$ degenerate levels corresponding to a particular fast eigenenergy ϵ_K. In this form, we clearly see that the rotation induces a hierarchy of Berry potentials in each K-band, of the generic form identical to eq.(8.177). This field is truly a gauge field. Indeed, any local rotation of the form $\psi_{KM} \rightarrow D_{MN}^K \psi_{KN}$ where D^K is a $(2K+1)$-dimensional matrix spanning the representation of rotation in the K-space can be compensated by a gauge transformation of the Berry potential $A^K \rightarrow D^K(\partial_t + A^K)D^{K\dagger}$ leaving S invariant. A potential which has this transformation property is called "invariant gauge potential"[36].

The structure of the Berry potential depends on the choice of the parametrization of the isorotation S (*i.e.*, gauge freedom). For the parametrization $S = a_4 + i\vec{a} \cdot \vec{\tau}$ with the unitary constraint $a \cdot a = 1$ (unitary gauge), we have

$$A^K = T_K^a A_K^a = T_K^a \left(g_K \, \frac{\eta_{\mu\nu}^a \, a_\mu da_\nu}{1 + a^2} \right) \tag{8.214}$$

where η is the 't Hooft symbol defined in Chapter 2 (see the Appendix) and g_K the induced coupling to be specified below. The T's refer to the K-representation of $SU(2)$, the group of isorotations. In the unitary gauge the Berry potential has the algebraic structure of a unit size instanton in isospace, *i.e.*, the space of the slow variables. Note that it is not the Yang-Mills instanton, since the above configuration is not self-dual due to the unitarity gauge constraint. This configuration is a nonabelian generalization of the monopole-like solution present in the diatomic molecular case.

To make our analogy more quantitative, let us refer to the Grassmannians c in the valence states by α's and those in the Dirac sea by β's. Clearly (8.212) can be rewritten in the form

$$
\begin{aligned}
S = &\sum_{KMN} \int dt \, \alpha^\dagger_{KM} \left[(i\partial_t - \epsilon_K)\mathbf{1}_{MN} - (A^K)_{MN} \right] \alpha_{KN} \\
+ &\sum_{KMN} \int dt \, \beta^\dagger_{KM} \left[(i\partial_t - \epsilon_K)\mathbf{1}_{MN} - (A^K)_{MN} \right] \beta_{KN} . \quad (8.215)
\end{aligned}
$$

Integrating over the Dirac sea *in the presence of valence quarks* yields the effective action

$$
\begin{aligned}
S = &\sum_{KMN} \int dt \, \alpha^\dagger_{KM} \left[(i\partial_t - \epsilon_K)\mathbf{1}_{MN} - (A^K)_{MN} \right] \alpha_{KN} \\
+ &\; i \mathrm{Tr} \ln \left((i\partial_t - \epsilon_K)\mathbf{1}_{MN} - (A^K)_{MN} \right) \quad\quad (8.216)
\end{aligned}
$$

where the Trace is over the Dirac sea states. The latter can be Taylor-expanded in the isospin velocities $\dot{a}_\mu$ in the adiabatic limit,

$$
i \mathrm{Tr} \ln \left((i\partial_t - \epsilon_k)\mathbf{1}_{MN} - (\mathcal{A}^K_\mu)_{MN}\dot{a}_\mu \right) = \int dt \, \frac{\mathcal{I}_q}{2} \dot{a}_\mu \dot{a}_\mu + \cdots \quad (8.217)
$$

We have exposed the velocity dependence by rewriting the form $A^K_{MN} = (\mathcal{A}^K_\mu)_{MN}\dot{a}_\mu$. Linear terms in the velocity are absent since the Berry phases in the sea cancel pairwise in the $SU(2)$ isospin case under consideration. For $SU(3)$ they do not and are at the origin of the Wess-Zumino term. The ellipsis in (8.217) refers to higher derivative terms. $\mathcal{I}_q$ is the moment of inertia of the bag, the explicit form of which is not needed for our consideration. We would like to point out that this term includes implicitly the valence quark effect, because the levels of the Dirac sea are modified due to the presence of the valence quarks.

To see the general motivation for studying the excited states via Berry phases, let us consider the case of the bag containing one valence quark in the $K = 1$ state. The action for the adiabatic motion of this quark is obtained from the above formulae and yields

$$S = \int dt \left[i\alpha_{1M}^\dagger \dot{\alpha}_{1M} - \epsilon_1 \alpha_{1M}^\dagger \alpha_{1M} + \frac{1}{2} \mathcal{I}_q \dot{a}_\mu \dot{a}_\mu \right.$$
$$\left. + \dot{a}_\mu (\mathcal{A}_\mu^1)_{MN} \alpha_{1M}^\dagger \alpha_{1N} \right]. \tag{8.218}$$

As we will see below, when canonically quantized, the generic structure of the resulting Hamiltonian is identical to (8.201) and shows that the excited quark system in the slow variable space behaves as a spinning charged particle coupled to an instanton-like gauge field centered in an S^3 sphere in the four-dimensional isospin space. This illustrates once again the universal character of the Berry phases.

(ii) Quantization

To quantize the system, we note that (8.218) describes the motion of a particle with nonabelian charges coupled to a gauge field on S^3. Since S^3 is isomorphic to the group manifold of $SU(2)$, it is convenient to use the left or right Maurer-Cartan forms as a basis for the vielbeins (one-form notation understood)

$$S^\dagger idS = -\omega_a \tau_a = -v_a^c(\theta) d\theta^c \tau^a \tag{8.219}$$

where we expressed the "velocity" forms ω in the basis of the vielbeins v_a^c, and θ denotes some arbitrary parametrization of the $SU(2)$, *e.g.* Euler angles. In terms of vielbeins, the induced Berry potential simplifies to

$$\mathcal{A}^c = g_K A_K^c = - g_K v_a^c(\theta) T^a \tag{8.220}$$

where T^a are the generators of the induced Berry structure in the K representation and g_K is the corresponding charge[27], the explicit form of which is not needed at this point and will be discussed later on, when we discuss baryonic excitations.

Note that we could have equally well used the right-invariant Maurer-Cartan form instead of the left-invariant Maurer-Cartan form (8.219). The field strength can be written in terms of A defined in eq.(8.220)

$$\mathcal{F}_K = dA_K - ig_K A_K \wedge A_K = -(1 - g_K/2)\epsilon^{mij} T_K^m v^i \wedge v^j. \quad (8.221)$$

$\mathcal{F}_K$ vanishes for $g_K = 2$, *i.e.* the Berry potential becomes a pure gauge. The vielbeins, and hence A and $\mathcal{F}$ are frame-dependent, but to quantize the system no specific choice of framing is needed. The canonical momenta are $p_a = \partial L/\partial\dot\theta_a$. Our system lives on S^3 and is invariant under $SO(4) \sim SU(2) \times SU(2)$. Right and left generators are defined as

$$
\begin{aligned}
R_a &= u_a^c\, p_c, \\
L_a &= D_{ab}(S)\, R_b
\end{aligned}
\qquad (8.222)
$$

where $u_i^a v_c^i = \delta_c^a$ and $D(S)$ spans the adjoint representation of the $SU(2)$. Following the standard procedure we get our Hamiltonian in terms of the generators

$$H^* = \epsilon_K \mathbf{1} + \frac{1}{8\mathcal{I}}\,(R_j - g_K T_{Kj})(R_j - g_K T_{Kj}). \quad (8.223)$$

As a result the Hamiltonian for a single excited quark takes the simple form

$$H^* = \epsilon_1 \mathbf{1} + \frac{1}{8\mathcal{I}}\left(\vec{R}^2 - 2g_K \vec{R}\cdot\vec{T}_K + g_K^2 \vec{T}_K^2\right). \quad (8.224)$$

The spectrum can be readily constructed if we notice that (8.224) can be rewritten solely in terms of the independent Casimir operators:

$$H^* = \epsilon_K \mathbf{1} + \frac{1}{2\mathcal{I}}\left[+\frac{g_K}{2}\vec{J}_K^{\,2} + (1 - \frac{g_K}{2})\vec{I}^2 - \frac{g_K}{2}(1 - \frac{g_K}{2})\vec{T}^2\right] \quad (8.225)$$

where $\vec{J}_K = \vec{R}/2 - \vec{T}_K$ and $\vec{I} = \vec{L}/2$. We recall that $\vec{L}^2 = \vec{R}^2$ on S^3. The set J_K^2, I^2, J_{K3} and I_3 forms a complete set of independent generators that commute with H^*. They can be used to span the Hilbert space of a single quark in the K band. We will identify J_K with the angular momentum and I with the isospin.

The spectrum associated with (8.225) will be analyzed both in the light-quark and heavy-quark sectors. Let us briefly mention here that in the light-quark system, the limit $g_K = 0$ for which the induced angular momentum $-\vec{T}_K$ decouples from the system (which would correspond to $(1 - \kappa) = 0$ in the diatomic molecule) is never reached. Neither is the abelian limit that will be attained for $g_K = 2$ physically relevant. On the other hand, as we will explicitly show in Chapter 9, the decoupling limit is attained in heavy-quark baryons because of the presence of heavy-quark symmetry that is present in QCD in the limit the heavy-quark mass becomes infinite.

8.3.4. *Baryon excitations*

In Chapter 7, we discussed what we classify as "ground-state" baryons and their rotational excitations. Thus in the light u- and d-quark sector, we encountered only the states with $J = I$ and in the $SU(3)$ sector, strange baryons were rotational excitations in the strange direction on top of the $J = I$ nonstrange baryons. To describe the excitations with $J \neq I$, one requires "vibrational" modes and Berry potentials associated with these modes.

In the next chapter, we shall consider the strangeness as a "vibration" with its frequency ω_K much "faster" than the rotational frequency discussed above and describe strange hyperons in terms of nonabelian Berry potentials induced when the vibrational mode is integrated out. As we will see, we will reproduce exactly the Callan-Klebanov spectrum for hyperons in this case.

Here let us simply assume that we can apply the same adiabaticity relation to the vibrational excitations in the light u- and d-sector. Again using the chiral-bag picture, we can think of the vibration in the meson sector to be the fluctuating pion field and in the quark sector to be the excitation of one of the quarks sitting in $K = 0^+$ level to higher K levels, the lowest of which is the $K = 1^+$ level. The relevant Hamiltonian is eq. (8.225):

$$H^* = \epsilon_K 1 + \frac{1}{2\Lambda}\left[+\frac{g_K}{2}\vec{J_K}^2 + (1 - \frac{g_K}{2})\vec{I}^2 - \frac{g_K}{2}(1 - \frac{g_K}{2})\vec{T}^2\right]. \quad (8.226)$$

This describes only one quark excited from the $K = 0^+$ level to a K level. One could generalize this to excitations of more than one quark with some simplifying assumptions (*i.e.*, *quasiparticles*) but this may not give a realistic description in the light-quark sector whereas in the heavy-quark sector considered in the next chapter, it is found to be quite accurate. Here we will not consider multi-quark excitations.

If one confines to one-quark excitations, then there are three parameters involved. They are all calculable in a specific model as described in the chiral bag model [37] but given the drastic approximations, one should not expect that the calculation would yield reasonable values. On the other hand, the generic structure of the mass formula suggests that given a set of parameters, the mass formula could make useful predictions. Suppose then that we write mass formulas eliminating the parameters ϵ_K, Λ and g_K. We give a few of the resulting mass formulas and see how they work. For instance, in the Roper channel, it follows from (8.226) that

$$P11 - N = P33 - \Delta \tag{8.227}$$

where here and in what follows we use particle symbols for their masses. Empirically the left-hand side is 502 MeV and the right-hand side is 688 MeV. In the odd-parity channel,

$$D13 - D35 + \Delta - N = -\frac{1}{4}(D35 - S31). \tag{8.228}$$

From the data, we get 116 MeV for the left side and 76 MeV for the right. Also

$$S31 - S11 = \frac{5}{2}(\Delta - N) - \frac{3}{2}(D35 - D13). \tag{8.229}$$

Empirically the left-hand side is 85 MeV and the right-hand side gives 125 MeV.

The agreement of the mass formulas is reasonable. But it is not very good either. This is not surprising as already warned above. Specifically the following should be noted. In the pure skyrmion description, we know that the vector mesons ρ, ω and possibly a_1 play an important role in describing some of the channels involved. For instance, to understand the phase shifts of the $S31$ and $S11$ channels, it is found that an anomalous parity (Wess-Zumino) term involving vector mesons consistent with hidden gauge symmetry is essential[38]. Now, in our framework what this means is that the quark excitations corresponding to the vector mesons ρ and ω etc. must be identified. This problem has not yet been fully understood in terms of induced gauge structure.

Appendix

(A_I) Conserved angular momentum in the presence of the monopole

For the purpose of understanding better the structure of the diatomic molecular system, we review here the known case of a particle coupled to an external gauge field of 't Hooft-Polyakov monopole[35] with a coupling constant g to which our nonabelian gauge field will correspond.

Asymptotically the magnetic field involved is of the form

$$\vec{B} = -\frac{\hat{r}(\hat{r} \cdot \mathbf{T})}{gr^2} \tag{A8.1}$$

which is obtained from the asymptotic form of the gauge field $\vec{A}$

$$A_i^\alpha = \epsilon_{\alpha ij} \frac{r_j}{gr^2}, \tag{A8.2}$$

$$\vec{B} = \vec{\nabla} \times \vec{A} - ig[\vec{A}, \vec{A}]. \tag{A8.3}$$

The Hamiltonian of a particle coupled to a 't Hooft-Polyakov monopole can therefore be written, for $g = 1$, in close analogy with eq.(8.183),

$$\begin{aligned} H &= \frac{1}{2M}(\vec{p} - \vec{A})^2 \\ &= \frac{1}{2M}\vec{D} \cdot \vec{D} \end{aligned} \tag{A8.4}$$

where $\vec{D} = \vec{p} - \vec{A}$.

One can see easily that the mechanical angular momentum $\vec{L}_m$ of a particle

$$\vec{L}_m = M\vec{r} \times \dot{\vec{r}} = \vec{r} \times \vec{D} \tag{A8.5}$$

does not satisfy the $SU(2)$ algebra after canonical quantization in eq.(8.182) and moreover it cannot be a symmetric operator that commutes with the Hamiltonian. The conventional angular momentum, $\vec{L}_o = \vec{r} \times \vec{p}$, satisfies the usual angular momentum commutation rule. However it does not commute with the Hamiltonian and hence cannot be a conserved angular momentum of the system. The conserved angular momentum can be constructed by modifying $\vec{L}_m$ to

$$\vec{L} = \vec{L}_m + \vec{Q}, \tag{A8.6}$$

where $\vec{\mathbf{Q}} = \vec{Q}^\alpha I^\alpha$ is determined as follows[39]. The first condition required for $\vec{\mathbf{Q}}$ is the consistency condition that $\vec{L}$ satisfy the $SU(2)$ algebra

$$[L_i, L_j] = i\epsilon_{ijk}L_k. \tag{A8.7}$$

This leads to an equation for $\vec{\mathbf{Q}}$,

$$\vec{r}(\vec{r} \cdot \mathbf{B}) + \vec{r}\vec{\mathcal{D}} \cdot \vec{\mathbf{Q}} - \vec{\mathcal{D}}(\vec{r} \cdot \vec{\mathbf{Q}}) = 0. \tag{A8.8}$$

where

$$\vec{\mathcal{D}} = \vec{\nabla} - i[\vec{\mathbf{A}}, \quad]. \tag{A8.9}$$

The second equation is obtained by requiring that $\vec{L}$ commute with H,

$$[\vec{L}, H] = 0. \tag{A8.10}$$

Equation (A8.10) can be replaced by a stronger condition

$$[L_i, \mathbf{D}_j] = i\epsilon_{ijk}\mathbf{D}_k, \tag{A8.11}$$

which leads to

$$\mathcal{D}_i\mathbf{Q}_j + \delta_{ij}\vec{r} \cdot \mathbf{B} - r_i\mathbf{B}_j = 0. \tag{A8.12}$$

It is obvious that $\vec{L}$ satisfying eq.(A8.11) or (A8.12) commutes with the Hamiltonian eq.(A8.4).

In the case of 't Hooft-Polyakov monopole, eqs.(A8.1) and (A8.2), it can be shown that

$$\vec{\mathbf{Q}} = \hat{r}(\hat{r} \cdot \mathbf{I}) \tag{A8.13}$$

satisfies eqs. (A8.8) and (A8.12). After inserting eq.(A8.13) into eq.(A8.6), we get

$$\begin{aligned}\vec{L} &= \vec{L}_m + \hat{r}(\hat{r} \cdot \mathbf{I}) \tag{A8.14} \\ &= \vec{r} \times \vec{p} + \vec{I}, \tag{A8.15}\end{aligned}$$

where

$$I_i = \delta_{i\alpha}I^\alpha. \tag{A8.16}$$

Equations (A8.15) and (A8.16) show clearly how the isospin-spin transmutation occurs in a system where a particle is coupled to a nonabelian monopole[40].

This analysis can be applied to the abelian $U(1)$ monopole just by replacing $\hat{r} \cdot \vec{I}$ by minus sign in eqs.(A8.13) and (A8.14). (We are considering a Dirac monopole with $e = g = 1$.) Then

$$\vec{Q} = \hat{r}, \tag{A8.17}$$

$$\vec{L} = m\vec{r} \times \dot{\vec{r}} - \hat{r}. \tag{A8.18}$$

This can be rewritten as

$$\vec{L} = \vec{r} \times \vec{p} - \vec{\Sigma}, \tag{A8.19}$$

where

$$\vec{\Sigma} = \left(\frac{(1 - \cos\theta)}{\sin\theta} \cos\phi, \frac{(1 - \cos\theta)}{\sin\theta} \sin\phi, 1 \right) \tag{A8.20}$$

which is the form frequently seen in the literature.

References

1. S. Nadkarni, H.B. Nielsen and I. Zahed, Nucl. Phys. **B253**, 308 (1984).
2. T.H.R. Skyrme, Nucl. Phys. **31**, 556 (1962).
3. E. Witten, Nucl. Phys. **B223**, 422, 433 (1983).
4. For discussions and references relevant to the present treatment, see G.E. Brown and M. Rho, Comments Part. Nucl. Phys. **18**, 1 (1988).
5. N. Dorey, J. Hughes and M.P. Mattis, Phys. Rev. Lett. **73**, 1211 (1994).
6. A.V. Manohar, Phys. Lett. **B336**, 502 (1994).
7. S. Nadkarni and H.B. Nielsen, Nucl. Phys. **B263**, 1 (1986).
8. S. Coleman, Phys. Rev. **D11**, 2088 (1975); S. Mandelstam, Phys. Rev. **D11**, 3026 (1975).
9. R.J. Perry and M. Rho, Phys. Rev. **D34**, 1169 (1986).
10. J. Schwinger, Phys. Rev. **128**, 2425 (1969).
11. H.B. Nielsen and A. Wirzba, in *Elementary Structure of Matter*, Springer Proceedings in Physics **26** (1988).
12. S.L. Adler, Phys. Rev. **177**, 2426 (1969); J.S. Bell and R. Jackiw, Nuovo Cimento **60A**, 107 (1969).
13. E. Witten, Comm. Math. Phys. **92**, 455 (1984).
14. M. Rho and I. Zahed, Phys. Lett. **B182**, 274 (1986); H. Falomir, M:A. Muschietti and E.M. Santangelo, Phys. Lett. **B205**, 93 (1988); Phys. Rev. **D37**, 1677 (1988). For a discussion in (3+1) dimensions, see Y.-X. Chen, R.-H. Yue and Y.-Z. Zhang, Phys. Lett. **B202**, 587 (1988).

15. T.H. Hansson and R.L. Jaffe, Phys. Rev. **D28**, 882 (1983); I. Zahed, Phys. Rev. **D30**, 2221 (1984).

16. We follow A.P. Balachandran, in *High-Energy Physics 1985*, vol. 1, ed. by M.J. Bowick and F. Gürrsey (World Scientific, Singapore, 1986).

17. H.B. Nielsen, M. Rho, A. Wirzba and I. Zahed, Phys. Lett. **B269**, 389 (1991).

18. J. Goldstone and R.J. Jaffe, Phys. Rev. Lett. **51**, 1518 (1983).

19. H.B. Nielsen, M. Rho, A. Wirzba and I. Zahed, Phys. Lett. **B281**, 345 (1992).

20. V.A. Novikov, M.A. Shifman, A.I. Vainshtein and V.I. Zakharov, Nucl. Phys. **B191**, 301 (1981).

21. P.H. Damgaard, H.B. Nielsen and R. Sollacher, Nucl. Phys. **B385**, 227 (1992); Phys. Lett. **B296**, 132 (1992).

22. K. Fujikawa, Phys. Rev. **D21**, 2848 (1980).

23. For a pedagogical discussion, see K. Huang, *Quarks, Leptons and Gauge Fields* (World Scientific, Singapore, 1982) p 152ff. Other useful references are C. Itzykson and J.-B. Zuber, *Quantum Field Theory* (McGraw-Hill Co., New York, 1985); P. Ramond, *Field Theory: A Modern Primer* (Addison-Wesley Publishing Co., Menlo Park, Calif., 1989).

24. P.H. Damgaard, H.B. Nielsen and R. Sollacher, Nucl. Phys. **B414**, 541 (1994).

25. See A. Shapere and F. Wilczek, eds., *Geometric phases in physics* (World Scientific, Singapore, 1989).

26. M. Bando, T. Kugo and K. Yamawaki, Phys. Repts. **164**, 217 (1988).

27. This section is based on the work by H.K. Lee, M.A. Nowak, M. Rho and I. Zahed, "Nonabelian Berry phases in baryons," Ann. of Phys. (NY), **227**, 175 (1993).

28. M. Stone, Phys. Rev. **D33**, 1191 (1986).

29. M.V. Berry, Proc. Roy. Soc. (London) **A392**, 45 (1984).

30. E. Rabinovici, A. Schwimmer and S. Yankielowicz, Nucl. Phys. **B248**, 523 (1984).

31. A. Bohm, B. Kendrick, M.E. Loewe and L.J. Boya, J. Math. Phys. **33**, 977 (1992); J. Moody, A. Shapere and F. Wilczek, Phys. Rev. Lett. **56**, 893 (1986); B. Zygelman, Phys. Rev. Lett. **64**, 256 (1990).

32. F. Wilczek and A. Zee, Phys. Rev. Lett. 52, 2111 (1984).

33. L.D. Landau and E.M. Lifshitz, *Quantum Mechanics* (Pergamon Press, 1976).

34. R. Casalbuoni, Nuov. Cim. **33A** (1976)389; A.P. Balachandran, Per Salomonson, Bo-Sture Skagerstam, and Jan-Olov Winnberg, Phys. Rev. **D15** (1977) 2308.

35. G. 't Hooft, Nucl. Phys. **B79**, 276 (1974); A. Polyakov, JETP Lett. **20** (1974) 194.

36. R. Jackiw, Phys. Rev. Lett. **56**, 2779 (1986).

37. H.K. Lee, M.A. Nowak, M. Rho and I. Zahed, Phys. Lett. **B255**, 96 (1991); Phys. Lett. **B272**, 109 (1991).

38. D. Masak, H. Walliser and G. Holzwarth, Nucl. Phys. **A536**, 583 (1992).

39. R. Jackiw, Acta Physica Austrica. Suppl. **XXXII**, 383 (1980); C.N. Yang, *Lectures on Frontiers in Physics* (Korean Physical Society, Seoul, 1980).

40. R. Jackiw and C. Rebbi, Phys. Rev. Lett. **36**, 1116 (1976); P. Hasenfratz and G. 't Hooft, Phys. Rev. Lett. **36**, 1119 (1976).

CHAPTER 9

STRANGE AND HEAVY BARYONS

9.1. Introduction

Thus far the quark masses in QCD have been treated as small perturbations around the massless (chiral) limit. As a result, chiral symmetry was used as a guiding principle for organizing the low-energy processes, involving chiefly pions and nucleons. While this point of view is well justified for the light-quark flavors (up and down), it may not be for the heavy flavors (strange, charm,...). In this chapter, we will take the opposite point of view. Instead, we will assume that the quark masses in QCD are infinitely heavy (heavy-quark limit). As a result, hadrons with one or several heavy quarks exhibit a new type of symmetry : invariance under spin flip of the heavy quarks. This heavy-quark symmetry can be used in conjunction with chiral symmetry to organize the low-energy structure of processes involving light and heavy hadrons.

The above dichotomy can be summarized in two limiting cases: (a) "bottom-up" and (b) "top-down." These two extremes describe the cases when (a) $m_q \ll \Lambda_\chi$ and (b) $m_Q \gg \Lambda_\chi$ where Λ_χ is the chiral symmetry scale, q denotes light quark and Q heavy quark. The systems studied in Chapters 7 and 8 belong to the first category and the charmed-quark and bottom-quark baryons described in this chapter belong to the latter. In the former regime, chiral symmetry is relevant; in the latter, heavy-quark symmetry[1,2,3] is relevant. Both are intrinsic symmetries of QCD that are however not immediately visible: chiral symmetry is "visible" in the QCD Lagrangian but realized in a hidden way (Goldstone mode); heavy-quark symmetry is not visible in the classical Lagrangian but realized in a visible way (heavy-quark multiplets).

After exposing the ambivalence (light/heavy) in the handling of the strange quarks at low energy, we shall proceed to introduce the concepts behind the heavy-quark limit. Borrowing on the arguments presented in the earlier chapters, we will show how the light and heavy degrees of freedom can be described quantitatively using the instanton model of the QCD vacuum, and generically following the dictates of chiral and heavy-quark symmetries. In particular, we show that the heavy-quark excitations acquire their quantum numbers through Berry phases. Our exposition will be along the "top-down" scenario, and will provide us with an important contrast to the "bottom-up" scenario used so far.

9.1.1. *Quark masses*

In QCD, the quark mass term is given by

$$\mathcal{L}_{mass}^{\text{QCD}} = -\bar{\psi}\mathcal{M}\psi \tag{9.1}$$

where ψ is the quark field $\psi^T = (u \; d \; s \cdots)$ and $\mathcal{M}$ the mass matrix which for our purpose could be taken diagonal and real (for zero θ angle)

$$\text{diag}\mathcal{M} = (m_u \; m_d \; m_s \; m_c \; m_b \; m_t). \tag{9.2}$$

These QCD masses (called "current quark" masses) renormalize. A microscopic theory which is presumably more fundamental than the standard model fixes them at some high-energy scale of the order of, say, 1 TeV. Their relation to the low-energy constants in effective theories of the standard model follows from renormalization. To lowest order in chiral perturbation theory, the renormalization of the quark masses is multiplicative and flavor independent [4]

$$m_{\pi+}^2 = -\frac{1}{f_\pi^2}(m_u + m_d)\langle\overline{u}u\rangle_0 + \mathcal{O}(m^2) + \mathcal{O}(e^2),$$

$$m_{K+}^2 = -\frac{1}{f_\pi^2}(m_u + m_s)\langle\overline{u}u\rangle_0 + \mathcal{O}(m^2) + \mathcal{O}(e^2),$$

$$m_{K^0}^2 = -\frac{1}{f_\pi^2}(m_d + m_s)\langle\overline{u}u\rangle_0 + \mathcal{O}(m^2) + \mathcal{O}(e^2). \tag{9.3}$$

The quark condensate is evaluated in the chiral limit with $\langle\overline{u}u\rangle_0 = \langle\overline{d}d\rangle_0 = \langle\overline{s}s\rangle_0$, and the $\mathcal{O}(m^2)$ corrections are from strong interactions and the $\mathcal{O}(e^2)$

corrections are from electromagnetic interactions. The results are just the Gell-Mann-Oakes-Renner relations of Chapter 6. Using Dashen theorem [5] for the electromagnetic self-energies, together with (9.3), gives [4]

$$\frac{m_d}{m_u} = \frac{m_{K^0}^2 - m_{K^+}^2 + m_{\pi^+}^2}{2m_{\pi^0}^2 + m_{K^+}^2 - m_{K^0}^2 - m_{\pi^+}^2} \approx 1.8,$$

$$\frac{m_s}{m_d} = \frac{m_{K^0}^2 + m_{K^+}^2 - m_{\pi^+}^2}{m_{K^0}^2 - m_{K^+}^2 + m_{\pi^+}^2} \approx 20.1. \tag{9.4}$$

The scale-dependence weakens in the ratios. Equation (9.4) shows that the quark masses are not at all degenerate. The down quark is about twice the up quark, and the strange quark about twenty times the down quark. There are higher-order corrections to (9.3) in both the strong and electromagnetic effects. The former may be large and of order $(m_K/4\pi f_\pi)^2 \sim 0.25$. The latter are small and of order $e^2/4\pi \sim 10^{-2}$. Hence, (9.4) should hold within 25%.

To get the absolute values of the current quark masses, extra assumptions are needed. An early estimate of these masses for a typical scale of 1 GeV is [6]

$$m_u \simeq 4\text{MeV}, \quad m_d \simeq 6\text{MeV}, \quad m_s \simeq 135\text{MeV} \tag{9.5}$$

which is in overall consistency with (9.4). More elaborate estimates are also available [7]. The masses for the c, b and t quarks are heavier than the chiral scale. We will come back to the c and b quarks later. While the validity of first order chiral perturbation theory for the up and down quarks is perhaps justified, this may not be the case for the strange quark. This throws some further uncertainty on the second ratio in (9.4) and may have important consequences on such issues as the strangeness content of the proton (see below) and kaon condensation (see Chapter 10). We will take (9.5) as the canonical values.

Without (9.1), the QCD Lagrangian is chirally invariant. Thus in effective Lagrangians we are interested in, the chiral symmetry-breaking pieces must therefore be proportional to the power of the mass matrix $\mathcal{M}$. If $\mathcal{M}$ is small, then the leading symmetry-breaking term in an effective Lagrangian is of the form

$$\mathcal{L}_{\text{SB}}^{eff} = f^2 B \operatorname{Tr}(\mathcal{M} U^\dagger + \mathcal{M}^\dagger U) \tag{9.6}$$

where U is the chiral meson field defined in Chapters 6 and 7, f is the constant related to the pion-decay constant f_π and B is a constant related to the quark condensate in the chiral limit (see Chapter 6). The QCD mass term implies that $\mathcal{M}$ transforms as $(3,\bar{3}) + (\bar{3},3)$ under chiral $SU(3) \times SU(3)$

$$\mathcal{M} \to L\mathcal{M}R^\dagger \tag{9.7}$$

with $L \in SU_L(N_f)$ and $R \in SU_R(N_f)$. Then higher-order terms in the quark mass can be constructed by writing terms involving higher powers of $\mathcal{M}$ that are invariant under (9.7).

Beyond lowest-order chiral perturbation theory, the quark masses are no longer multiplicatively renormalizable. In particular they receive an additive renormalization proportional to $m_d m_s$ for the up quark, $m_s m_u$ for the down quark and $m_u m_d$ for the strange quark. These terms are induced in the renormalization group evolution by the small size instantons. Their determination is infrared-sensitive. In the effective Lagrangian approach this issue translates to a non-unique contribution when next-to-leading-order effects in $\mathcal{M}$ are considered. The problem is that the matrix

$$\mathcal{M}' = \alpha_1 \mathcal{M} + \alpha_2 (\mathcal{M}^\dagger)^{-1} \det \mathcal{M} \tag{9.8}$$

with arbitrary constants α_i transforms under (9.7) in the same way as $\mathcal{M}$. Specifically this means for a diagonal matrix $\mathcal{M}$

$$m_u' = \alpha_1 m_u + \alpha_2 m_d m_s \quad (\text{cyclic } u \to d \to s \to u). \tag{9.9}$$

Therefore if $\mathcal{L}(U, \partial U, \mathcal{M})$ is an effective Lagrangian consistent with chiral symmetry, then so is $\mathcal{L}(U, \partial U, \mathcal{M}')$. This ambiguity was noted by Kaplan and Manohar [8]. Obviously this is not a symmetry of proper QCD but a hidden symmetry in effective theories [9]. It does not affect the considerations below.

9.1.2. *Strangeness content of the proton*

With the advent of precise polarization experiments on nucleon targets at high energy, the theoretical curiosity regarding the amount of strangeness in the nucleon state has turned into a physical reality. This issue is important for understanding the relationship between the nucleon axial-singlet charge and its spin, and may have fundamental consequences for kaon-nucleon interactions and kaon condensation in dense neutron-star matter.

To assess this question in the present section we will rely on some of the methods and models discussed in the preceding chapters.

- Skyrmions

To estimate the amount of the nucleon mass due to strangeness, we assume as before that chiral $SU(3) \times SU(3)$ is a good symmetry and treat the mass term as perturbation. In QCD variables, the quantity we are interested in is

$$Y \equiv \frac{\langle \bar{s}s \rangle_P}{\langle \bar{u}u + \bar{d}d + \bar{s}s \rangle_P}. \tag{9.10}$$

The expectation values are connected

$$\langle \bar{q}q \rangle_N \equiv \langle N | \left(\bar{q}q - \langle 0|\bar{q}q|0\rangle \right) |N\rangle \tag{9.11}$$

with the normalization $\langle N|N \rangle = 1$. In terms of the chiral field U, the relevant matrix elements are

$$\langle \bar{u}u + \bar{d}d \rangle_N \propto \quad \langle N| \operatorname{Tr}[P(U + U^\dagger - 2)|N\rangle$$
$$\propto \quad \langle N|[2 + D_{88}(S)]|N\rangle,$$

$$\langle \bar{s}s \rangle_N \propto \quad \langle N| \operatorname{Tr}[Q(U + U^\dagger - 2)]|N\rangle$$
$$\propto \quad \langle N|[1 - D_{88}(S)]|N\rangle \tag{9.12}$$

where $P = \operatorname{diag}(1,\ 1,\ 0)$, $Q = \operatorname{diag}(0,\ 0,\ 1)$, $D_{ab}(S) = \frac{1}{2}\operatorname{Tr}(\lambda_a S \lambda_b S^\dagger)$ and S is the $SU(2)$ collective rotation matrix defined in Chapter 7. The overall constants are not needed for our arguments. They drop out in the ratios.

If we ignore the effects of the current quark masses on the nucleon wave-function, the ratio (9.10) may be readily evaluated. Indeed, using the $SU(3)$ wave functions discussed in Chapter 7, we obtain

$$Y = \frac{\frac{7}{10}}{\frac{12}{10} + \frac{11}{10} + \frac{7}{10}} = \frac{7}{30} \approx 0.23 \tag{9.13}$$

where the matrix elements of (9.10) are written – apart from an overall constant – in the order they appear in the definition of Y, eq.(9.10). The strangeness contribution to the nucleon state is comparable to the up and down contributions. From (9.4) it follows that the net contribution to the nucleon mass is about twenty times larger!

Let us now see what happens to Y if one takes into account the effects of the quark masses on the nucleon state viewed as an $SU(3)$ skyrmion. In the Yabu-Ando approach described in Chapter 7, the wave function has the representation mixing. Taking for illustration the perturbative form (4.107), Chapter 7, we get

$$Y \approx \frac{7}{30} - \frac{43}{2250} \left(\frac{b\gamma}{4} \right) + \cdots. \tag{9.14}$$

Hyperon spectra suggest that $b\gamma/4 \approx 3.3$. With this we again find a substantial admixture of strangeness

$$Y \approx 0.17. \tag{9.15}$$

The exact diagonalization reduces this number somewhat further to about 0.12. This is about half the value found in (9.13), still substantial. So, we conclude that the $SU(3)$ soliton description of the nucleon supports a large strangeness contribution to the nucleon mass, and perhaps to other observables as well.

- Gell-Mann-Okubo formula

To contrast the above result, consider [10] the πN sigma term discussed in Chapter 5:

$$\Sigma_{\pi N} \equiv \hat{m} \langle \bar{u}u + \bar{d}d \rangle_P = 45 \pm 10 \text{ MeV} \tag{9.16}$$

with

$$\hat{m} = \frac{1}{2}(m_u + m_d) \tag{9.17}$$

where the last equality is the presently accepted value with a large theoretical error bar. There is still a considerable uncertainty on the precise value of the sigma term, but for our purpose it is not important what the precise value is. Now consider the operator

$$\begin{aligned} O_8 &\equiv \hat{m}(\bar{u}u + \bar{d}d - 2\bar{s}s) \\ &= \frac{\hat{m}}{m_s - \hat{m}}(m_s - \hat{m})(\bar{u}u + \bar{d}d - 2\bar{s}s). \end{aligned} \tag{9.18}$$

Using the current algebra value of the quark mass ratio

$$\frac{m_s + \hat{m}}{\hat{m}} \approx \frac{m_s}{\hat{m}} \approx 25 \tag{9.19}$$

and the Gell-Mann-Okubo mass formula, which follows from the assumption that the octet mass splitting is governed by the operator O_8 (see the caveat on this later)

$$\langle O_8 \rangle_P = \frac{3\hat{m}}{m_s - \hat{m}}(M_\Xi - M_\Lambda) \approx 26 \text{ MeV} \tag{9.20}$$

we obtain from eqs.(9.16) and (9.18)

$$Y \approx 0.17 \tag{9.21}$$

which is in agreement with the second $SU(3)$ skyrmion value (9.15). This would seem to indicate again that there is a substantial amount of strangeness flavor mixed in the nucleon, which would be at odds with the usual quark-model picture that for instance, the proton is made up of two up and one down quarks.

- Lattice and χPT

Lattice simulations in quenched approximation[11] give $Y \approx 0.2$, so such a large strangeness in the proton cannot be ruled out. In any case, such a large Y will have a rather dramatic effect on kaon condensation in dense hadronic matter discussed in the next chapter.

Chiral perturbation theory that includes terms of $O(m_s^{3/2})$ predicts about $Y \approx 0.09$ with a large error bar which makes even $Y \sim 0$ compatible[12]. If Y were close to zero, it would not be reasonable to treat the strange quark as "light" in the sense that the up and down quarks are light. We will see later that a model that treats the s quark as "heavy" with $Y \sim 0$ (*i.e.*, the Callan-Klebanov model[13]) describes hyperons surprisingly well.

So far we have assumed that the octet operator O_8 is directly responsible for the mass splitting. But this assumption could be invalidated by the trace anomaly as pointed out by Ball, Forte and Tigg.[14] Hence, it is not at all impossible that the relation between the sigma term and the mass splitting used above, could be significantly affected by non-perturbative effects. At the moment it is not possible to make a reliable estimate of the latter but an approximate model-dependent estimate shows that the gluon contribution is not purely flavor-singlet and can be quantitatively important.

9.1.3. *KN sigma term*

The KN sigma term defined as

$$\Sigma_{KN} \equiv \frac{1}{2}(\hat{m} + m_s)\langle \bar{u}u + \bar{s}s \rangle_P \qquad (9.22)$$

figures in the nucleon mass as well as in kaon condensation in dense nuclear matter described in the next chapter. As in the case of the πN sigma term, one could eventually extract this quantity from accurate threshold KN scattering data. At present this is not possible. Here we will make a theoretical estimate based on the previous analysis on the strangeness content of the proton.

To do this, let us consider three cases: $Y = 0, 0.1, 0.2$. We take the accepted value $\Sigma_{\pi N} \approx 45$ MeV, and the quark masses $\hat{m} \approx 5$ MeV, $m_s \approx 135$ MeV. The resulting Σ_{KN} then is

$$\Sigma_{KN} \approx (2.3, \ 2.8, \ 3.4)m_\pi \quad \text{for} \quad Y = 0, \ 0.1, \ 0.2. \qquad (9.23)$$

The lattice result[11], $\Sigma_{KN} = (450 \pm 30)$ MeV, supports the larger strangeness content of the proton. The kaon-nucleon scattering data analyzed in Chapter 10 are consistent with this value.

9.2. Heavy Quark Hadrons

We have seen above that strangeness could actually be viewed as heavy rather than light. If that is the case, perhaps the best strategy is to start from the infinitely heavy-quark limit, instead of the chiral limit. The aim of this section is to develop some insights into this new limit, with the understanding that it is perhaps more relevant for heavy flavors such as top and bottom rather than strange. Although, a priori, the approach looks irrelevant to the issues so far discussed, we will see as the exposition proceeds that this is not the case. In fact, we will borrow from most of the concepts introduced so far (effective action, solitons, bags, Berry phases) to describe the heavy-quark systems. In a way, this emphasizes the generic character of these concepts for soft QCD processes, showing their inter-complementarity in understanding the physics in these systems.

In this section we will first discuss the basic phenomenological aspects of heavy quarks and hadrons, and their generic aspects through a bag model description. We then proceed to discuss the heavy-quark expansion in the context of the instanton model to the QCD vacuum discussed in Chapter 2.

This model captures some of the essential soft effects on the heavy hadrons. The resulting effective interactions between the soft light and heavy quarks will be discussed. Using approximate bosonization, we rephrase the effective heavy-quark action in terms of heavy meson degrees of freedom. In the large N_c limit, the effective action supports heavy soliton configurations (heavy baryons) in total analogy with the "top-down" framework used for the light hadrons. As discussed in Chapter 8, we will show that the heavy solitons acquire their quantum numbers through Berry phases, much like in the Callan-Klebanov [13] scheme of Chapter 7. In this way the analogy with the "bottom-up" approach is total.

9.2.1. *Generalities*

Heavy hadrons are mesons and baryons with at least one heavy quark. Although we will be mostly interested in taking the heavy-quark mass to infinity, we will be mainly interested in the the top quark t with a mass of $176 - 199$ GeV, the bottom quark b with a mass of $4.7 - 5.3$ GeV, the charmed quark c with a mass of $1.3 - 1.7$ GeV, and to some extent the strange quark s with a mass of $100 - 300$ MeV.

Heavy mesons $Q\bar{q}$ may be organized in their ground state into multiplets with $I(J^P) = \frac{1}{2}(0^-, 1^-)$. In the heavy-quark limit, the multiplets are invariant representations of the heavy-quark symmetry group (essentially left spin rotation). Empirically $(K, K^*) = (493, 892)$ MeV, $(D, D^*) = (1869, 2010)$ MeV and $(B, B^*) = (5278, 5324)$ MeV.

The heavy baryons with one heavy quark will be of the type Qqq, with the conventions being $\Lambda_Q = 0\frac{1}{2}^+$, $\Sigma_Q = 1\frac{1}{2}^+$ and $\Sigma_Q^* = 1\frac{3}{2}^+$. The ones with two heavy quarks will be of the type QQq. A few heavy baryons have already been observed. Organizing them in multiplets invariant under heavy-quark symmetry, we have $\Lambda_s(\Sigma_s, \Sigma_s^*) = 1116\,(1195, 1385)$ MeV, $\Lambda_c(\Sigma_c, \Sigma_c^*) = 2284\,(2455, 2530 \pm 5 \pm 5)$ MeV, and $\Lambda_b(\Sigma_b, \Sigma_b^*) = 5656\,(5829, 5885)$ MeV. Other measured heavy (charmed) baryons include $(\Xi_c, \Xi_c^*) = (2468, 2642.8 \pm 2.2\,[15])$ MeV and $\Omega_c = 2704$ MeV. Heavy baryons with more than one heavy quark have not been found yet. The issue of multiquark states (exotics) with heavy flavor is still empirically debated.

9.2.2. *Bag model estimate*

The basic principles at work in a heavy-light system are best illustrated by using a simple bag model description, as discussed in Chapter 8.

If we were to insert a heavy source in a spherical cavity of radius R (without any cloud), then the total ground state energy could be organized using the bare heavy-quark mass m_Q as $E = m_Q + E_0 m_Q^0 + E_1 m_Q^{-1} + \cdots$. The contribution E_0 refers to the energy of the light quarks present in the cavity and is standard (see Chapter 8). The contribution E_1 corresponds to

$$\frac{E_1}{m_Q} = \frac{(\pi/R)^2}{2m_Q} + \frac{\vec{\mu}^a \cdot \vec{B}^a}{2m_Q} \tag{9.24}$$

where the first term is the recoil of the heavy quark, and the second term is the magnetic interaction between the average magnetic field induced by the light quark at the center of the bag, and the magnetic moment of the heavy quark. While schematic, (9.24) captures the essence of the $1/m_Q$ corrections in heavy-quark physics. Using standard bag model parameters [16], we have for charmed mesons (recoil, spin) $\sim$ (400, 20) MeV, while for bottom mesons (recoil, spin) $\sim$ (100, 5) MeV [17].

9.2.3. *Heavy quark expansion*

To understand the soft effects on the heavy-quark dynamics, we will use the instanton model to the QCD vacuum as discussed in Chapter 2. Our arguments will parallel recent considerations by Chernyshev *et al.* [18] They are generic, and should extend to any constituent quark model (e.g. Nambu-Jona-Lasinio model with heavy quarks) provided that the QCD symmetries are respected.

Let $q(x)$ be the light-quark field, and $\psi(x)$ the heavy-quark field. For m_Q much larger than the typical scale of the problem, Λ_{QCD}, one may use Λ_{QCD}/m_Q expansion to analyze the heavy-light correlation functions as was done in Chapter 2. This program can be implemented systematically using a Foldy-Wouthuysen transformation[19] of the heavy-quark field [20],

$$\psi(x) \sim e^{-i\gamma_0 m_Q t} \, e^{-i\sigma_{0i}[\nabla^0, \nabla^i]/4m_Q^2} \, e^{-i\vec{\gamma}\cdot\vec{\nabla}/2m_Q} \, Q(x) \tag{9.25}$$

where $\nabla = \partial - iA$ and with g (the gauge coupling) set to one. The first transformation rescales the momenta, the second eliminates the odd parts, and the third removes the mass term. The successive transformations associated to (9.25) are vector-like, unitary and gauge-covariant. In terms of (9.25) the QCD part of the action for the heavy field ψ becomes

$$\mathcal{L}_\psi \sim \overline{Q} i\gamma^0 \nabla^0 Q + \overline{Q} \hat{O}_1 Q + \mathcal{O}(m_Q^{-2}) \tag{9.26}$$

where the operator $\hat{O}_1$ is defined as

$$\hat{O}_1 = -\frac{\vec{\nabla}^2}{2m_Q} - \frac{\vec{\sigma} \cdot \vec{B}}{2m_Q} \qquad (9.27)$$

with $B^i = -i\epsilon^{ijk}[\nabla^i, \nabla^j]$. Equation (9.26) has the expected FW form to order m_Q^{-1}. The first term in (9.27) is the heavy quark recoil, and the second term is the spin-magnetic contribution familiar from Ampere's law.

9.2.4. *Heavy quark propagator*

In terms of (9.26), the heavy-quark propagator to order $\mathcal{O}(1/m_Q^2)$ reads

$$S_Q = S_\infty + S_\infty \hat{O}_1 S_\infty + \mathcal{O}(m_Q^{-2}) \qquad (9.28)$$

with $S_\infty = \gamma_4/i\nabla_4$ being the free part. This construction can be carried out to arbitrary orders in $1/m_Q$ [18,20], given that the heavy quark expansion is not upset by renormalization [20].

The heavy-quark propagator (9.28) may be analyzed in a random instanton gas. In the planar approximation discussed in Chapter 2, it satisfies the integral equation [18,21]

$$S_\infty^{-1} = S_*^{-1} - \sum_{I,\bar{I}} \langle \left(A\!\!\!/_{4,I}^{-1} - S_\infty \right)^{-1} \rangle \qquad (9.29)$$

where $A\!\!\!/_4 = \gamma^4 A^4$ and $S_* = i\gamma^4 \partial^4$. The sum is over all instantons and anti-instantons and the averaging is over the position z_I and the $SU(N_c)$ color orientation U_I, as discussed in Chapter 2. For a low instanton density $n = N/V_4 \sim 1$ fm^{-4}, we can iterate (9.29) in powers of n. The result is

$$S_\infty^{-1} = S_*^{-1} + in \left(\Theta_0 + \frac{1}{\rho m_Q} \Theta_1 + \mathcal{O}\left(\frac{1}{\rho^2 m_Q^2}\right) \right) + \mathcal{O}(n^2). \qquad (9.30)$$

Substituting (9.30) into (9.29), we obtain

$$\Theta_0 = \int d^4 z_I \, \text{Tr}_c \left(S_*^{-1} \left(\frac{1}{i\gamma^4 \nabla_{4,I}} - S_* \right) S_*^{-1} + I \to \bar{I} \right) \qquad (9.31)$$

and

$$\Theta_1 = \int d^4 z_I \, \text{Tr}_C \left(S_*^{-1} \left(\frac{1}{i\gamma^4 \nabla_{4,I}} - S_* \right) \mathcal{O}_1 \right.$$
$$\left. \times \left(\frac{1}{i\gamma^4 \nabla_{4,I}} - S_* \right) S_*^{-1} + I \to \bar{I} \right) \qquad (9.32)$$

where $\mathcal{O}_1 = -\vec{\nabla}^2/2 - \vec{\sigma}\cdot\vec{B}/2$. Both Θ_0 and Θ_1 are τ-dependent. To proceed further, we note that in coordinate space, the heavy-quark propagator in the one-instanton background reads

$$\langle x|\frac{1}{i\nabla_{4,I}}|0\rangle = \delta(\vec{x})\,\theta(\tau)\,\frac{1+\gamma^4}{2}\,\mathbf{P}e^{i\int_0^\tau ds A_4(x_s - z_I)} \qquad (9.33)$$

with $x_s = (s, \vec{x})$. Inserting (9.33) into (9.31, 9.32) and using the one-instanton configuration A_μ^a of Chapter 2 (singular gauge) yields for large times

$$\langle x_{-\infty}|(n\Theta_0, \frac{n}{\rho m_Q}\Theta_1)|x_{+\infty}\rangle \sim (\Delta_0 M_Q, \Delta_1 M_Q). \qquad (9.34)$$

The shift in the heavy-quark mass is $\Delta_0 M_Q \sim 70\,\text{MeV}$ [21] from Θ_0 plus $\Delta_1 M_Q \simeq 16\,\text{MeV}$ from Θ_1 for a c-quark with $m_c = 1350\,\text{MeV}$ [18]. The recoil effect $\Delta_1 M_Q$ is an order of magnitude down compared to the naive bag estimate (9.24). This can be understood by noting that in the presence of instantons, the energy of a heavy quark can be rewritten schematically as

$$E = 32\pi \times n\rho^4 \times \left(1/\rho + \frac{1/\rho^2}{m_Q} + ...\right) \qquad (9.35)$$

where the factors follow from (9.30) and (9.34). Equation (9.35) is the analog of (9.24). For the instanton parameters used and a charmed quark, (9.35) yields

$$E \sim 32\pi \times 10^{-3} \times \left(600 + \frac{1}{2}600 + ...\right) \sim \left(60 + 30 + ...\right)\,\text{MeV} \qquad (9.36)$$

which shows that the zeroth order shift in the mass is about 60 MeV, while the recoil effect is about 30 MeV, as expected.

In the arguments to follow in this chapter, we will use the fact that the light-quark propagator is $S^{-1} \sim S_0^{-1} + i\sqrt{n}\Sigma(x)$ with an average light-quark mass shift $\Delta M_q \sim 420\,\text{MeV}$ [18,22], for an instanton density $n \sim 1\,\text{fm}^{-4}$. This mass shift is slightly larger than the one discussed in Chapter 2.

9.2.5. *Effective interactions*

To characterize the effective interactions between the soft parts of the light and heavy-quark fields immersed in a dilute and random ensemble of instantons and anti-instantons, consider the heavy-light correlator C. Using arguments similar to those developed in Chapter 2, and to leading order in the instanton density n, we have [18]

$$C^{-1} \sim S^{-1} \otimes S_{\infty}^{T,\,-1} - n \int d^4 z_I \, \mathrm{Tr}_c \left([\, L\,]_I \otimes [\, H\,]_I + I \to \bar{I} \right) \quad (9.37)$$

with

$$[\, L\,]_I = S_0^{-1} \left(\frac{|\Phi_0\rangle\langle\Phi_0|}{i\sqrt{n\Sigma_0}} - S_0 \right) S_0^{-1} \quad (9.38)$$

$$[\, H\,]_I = S_*^{-1} \left(\frac{1}{i\gamma^4\nabla_{4,I}} - S_* \right) S_*^{-1}$$

$$+ \frac{1}{m_Q} S_*^{-1} \left(\frac{1}{i\gamma^4\nabla_{4,I}} - S_* \right) \mathcal{O}_1 \left(\frac{1}{i\gamma^4\nabla_{4,I}} - S_* \right) S_*^{-1}$$

where $\Phi_0(x)$ is the light zero mode state (2.18) and $\Sigma_0 = \langle\Phi_0|\Sigma|\Phi_0\rangle \sim (240 \text{ MeV})^{-1}$ is the induced soft scale in the instanton ensemble. In the long-wavelength limit, the instanton size is small, and a local interaction between the effective fields $\mathbf{Q}$ and $\mathbf{q}$ can be derived much like the 't Hooft interaction between the light effective fields $\mathbf{q}$ [23,24] discussed in Chapter 2. Indeed, the Bethe-Salpeter equation associated to (9.37) leads to the effective vertex

$$\Gamma_{bd}^{ac}(x, y, \, x', y') = -inN_c \int d^4 z_I \int dU_I$$

$$\times \left(U_i^a \langle x|[\, L_I\,]_j^i|x'\rangle U^{\dagger j}_{\;\;b} \otimes U_k^c \langle y|[\, H_I\,]_l^k|y'\rangle U^{\dagger l}_{\;\;d} + I \to \bar{I} \right) \quad (9.39)$$

where the color matrices have been explicitly displayed. This vertex function gives rise to an effective action $\mathcal{S}_I$

$$\Gamma_{bd}^{ac}(x, y, \, x', y') = \frac{\partial^4 \mathcal{S}_I}{\partial q^a(x)\, \partial q_b^\dagger(x')\, \partial Q^c(y)\, \partial Q_d^\dagger(y')} \quad (9.40)$$

which is essentially non-local. In the long-wavelength approximation the kernel (9.39) factorizes into two independent kernels as $x \to x'$ and $y \to y'$.

The corresponding effective action reads

$$
\mathcal{S}_I = -in N_c \int d^4 z_I \int dU_I
$$

$$
\times \left[\int d^4 x \, \mathbf{q}^\dagger(x) \, U_I \, \langle x | \, L_I \, | x \rangle \, U_I^\dagger \, \mathbf{q}(x) \right.
$$

$$
\left. \times \int d^4 y \, \mathbf{Q}^\dagger(y) \, U_I \, \langle y | \, H_I \, | y \rangle \, U_I^\dagger \, \mathbf{Q}(y) \right] + I \to \bar{I} \qquad (9.41)
$$

and yields the effective interaction in Euclidean space (to leading order in $1/N_c$)

$$
\mathcal{L}_{qQ}^E = n \left(-\frac{16\pi\rho^3 I_Q}{N_c} \right) \left(\frac{4\pi^2\rho^2}{\sqrt{n}\Sigma_0} \right)
$$

$$
\times \left(i\mathbf{Q}^\dagger \frac{1+\gamma_4}{2} \mathbf{Q} \, i\mathbf{q}^\dagger \mathbf{q} + \frac{1}{4} i\mathbf{Q}^\dagger \frac{1+\gamma_4}{2} \lambda^a \mathbf{Q} \, i\mathbf{q}^\dagger \lambda^a \mathbf{q} \right). \qquad (9.42)
$$

The first bracket in (9.42) arises from the heavy-quark part and the second bracket from the light-quark part. Wick-rotating to Minkowski space gives

$$
\mathcal{L}_{qQ} = -\left(\frac{\Delta M_Q \Delta M_q}{2n N_c} \right) \left(\overline{\mathbf{Q}} \frac{1+\gamma^0}{2} \mathbf{Q} \, \overline{\mathbf{q}} \mathbf{q} + \frac{1}{4} \overline{\mathbf{Q}} \frac{1+\gamma^0}{2} \lambda^a \mathbf{Q} \, \overline{\mathbf{q}} \lambda^a \mathbf{q} \right) (9.43)
$$

which is to be compared with the 't Hooft vertex for two light flavors $\mathbf{q} = (\mathbf{u}, \mathbf{d})$

$$
\mathcal{L}_{qq} = \left(\frac{\Delta M_q^2}{n N_c} \right) \left(\det \overline{\mathbf{q}}_R \mathbf{q}_L + \det \overline{\mathbf{q}}_L \mathbf{q}_R \right). \qquad (9.44)
$$

Equation (9.44) is the local version of the 't Hooft vertices discussed in Chapter 2. The interaction (9.43) is dominated by the Coulomb-like second term and has a proper heavy-quark spin symmetry. The recoil effect renormalizes the strength of the interaction through $\Delta M_Q = \Delta_0 M_Q + \Delta_1 M_Q \sim 86 \mathrm{MeV}$. The spin part gives rise to a chromomagnetic interaction

$$
\mathcal{L}_{qQ}^{spin} = \frac{\Delta M_q \, \Delta M_Q^{spin}}{2n N_c} \frac{1}{4} \overline{\mathbf{Q}} \frac{1+\gamma^0}{2} \lambda^a \sigma^{\mu\nu} \mathbf{Q} \, \overline{\mathbf{q}} \lambda^a \sigma^{\mu\nu} \mathbf{q} \qquad (9.45)
$$

with $\Delta M_Q^{spin} \sim n\rho^2/m_Q \sim 3$ MeV for a charmed quark [18]. As a Coulomb-like term in (9.43), it has a smooth N_c^0 limit for the large N_c and is attractive in the spin zero, color-singlet channel.

Similar arguments may be applied to the heavy mesons $\overline{Q}Q$ as well. To order $1/m_Q$ and in the planar approximation, the effective interaction among the heavy quarks is given by

$$\mathcal{L}_{QQ} = -\left(\frac{\Delta M_Q \Delta M_Q}{2nN_c}\right)$$
$$\times \left(\overline{\mathbf{Q}}\frac{1+\gamma^0}{2}\mathbf{Q}\,\overline{\mathbf{Q}}\frac{1+\gamma^0}{2}\mathbf{Q} + \frac{1}{4}\overline{\mathbf{Q}}\frac{1+\gamma^0}{2}\lambda^a\mathbf{Q}\,\overline{\mathbf{Q}}\frac{1+\gamma^0}{2}\lambda^a\mathbf{Q}\right). \quad (9.46)$$

The recoil effects appear to first order in $1/m_Q$ and renormalize ΔM_Q. The spin effects are of second order in $1/m_Q$, and result in

$$\Delta\mathcal{L}_{QQ}^{spin} = \frac{1}{4}\left(\frac{[\Delta M_Q^{spin}]^2}{2nN_c}\right)\overline{\mathbf{Q}}\frac{1+\gamma^0}{2}\lambda^a\sigma_1^{\mu\nu}\mathbf{Q}\,\overline{\mathbf{Q}}\frac{1+\gamma^0}{2}\lambda^a\sigma_2^{\mu\nu}\mathbf{Q}. \quad (9.47)$$

For heavy baryons of the type qqQ we have

$$\mathcal{L}_{qqQ} = -\left(\frac{\Delta M_Q \Delta M_q^2}{2n^2 N_c^2}\right)\left(\overline{\mathbf{Q}}\frac{1+\gamma^0}{2}\mathbf{Q}\,(\det\overline{\mathbf{q}}_L\mathbf{q}_R + \det\overline{\mathbf{q}}_R\mathbf{q}_L)\right.$$
$$\left.+\frac{1}{4}\,\overline{\mathbf{Q}}\frac{1+\gamma^0}{2}\lambda^a\mathbf{Q}\,(\det\overline{\mathbf{q}}_L\lambda^a\mathbf{q}_R + \det\overline{\mathbf{q}}_R\lambda^a\mathbf{q}_L)\right) \quad (9.48)$$

and to second order in $1/m_Q$

$$\mathcal{L}_{qqQ}^1 = -\left(\frac{\Delta M_Q^{spin}\,\Delta M_q^2}{n^2 N_c^2}\right)\frac{1}{4}\,\overline{\mathbf{Q}}\frac{1+\gamma^0}{2}\lambda^a\sigma_{\mu\nu}\mathbf{Q}$$
$$\times\left(\det\overline{\mathbf{q}}_L\lambda^a\sigma_{\mu\nu}\mathbf{q}_R + \det\overline{\mathbf{q}}_R\lambda^a\sigma_{\mu\nu}\mathbf{q}_L\right). \quad (9.49)$$

The phenomenological implications of these interactions on heavy-light spectra will be discussed next.

9.2.6. *Heavy hadron spectra*

The effective interactions between the soft components of the light and heavy quarks can be analyzed either directly as in a constituent quark model, or indirectly using bosonization techniques in the long-wavelength limit. In this section we will give a brief description of the former, while in the next section we will formulate and use the latter.

The contributions of the various instanton interactions derived above to the heavy hadron spectra will be estimated using a variational approach.

For mesons, the generic Hamiltonian is

$$H = \frac{\vec{p}_q^2}{2m_q} + \frac{\vec{p}_Q^2}{2m_Q} + \frac{1}{2}M\omega^2|\vec{r}_q - \vec{r}_Q|^2 + H^{(2)} \qquad (9.50)$$

where M is the reduced mass of the heavy-light system, with $m_q = \Delta M_q \sim$ 420MeV and $m_Q = m_c + \Delta M_Q \sim (1350+86)$ MeV. The harmonic potential provides a simple mechanism of confinement. The instanton-induced interaction $H^{(2)}$ derived from eqs. (9.43-9.48) will be treated as a perturbation. The trial wave function is

$$\psi(\chi) = \left(\frac{2\alpha}{\pi}\right)^{3/4} e^{-\alpha\chi^2} \qquad (9.51)$$

where $\vec{\chi} = \frac{1}{\sqrt{2}}(\vec{r}_q - \vec{r}_Q)$. Minimizing the expectation value of (9.50) in (9.51) with respect to α yields $\alpha = \frac{1}{2}M\omega$ with the confining energy $\mathcal{E}_\alpha = \frac{3}{2}\omega$ as expected. Since the size $r = \sqrt{\frac{1}{2\alpha}}$ of the ground state is a function of the reduced mass M, we fix our parameters by the size of the heavy-light system $r_{qQ} = 0.6\,\mathrm{fm}$. Then the size of the heavy-heavy system is $r_{QQ} \simeq 0.4\,\mathrm{fm}$, and the confining energy is about $\mathcal{E}_\alpha \simeq 250\mathrm{MeV}$ for both of them.

- *Mesons*

For heavy mesons the relevant two-body instanton-induced interactions are for $Q\bar{q}$

$$H_{qQ}^{(2)} = \left(\frac{\Delta M_Q \Delta M_q}{2nN_c}\right)\left(1 + \frac{1}{4}\lambda_q^a\lambda_Q^a\right)\delta^3(\vec{r}_q - \vec{r}_Q), \qquad (9.52)$$

and for $Q\bar{Q}$,

$$H_{QQ}^{(2)} = \left(\frac{\Delta M_Q \Delta M_Q}{2nN_c}\right)\left(1 + \frac{1}{4}\lambda_{Q_1}^a\lambda_{Q_2}^a\right)\delta^3(\vec{r}_{Q_1} - \vec{r}_{Q_2}). \qquad (9.53)$$

The induced spin-interaction in the $Q\bar{q}$ configuration is given by

$$H_{qQ}^{2,s} = -\left(\frac{\Delta M_Q^{spin}\Delta M_q}{2nN_c}\right)\frac{1}{4}\vec{\sigma}_q \cdot \vec{\sigma}_Q \, \lambda_q^a \cdot \lambda_Q^a \, \delta^3(\vec{r}_q - \vec{r}_Q). \qquad (9.54)$$

We recall that ΔM_Q^{spin} is suppressed by one power of $1/m_Q$. Using (9.52) and the trial wave function (9.51), we have

$$\langle H_{qQ}^{(2)}\rangle \sim -C_F\left(\frac{\Delta M_Q \Delta M_q}{2nN_c}\right)|\psi(\vec{0})|^2$$

$$\sim -\frac{N_c}{2} \left(\frac{\Delta M_Q \Delta M_q}{2n N_c} \right) \left(\frac{1}{\sqrt{\pi} r_{qQ}} \right)^3 \tag{9.55}$$

and similarly for (9.53). Thus $\langle H_{qQ}^{(2)} \rangle \sim -183$ MeV and $\langle H_{QQ}^{(2)} \rangle \sim -103$ MeV. These numbers should be compared respectively to -140 MeV and -70 MeV, as quoted in ref.[18] using qualitative arguments. The spin corrections are of order $\langle H_{qQ}^{2,s} \rangle \simeq -42$ MeV, a result that is consistent with the constituent quark model estimate of ~ -27 MeV [25]. The spin-induced interaction in heavy systems such as charmonium is tiny, $\langle H_{QQ}^{2,s} \rangle \sim -0.7$ MeV.

To evaluate the spectrum in heavy-light and heavy-heavy systems, we use the mass formulae

$$M_{qQ} = \langle H^{(0)} + H^{(1)} + H^{(2)} \rangle. \tag{9.56}$$

Here $H^{(0)}$ is the sum of the binding energy $\mathcal{E}_\alpha$ and the current masses $m_s = 150$MeV, $m_c = 1350$MeV, $m_b = 4700$MeV (we take all light flavors to be massless). $H^{(1)}$ stands for the induced instanton mass for the light $\Delta M_q \sim 420$MeV and heavy $\Delta M_Q \sim 86$MeV quarks. $H^{(2)}$ provides the extra Coulomb binding energy discussed above and the $1/m_Q$ hyperfine splitting within the multiplets. The results are summarized in Table 9.1. They are in overall agreement with experiment and the constituent quark model [25].

Table 9.1. Mesonic spectrum

	$I(J)^P$	Exp. value	Instantons [18]	QM [25]
D^0	$\frac{1}{2}(0^-)$	1869	1881	$1800 - 1860$
D^*	$\frac{1}{2}(1^-)$	2010	1965	$1930 - 1990$
$D_s^{\pm}$	$0(0^-)$	1969	2031	1975
$D_s^{\pm}$	$0(1^-)$	2110	2115	2061
B^0	$\frac{1}{2}(0^-)$	5278	5231	

• *Baryons*

Heavy baryons may be analyzed in a similar fashion using the induced interactions discussed at the end of Section 3. First, we note that the analog of the one-gluon exchange in our case is the induced two-body interaction (9.43), that is

$$H_{qq}^{(2)} = \left(\frac{\Delta M_q \Delta M_q}{n N_c} \right) \left(1 + \frac{3}{32} \lambda_1^a \cdot \lambda_2^a - \frac{9}{32} \sigma_1 \cdot \sigma_2 \, \lambda_1^a \cdot \lambda_2^a \right) \delta^3(\vec{r}_1 - \vec{r}_2) \tag{9.57}$$

and similarly for baryons with two heavy quarks. This interaction is expected to be attractive and hence binding. We note that it scales like N_c in baryonic configurations. Instantons in heavy baryons induce also a three-body interaction (9.48),

$$\begin{aligned}
H_{qqQ}^{(3)} = & -\left(\frac{\Delta M_Q \Delta M_q^2}{2n^2 N_c^2} \right) \sum_{i<j} \delta^3(\vec{r}_i - \vec{r}_j) \, \delta^3(\vec{r}_j - \vec{r}_Q) \\
& \times \left\{ 1_Q \cdot \left[1 + \frac{3}{32} \lambda_i^a \cdot \lambda_j^a - \frac{9}{32} \sigma_i \cdot \sigma_j \lambda_i^a \cdot \lambda_j^a \right] \right. \\
& + \frac{1}{4} \lambda_Q^a \cdot \left[\lambda_j^a + \frac{3}{32} \lambda_i^b \cdot (\lambda^b \lambda^a)_j - \frac{9}{32} \sigma_i \cdot \sigma_j \, \lambda_i^b \cdot (\lambda^b \lambda^a)_j \right. \\
& \left. \left. + (i \leftrightarrow j) \right] \right\}.
\end{aligned} \tag{9.58}$$

This interaction scales as N_c^0. Although subleading in our previous book-keeping arguments, we will keep it in our $N_c = 3$ considerations.

For baryons, the trial wave functions will be chosen in the form

$$\psi(\chi, \eta) = \left(\frac{2\alpha}{\pi} \right)^{3/2} e^{-\alpha(\chi^2 + \eta^2)} \tag{9.59}$$

where $\vec{\chi} = \frac{1}{\sqrt{2}}(\vec{r}_1 - \vec{r}_2)$ and $\vec{\eta} = \sqrt{\frac{1}{6}}(\vec{r}_1 + \vec{r}_2 - 2\vec{r}_Q)$ are the standard Jacobi coordinates. Here we choose $r_{qqQ} = 1 \, \text{fm}$ for the size of the heavy-light baryons. The size of the heavier configurations will be set to $r_{qQQ} \simeq 0.86$ fm for Qqq and $r_{QQQ} \simeq 0.7$ fm for QQQ. The confining potential energy is about $\mathcal{E}_\alpha \sim 500$ MeV for all of them.

The addition of one more quark in the baryonic configurations brings in the three-body contribution to the energy an additional overall factor of $\mathcal{R}_q$ for a light quark, and $\mathcal{R}_Q$ for a heavy quark, in comparison to (9.57). Specifically,

$$\mathcal{R}_{q,Q} = -2 \left(\frac{\Delta M_{q,Q}}{2n N_c} \right) \left(\frac{1}{\sqrt{\pi} r} \right)^3 .$$

Table 9.2. Baryonic spectrum

	I	J	Exp. value	Instantons [18]	QM [25]	Ref. [26]
Λ_c	0	$\frac{1}{2}$	(2285)	2376	2200	2170
Σ_c	1	$\frac{1}{2}$	(2453)	2502	2360	2421
Ξ_c	$\frac{1}{2}$	$\frac{1}{2}$	2468	2652	2420	2421
Ω_c	0	$\frac{1}{2}$	(2704)	2802	2680	2645
Ξ_{cc}	$\frac{1}{2}$	$\frac{1}{2}$	?	3558	3550	3510
Ω_{cc}	0	$\frac{1}{2}$	?	3708	3730	3698
Ω_{ccc}	0	$\frac{3}{2}$	?	4808	4810	4784

The three-body contribution is repulsive, whereas the two-body contribution is attractive. Thus, there is a subtle interplay between two- and three-body interactions in the determination of the overall energy of the heavy-light baryonic systems. For a baryon size $r = 1$ fm, $\mathcal{R}_q \sim -0.75$ and $\mathcal{R}_Q \sim -0.06$. We note that the results are very sensitive to the size of the hadron. Indeed, for $r = 0.9$ fm, $\mathcal{R}_q \sim -1$, and for $r = 0.4$ fm, $\mathcal{R}_Q \sim -1$. In other words, two- and three-body contributions become comparable in strength. The size r is fixed by the choice of the potential (9.57) and is independent of the character of the induced interaction in our discussion. Here, we chose to work with $r_{qQ} = 0.6$ fm for the heavy-light mesons, and $r_{qqQ} \sim 1$ fm for the heavy baryons. A smaller size, say $r_{qqQ} \sim 0.5$ fm, would not fit the spectrum. Of course, other choices may also be possible, with other choices of the potential in (9.50).

We present the spectra for heavy baryons in Table 9.2. The results are in a reasonable agreement with experiment and other models. Major uncertainty comes from the $1/N_c$ corrections (non-planar graphs) and the approximation for the ground-state wave functions. As we have seen, the first-order corrections came up to 25% of the leading order. Since we are making two expansions, Λ_{QCD}/m_Q and $1/N_c$, one would also expect substantial contributions coming from the first order corrections in $1/N_c$ for physical $N_c = 3$.

• *Exotics*

The method of constructing two- and three-body interactions using instanton induced effects can be extended to multi-quark configurations.

For example, for $QQqq$ configurations, the four-body interaction is of the form

$$
\begin{aligned}
\mathcal{L}_{qqQQ} =&-nN_c\left(\frac{\Delta M_q}{nN_c}\right)^2\left(\frac{\Delta M_Q}{2nN_c}\right)^2 \\
&\times\Bigg(\overline{\mathbf{Q}}\frac{1+\gamma^0}{2}\mathbf{Q}\ \overline{\mathbf{Q}}\frac{1+\gamma^0}{2}\mathbf{Q}\ (\det\overline{\mathbf{q}}_L\mathbf{q}_R + \det\overline{\mathbf{q}}_R\mathbf{q}_L) \\
&+\frac{1}{4}\ \overline{\mathbf{Q}}\frac{1+\gamma^0}{2}\lambda^a\mathbf{Q}\ \overline{\mathbf{Q}}\frac{1+\gamma^0}{2}\mathbf{Q}\ (\det\overline{\mathbf{q}}_L\lambda^a\mathbf{q}_R + \det\overline{\mathbf{q}}_R\lambda^a\mathbf{q}_L) \\
&+\frac{1}{4}\ \overline{\mathbf{Q}}\frac{1+\gamma^0}{2}\lambda^a\mathbf{Q}\ \overline{\mathbf{Q}}\frac{1+\gamma^0}{2}\lambda^a\mathbf{Q}\ (\det\overline{\mathbf{q}}_L\mathbf{q}_R + \det\overline{\mathbf{q}}_R\mathbf{q}_L)\Bigg).
\end{aligned}
$$
(9.60)

The overall sign is consistent with the naive expectation, that the n-body interaction follows from the (n+1)-body interaction by contracting a light quark line, resulting in an overall minus sign (quark condensate).

We recall that for each extra light quark, the penalty factor in the energy is $\mathcal{R}_q$. Starting with $r = 1$ fm for three quark states (whether heavy or light), we find that the radius r shrinks to 0.93 fm for one additional light quark (four-quark state), to 0.88 fm for an extra one (five-quark state), and to 0.84 fm for still another one (six-quark state). For $r = 1$ fm, $\mathcal{R}_q \sim -0.72$, while for $r = 0.9$ fm, $\mathcal{R}_q = -1$. It follows that the three-body interaction (repulsive) will tend to overcome the binding energy provided by the two-body interaction (assumed attractive) in the multiquark configurations of the type $(\bar{q}\bar{q}qq)$, $(\bar{Q}\bar{q}qq)$, $(\bar{q}q\,qqq)$, $(\bar{Q}q\,qqq)$ and $(qqq\,qqq)$. As a result, the H-dibaryon viewed as a six-light-quark state $(qqq\,qqq)$, will be unbound by the three body-forces induced by instantons, in agreement with the conclusion reached by Oka and Takeuchi [27] *

Adding a heavy quark brings about a penalty factor $\mathcal{R}_Q$ in the energy. This factor is 0.06 for $r = 1$ fm, and 1 for $r = 0.4$ fm. Using the harmonic potential with two light quarks and four heavy quarks yields $r = 0.8$ fm and $\mathcal{R}_Q = 0.1$. Thus, for heavy six-quark states, the three-body interaction is 10% of the two-body interaction, hence small. In this respect, if we were to think about the H-dibaryon as a six-heavy-light-quark state $(Qqq\,Qqq)$, it will *not* be unbound by the three-body interaction. Similarly, the four-body-interaction will be expected to be about 1% of the two-body, about

---

*In ref. [27] the radius $r = 0.5$ fm was used in comparison with $r = 0.84$ fm in the present case.

the same order of magnitude as the hyperfine splitting discussed above. We therefore conclude that the multi-body effects are only important for multi-quark states near threshold.

Specific calculations for exotics containing two and three heavy quarks of the type $(\bar{Q}\bar{Q}\,qq)$ and $(\bar{Q}Q\,Qqq)$ can also be carried out. With the above choice of parameters, the configurations are found to be stable against strong decay through $\bar{Q}\bar{Q}\,qq \to \bar{Q}q + \bar{Q}q$ and $\bar{Q}Q\,Qqq \to \bar{Q}Q + Qqq$ or $\bar{Q}Q\,Qqq \to \bar{Q}q + QQq$. In both cases the binding energy is of the order of 10 MeV, in agreement with other models [26,28].

9.2.7. *Heavy meson effective action*

In the above discussion of the hadronic spectra, we have made explicit use of the constituent quarks through a potential model. In the spirit of Chapter 2, we can also bosonize [29] directly the effective interactions (9.43-9.49), to generate an effective action for the long-wavelength description of the soft heavy-light hadronic physics. This will be discussed in this section, following arguments by Nowak and Zahed [30], and also Bardeen and Hill[31].

Generically, a Fierz rearrangement of the four fermion interaction (9.43), say, gives

$$
\begin{aligned}
\mathcal{L} \sim &+(\bar{\mathbf{Q}}\mathbf{q})(\bar{\mathbf{q}}\mathbf{Q}) + \bar{\mathbf{Q}}\gamma_5\mathbf{q}(\bar{\mathbf{q}}\gamma_5\mathbf{Q}) \\
&+(\bar{\mathbf{Q}}\gamma_\mu \frac{1-\not{p}}{2}\mathbf{q})(\bar{\mathbf{q}}\gamma_\mu \frac{1-\not{p}}{2}\mathbf{Q}) \\
&+(\bar{\mathbf{Q}}\gamma_\mu \frac{1-\not{p}}{2}\gamma_5\mathbf{q})(\bar{\mathbf{q}}\gamma_\mu \frac{1-\not{p}}{2}\gamma_5\mathbf{Q})
\end{aligned}
\tag{9.61}
$$

where the bold-faced $\bar{\mathbf{q}}\mathbf{Q}$ combinations (indicated in parentheses) respectively carry $0^+, 0^-, 1^-$ and 1^+ spin-parity assignments (with tensor mesons omitted). These combinations can be bosonized in the standard way using the multiplets $H = (0^-, 1^-)$ and $G = (0^+, 1^+)$

$$
\begin{aligned}
H &= \frac{1+\not{p}}{2}(\gamma^\mu P_\mu^* + i\gamma_5 P), \\
G &= \frac{1+\not{p}}{2}(\gamma^\mu \gamma_5 Q_\mu^* + Q).
\end{aligned}
\tag{9.62}
$$

Empirically, the multiplet with $I(J^P) = \frac{1}{2}(0^-, 1^-)$ corresponds to $(K, K^*) = (495, 892)$ MeV for the "heavy" quark s, $(D, D^*) = (1869, 2010)$ MeV for the c quark and $(B, B^*) = (5278, 5324)$ MeV for the b quark. The unobserved multiplet $(\tilde{D}, \tilde{D}^*)$ has $I(J^P)$ assignment $(\frac{1}{2}(0^+, 1^+))$ and corresponds to the parity-even partners of (D, D^*). The heavy-vector fields are transverse, $v^\mu P^*_\mu = v^\mu Q^*_\mu = 0$. The "bosonized" version of (9.61) is

$$S = \sum_v \int d^4x \overline{\chi} [\mathbf{1}_2 (i\nabla_L - \Delta M_q)\gamma_5^- + \mathbf{1}_2(i\nabla_R - \Delta M_q)\gamma_5^+$$

$$+ \mathbf{1}_3(i\not{v}v\cdot\partial - \Delta M_Q) + H + \overline{H} + G + \overline{G}]\chi \qquad (9.63)$$

where the covariant L, R derivatives are : $\nabla_L = \partial - iL$ and $\nabla_R = \partial - iR$. $\overline{H}$ and $\overline{G}$ are related, respectively, to H and G through

$$\overline{H} = \gamma^0 H^+ \gamma^0, \qquad \overline{G} = \gamma^0 G^+ \gamma^0. \qquad (9.64)$$

Here L_μ and R_μ are the light-vector fields, valued in $U_L(2)$ and $U_R(2)$ respectively, $\mathbf{1}_2 = \text{diag}(1,1,0)$, $\mathbf{1}_3 = \text{diag}(0,0,1)$ are the projectors onto the light and heavy sectors, respectively, and we are using the short-hand notation $\gamma_5^\pm \equiv \frac{1}{2}(1\pm\gamma_5)$. The quark field χ in (9.63) stands for $\psi = (q; Q_v)$. The $\hat{P}$'s are off-diagonal in flavor space, so that $H = H_a T^a$ with $a = 1, 2$ and $T^1 = (\lambda_4 - i\lambda_5)/2$, $T^2 = (\lambda_6 - i\lambda_7)/2$. For simplicity we consider here the third s quark as "heavy". The generalization for N_q light quarks and N_Q heavy quarks, and also to other variants of (9.61), are straightforward.

The light-light and heavy-light quark dynamics follows from the dressed action (9.63) through a derivative expansion. The momenta are bounded from above by the cutoff Λ, corresponding to the scale of the chiral symmetry breakdown. Our expansion of (9.63) will be understood in the sense of $m_Q/\Lambda \to \infty$. To second order, the heavy-light induced action reads

$$S_H = N_c \text{Tr} \left(\mathbf{1}_2 \Delta_l H \mathbf{1}_3 \Delta_h \overline{H} \right)$$

$$- N_c \text{Tr} \left(\mathbf{1}_2 \Delta_l (\not{V}\Delta_l + \not{A}\Delta_l\gamma_5) H \mathbf{1}_3 \Delta_h \overline{H} \right) + \cdots \qquad (9.65)$$

where the functional trace includes tracing over the space, flavor and spin indices. We have denoted $\Delta_l = (i\not{\partial} - \Delta M_q)^{-1}$ and $\Delta_h = (i\not{v}v\cdot\partial - \Delta M_Q)^{-1}$, and

$$\gamma_5^+ \not{R} + \gamma_5^- \not{L} = \not{V} + \gamma_5\not{A}. \qquad (9.66)$$

The ellipsis in (9.65) stands for higher insertions of vectors and axials. A similar action is expected for the heavy chiral partners G. To the order considered, there are also cross terms generated by the axial current.

After carrying out the trace over space in (9.65) and renormalizing the heavy-quark fields, $H \to H/\sqrt{Z_H}$ and $G \to G/\sqrt{Z_G}$, we obtain to leading order in the gradient expansion for the H's ($s_l^{P_l} = \frac{1}{2}^-$),

$$
\begin{aligned}
\mathcal{L}_v^H =& -\frac{i}{2}\mathrm{Tr}(\bar{H}v^\mu\partial_\mu H - v^\mu\partial_\mu\bar{H}\,H) \\
&+\mathrm{Tr}V_\mu\bar{H}Hv^\mu - g_H\mathrm{Tr}A_\mu\gamma^\mu\gamma_5\bar{H}H + m_H\mathrm{Tr}\bar{H}H
\end{aligned}
\tag{9.67}
$$

and for the G's ($s_l^{P_l} = \frac{1}{2}^+$) †

$$
\begin{aligned}
\mathcal{L}_v^G =& +\frac{i}{2}\mathrm{Tr}(\bar{G}v^\mu\partial_\mu G - v^\mu\partial_\mu\bar{G}\,G) \\
&-\mathrm{Tr}V_\mu\bar{G}Gv^\mu - g_G\mathrm{Tr}A_\mu\gamma^\mu\gamma_5\bar{G}G + m_G\mathrm{Tr}\bar{G}G.
\end{aligned}
\tag{9.68}
$$

The parameters in the above formulae are given by ($P_l = \pm$ is the parity of the light part of H, G)

$$
\begin{aligned}
Z_{P_l} =& N_c\int_0^\Lambda\frac{d^4Q}{(2\pi)^4}\left[-\frac{1}{v\cdot Q + \Delta M_Q + i\epsilon}\cdot\frac{1}{Q^2 - \Delta M_q^2 + i\epsilon}\right. \\
&\left.+ P_l\frac{2\Delta M_q}{Q^2 - \Delta M_q^2 + i\epsilon}\right] \\
g_{P_l} =& \frac{N_c}{Z_{P_l}}\int_0^\Lambda\frac{d^4Q}{(2\pi)^4}\frac{1}{v\cdot Q + \Delta M_Q + i\epsilon}\cdot\frac{\Delta M_q^2 - Q^2/3}{(Q^2 - \Delta M_q^2 + i\epsilon)^2} \\
m_{P_l} =& \frac{N_c}{Z_{P_l}}\int_0^\Lambda\frac{d^4Q}{(2\pi)^4}\left[\frac{1}{Q^2 - \Delta M_q^2 + i\epsilon}\right. \\
&\left.+ P_l\frac{1}{v\cdot Q + \Delta M_Q + i\epsilon}\cdot\frac{\Delta M_q}{Q^2 - \Delta M_q^2 + i\epsilon}\right].
\end{aligned}
\tag{9.69}
$$

Note that m_H and m_G are of order m_Q^0. In the double limit $m_Q/\Lambda \to \infty, \Sigma/\Lambda \to 0$ and for $\Delta M_Q = 0$, both coupling constants g_G, g_H take the value $1/3$.[32] Note, however, that the effect of the chiral symmetry breakdown ($\Sigma \neq 0$) is dramatically different for the coupling constants. Whereas g_H stays equal to $1/3$ in the broad range of the x, g_G grows rapidly with x, reaching the value around 1 for the physically reasonable value of $x \sim 1/3$.‡

---

†We label H and G by the total angular momentum s_l and parity P_l of the *light* degrees of freedom.

‡The recent CLEO collaboration data seem to indicate an empirical value $g_H \approx .58$ with a large error bar. See Cho and Georgi[33] for an analysis.

This prediction has to be contrasted with the estimates of the nonrelativistic quark model, which gives $g_H = g_A$, $g_G = g_A/3$ [34], where $g_A = 0.75$ [35].

The interaction between the H's and the G's is given by

$$\mathcal{L}_v^{HG} = +\sqrt{\frac{Z_H}{Z_G}}\,\mathrm{Tr}(\gamma_5 \overline{G} H v^\mu A_\mu) - \sqrt{\frac{Z_G}{Z_H}}\,\mathrm{Tr}(\gamma_5 \overline{H} G \gamma^\mu A_\mu). \qquad (9.70)$$

Notice that $Z_G/Z_H = g_H/g_G$ reduces to 1 in the heavy-quark limit and $x \to 0$. The mass splittings between the H's ($D, D^*, ...$) and their chiral partners G's ($\tilde{D}, \tilde{D}^*, ...$) imply, to order $(m_Q^0 N_c^0)$, the mass relations

$$m(\tilde{D}^*) - m(D^*) = m(\tilde{D}) - m(D) = O(\Sigma) \qquad (9.71)$$

where again Σ is the constituent mass of the light quarks in the chiral limit. This is expected since the D, D^* are S-wave mesons, while the $\tilde{D}$, $\tilde{D}^*$ are P-wave mesons (not yet observed)§ In the nonrelativistic quark model, the difference is centrifugal and of order $m_Q^0{}^2$. For the broad range of the value of x, the mass difference is constant and is equal to $\sim 0.1\Lambda$. This difference is smaller than the naive estimate Σ, due to the presence in (9.69) of large logarithmic corrections $O(x \ln x)$.

The preceding arguments suggest that in the heavy-quark and chiral limit, a similar effective action should involve the chiral partners of the heavy pseudoscalars and vectors, here denoted by G. Their role in the soliton scenario might be important for the description of opposite-parity heavy-baryon states. They also allow for a qualitative estimation for the coefficients involved, and provide a rationale for the systematic expansion in both $k_\pi/(\sqrt{N_c}\Lambda)$ (derivative or $1/N_c$ expansion) and Λ/m_Q (heavy-quark expansion).

Before we move to the description of heavy baryons in the effective action context, we note that the excited states of the heavy mesons, corresponding to the empirical states $(D_1, D_2^*) = (2420, 2460)$ MeV with the assignment $\frac{1}{2}(1^+, 2^+)$, can also be described in a similar way [30].

---

§The observed $D_1(2400)$ and $D_2^*(2460)$ are components of the $s_l^{P_l} = \frac{3}{2}^+$ multiplet.

9.3. Heavy Solitonic Baryons

9.3.1. *Heavy solitons*

The scaling with N_c of the overall heavy-light effective action (9.65) before renormalization suggests that the heavy-light system should be entrusted with the same weight as the light-light counterpart (which is well known and hence omitted in our discussion) in the large N_c limit, implying that a soliton description for heavy baryons is perhaps justified [36]. The large N_c limit appears to be compatible with the heavy-quark limit in the meson sector. This point is *a priori* not obvious since terms of the form $m_Q/N_c\Lambda$ and others cannot be ruled out on general grounds. Thus our approximate bosonization version of the effective interactions inherited from the random instanton model yields a long-wavelength description that appears to be consistent with the heavy-quark symmetry encoded in QCD[37]. We note that in the meson sector the $HH\pi$ interaction is of order $1/\sqrt{N_c}$ as expected. In the soliton sector, however, the effect of this interaction is in *principle* of order N_c, since we expect $H \sim \sqrt{N_c}$ and $\pi \sim \sqrt{N_c}$ (*i.e.*, semi-classical π field). However, in the heavy-quark limit, the heavy meson field is *in general* localized over a range of the order of $1/(N_c m_Q)$ affecting the energy to order N_c^0. Hence, one would expect heavy baryons composed of pions and P, P^* (or Q, Q^*) to emerge to order N_c^0 or lower. Let us see how this scenario works in detail.

Our starting point is the effective action (9.67) for heavy-light mesons in the infinite quark mass limit. If we denote by

$$H = \frac{1+\gamma^0}{2}(-\gamma_i P_i^* + i\gamma_5 P) \qquad \text{and} \qquad \overline{H} = \gamma^0 H^\dagger \gamma^0 \qquad (9.72)$$

the $(0^-, 1^-)$ degenerate doublet in the rest frame of the heavy quark, then

$$\mathcal{L}_H = -i\mathrm{Tr}(\partial_t H\overline{H}) + \mathrm{Tr}HV^0\overline{H} - g_H\mathrm{Tr}HA^i\sigma^i\overline{H} + m_H\mathrm{Tr}H\overline{H} \quad (9.73)$$

where the vector and axial currents are entirely pionic and read, defined here slightly different from previously, as

$$V_\mu = +\frac{i}{2}\left(\xi\partial_\mu\xi^\dagger + \xi^\dagger\partial_\mu\xi\right),$$
$$A_\mu = +\frac{i}{2}\left(\xi\partial_\mu\xi^\dagger - \xi^\dagger\partial_\mu\xi\right). \qquad (9.74)$$

The pion field $\xi = \sqrt{U}$ is described by the usual Skyrme type action. With our conventions, $\vec{\sigma} = \gamma_5 \gamma^0 \vec{\gamma}$.

The effective action following from (9.73) is invariant under local $SU_V(2)$ symmetry (h), in which V transforms as a gauge field, A transforms covariantly and $H \to Hh^\dagger$ and $\overline{H} \to h\overline{H}$. It is also invariant under heavy-quark symmetry $SU_Q(2)$ (S), $H \to SH$ and $\overline{H} \to \overline{H}S^\dagger$. Specifically,

$$\delta \vec{P}^* = \vec{\alpha} P + (\vec{\alpha} \times \vec{P}^*), \qquad \delta P = -\vec{\alpha} \cdot \vec{P}^* \qquad (9.75)$$

in infinitesimal form. We note that the $\vec{P}^* = \vec{0}$ part of (9.73) relates to the effective action used by Callan and Klebanov [13] in the infinite kaon mass limit following the nonrelativistic field substitution $P \to e^{-im_Q t} P / \sqrt{m_Q}$.

In the soliton sector, the pion field is in the usual hedgehog configuration. In this case it is useful to organize the H field in K-partial waves. Generically

$$H(x,t) = \sum_{KM} a_{KM}(t) H_{KM}(x) \qquad (9.76)$$

where the a's annihilate H particles with good K-spin where $\mathbf{K} = \mathbf{I} + \mathbf{J} = \mathbf{K}_L + \mathbf{S}_Q$ with $\mathbf{I}$ and $\mathbf{J}$ the total isospin and angular momentum of the H-soliton system, and $\mathbf{S}_Q$ the spin of the heavy quark.[¶] In the original approach of Callan and Klebanov [13], the $K^\pi = \frac{1}{2}^+$ state in the kaon channel was found to bind to the soliton. The binding was strong in the presence of the Wess-Zumino-Witten term and weak in its absence. In the heavy-quark limit, the Wess-Zumino-Witten term decouples from the heavy mesons. Thus the mechanism of binding in the heavy-quark limit depends solely on the background pionic potential and is therefore model-dependent. This notwithstanding, we define the classical fields to be

$$\xi_0(\vec{x}) = e^{i\tau \cdot \hat{x} F(r)/2}, \qquad H(\vec{x}) = G(\vec{x}) \tau \cdot \hat{x}. \qquad (9.77)$$

The pion field is in the usual hedgehog configuration $K^\pi = 0^+$, while the H multiplet has $K^\pi = \frac{1}{2}^+$. The generalization to higher K-partial waves is straightforward. Since we can construct three different grand spin states for $K = \frac{1}{2}$ (for the broader discussion and an explicit construction, see Min *et al.*, [38]), to order $(m_Q^0 N_c^0)$, the Hamiltonian following from (9.73) simplifies

---

[¶]Note that one could equally use $\mathbf{K}_L = \mathbf{I} + \mathbf{J}_L = \mathbf{K} - \mathbf{S}_Q$ to organize the H field, since it is also a symmetry of the classical action.

to

$$H_0^{\frac{1}{2}} = M_S + m_Q - m_H \text{Tr}(G\overline{G}) + g_H F'(0) \begin{pmatrix} 0 & \frac{\sqrt{3}}{2} & 0 \\ \frac{\sqrt{3}}{2} & -1 & 0 \\ 0 & 0 & \frac{1}{2} \end{pmatrix} \quad (9.78)$$

where we have assumed that the heavy mesons are infinitely peaked in space (zero kinetic energy) with a range proportional to the heavy-quark Compton wavelength, so that $H\xi_0 \sim H(i\tau \cdot \hat{x})$ with $F(0) = \pi$. After diagonalization, we see that two states with eigenvalue $-g_H F'(0)/2$ are not bound and the third is bound. Restricting to the latter in the $K^\pi = \frac{1}{2}^+$ band simplifies (9.78) to

$$H_0^{\frac{1}{2}} = M_S + m_Q + m_H + \frac{3}{2} g_H F'(0). \quad (9.79)$$

Clearly the character of the binding depends on the partial wave considered and also on the terms of order $m_Q^0 N_c^0$ kept in the effective action. There is binding in the $K = \frac{1}{2}^+$ channel provided that $g_H > 0$. The binding energy is $\Delta E = 3g_H F'(0)/2$.[||]

To quantize (9.77), we introduce the isospin collective coordinate $S(t)$ and rotate the bound state as a *whole*

$$\xi \to S(t)\xi_0 S(t)^\dagger, \qquad H \to HS(t)^\dagger. \quad (9.80)$$

In this way H refers to the heavy mesons in the (isospin) co-moving frame. To order $m_Q^0 N_c^{-1}$, the action is given by

$$S_A = -\Omega \text{Tr}(S^\dagger \dot{S})^2 + i \text{Tr}(GS^\dagger \dot{S}G). \quad (9.81)$$

Aside from the rotator action due to the soliton, the cranking induces (as we know from Chapter 8) a Berry phase with a normalization that is fixed by the underlying $SU_V(2)$ isospin symmetry. For the $H-$ multiplet, the Berry phase is found to vanish *identically* in the $K = \frac{1}{2}^+$ shell

$$\vec{\Phi}(\infty) = -\text{Tr}(G\vec{I}\overline{G}) = \frac{1}{2}\text{Tr}\left(P\vec{I}P^\dagger + P_j^* \vec{I} P_j^{*\dagger}\right) = \frac{1}{32}\left(\vec{\tau}_{\frac{1}{2}} - \vec{\tau}_{\frac{1}{2}}\right) = \vec{0} \quad (9.82)$$

where we have used $S^\dagger \dot{S} = i\tau \cdot \omega/2$ and τ_K refers to the embedded form of τ in K-space. The contribution of P is *exactly* balanced by the contribution

---

[||]With our conventions, $F'(0) < 0$ and the baryon number carried by the soliton is $+1$.

of P^* [39,40]. The cancellation can be understood if we note that in P^* the isospin is anti-coupled to the spin to form a $K = \frac{1}{2}^+$ state. Since (9.82) simply counts the isospin in H, the cancellation follows. The resulting spectrum is simple as we now show.

The isospin operator follows from the transformation

$$S \to e^{i\beta \cdot I} S, \qquad H \to H, \qquad (9.83)$$

as H is blind to isospin in the (isospin) co-moving frame. The isospin generator follows from the conventional Noether construction. It is

$$\mathbf{I}^a = \frac{1}{2}\mathrm{Tr}(\tau^a S \tau^b S^\dagger)\left(\Omega\omega^b + \mathbf{\Phi}^b(\infty)\right)$$
$$= -\frac{1}{2}\mathrm{Tr}(\tau^a S \tau^b S^\dagger)\mathbf{J}_R^b = -\mathbf{R}^{ab}(S)\mathbf{J}_R^b \qquad (9.84)$$

with $\mathbf{J}_R$ the angular momentum of the rotator. Note that (9.84) acts only on the wave function of the rotator. The angular momentum operator $\mathbf{J}$ can be obtained through rotations in space. Since H carries an internal spin through P^*, care is required. Recalling that the grand spin is $\mathbf{K} = \mathbf{I} + \mathbf{J}$, we find (since ξ_0 carries zero K-spin)

$$Se^{+i\beta \cdot J}\xi_0 e^{-i\beta \cdot J}S^\dagger = Se^{-i\beta \cdot I}\xi_0 e^{+i\beta \cdot I}S^\dagger \qquad (9.85)$$

and hence

$$S \to Se^{-i\beta \cdot I} \qquad (9.86)$$

and also

$$e^{+i\beta \cdot J}He^{-i\beta \cdot J}S^\dagger = e^{+i\beta \cdot K}e^{-i\beta \cdot I}He^{-i\beta \cdot K}\left(Se^{-i\beta \cdot I}\right)^\dagger. \qquad (9.87)$$

The Noether current following from this transformation is

$$\vec{\mathbf{J}} = -\Omega\vec{\omega} + \mathrm{Tr}([G, \vec{\mathbf{I}}]\overline{G}) - \mathrm{Tr}([G, \vec{\mathbf{K}}]\overline{G}) = \vec{\mathbf{J}}_R - \mathrm{Tr}([G, \vec{\mathbf{K}}]\overline{G}). \quad (9.88)$$

The H particle has transmuted its "K-spin" to angular momentum. This transmutation is important for the correct quantum-number identification. For strange hyperons, only kaons are relevant, and the transmutation involves only the isospin of the kaon. In the present case the spin of P^* is also relevant. Finally, in the heavy-quark limit, the spin of the heavy quark is a good quantum number. This can be seen by noting that (9.73) is invariant under

$$H \to e^{+i\beta \cdot \mathbf{S}} H. \tag{9.89}$$

The Noether current associated with this symmetry is given by

$$\vec{\mathbf{S}}_Q = +\text{Tr}(\overline{G}\vec{\mathbf{S}}G) = -i \int d^3x \left(\vec{P}^* P^\dagger - P\vec{P}^{*\dagger} + \vec{P}^* \times \vec{P}^{*\dagger} \right) \tag{9.90}$$

mixing P and P^* as expected. For completeness, note that the light quark in H carries a spin $\mathbf{S}_q$ given by

$$\vec{\mathbf{S}}_q = -\text{Tr}(G\vec{\mathbf{S}}\overline{G}) = +i \int d^3x \left(\vec{P}^* P^\dagger - P\vec{P}^{*\dagger} - \vec{P}^* \times \vec{P}^{*\dagger} \right). \tag{9.91}$$

The combination $\vec{\mathbf{S}}_H = -\text{Tr}([G, \vec{\mathbf{S}}]\overline{G}) = \vec{\mathbf{S}}_Q + \vec{\mathbf{S}}_q$ is the total spin in the H particle. In the $K = \frac{1}{2}^+$ shell, it reduces to $\vec{\sigma}/2$. The H particle bound to the skyrmion resembles a heavy fermion with spin 1/2. A similar transmutation occurs in the Callan-Klebanov construction [13]. This fermionization of the original bosonic degrees of freedom through the hedgehog structure is what makes skyrmions so remarkable. This result carries to higher K-shells.

To order $m_Q^0 N_c^{-1}$ the Hamiltonian follows from (9.81) through canonical rules. In the $K = \frac{1}{2}^+$ shell, it is of the form

$$H_1^{\frac{1}{2}} = \frac{(\vec{\mathbf{J}}_R - \vec{\Phi}(\infty))^2}{2\Omega} = \frac{\mathbf{J}_R^2}{2\Omega}. \tag{9.92}$$

Since the G's refer to the S-wave multiplet, we have, using (9.82),

$$\vec{\mathbf{J}}_R = \vec{\mathbf{J}} + \text{Tr}([G, \vec{\mathbf{K}}]\overline{G}) = \vec{\mathbf{J}} - \vec{\mathbf{S}}_H. \tag{9.93}$$

Due to the fermionization of the bosonic degrees of freedom, $\mathbf{J}_R$ plays the role of the effective angular momentum $\mathbf{J}_L$ carried by the light degrees of freedom. We conclude that to order $m_Q^0 N_c^{-1}$,

$$H_1^{\frac{1}{2}} = \frac{\mathbf{I}^2}{2\Omega} = \frac{(\vec{\mathbf{J}} - \vec{\mathbf{S}}_H)^2}{2\Omega}. \tag{9.94}$$

Thus to order $m_Q^0 N_c^{-1}$ the Σ and Σ^* are degenerate. If we recall that $(|I, J^\pi, J_L >)$

$$|\Lambda_Q\rangle = |0, \frac{1}{2}^+, 0\rangle,$$

$$|\Sigma_Q\rangle = |1, \frac{1}{2}^+, 1\rangle,$$

$$|\Sigma_Q^*\rangle = |1, \frac{3}{2}^+, 1\rangle, \tag{9.95}$$

then we find that

$$\left(m_{\Sigma_Q} - m_{\Lambda_Q}\right) = \frac{2}{3}\left(m_\Delta - m_N\right). \tag{9.96}$$

The Hilbert space in our case is a product space of the type $|H> \otimes |S>$ with explicit S_Q symmetry. The latter is absent in the spectrum discussed by Callan and Klebanov [13]. To order $m_Q^0 N_c^0$, there is a huge degeneracy, part of which is lifted by H_1 to order $m_Q^0 N_c^{-1}$.

To summarize, we note that the low-energy effective action in the heavy-quark limit involves not only the kaons but their heavy-quark symmetry partners, P^*. In this limit, the rotator angular momentum is just the angular momentum carried by the light quarks in the bound H-soliton system, as the Berry phase is found to vanish identically[39,40]. If we were to depart from the latter, then we would expect the Berry phase to receive unequal contributions from P and P^*, so that

$$\vec{\Phi}(m_Q) = C_P(m_Q)\text{Tr}(P\vec{I}P^\dagger) + C_{P^*}(m_Q)\text{Tr}(P_j^*\vec{I}P_j^{*\dagger}). \tag{9.97}$$

The coefficients C_{P,P^*} are model-dependent for finite m_Q. For $m_Q \to \infty$, we have $C_P(\infty) = -C_{P^*}(\infty) = 1$, the Berry phase vanishes and one recovers the heavy-quark symmetry in the hyperon spectrum. This result is model-independent. In these terms, the Hamiltonian (9.94) has the suggestive form

$$H_1 = \frac{1}{2\Omega}\left((\vec{J} - \vec{S}_H) - (1 - C_P)\text{Tr}(P\vec{I}P^\dagger) - (1 - C_P^*)\text{Tr}(P_j^*\vec{I}P_j^{*\dagger})\right)^2. \tag{9.98}$$

If we were to ignore the contributions from P^* for any finite m_Q, then (9.98) would reduce to

$$H_1 = \frac{1}{2\Omega}\left(\vec{J} - (1 - C_P)\text{Tr}(P\vec{I}P^\dagger)\right)^2. \tag{9.99}$$

This is the Hamiltonian discussed originally by Callan and Klebanov [13]. This would reduce to (the incorrect) $H_1 = \mathbf{J}^2/2\Omega$ as opposed to (the correct) $H_1 = \mathbf{I}^2/2\Omega$ in the heavy-quark limit. It is amusing to note that

(9.99) implies the following model-independent mass relation

$$\frac{1}{3}(2m_{\Sigma^*} + m_{\Sigma}) - m_{\Lambda} = \frac{2}{3}(m_{\Delta} - m_N) \qquad (9.100)$$

reducing to (9.96) if we were to set by hand $m_{\Sigma^*} = m_{\Sigma}$. This suggests that the model-independent relations following from the conventional approach (ignoring heavy-quark symmetry) may still be consistent for heavy baryons provided that the proper degenerate multiplets are identified [39,40].

To conclude, the construction we have presented parallels the construction that Callan and Klebanov [13] used in their original formulation. We have assumed that the $K^{\pi} = \frac{1}{2}^+$ bound state survives the heavy-quark limit (*albeit* through a model-dependent pionic potential) and quantized it. (Higher-K^{π} bound states are not ruled out. However they yield higher spin hyperon states and are unlikely to occur in nature.) In this respect the present quantization differs from the one used by Manohar and collaborators [41]. In the present approach, the P and P^* are defined in the (isospin) co-moving frame making their quantization simpler for the bound-state problem since they do not carry good isospin (they carry good K-spin). The *dressed* P and P^* used by Manohar and collaborators are defined in the laboratory frame and their quantization is simpler for the scattering problem since they carry good isospin (as asymptotics P and P^* do). The two descriptions are related by a global isospin rotation. In the former, $\mathbf{K} = \mathbf{I} + \mathbf{J}$ is used to expand the heavy-quark field while in the latter $\mathbf{K}_L = \mathbf{K} - \mathbf{S}_Q$ is used.

9.3.2. *Hyperon spectrum*

So far, we have considered one heavy quark in the system. How can one describe the baryons with one or more heavy quarks such as Ξ and Ω? For this, we need to make an assumption on the interaction between heavy mesons. For simplicity, we will assume that mesons interact weakly so that higher-order terms in meson field can be ignored. This is essentially a quasiparticle approximation and was of course the basic premise with which we have started. This approximation is exact in the large N_c limit. Now two or more heavy mesons, each of which undergoes spin-isospin transmutation, can then be combined to give the states of total flavor quantum number and angular momentum with the $SU(2)$ soliton. It has been shown that the hyperon octet and decuplet do come out correctly in agreement with experiments and with quark models[42]. Table 7.3 in Chapter 7 shows the quantum number identification of strange and charmed hyperons. A straightforward calculation gives the fine- and hyperfine-structure mass formula for the baryons containing up to two heavy-quark flavors[36]

$$M(I, J, n_1, n_2, J_1, J_2, J_m) = M_{\text{sol}} + n_1\omega_1 + n_2\omega_2 + \frac{D}{2I} \qquad (9.101)$$

where

$$D = I(I+1) + (c_1 - c_2)[c_1 J_1(J_1+1) - c_2 J_2(J_2+1)] + c_1 c_2 J_m(J_m+1)$$
$$+ [J(J+1) - J_m(J_m+1) - I(I+1)]$$
$$\times \left[\frac{c_1 + c_2}{2} + \frac{c_1 - c_2}{2} \frac{J_1(J_1+1) - J_2(J_2+1)}{J_m(J_m+1)} \right].$$

Here the subscript i in J_i and c_i represents the heavy-meson flavor and J_m stands for all possible vector addition of the two angular momenta $\vec{J}_1$ and $\vec{J}_2$ labeling different states. The mass formula (9.101) for fine and hyperfine splittings should be compared to the mass relations discussed in the dual quark picture of section 9.2.6 (baryons) when $1/N_c$ counting arguments are used. It is reassuring to note that both descriptions are consistent with each other and the data.

9.3.3. *Connection to Berry structure*

In Chapter 8, we introduced the notion that a nonabelian Berry potential emerges when a "particle-hole" excitation corresponding to a $Q\bar{q}$ configuration is eliminated in the presence of a slow rotational field, *i.e.*, the class B case. We now show that the corresponding excitation spectrum is identical to what we obtained above, (9.101), when a pseudoscalar meson K is bound to the soliton as in the Callan-Klebanov scheme. For this it suffices to notice that the angular momentum lodged in the rotating soliton is J_l and that lodged in the Berry potential is J_K. The coefficient c is just the Berry charge $(1 - \kappa)$ in the diatomic-molecular case and g_K in the case of the chiral bag. Although eq.(9.101) is derived in the Callan-Klebanov framework, the resulting spectrum is quite generic with a general interpretation in terms of a Berry potential.

Figure 9.1 shows the hyperon spectrum for a value c fixed by the empirical relation

$$c = \frac{2(M(\Sigma^*) - M(\Sigma))}{2M(\Sigma^*) + M(\Sigma) - 3M(\Lambda)} \approx 0.62. \qquad (9.102)$$

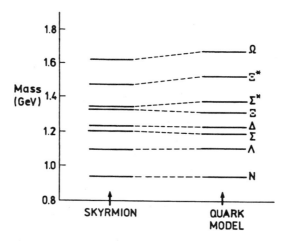

Fig. 9.1. Strange hyperons: the parameter $c_K = 0.62$ is fixed by eq.(9.102) and the $\omega_K = 223$ by fitting the Λ for the given c.

Now given an effective Lagrangian, one can calculate the meson frequency ω_i and the hyperfine coefficients c_i and use eq.(9.101) to predict the spectrum. For strange hyperons, this is known to work fairly well[13]. Given the generic form, it is tempting to extend the mass formula (9.101) to heavier-quark systems such as charmed and bottom baryons by taking the ω_i's and c_i's as parameters and by assuming one or more bound mesons as *quasiparticles* (*i.e.*, independent particles) which allows us to use the formula (9.101). Figure 9.1 shows how well strange hyperons are described when ω_K and c_K are taken as parameters. (The ground-state parameters $\mathcal{I}$, f_π and e (or g)– in the case of the Skyrme term – are known from that sector.) This is a remarkable agreement with experiment. The parameters so determined are rather close to the Callan-Klebanov prediction with the standard Skyrme model with a suitable symmetry-breaking term implemented.

In Figure 9.2 the spectrum of charmed baryons containing one or more charm quarks and strange quarks are given. The only parameters of the model, ω_D and c_D, are fixed by fitting Λ_c and Σ_c. The prediction is compared with quark-model predictions. The agreement with quark models is impressive, much better than one would have the right to expect, considering that the mass of the charmed quark is not so heavy. Similar results are obtained for b-quark baryons.

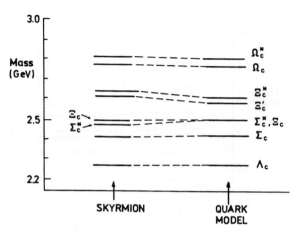

Fig. 9.2.   Charmed hyperons: the two parameters $\omega_D = 1418$ MeV and $c_D = 0.14$ are obtained from fitting Λ_c and adjusting c_D.

9.3.4. Connection to large N_c QCD

The general characteristics of skyrmions, both the ground state (light-quark baryons) and excited states (heavy-quark baryons), can be understood in terms of large N_c QCD as discussed by Dashen and Manohar[43] and explained in Chapter 3. (We will refer to the large N_c constraints discussed by these authors as Dashen-Manohar constraints.) The leading non-vanishing correction to certain ratios as g_A/μ_V (where μ_V is the isovector magnetic moment of the nucleon) are of order $1/N_c^2$. This can happen only if there is an infinite tower of degenerate states contributing in the intermediate state, their contributions mutually canceling each other in the limit of large N_c. The degeneracy is lifted to order $1/N_c$. This can be seen by observing that the Born term for baryon-pion scattering violates both large N_c constraint and unitarity, and hence must be canceled – among the tower of intermediate states – to $O(1/N_c)$ or alternatively by assuring that one loop chiral perturbation corrections to pion-baryon scattering do not violate the large N_c constraint and unitarity.

A consequence of the Dashen-Manohar QCD constraints in light-quark baryons is found to be that baryon splittings must be proportional to $1/N_c$ and $\vec{J}^2$ where $\vec{J}$ is the angular momentum of the baryon [44]. This is precisely the structure of the light-quark spectrum we obtained in the skyrmion picture. This can be understood as a general consequence of large N_c QCD which predicts that the spin-dependent terms in the energy vanish as $N_c \to \infty$. Now if baryons contain one heavy quark Q plus two light quarks, heavy-quark symmetry requires that the hyperfine splitting in the heavy baryons must be inversely proportional to the heavy-quark mass m_Q. Thus large N_c QCD predicts the splitting to be $\sim (N_c m_Q)^{-1}$. Furthermore from the Dashen-Manohar constraints, it follows [45] that the hyperfine splitting for, say, $\Sigma_Q^* - \Sigma_Q$ must be of the form, $\sim \vec{J_l} \cdot \vec{J_Q}$, where J_l is the skyrmion angular momentum and J_Q the heavy-quark angular momentum. This is identical to what we found in the heavy baryon spectrum (see (9.99))

$$\Delta H \sim \frac{1}{\mathcal{I}} c_\Phi \vec{J_l} \cdot \vec{J_\Phi} \qquad (9.103)$$

where J_Φ is the angular momentum lodged in the Berry potential generated from the heavy meson Φ containing the heavy quark Q when the latter is integrated out. (The moment of inertia $\mathcal{I}$ is of $O(N_c)$ and c is of $O(m_Q^{-1})$, so (9.103) is of $O([N_c m_Q]^{-1})$.)

The same argument applies to non-relativistic quark models as one expects from the large N_c equivalence between the skyrmion model and non-relativistic quark model[46].

References

1. N. Isgur and M.B. Wise, Phys. Rev. Lett. **66**, 1130 (1991) and references given therein; M.B. Wise, "New symmetries of the strong interactions," Lake Louise lectures, CALT-68-1721 (1991).
2. E.V. Shuryak, Phys. Lett. **93B**, 134 (1980); Nucl. Phys. **B198**, 83 (1982); M.B. Voloshin and M.A. Shifman, Yad. Fiz. **45**, 463 (1987); **47**, 801 (1988).
3. For recent reviews see: M.B. Wise, Lectures given at the CCAST Symposium on Particle Physics at the Fermi Scale, May – Jun. 1993, Caltech preprint CALT-68-1860 (1993); H. Georgi, Preprint HUPT-91-A039 [Published in 1991 TASI Proceedings]; B. Grinstein, Ann. Rev. Nucl. Part. Sci. **42**, 101 (1992); M. Neubert, Phys. Repts. **245**, 259 (1994).
4. S. Weinberg, Transactions of the New-York Academy of Sciences, Vol. **38**,185 (1977).
5. R. Dashen, Phys. Rev. **183**, 1245 (1969).
6. J. Gasser and H. Leutwyler, Nucl. Phys. **B94**, 269 (1975).
7. J. Bijnens, J. Prades and E. de Rafael, Phys. Lett. **B348**, 226 (1995).
8. D.B. Kaplan and A.V. Manohar, Phys. Rev. Lett. **56**, 2004 (1990).

9. For a further discussion on this matter, see H. Leutwyler, "Chiral Effective Lagrangians," in *Proc. of the Theoretical Advanced Study Institute 1991*, ed. R.K. Ellis, C.T. Hill and J.D. Lykken (World Scientific, Singapore, 1992).

10. J.F. Donoghue and C.R. Nappi, Phys. Lett. **B168**, 105 (1986).

11. S.-J. Dong and K.F. Liu, "$\pi N \sigma$ term and quark spin content of the nucleon," hep-lat/9412059; M. Fukugita, Y. Kuramashi, M. Okawa and U. Ukawa, Phys. Rev. **D51**, 5319 (1995).

12. E. Jenkins and A.V. Manohar, Phys. Lett. **B281**, 336 (1992).

13. C.G. Callan and I. Klebanov, Nucl. Phys. **B262**, 365 (1985); N.N. Scoccola, H. Nadeau, M.A. Nowak and M. Rho, Phys. Lett. **B201**, 425 (1986);C.G. Callan, H. Hornbostel and I. Klebanov, Phys. Lett. **B202**, 425 (1986).

14. R.D. Ball, S. Forte and J. Tigg, Nucl. Phys. **B428**, 485 (1994).

15. CLEO Collaboration, P. Avery *et al.*, report CLNS95/1352, hep-ex/9508010.

16. F.E. Close, *An introduction to quarks and partons* (Academic Press, London, 1979).

17. I. Zahed, in *Proceedings of XXXIII Cracow School of Theoretical Physics, Zakopane 1993*, Acta Phys. Pol.**B25**, 151 (1994).

18. S. Chernyshev, M.A. Nowak and I. Zahed, Phys. Lett. **B350**, 238, (1995); S. Chernyshev, M.A. Nowak and I. Zahed, SUNY preprint, SUNY-NTG-95-33.

19. L.L. Foldy and S.A. Wouthuysen, Phys. Rev. **78**, 29 (1950).

20. J. G. Körner and G. Thompson, Phys. Lett. **B 264**, 185 (1991; S. Balk, J. G. Körner and D. Pirjol, Nucl. Phys. **B428**, 499 (1994) and Ahrenshoop Symp. 1993, p. 315.

21. D.I. Diakonov, V.Yu. Petrov, and P.V. Pobylitsa, Phys. Lett. **B226**, 372 (1989).

22. P. V. Pobylitsa, Acta Phys. Pol. **B22**, 175 (1991). (Note that the mass shift for light quarks obtained in this paper differs from the one discussed in Chapter 2, (2.83), but numerically the difference is small.)

23. M.A. Shifman, A.I. Vainshtein, and V.I. Zakharov, Nucl. Phys. **B163**, 46 (1980).

24. M.A. Nowak, J.J.M. Verbaarschot and I. Zahed, Nucl. Phys. **B324**,1 (1989).

25. A. De Rujula, H. Georgi, and S.L. Glashow, Phys. Rev. **D12**, 147 (1975).

26. D.O. Riska and N.N. Scoccola, Phys. Lett. **B299**, 388 (1993).

27. M. Oka and S. Takeuchi, Nucl. Phys. **A 524**, 649 (1991).

28. A.V. Manohar and M.B. Wise, Nucl. Phys. **B399**, 17 (1993).

29. See, *e.g.*, I. Zahed and G. Brown, Phys. Rep. **142**, 1 (1986); R.D. Ball, "Desperately seeking mesons", in *Proc. of the Workshop on Skyrmions and Anomalies*, ed. by M. Jeżabek and M. Praszałowicz, (World Scientific 1987), p. 54; D. Ebert and H. Reinhardt, Phys. Lett. **B173**, 153 (1986); A. Dhar and S. R. Wadia, Phys. Rev. Lett. **52**, 959 (1984); R.T. Cahill, J. Praschifka and C. D. Roberts, Phys. Rev. **D36**, 209 (1987); Ann. Phys. (NY) **188**, 20 (1988).

30. M. A. Nowak and I. Zahed, Phys. Rev. **D48**, 356 (1993).

31. W. A. Bardeen and C. T. Hill, Phys. Rev. **D49**, 409 (1994).

32. M. A. Nowak, M. Rho and I. Zahed, Phys. Rev. **D48**, 4370 (1993).
33. P. Cho and H. Georgi, Phys. Lett. **B296**, 408 (1992); **B300**, 410 (1993).
34. A.F. Falk and M. Luke, Phys. Lett. **B292**, 119 (1992).
35. A. Manohar and H. Georgi, Nucl. Phys. **B234**, 189 (1984).
36. M. Rho. D.O. Riska and N.N. Scoccola, Phys. Lett. **B251**, 597 (1990); Z. Phys. **A341**, 343 (1992);Y. Oh, D.-P. Min, M. Rho and N.N. Scoccola, Nucl. Phys. **A534**, 493 (1991).
37. M.B. Wise, Phys. Rev. **D45**, R2118 (1992); T.-M. Yan *et al*, Phys. Rev. **D46**, 1148 (1992); G. Burdman and J.F. Donoghue, Phys. Lett. **B280**, 287 (1992).
38. D.-P. Min, Y. Oh, B.-Y. Park and M. Rho, Int. Jour. Mod. Phys. **E4**, 47 (1995).
39. M. A. Nowak, M. Rho and I. Zahed, Phys. Lett. **B303**, 130 (1993). (A dependence on the relative phase between the pseudoscalar and the vector mentioned in this paper is an artefact of the ansatz and has no physical significance.)
40. D.-P. Min, Y. Oh, B.-Y. Park and M. Rho, "Soliton structure of heavy baryons," SNU preprint SNUTP-92-78, unpublished.
41. E. Jenkins, A.V. Manohar and M.B. Wise, Nucl. Phys. **B396**, 27 (1993); Z. Guralnik, M. Luke and A.V. Manohar, Nucl. Phys. **B390**, 474 (1993); E. Jenkins and A.V. Manohar, Phys. Lett. **B294**, 273 (1992).
42. C.C. Callan and I. Klebanov[13] ; N.N. Scoccola and A. Wirzba, Phys. Lett. **B66**, 1130 (1991).
43. R. Dashen and A.V. Manohar, Phys. Lett. **B315**, 425, 438 (1993).
44. E. Jenkins, Phys. Lett. **B315**, 441 (1993).
45. E. Jenkins, Phys. Lett. **B315**, 447 (1993).
46. A.V. Manohar, Nucl. Phys. **B248**, 19 (1984).

CHAPTER 10

BARYONIC MATTER

10.1. Introduction

We have discussed so far the structure of hadrons and their interactions in isolation, focusing on the structure of the vacuum and the characteristics of elementary excitations on top of it. In this chapter, we shall exploit what we learned on the properties of elementary excitations to describe many-baryon systems, that is, complex nuclei, relying as much as possible on the generic feature of chiral symmetry and its realization discussed in the preceding chapters. We will focus here mainly on many-body systems with light quarks with the strange quark considered on the same footing.

The physics of nuclei and nuclear matter is the physics of *strong* interactions in the sense described in this book. One might naively think that implementing what we discussed in the preceding chapters in the physics of nuclei would be infeasible, if not impossible. The reason is this: in terms of effective degrees of freedom, the pion-nucleon coupling

$$\mathcal{L}_{\pi N} \sim \frac{g_{\pi NN}}{2m_N} \overline{N} \gamma^\mu \gamma_5 \tau^a N \ \partial_\mu \pi^a \tag{10.1}$$

with $g_{\pi NN}^2/4\pi \sim 14$ is strong, and clearly a perturbative expansion in $g_{\pi NN}$ would be doomed. On the other hand, it is now very well established that the notion of the "quasiparticle" works extremely well in nuclear physics as we have learned from the success of shell model: A nucleus is not a soup of pions and nucleons nor of quarks and gluons but a system consisting of well-defined single-particle (that is, quasiparticle) excitations. Although various phenomenological descriptions using the interaction (10.1) at low-order expansion are found to be quite successful, this does not mean that a low-order perturbation theory with the strong coupling constant "miraculously" works. In this section, we will present two complementary (and most likely related) points of views that are not rooted in conventional perturbation theory that offer a modern way of describing nuclei based on chiral symmetry: A semi-classical expansion and an application of chiral perturbation theory.

In the semi-classical approach, $\hbar \sim 1/\sqrt{N_c}$, $g_{\pi NN} \sim (\sqrt{N_c})^3$ and $m_N \sim N_c$; thus the pion-nucleon interaction is strong $g_{\pi NN}/m_N \sim \sqrt{N_c} \sim 1/\hbar$. In this case, the nucleon is viewed as a soliton, and pion-nucleon and nucleon-nucleon interactions are organized in $1/\hbar$. To lowest order in $1/\hbar$, the nucleon is non-relativistic and extended. The accuracy of the description, therefore, depends on the accuracy of the starting effective Lagrangian as well as the reliability of the semi-classical expansion. This point of view is well motivated by large-N_c QCD arguments, although in nature $N_c = 3$.

In chiral perturbation theory, chiral power counting is introduced so that the scattering amplitudes and form factors involving a soft momentum scale $\sqrt{s} \sim m_\pi$ are expanded using

$$\frac{g_{\pi NN}\sqrt{s}}{4\pi m_N} \sim \frac{g_A \, m_\pi}{4\pi f_\pi} \tag{10.2}$$

where the Goldberger-Treiman relation has been used. Since $g_A \sim 1$, the expansion is well-defined if $\sqrt{s} \ll \Lambda_\chi \sim 4\pi f_\pi$. This expansion can be effected using the effective Lagrangian approach given in Chapters 5 and 6. In this approach, the nucleon is a point-like object with its extended structure emerging through pion loops. The accuracy of this description relies on the character of the chiral expansion. This point of view is empirically motivated by a number of low energy relations as discussed in Chapter 5, including the Goldberger-Treiman relation.

The soliton approach, while well motivated by the large N_c arguments of Chapter 3, is intrinsically limited to the extent that we do not really know what the realistic large N_c QCD effective Lagrangian is. Furthermore as discussed by several authors (see *e.g.*, Coriano *et al.* [1]), it is not completely obvious whether the S-matrix is fully consistent with the nucleon treated as a soliton, not to mention nuclear systems as solitons. Similarly, while chiral perturbation theory as applied to nucleons is well motivated by the success of current algebra – and more systematic in scope, the approach is limited in practice to only few-loop orders due to the fact that there are usually not enough data to fix the constants ("counterterms" introduced in Chapters 5 and 6). In many-body system, this means that the treatment can be limited to only long-distance interactions between two or more nucleons while in nuclear interactions, all ranges from a pion Compton wavelength down to the hard-core size are probed.

Although different in their systematics, the two approaches mentioned above bear much in common and are most likely related. Indeed recent works[2] suggest rather strongly that such a relation can be established, at least in the one-baryon sector in the large N_c limit. It is an open problem to find the bridge but we will assume here that we can map, in nuclear systems as in the elementary nucleon, the semi-classical (soliton) description to the description in which baryons are introduced as local massive matter fields. They both start from empirically motivated chiral effective Lagrangians.

In Section 2 of this chapter, we present the semi-classical approach. We assume that we are given a large N_c effective Lagrangian as discussed in Chapter 7, and develop many-body dynamics therefrom, suitably extending what we know of the single skyrmion structure. The mathematics of multi-skyrmion structures, while theoretically interesting, remains still in its infancy, and much work is still needed to understand beyond baryon number 2. We shall therefore limit our discussion to the two-nucleon system, *i.e.*, the deuteron and the two-nucleon force.

The remaining (main) part of this chapter will be devoted to the approach based on chiral perturbation theory with nucleons. We start from an effective Lagrangian that consists of mesons and baryons and guided by chiral symmetry and other symmetries of the strong interactions, develop chiral perturbation theory for nuclear dynamics. The main task then would be to understand nuclear forces from the point of view of chiral symmetry and then develop nuclear response functions in the same scheme. This approach will be discussed with applications to both finite nuclei and infinite nuclear matter.

When the many-body system is relatively dilute (say, up to nuclear matter density ρ_0), both of the above-mentioned approaches are tractable by means of virial-like expansions. For very dense matter ($\rho \gg \rho_0$), the situation is a lot more complex as higher-order density effects become important. Consequently this regime is more or less unknown as there is little guidance from experiments. In this chapter, only those systems where the complications are minimized will be discussed.

10.2. Nuclei As Skyrmions

Given a realistic effective Lagrangian of meson fields, the topological soliton sector should in principle give rise not only to the winding number one ($W = 1$) objects considered above but also to arbitrary winding-number objects. In the strong interactions, the winding number is the baryon number n_B and hence we are to obtain nuclei with the baryon number $n_B = A$ where A is the mass number of the nucleus when properly quantized. As mentioned, although there are attempts to describe $A \geq 3$ nuclei in terms of multi-winding-number skyrmions[3], our understanding of the mathematics of multi-skyrmion configuration remains rather poor. Given the approximations that one is forced to make in solving the problem, we do not know how close the multi-winding-number solution is to real nuclei. In this section, therefore, we shall limit our discussion to the simplest nuclear system, the deuteron and the nucleon-nucleon potential. Also instead of working with a general, vector-meson-implemented Lagrangian, we will consider the simplest one, the original Skyrme Lagrangian discussed in Chapter 7:

$$\mathcal{L}_{skyrme} = \frac{f_\pi^2}{4} \operatorname{Tr} \left[\partial_\mu U \partial^\mu U^\dagger\right] \; + \; \frac{1}{32g^2} \operatorname{Tr} \left[U^\dagger \partial_\mu U, U^\dagger \partial_\nu U\right]^2$$
$$+ \; r \operatorname{Tr} \left(\mathcal{M}U + h.c.\right) + \cdots \quad (10.3)$$

where we have put $f = f_\pi$ and $r = f^2 B_0/2$ (see Chapter 6). The last term in (10.3) is the leading chiral symmetry-breaking term of $O(\partial^2)$ in the "chiral counting" (as discussed in Chapters 5 and 9). In dealing with nonstrange nuclear systems, U will be the chiral field in $SU(2)$, so the Wess-Zumino term is missing from (10.3). Later we will consider the chiral field valued in $SU(3)$ in dealing with kaon-nuclear interactions.

In what follows, we will understand the quartic and higher derivative terms in (10.3) as the result of integration over higher meson-resonances. In this way, the possible acausality problems inherent to higher derivative actions[4] are avoided. In this sense, the Skyrme quartic term is understood to capture in some average way all effects of heavy degrees of freedom that are "integrated out." Thus much of short-distance physics is taken into account. More sophisticated Lagrangians with vector mesons and baryon resonances bring about quantitative improvements but overall no qualitative changes on the bulk properties. This was quite clear in hyperon phenomenology discussed in the preceding chapter.

10.2.1. *Deuteron*

The simplest nucleus, the deuteron, is a loosely bound proton-neutron system with a large quadrupole moment: a binding energy of 2.2 MeV and a root-mean-square radius of 2.095 fm. Since the two nucleons are on the average widely separated, we could start with the product *ansatz* for the chiral field with winding number 2, and ignore quantum effects. Let us put the centers of the two skyrmions symmetrically about the origin along the x axis. When the two nucleons are widely separated, a reasonable configuration for the system is the product form (see later for a more realistic form),

$$U_1(\vec{r} + s\hat{x})SU_1(\vec{r} - s\hat{x})S^\dagger \qquad (10.4)$$

where U_1 is the hedgehog with baryon number 1 and S is a constant (global) $SU(2)$ matrix describing the relative isospin orientation of the two skyrmions. (In the literature, one frequently uses A for the global $SU(2)$ rotation. Here as in Chapter 7, we reserve A for the induced axial current.) The two skyrmions are separated along the x axis by the distance of $2s$. It turns out from the pioneering work of Jackson, Jackson and Pasquier [5] that the most attractive configuration is obtained when the isospin of one skyrmion is rotated by an angle of 180° about an axis perpendicular to the x axis (*i.e.*, the axis of separation). Choosing the z axis for the rotation,

$$A = e^{i\pi\tau_3/2} = i\tau_3 \qquad (10.5)$$

the $n_B = 2$ chiral field takes the form

$$U_s(x, y, z) = U_1(x + s, y, z)\tau_3 U_1(x - s, y, z)\tau_3. \qquad (10.6)$$

Let us see what symmetries this *ansatz* has. With the explicit form for the single hedgehog $U_1 = \exp(i\vec{\tau}\cdot\hat{r}F(r))$, it is easy to verify that it satisfies the following discrete symmetry relations

$$U_s(-x, y, z) = \tau_2 U_s^\dagger(x, y, z)\tau_2, \qquad (10.7)$$
$$U_s(x, -y, z) = \tau_1 U_{-s}^\dagger(x, y, z)\tau_1, \qquad (10.8)$$
$$U_s(x, y, -z) = U_{-s}^\dagger(x, y, z). \qquad (10.9)$$

It also has the symmetry under the parity transformation

$$U_s^\dagger(-x, -y, -z) = \tau_3 U_s(x, y, z)\tau_3. \tag{10.10}$$

These discrete symmetries can be summarized as follows

$$U_s(x, -y, -z) = \tau_1 U_s(x, y, z)\tau_1. \tag{10.11}$$

What this says is that a spatial rotation about the axis of separation by angle of π is equivalent to an isospin rotation about the x axis by π. This symmetry forbids the quantum number $I = J = 0$ where I is the isospin and J the angular momentum of the state. This is consistent with Pauli exclusion principle. However the product ansatz is not a solution of the equation of motion and in addition there is nothing to prevent another deuteron-like state (say, d') with $I = 0$ and $J = 1$ to appear nearly degenerate with the deuteron. Indeed a simple analysis does show that a d' is present when the product ansatz is used. In nature, there is only one bound state, so while the product ansatz describes the deuteron property (and more generally static two-nucleon properties [6]), in qualitative agreement with experiments, it does not have enough symmetry of nature. In addition to the symmetry (10.11),

$$U_2(x, -y, -z) = \tau_1 U_2(x, y, z)\tau_1, \tag{10.12}$$

other discrete symmetries not present in the product ansatz but needed to eliminate the spurious d' are

$$U_2(-x, y, -z) \;=\; \tau_1 U_2(x, y, z)\tau_1, \tag{10.13}$$

$$U_2(-x, -y, z) \;=\; U_2(x, y, z). \tag{10.14}$$

(We are using the subscript 2 to indicate the winding number-2 configuration.) In fact the last symmetry (10.14) is a special case (*i.e.*, $\alpha = \pi$) of the continuous cylindrical symmetry

$$U_2(\rho, \phi + \alpha, z) = e^{-i\alpha\tau_3} U_2(\rho, \phi, z)e^{i\alpha\tau_3} \tag{10.15}$$

where the z axis is taken to be the symmetry axis. This has a toroidal geometry. A $n_B = 2$ skyrmion with the symmetries (10.12), (10.13), (10.14) and (10.15) is found to be the lowest energy solution of the skyrmion equation[7]. With this configuration, there is no bound d' state as it is pushed up by rotation. We will refer to this as toroidal skyrmion. To be more quantitative, let us see what the classical energies of various configurations come out to be[8]. For this, we put the length in units of $(gf)^{-1}$ and the energy in units of $f/2g$. In these units, the energy of two infinitely separated skyrmions is $1.23 \times 24\pi^2$ whereas the toroidal structure with coincident skyrmions comes slightly lower, at $1.18 \times 24\pi^2$. The latter corresponds to the lowest energy solution of the $n_B = 2$ skyrmion. The $n_B = 2$ spherical hedgehog configuration has much higher energy, $1.83 \times 24\pi^2$.

The question is: What is the deuteron seen in nature? While lowest in energy, the toroidal configuration cannot be the dominant component of the deuteron. For with that configuration, the size – and hence the quadrupole moment – of the deuteron would be much smaller (see Braaten and Carson in [7]), the calculated size being $\sqrt{\langle r^2 \rangle_d} \approx 0.92$ fm while experimentally it is 2.095 fm and the calculated quadrupole moment being $Q \approx 0.082$ fm^2 while experimentally it is 0.2859 fm^2. This compactness of the structure is easily understandable in terms of much larger binding energy – about an order of magnitude too large – associated with this configuration. Clearly then this configuration must constitute a small component of the deuteron wave function. Instead the size and quadrupole moment indicate that the widely separated configuration should be closer to nature. Even so, the deuteron seems to share the torus symmetry in a variety of ways as observed in its static electromagnetic properties and other moments. In particular, short-distance properties of the deuteron could be governed by the toroidal configuration.

Even though reducing from a field theory involving solitons to a quantum mechanics in the space spanned by the collective coordinates ("moduli space") brings in an enormous simplification, the reality must still be rather complex in terms of skyrmion configurations. A realistic treatment[8,9] must encompass the unstable spherical hedgehog with winding number 2, the manifold of which is twelve-dimensional including six unstable modes, the infinitely separated two skyrmion configuration – which is also twelve-dimensional – and the eight-dimensional toroidal configuration. This is a very difficult problem and it is highly unlikely that full analytical solutions will be forthcoming in the near future. It is however expected that numerical simulations will provide accurate solutions in the near future. Indeed an initial "experiment" of that type has been recently performed. While the result is only preliminary and semiquantitative at best, it is nonetheless a remarkable development. Since the presently available result will undoubtedly be superseded soon by better numerical simulations, we will just summarize what has been achieved so far[10].

In the analysis of Crutchfield *et al.* [10], initially two widely separated skyrmions (which are not a solution on a finite lattice which they are using) are prepared by relaxation. The continuum solution is imposed on the lattice. As the skyrmions relax, they approach each other through the dissipation of their orbital energy. At the point of closest approach, the two skyrmions merge into the optimal toroidal configuration and then scatter at angle $\pi/2$, a feature generic of two solitons scattering at low energy, such as monopole-monopole, skyrmion-skyrmion and vortex-vortex scattering. This process is found to repeat, with the skyrmions falling into the toroidal form and scattering at right angles. At each cycle, energy is transferred from the orbital motion to a radial excitation of the individual skyrmion. Quantizing the nearly periodic motion, it is found that for large g^2 (for which the Skyrme quartic term emerges from the ρ meson kinetic energy term in a more realistic hidden gauge symmetry Lagrangian[11]) there is only one bound state with $J = 1$ and $I = 0$, with an unbound $J = 0$ and $I = 1$ state nearby. (There are instead many bound states for weak g^2. But for a weak g^2, there is no reason why the Skyrme Lagrangian would be a good approximation.) The bound wave function describes a state peaked at a configuration with two nucleons widely separated, hence elongated, with the size about twice that of the static toroidal structure and a quadrupole moment about 4 times while the short-range correlations are governed by the toroidal configuration. These are all in semiquantitative agreement with experiments.

10.2.2. *Nuclear forces*

The static minimum-energy configuration is known to be tetrahedron for $n_B = 3$, octahedron for $n_B = 4$ etc.[12] As in the case of torus, these configurations will make up only a small component of the wave functions of ^{3}He, ^{4}He etc. even though symmetries may be correctly described by – and short-distance physics may be sensitive to – them. The alternative to this – which is definitely more economical in computation and more predictive– is to construct a nucleon-nucleon potential from skyrmion structure and treat the nuclei in the conventional way. Furthermore, constrained static solutions that are required for constructing NN potentials are much easier to work out than the time-dependent problem mentioned above. The question is: Can one derive a nucleon-nucleon potential which can be compared with, say, the Paris potential[13] (or other realistic potentials)?

The $n_B = 2$ baryon has twelve collective degrees of freedom: three for translation (or center-of-mass position), three for orientations in space and isospin space, three for the spatial separation and three for the relative orientation in isospin. For the asymptotic configuration where two skyrmions are widely separated, the twelve degrees of freedom correspond to the coordinate $\vec{r}_i$ and the isospin orientation S_i for each skyrmion $i = 1, 2$. When quantized, they represent two free nucleon states. The product ansatz possessing twelve collective degrees of freedom is close to two free nucleons when widely separated but lacks the correct symmetry at short distance. When two skyrmions are on top of each other, as mentioned above, the lowest energy configuration is that of a torus (with axial symmetry) with eight collective degrees of freedom. Thus four additional collective degrees of freedom are needed for describing general $n_B = 2$ configurations. This constitutes the moduli space that is involved for the problem.

For general $n_B = 2$ configurations, global collective coordinates do not affect the energy, so for considering a potential, we may confine ourselves to the separation r of the two skyrmions and the relative isospace orientation which can be expressed by three Euler angles α, β and γ

$$C = S_1^\dagger S_2 = e^{i\tau_3\alpha/2}e^{i\tau_2\beta/2}e^{i\tau_3\gamma/2}. \tag{10.16}$$

When global collective coordinates are quantized, the resulting Hamiltonian involves a potential which will then depend only on the separation r which we choose here to be along the z axis and the three Euler angles. The potential therefore must be expressible in terms of the Wigner $\mathcal{D}$-functions as[14]

$$V(r, C) = \sum_{j=0}^{\infty} \sum_{m=-j}^{j} \sum_{n=-j}^{j} V_{jmn}(r)\mathcal{D}_{m,n}^{(j)}(C). \tag{10.17}$$

This is the most general form since the $\mathcal{D}$ functions form a complete set over $SU(2)$ for integer j. This can be further reduced by using the symmetry associated with rotation about the z axis – therefore reducing the Euler angles to two, namely, β and $(\alpha - \gamma)$ –and the relation $V(r, C) = V(r, C^\dagger)$ due to the $SU(2)$ symmetry, to the form

$$V(r, C) = \sum_{j=0}^{\infty} \sum_{m=0}^{j} V_{jm} \frac{1}{2} \left[\mathcal{D}_{m,m}^{(j)}(C) + \mathcal{D}_{-m,-m}^{(j)}(C) \right]. \tag{10.18}$$

The nucleon-nucleon potential is then defined by taking the matrix element with the asymptotic two-nucleon wave function

$$\Psi(1,2) \sim \mathcal{D}_{i_1,-s_1}^{(1/2)}(S_1)\mathcal{D}_{i_2,-s_2}^{(1/2)}(S_2). \qquad (10.19)$$

The potential so defined is guaranteed to give the correct asymptotic behavior shared by the product ansatz and the correct short-distance behavior shared by the toroidal configuration. Although there is no rigorous proof, only the partial waves $j \leq 1$ turn out to contribute significantly. Ignoring the higher partial waves $j \geq 2$, we have

$$\begin{aligned} V(r,C) &= V_{00}(r)\mathcal{D}_{0,0}^{(0)}(C) + V_{10}\mathcal{D}_{0,0}^{(1)}(C) \\ &\quad + V_{11}(r)[\mathcal{D}_{1,1}^{(1)}(C) + \mathcal{D}_{-1,-1}^{(1)}(C)] \end{aligned} \qquad (10.20)$$

which, when sandwiched with the wave function (10.19), gives

$$\begin{aligned} V(1,2) = V_C(r) &+ V_{SS}(r)\vec{\tau}_1 \cdot \vec{\tau}_2\vec{\sigma}_1 \cdot \vec{\sigma}_2 \\ &+ V_T(r)\vec{\tau}_1 \cdot \vec{\tau}_2(3\vec{\sigma}_1 \cdot \hat{r}\vec{\sigma}_2 \cdot \hat{r} - \vec{\sigma}_1 \cdot \vec{\sigma}_2) \end{aligned} \qquad (10.21)$$

with

$$V_C = V_{00}, \quad V_{SS} = 25(V_{10} + V_{11})/247, \quad V_T = 25(2V_{10} - V_{11})/486 \qquad (10.22)$$

with the potential functions V_i given by the skyrmion solution of the classical equation of motion. The V's are obtained numerically.

There is considerable uncertainty in defining the separation r between two skyrmions because of their extended structure. A convenient definition is

$$\left(\frac{r}{2}\right)^2 = \frac{1}{n_B} \int d^3x \, \vec{x}^2 \mathcal{B}^0(\vec{x}) \qquad (10.23)$$

where $\mathcal{B}^0$ is the baryon number density. This reduces to the usual separation $r = |\vec{r}_1 - \vec{r}_2|$ if one takes the form

$$\mathcal{B}^0(\vec{x}) = \delta^3(\vec{x} + \frac{1}{2}r\hat{e}_3) + \delta^3(\vec{x} - \frac{1}{2}r\hat{e}_3).$$

It may be possible to circumvent the above ambiguity. To do so, a full quantum skyrmion-skyrmion configuration, including the collective variables associated to the kinetic motion, has to be considered. Take, for instance, the ansatz

$$U(\vec{x},t) = S_1 U_1(\mathbf{x} - \mathbf{X} - \frac{\mathbf{r}}{2})S_1^\dagger U_\pi(\mathbf{x} - \mathbf{X}, t)S_2 U_1(\mathbf{x} - \mathbf{X} - \frac{\mathbf{r}}{2})S_2^\dagger \qquad (10.24)$$

where $\mathbf{X}$ is the center-of-mass collective coordinate, S_1 and S_2 the collective isospin coordinates with $S = S_1^\dagger S_2$ (see eq.(10.16)), and $\mathbf{r}$ the relative separation between the two skyrmions. Here U_π are the quantum pions, that is, the pion cloud. When promoted to collective coordinates, S_1, S_2, $\mathbf{X}$ and $\mathbf{r}$ intertwine. Now the semi-classical expansion in the two-skyrmion configuration requires a double expansion in $\hbar \sim 1/\sqrt{N_c}$ and the range of the interaction $e^{-m_\pi r}$ [15]. Starting from (10.24), it can be shown that in the one-pion range, the interaction is ansatz-independent, while in the two-pion range, the ansatz-dependence is largely canceled (either through the shape of U_1 or the definition of $\mathbf{r}$) by the iteration of the one-pion fluctuations. The latter provides a potential source of attraction in the central channel as discussed by Thorsson and Zahed [16]. The validity of the double expansion holds for ranges that are larger than or equal to 1.4 fm. To probe the potential in the three-pion range and shorter requires further iterations of the pion fluctuations (higher-loops) in the context of the double expansion. In a way, the approach and the ensuing result are reminiscent of the Born-Oppenheimer construction in which the nuclear potential involves not only direct interactions but also self-consistently induced potentials through the pion cloud. It can provide a systematic ansatz-free analysis of the two- and many-nucleon problem.

In summary, when suitable $1/N_c$ corrections are minimally implemented by taking into account the state mixing between NN, $N\Delta$ and $\Delta\Delta$ intermediate states as described by Amado *et al.* [17,18], skyrmions provide a semi-quantitative account of the long- and medium-range interaction between two nucleons, in a manner that is ambiguity-free [16,18]. The potentials obtained by Walhout and Wambach and Amado *et al.* are shown in Figure 10.1. The short-distance repulsion and tensor forces also seem to be well described in the model, although the ansatz-dependence has to be removed as described above. Considering the enormous simplicity of the effective Lagrangian used, this is truly remarkable. Does this imply that nature vindicates the description of nucleons and nuclei in terms of chiral solitons? Do we live in a world for which $N_c = 3 \gg 1$? We cannot answer this question. In what follows, we shall leave this picture and turn to the description based on chiral perturbation theory with a Lagrangian that contains the baryons as (point-like) matter fields.

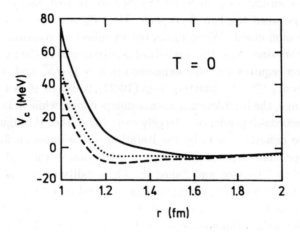

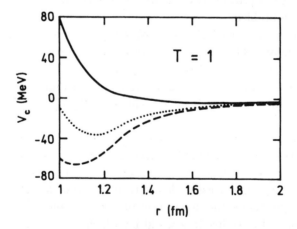

Fig. 10.1. Nucleon-nucleon central potential V_C^T derived from baryon number 2 skyrmion compared with the results of a phenomenological (Reid soft-core) potential. The solid line represents the nucleon-only (leading N_c) result of Walhout and Wambach[14], the dashed line the full Born-Oppenheimer diagonalization of the $N\Delta$ state mixing of Walet and Amado[17] and the dotted line the Reid soft-core potentials.

10.3. Nuclear Effective Chiral Lagrangians

10.3.1. *Chiral symmetry in complex nuclei*

Chiral symmetry in the form of effective Lagrangians we shall consider here has played an extremely important and powerful role in nuclear physics since many years [19], although the nuclear physics community has not paid much attention to it until recently. It is presently one of the major preoccupations of many nuclear theorists as it has become clear that the main ingredient of QCD relevant to nuclear dynamics is the spontaneously broken chiral symmetry, relegating confinement and asymptotic freedom – other important facets of QCD – to a higher energy regime. Though at low energy chiral symmetry addresses primarily to pions, it turns out to play a crucial role where other degrees of freedom than pions intervene. The reason for this, as we shall learn shortly, is that *pions dominate whenever they can and if they do not, then there are reasons for it, which also have something to do with chiral symmetry.* This will be later referred to as "chiral filtering" phenomenon.

In dealing with nuclei, one confronts several facets of hadron structure. The first is the property of the basic building block of the nucleus, which is of course the nucleon which is in turn made of quarks and gluons: We thus have to understand what the nucleon is in order to understand how two or more nucleons interact. Given our understanding of the nucleon, we have to understand how they interact – following what symmetries and dynamics – and how the individual property of the nucleon is modified by the change of environment. In all these, we would like that the same guiding principles apply for both the basic structure of the nucleon and the residual interaction between them. This is where the chiral perturbation approach described in Chapter 6 implemented with baryons enters. We start with how the baryons are incorporated into chiral Lagrangians.

10.3.2. *Chiral Lagrangians for baryons and Goldstone bosons*

In Chapter 7, we dealt with chiral Lagrangians with mesons only, a structure implied by large-N_c arguments. We discussed there how baryons could emerge from meson fields as solitons (skyrmions). One day it may be possible to describe, say, ^{208}Pb with a Lagrangian whose parameters are entirely fixed in the zero-baryon sector. In fact we might be able to expect that the Pb nucleus is understood *because* the nucleon is understood, through a concept akin to the tumbling of scale hierarchies in particle spectrum as discussed by Nambu[20].

Here we will be less ambitious but more systematic. We will treat the nucleon as infinitely heavy, as suggested by the large N_c arguments, but point-like instead. Meson-nucleon dynamics will be sought in the most general effective Lagrangian consistent with chiral symmetry, perturbative unitarity and causality, and organized using chiral counting arguments as advocated by Weinberg[21] (see also Chapter 6), and discussed more recently by many authors[22]. Since the mass of the baryon m_B is large, the expansion parameters will be sought in the form : $(\partial, \mathcal{M})/(m_B, \Lambda_\chi)$ (where $\mathcal{M}$ is the quark mass matrix).

Although for the moment we will be mostly concerned with the up- and down-quark $(SU_f(2))$ sector, we shall write the Lagrangian for the $SU(3)$ flavor as it will be needed for kaon-nuclear interactions and kaon condensation in the later part of this chapter. As explained in Chapters 1 and 2, the $U_A(1)$ symmetry is broken by anomaly, so the associated ninth pseudoscalar meson η' is massive. It will not concern us here. The scale invariance is broken by the trace anomaly with the associated scalar χ playing an important role for scale properties of the hadrons. This feature will be discussed later. Let the octet pseudoscalar field $\pi = \pi^a T^a$ be denoted by

$$\sqrt{2}\pi = \begin{pmatrix} \frac{1}{\sqrt{2}}\pi^0 + \frac{1}{\sqrt{6}}\eta & \pi^+ & K^+ \\ \pi^- & -\frac{1}{\sqrt{2}}\pi^0 + \frac{1}{\sqrt{6}}\eta & K^0 \\ K^- & \bar{K}^0 & -\frac{2}{\sqrt{6}}\eta \end{pmatrix} \qquad (10.25)$$

and the octet baryon field $B = B^a T^a$ be given by

$$B = \begin{pmatrix} \frac{1}{\sqrt{2}}\Sigma^0 + \frac{1}{\sqrt{6}}\Lambda & \Sigma^+ & p \\ \Sigma^- & -\frac{1}{\sqrt{2}}\Sigma^0 + \frac{1}{\sqrt{6}}\Lambda & n \\ \Xi^- & \Xi^0 & -\frac{2}{\sqrt{6}}\Lambda \end{pmatrix}. \qquad (10.26)$$

We write the vector current V_μ and axial-vector current A_μ (with a sign opposite to that defined in Chapter 7) as A_μ as

$$\begin{pmatrix} V^\mu \\ -iA^\mu \end{pmatrix} = \frac{1}{2}\left(\xi^\dagger \partial^\mu \xi \pm \xi \partial^\mu \xi^\dagger\right) \qquad (10.27)$$

where $U = \xi^2 = \exp(i\pi^a T^a/f)$ with $\mathrm{Tr}T^a T^b = \frac{1}{2}\delta_{ab}$ (*i.e.*, $T^a \equiv \lambda^a/2$ where λ^a are the usual Gell-Mann matrices). Since U transforms $U \rightarrow$

$LUR^\dagger$ under chiral transformation with $L \in SU_L(3)$ and $R \in SU_R(3)$, ξ transforms $\xi \to L\xi g^\dagger(x) = g(x)\xi R^\dagger$ with $g \in SU(3)$ a local, nonlinear function of L, R and π. Under the local transformation $g(x)$, V_μ transforms as a gauge field

$$V_\mu \to g(V_\mu + \partial_\mu)g^\dagger \qquad (10.28)$$

while A_μ transforms covariantly

$$A_\mu \to gA_\mu g^\dagger. \qquad (10.29)$$

The baryon field can be considered as a matter field just as the vector meson fields ρ are. The general theory of matter fields interacting with Goldstone boson fields was formulated by Callan, Coleman, Wess and Zumino many years ago[23]. Since matter fields transform as definite representations only under the unbroken diagonal subgroup $SU_{L+R}(3)$, one may choose any representation that reduces to this subgroup. Different choices give rise to the same S-matrix. A convenient choice turns out to be the one that has the baryon field transform

$$B \to gBg^\dagger. \qquad (10.30)$$

Under vector transformation $L = R = V$, we have $g = V$ so that eq.(10.30) reduces to the desired vector (octet) transformation $B \to VBV^\dagger$. With this representation, the pion field has only derivative coupling to the baryons which facilitates the chiral counting developed below[24]. With the "gauge" field V_μ, we write the covariant derivative

$$D^\mu B = \partial^\mu + [V^\mu, B] \qquad (10.31)$$

which transforms $D^\mu B \to gD^\mu Bg^\dagger$. The Lagrangian containing the lowest derivative involving baryon fields is then

$$\begin{aligned}
\mathcal{L}_B &= i\,\mathrm{Tr}(\bar{B}\gamma^\mu D_\mu B) - m_B\,\mathrm{Tr}(\bar{B}B) \\
&+ D\,\mathrm{Tr}(\bar{B}\gamma_\mu\gamma_5\{A^\mu, B\}) + F\,\mathrm{Tr}(\bar{B}\gamma_\mu\gamma_5[A^\mu, B]) + O(\partial^2)
\end{aligned} \qquad (10.32)$$

where D and F are constants to be determined from experiments as we will see later. The mesonic sector is of course described by eq. (10.3), the chirally invariant part of which is, in terms of the chiral field U alone,

$$\mathcal{L}^U = \frac{f_\pi^2}{4}\,\mathrm{Tr}(\partial_\mu U \partial^\mu U^\dagger) + O(\partial^4). \qquad (10.33)$$

In eqs.(10.32) and (10.33), a truncation is made at the indicated order of derivatives.

Next we need to introduce chiral symmetry breaking terms since the symmetry is indeed broken by the quark mass terms. As in Chapter 9, we assume as suggested by QCD Lagrangian that the quark mass matrix $\mathcal{M}$ transforms as $(3, \bar{3}) + (\bar{3}, 3)$, namely, as

$$\mathcal{M} \rightarrow L\mathcal{M}R^{\dagger}. \tag{10.34}$$

Writing terms involving $\mathcal{M}$ that transform invariantly under chiral transformation, we have

$$\begin{aligned}
\mathcal{L}_{\chi SB} &= r\,\mathrm{Tr}\,(\mathcal{M}U + h.c.) + a_1\,\mathrm{Tr}\bar{B}\,(\xi\mathcal{M}\xi + h.c.)\,B \\
&\quad + a_2\,\mathrm{Tr}\bar{B}B\,(\xi\mathcal{M}\xi + h.c.) + a_3\,\mathrm{Tr}\bar{B}B\,\mathrm{Tr}\,(\mathcal{M}U + h.c.)\,.
\end{aligned} \tag{10.35}$$

In writing this, only CP-even terms are kept. In general, without restriction on CP, the $h.c.$ term would be multiplied by complex conjugation of the coefficient which would in general be complex. The mass matrix $\mathcal{M}$ is formally $O(\partial^2)$, so to the order involved in (10.35), we have to implement with a Lagrangian that contains two derivatives. We will call it $\delta\mathcal{L}_B$. It will be specified below.

As given, the baryon Lagrangian (10.32) does not lend itself to a straightforward chiral expansion[25]. This is because the baryon mass m_B is not small compared with the chiral scale $\Lambda_\chi \approx 1$ GeV: as explained in more detail below, while the space derivative acting on the baryon field can be considered to be small, the time derivative cannot be, since it picks up the baryon mass term. It is therefore more appropriate to consider the baryon to be heavy and make a field redefinition so as to eliminate the mass term from the Lagrangian. This leads to the heavy-fermion formalism (HFF). To do this we let[26]

$$B \rightarrow e^{im_B v \cdot x} B_+ \equiv e^{im_B v \cdot x} \frac{1 + \gamma \cdot v}{2} B_v \tag{10.36}$$

where v_μ is the velocity four-vector of the baryon with $v^2 = 1$. Each field is labeled by v, which in the $m_B \rightarrow \infty$ constitutes a super-selection rule. The rescaling (10.36) eliminates the centroid mass dependence m_B from the baryon Lagrangian $\mathcal{L}_B$ and relegate the negative-energy Dirac component

$\frac{1}{2}(1 - \gamma \cdot v)B$ to a $1/m_B$ correction. Denoting B_+ simply by B (with the label v understood), we have

$$\mathcal{L}_B = \text{Tr}\bar{B}iv \cdot DB + 2D\,\text{Tr}\bar{B}S^\mu\{A_\mu, B\} + 2F\,\text{Tr}\bar{B}[A_\mu, B] + \cdots \quad (10.37)$$

and $\delta\mathcal{L}_B$ takes the form

$$\begin{aligned}
\delta\mathcal{L}_B = \ & c_1\,\text{Tr}\bar{B}D^2B + c_2\,\text{Tr}\bar{B}(v \cdot D)^2B + d_1\,\text{Tr}\bar{B}A^2B \\
& + d_2\,\text{Tr}\bar{B}(v \cdot A)^2B + f_1\,\text{Tr}\bar{B}(v \cdot D)(S \cdot A)B \\
& + f_2\,\text{Tr}\bar{B}(S \cdot D)(v \cdot A)B + f_3\,\text{Tr}\bar{B}[S^\alpha, S^\beta]A_\alpha A_\beta B \\
& + \cdots
\end{aligned} \quad (10.38)$$

where the ellipsis stands for other terms of the same class which we do not need explicitly. Here S_μ is the spin operator defined as $S_\mu = \frac{1}{4}\gamma_5[\not{v}, \gamma_\mu]$ constrained to $v \cdot S = 0$ and the constants a_i, c_i, d_i, f_i and r are parameters to be fixed from experiments.

In many-body systems like nuclei, we have to include also higher order terms in baryon field, such as[27]

$$\mathcal{L}_4 = \sum_i \text{Tr}\left(\bar{B}\Gamma_i B\right)^2 + \cdots \quad (10.39)$$

where Γ_i are matrices constrained by flavor and Lorentz symmetries which we assume do not contain derivatives. The ellipsis stands for four-fermion interactions containing derivatives and also for higher-fermion interaction terms. Such terms with quartic fermion fields can be thought of as arising when meson exchanges with the meson mass greater than the chiral scale are eliminated. We can also have derivatives multiplying Γ_i. We will encounter such derivative-dependent four-fermion terms later on – as counterterms – when we consider nuclear responses to electroweak interactions.

In the form given above for $\mathcal{L}_B$ and $\delta\mathcal{L}_B$, derivatives on the baryon field pick up small residual four-momentum k_μ

$$p^\mu = m_B v^\mu + k^\mu. \quad (10.40)$$

In this form, a derivative on the baryon field is of the same order as a derivative on a pseudoscalar meson field or the mass matrix $\sqrt{\mathcal{M}}$. Also, any order in the chiral expansion calls for a specific dependence on v. A change in v in (10.40) involves different chiral orders but the physics is unchanged thanks to reparametrization invariance[28].

The Lagrangian we will work with then is the sum of (10.33), (10.37), (10.38) and (10.39)

$$\mathcal{L} = \mathcal{L}^U + \mathcal{L}_B + \delta\mathcal{L}_B + \mathcal{L}_4. \tag{10.41}$$

Now given this Lagrangian for three-flavor system, it is easy to reduce it to one for two flavors – which is what we need for most of nuclear physics applications. The pseudoscalar field is then a 2×2 matrix

$$\sqrt{2}\pi = \begin{pmatrix} \frac{1}{\sqrt{2}}\pi^0 & \pi^+ \\ \pi^- & -\frac{1}{\sqrt{2}}\pi^0 \end{pmatrix} \tag{10.42}$$

and the baryon field becomes a simple doublet

$$N = \begin{pmatrix} p \\ n \end{pmatrix} \tag{10.43}$$

with the resulting Lagrangian

$$\begin{aligned}
\mathcal{L} &= N^\dagger i\partial_0 N + iN^\dagger V_0 N - g_A N^\dagger \vec{\sigma} \cdot \vec{A} N \\
&+ bN^\dagger \left(\xi \mathcal{M}\xi + h.c.\right) N + h'N^\dagger N \operatorname{Tr}\left(\mathcal{M}U + h.c.\right) + \cdots \\
&+ \sum_i (N^\dagger \Gamma_i N)^2 + \delta\mathcal{L}_B \cdots,
\end{aligned} \tag{10.44}$$

where we have replaced $(F + D)$ by g_A, relevant for neutron beta decay, a procedure which is justified at the tree order. This relation no longer holds, however, when loop corrections are included. We should note that in HFF, we do not use the scalar density $\bar{N}N$, since this differs from the density $N^\dagger N$ by the "$1/m$" correction. Later when we embed the strong interaction system in dense matter, we will be dealing with $\langle N^\dagger N\rangle$, not with $\langle \bar{N}N\rangle$.

10.3.3. *Chiral power counting*

In this chapter, we are concerned with an amplitude that involves E_N nucleon lines (sum of incoming and outgoing lines), E_π pion lines and at most one electroweak current. The chiral Lagrangian approach is relevant at low energy and low momentum, so we are primarily concerned with slowly varying electroweak fields. In dealing with chiral power counting, the pion field and its coupling are simple because chiral symmetry requires derivative coupling. As mentioned, the chiral field U can be counted as $O(1)$

in the chiral counting, a derivative acting on the pion field as $O(Q)$ where Q is a characteristic pion four-momentum involved in the process, assumed to be of $O(m_\pi)$ or less – and in any case much less than the chiral scale of order 1 GeV – and $m_q \sim m_\pi^2 \sim O(Q^2)$. The situation with the baryon as given in (10.32) is somewhat more complicated because the baryon mass is of $O(1)$ relative to the chiral scale. Indeed, one should count $m_B \sim O(1)$, the generic baryon field $\psi \sim O(1)$ and hence $\partial^\mu \psi \sim O(1)$. This means that

$$(i\gamma_\mu \partial^\mu - m_B)\psi = iv_\mu \partial^\mu N \sim O(Q). \tag{10.45}$$

However the three-momentum of the baryon – and hence the space derivative of the baryon field – should be taken to be $O(Q)$ while its time derivative should be taken to be $O(1)$. The heavy-fermion formalism takes care of this problem of the heavy baryon mass.

In deducing the chiral counting rule, we must distinguish two types of graphs: one, *reducible* graphs and the other, *irreducible* graphs. Chiral expansion is useful for the second class of diagrams. An irreducible graph is a graph that cannot be made into disconnected graphs by cutting any intermediate state containing $E_N/2$ nucleon lines and either all the initial pions or all the final pions. Reducible graphs involve energy denominators involving the difference in nucleon kinetic energy which is much smaller than the pion mass whereas irreducible graphs involve energy denominators of $O(Q)$ or the pion mass. We must apply chiral perturbation theory to the irreducible graphs, but not to the reducible ones. The reducible ones involve infrared divergences and are responsible for such nonperturbative effects as binding or phase transitions. In the language of effective field theories [29], it is the effect of reducible graphs that transform "marginal" or "irrelevant" terms into "relevant" leading to interesting phenomena such as superconductivity. In the present context, the reducible graphs are included when one solves Lippmann-Schwinger equation or Schrödinger equation. We need therefore to confine ourselves to the irreducible graphs only.

Furthermore the counting rule is much more clearly defined without vector mesons than with vector mesons. The vector mesons with their mass comparable to the chiral scale bring in additional scale to the problem.

In what follows, we rederive and generalize Weinberg's counting rule applicable to *irreducible* graphs [30] using HFF. Although we shall not consider explicitly the vector-meson degrees of freedom, we include them here in addition to pions and nucleons. (We will treat vector mesons as very heavy. However we shall not use the heavy-meson formalism of Chapter 9 since the vector mesons we are considering are highly off-shell.) Much of what we obtain later turn out to be valid in the presence of vector mesons. Now in dealing with them, their masses will be regarded as comparable to the characteristic chiral mass scale of QCD $\sim 4\pi f_\pi$ which can be considered heavy compared to Q – say, scale of external three momenta or m_π. There is one subtlety we will encounter shortly and that is that in some situation the vector mesons have to be considered "light." The best way to handle this subtlety is to use hidden gauge symmetry[11].

In establishing the counting rule, we make the following key assumptions: Every intermediate meson (whether heavy or light) carries a four-momentum of order of Q while every intermediate nucleon carries a four-momentum of order of $m_N \sim 4\pi f_\pi$. In addition we assume that for any loop, the effective cutoff in the loop integration is of order of Q. We will be more precise as to what this means physically when we discuss specific processes, for this clarifies the limitation of the chiral expansion scheme.

An arbitrary Feynman graph can be characterized by the number $E_N(E_H)$ of external – both incoming and outgoing – nucleon (vector-meson) lines, the number L of loops, the number $I_N(I_\pi, I_H)$ of internal nucleon (pion, vector-meson) lines and the number C of separated pieces. Each vertex can in turn be characterized by the number d_i of derivatives and/or of m_π factors and the number n_i (h_i) of nucleon (vector-meson) lines attached to the vertex. For a nucleon intermediate state of momentum $p^\mu = mv^\mu + k^\mu$ where $k^\mu = \mathcal{O}(Q)$, we acquire a factor Q^{-1}, an internal pion line contributes a factor Q^{-2} while a vector-meson intermediate state contributes Q^0 ($\sim O(1)$). Finally a loop contributes a factor Q^4 because its effective cutoff is assumed to be of order of Q. We thus arrive at the counting rule that an arbitrary graph is characterized by the factor Q^ν (times a slowly varying function $f(Q/\mu)$ where $\mu \sim \Lambda_\chi$) with

$$\nu = -I_N - 2I_\pi + 4L + 4 - 4C + \sum_i d_i \qquad (10.46)$$

where the sum over i runs over all vertices of the graph. Using the identities, $I_\pi + I_H + I_N = L + V - C$, $I_H = \frac{1}{2}\sum_i h_i - E_H/2$ and $I_N = \frac{1}{2}\sum_i n_i - E_N/2$, we can rewrite the counting rule

$$\nu = 4 - 2C - \frac{E_N + 2E_H}{2} + 2L + \sum_i \nu_i, \quad \nu_i \equiv d_i + \frac{n_i + 2h_i}{2} - 2. \qquad (10.47)$$

We recover the counting rule derived by Weinberg[30] if we set $E_H = h_i = 0$.

The situation is different depending upon whether or not there is external gauge field (*i.e.*, electroweak field) present in the process. In its absence (as in nuclear forces), ν_i is non-negative

$$d_i + \frac{n_i + 2h_i}{2} - 2 \geq 0. \tag{10.48}$$

This is guaranteed by chiral symmetry. This means that the leading order effect comes from graphs with vertices satisfying

$$d_i + \frac{n_i + 2h_i}{2} - 2 = 0. \tag{10.49}$$

Examples of vertices of this kind are: πNN $(d_i = 1,\ n_i = 2,\ h_i = 0)$, hNN $(d_i = 0,\ n_i = 2,\ h_i = 1)$, $(\overline{N}\Gamma N)^2$ $(d_i = 0,\ n_i = 4,\ h_i = 0)$, $\pi\pi NN$ $(d_i = 1,\ n_i = 2,\ h_i = 0)$, $\rho\pi\pi$ $(d_i = 1,\ n_i = 0,\ h_i = 1)$, etc. In NN scattering or in nuclear forces, $E_N = 4$ and $E_H = 0$, and so we have $\nu \geq 0$. The leading order contribution corresponds to $\nu = 0$, coming from three classes of diagrams; one-pion-exchange, one-vector-meson-exchange and four-Fermi contact graphs. In πN scattering, $C = 1$, $E_N = 2$ and $E_H = 0$, we have $\nu \geq 1$ and the leading order comes from nucleon Born graphs, seagull graphs and one-vector-meson-exchange graphs. (We note here that scalar glueball fields χ which will be introduced later through conformal anomaly play only a minor role in πN scattering because the $\chi\pi\pi$ vertex $(d_i = 2,\ n_i = 0,\ h_i = 1)$ acquires an additional Q power.)

In the presence of external fields, the condition becomes[31,32]

$$\left(d_i + \frac{n_i + 2h_i}{2} - 2 \right) \geq -1. \tag{10.50}$$

The difference from the previous case comes from the fact that a derivative is replaced by a gauge field. The equality holds only when $h_i = 0$. We will later show that this is related to what is called the "chiral filter" phenomenon. The condition (10.50) plays an important role in determining exchange currents in nuclei. Apart from the usual nucleon Born terms which are in the class of "reducible" graphs and hence do not enter into our consideration, we have two graphs that contribute in the leading order to the exchange current: the "seagull" graphs and "pion-pole" graphs in standard jargon in the literature, both of which involve a vertex with $\nu_i = -1$. On the other hand, a vector-meson-exchange graph involves a

$\nu_i = +1$ vertex. This is because $d_i = 1$, $h_i = 2$ at the $J_\mu hh$ vertex. Therefore vector-exchange graphs are suppressed by power of Q^2. This counting rule is the basis for chiral filtering.

The key point of chiral perturbation theory is that for small enough Q with which the system is probed, physical quantities should be dominantly given by the low values of ν. The leading term is therefore given by tree diagrams with the next-to-leading order being given by one-loop diagrams with vertices that involve lowest derivative terms. The divergences present in one-loop graphs are then to be canceled by counterterms in the next order derivative terms and finite counterterms are to be determined from empirical data. This can be continued to an arbitrary order in a well-defined manner. A systematic application of this strategy to $\pi\pi$ scattering has met with an impressive success[33] as we discussed in Chapters 5 and 6.

10.3.4. *Effective theories in nuclear medium*

Now given an effective Lagrangian constructed above, how does one "embed" it in the many-particle system or medium? This issue – a long standing problem in nuclear physics – is still not resolved satisfactorily. In this section, we address this problem in a somewhat heuristic way invoking several (perhaps drastic) simplifying assumptions. It is not obvious that *all* the assumptions are correct and our hope is that future experiments will help validate or invalidate some of those assumptions.

The key argument that we shall employ is borrowed from the Wilsonian renormalization group strategy used in condensed matter and particle physics[29,34]. Consider a generic action $S(\phi)$ defined with a bare cutoff Λ. We decompose the field ϕ into "hard" (H) and "soft" (S) components, $\phi = \phi_H + \phi_S$, and rewrite the generating functional as

$$\int [d\phi_S][d\phi_H]e^{iS(\phi_S,\phi_H)} = \int [d\phi_S]e^{iS_\Lambda(\phi_S)} \qquad (10.51)$$

with

$$e^{iS_\Lambda(\phi_S)} \equiv \int [d\phi_H]e^{iS(\phi_S,\phi_H)}. \qquad (10.52)$$

The effective action S_Λ is given by the sum of all the terms allowed by the symmetries of the fundamental theory

$$S_\Lambda = \int d^4x \sum_i g_i O_i \qquad (10.53)$$

where g_i are cutoff-dependent couplings, and O_i cutoff-dependent operators. The effective actions we have discussed so far are a variant of (10.53) for the QCD action. How are these arguments modified in matter?

In dilute matter, for which the Fermi momentum $k_F \ll \Lambda$ (or $\rho \ll \rho_0$), it is reasonable to include the matter effects in the soft part of the effective action. In this case, the coefficients in (10.53) are expected to be independent of k_F. The scale separation used is that of the vacuum, and all what we said using either $1/N_c$ arguments or chiral perturbation theory should apply.

In a dense hadronic matter with $k_F \lesssim \Lambda$, the coefficients in (10.53) will surely carry a k_F dependence. One therefore expects that

$$S_\Lambda \to S_\Lambda^\star = \int d^4x \sum_i g_i^* O_i^* \qquad (10.54)$$

where the star indicates the dependence on k_F. The key point here is that (10.54) in medium is assumed to preserve – in the spirit of effective field theories – the same symmetry structure of the matter-free space. This assumption has been extensively used in condensed matter physics [34].

Scaled parameters

Although conceptually clear, the above arguments are difficult in practice to implement systematically starting from a given chiral Lagrangian. At a few times nuclear matter density, however, $k_F \sim \Lambda$, so a scale separation for assessing the k_F dependence of the coefficient functions can become difficult.* At these densities, the approach leading to (10.54) and thus the effective actions we have discussed before, may break down: In the skyrmion picture, higher-order terms in the effective action could become dominant, while in chiral perturbation theory, higher-loop effects and higher derivative terms could become so. This is a domain where lattice simulations would be most welcome, given the paucity of non-perturbative approaches at low energy. Finding the method to obtain an expansion of the type (10.54) for QCD in matter remains a challenge for nuclear theorists.

In this subsection, we present arguments that lead to interesting predictions for the scaled parameters that can be used in mean-field approximations. The relevant degrees of freedom – nucleons and mesons – that are

---

*See also Chapter 11 for a similar discussion on the temperature effects.

involved here could be considered as quasiparticles, in the sense that residual interactions are ignored in the first approximation. We can gain some idea as to how the low-energy constants $g^\star$ (including masses) in (10.54) depend on medium from the Nambu-Jona-Lasinio model[35]. Aside from the Goldstone modes which are protected by chiral symmetry up to the critical density, all hadronic masses are given by three times (baryons) or two times (mesons) the constituent quark mass[†] m_Q leading to

$$\frac{m_H^\star}{m_H} \approx \frac{m_Q^\star}{m_Q} \equiv \Phi(\rho). \tag{10.55}$$

The prediction of the NJL model for (10.55) is shown in Fig. 10.2. The same model predicts

$$f_\pi^\star/f_\pi \approx \Phi(\rho), \tag{10.56}$$

$$g_{\pi NN}^\star/g_{\pi NN} \approx 1. \tag{10.57}$$

Both the Gell-Mann-Oakes-Renner relation and the Goldberger-Treiman relation are maintained. The former implies that the chiral condensate scales with the constituent quark mass, while the latter $g_A^\star/g_A \sim 1$.

The scaling relations (10.55) and (10.56) have been encapsulated into a "universal scaling" for QCD in medium at low energy by Brown and Rho [36,37,38] (Brown-Rho scaling) and used to investigate a variety of medium phenomena, as we will discuss below. Although the reasoning is valid in mean field and hence does not lend readily to systematic "higher-order" corrections, the results are generic and are expected to hold in a more rigorous approach. Indeed, one can arrive at the same scaling starting from a chiral Lagrangian that contains n-Fermi interactions for $n \geq 4$ that leads to Walecka mean-field theory in nuclear matter[39,40]. This approach could offer a possible systematic calculational scheme. We shall not dwell on this approach but instead sketch the original arguments given in 1991[36].

The key ingredient for the argument is that while there is no low-lying local scalar degree of freedom in medium-free space, there can be one in medium – called "sigma (σ)" – at about 500 MeV at nuclear matter density. This is the sigma field that plays an important role in Walecka theory. We

---

[†]In the NJL model, most of the hadronic excitations are unstable against QQQ-decays (baryons) and $Q\overline{Q}$-decays (mesons), due to the absence of confinement. The exception is the Goldstone modes. In this sense, the issue of whether the vector mesons or baryons are bound in the model or its variant is moot.

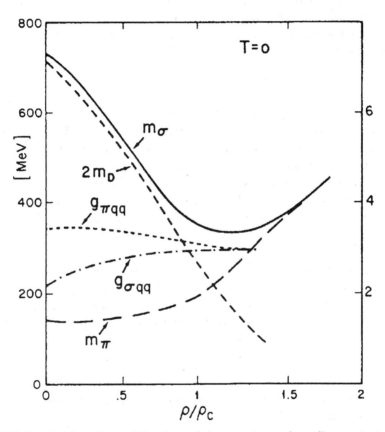

Fig. 10.2. Density dependence of the pion and sigma masses, and coupling constants in the Nambu-Jona-Lasinio model [35].

assume that this field is a chiral-singlet. We also assume that at some
high density, as suggested by Weinberg[41], there is a "mended symmetry"
which brings a scalar meson into a multiplet with the triplet of the pion.
Following Beane and van Kolck[42], we associate this scalar with the quarkish
(or dilaton) component of the scalar χ_D associated with the scale anomaly
discussed in Chapter 1. By continuity, it is natural to associate the scalar
χ_D with the Walecka scalar σ as was done in [39].

To incorporate the above observation into a scaling law, we introduce
the scalar field χ associated with the trace anomaly

$$\theta^\mu_\mu \equiv \partial_\mu D^\mu = \frac{\beta(g)}{2g} G^2 \equiv \chi^4 \tag{10.58}$$

where we ignored the quark mass contribution. We wish to write an effective
Lagrangian that is to be treated in mean field for describing physics in
medium, much in the same spirit as Walecka mean field theory. In doing
this, we use the fact that the trace anomaly is entirely given – apart from
a quark-mass term – by the χ^4 in (10.58). Therefore we write the effective
Lagrangian as a term which is scale-invariant $\mathcal{L}_{inv}$ and a potential term
$\mathcal{V}(\chi)$ which breaks the scale in the way dictated by QCD:

$$\mathcal{L}^{eff} = \mathcal{L}_{inv} + \mathcal{V} \tag{10.59}$$

with $(-4 + \chi \frac{\partial}{\partial \chi})\mathcal{V} \equiv \mathcal{D}\mathcal{V} = \chi^4$. We will assume that the potential contains
other terms that are invariant under scaling and that it determines the
"vacuum" expectation value $\chi^\star \equiv \langle \chi \rangle^\star$ for a given density ρ of the system.

We can now write down the Lagrangian $\mathcal{L}_{inv}$ using the U and χ fields.
Consider the current algebra term

$$\frac{f^2}{4} \text{Tr}(\partial_\mu U \partial^\mu U^\dagger).$$

We want this term to be zero under the $\mathcal{D}$ operation, so the appropriate
scale-invariant form is

$$\frac{f^2}{4} (\frac{\chi}{\chi_0})^2 \text{Tr}(\partial_\mu U \partial^\mu U^\dagger) \tag{10.60}$$

where χ_0 is the matter-free vacuum expectation value of the χ field. Now
in a medium defined by $\chi^\star$ for a given medium density, we can shift the χ
field in the usual manner $\chi \rightarrow \chi^\star + \chi$ and extract the current algebra term

$$\frac{(f^\star)^2}{4} \text{Tr}(\partial_\mu U \partial^\mu U^\dagger) \tag{10.61}$$

with $f^\star \chi_0 \equiv f\chi^\star$.

Next we consider the Skyrme quartic term

$$\frac{1}{32g^2} \, \mathrm{Tr}[U^\dagger \partial_\mu U, U^\dagger \partial_\nu U]^2 \tag{10.62}$$

which, as mentioned above, can be considered as coming from a massive vector meson when the latter is integrated out. This term is scale-invariant by itself, so no χ field needs to be multiplied into it. It follows immediately that

$$g^\star = g. \tag{10.63}$$

When the vector mesons ρ and ω are treated as hidden gauge bosons as done by Bando *et al.* [11], the coupling g is related to the vector-meson (V) mass by the KSRF relation $m^{\star 2}_V = 2f^{\star 2}g^2$ from which follows

$$\frac{m_V^\star}{m_V} \approx \frac{f^\star}{f} = \Phi(\rho). \tag{10.64}$$

Next from the skyrmion description of the baryon given in Chapter 7, we deduce that $m_B^\star \propto f^\star/g$ and hence [‡]

$$\frac{m_B^\star}{m_B} \approx \Phi(\rho). \tag{10.65}$$

Identifying the dilaton degree of freedom as given by Beane and van Kolck[42] with the quarkish component of χ, χ_D, we also obtain

$$\frac{m_\sigma^\star}{m_\sigma} \approx \Phi(\rho). \tag{10.66}$$

As mentioned above, the pion mass will be protected by chiral symmetry up to $\rho \sim \rho_0$ as in the case of the Nambu-Jona-Lasinio model.

It should be noted that the scaling of the parameters is governed by the dilatonic (quarkish) component of the scalar field, not by the gluonic component which, we know, does not change much as a function of density

---

[‡] As discussed elsewhere[38], there is an important short-range correction in nuclear matter for the nucleon mass in comparison with the meson mass, that is effective up to nuclear matter density. This correction gives the scaling $\frac{m_B^\star}{m_B} \approx \sqrt{g_A^\star/g_A}\,\Phi(\rho)$. As shown below, at nuclear matter density, $\sqrt{g_A^\star/g_A} \approx 0.89$. This compares with the QCD sum-rule result given below, $m_N^\star/m_N = 0.86\Phi$, at nuclear matter density.

or temperature. This was emphasized by Adami and Brown[37] and clarified further in the Beane-van Kolck formalism by Brown and Rho[38].

Although there are no lattice calculation with density, various QCD sum-rule calculations have been carried out for the ρ meson and nucleon in medium. Confirming an earlier calculation by Hatsuda and Lee[43] for the ρ mass, Jin and collaborators[44,45] obtained

$$m_\rho^\star(\rho_0)/m_\rho \;=\; 0.78 \pm 0.08, \tag{10.67}$$

$$m_N^\star(\rho_0)/m_N \;=\; 0.67 \pm 0.05. \tag{10.68}$$

This implies that $\Phi(\rho_0) \approx 0.78$ and $m_N^\star/m_N \approx 0.86\Phi(\rho_0)$.

10.3.5. *Applications*

(i) Nuclear forces

- *Two-body forces*

As a first application of the strategy developed above, we discuss Weinberg's approach to nuclear forces[30], in particular the consequences on n-body nuclear forces (for n=2,3,4) of the leading order chiral expansion with (10.44). From eq.(10.47), one can see that we want $L = 0$ (or tree approximation) and for nuclear forces,

$$d_i + \frac{1}{2}n_i - 2 \geq 0, \tag{10.69}$$

hence we want $d_i + \frac{1}{2}n_i - 2 = 0$ which can be satisfied by two classes of vertices: the vertices with $d_i = 1$ and $n_i = 2$ corresponding to the pseudovector πNN coupling and the vertices with $d_i = 0$ and $n_i = 4$. The former gives rise to the standard one-pion exchange force and the latter, coming from the four-nucleon interaction in (10.44), gives rise to a short-ranged δ-function force. For two-body forces, we have[§]

$$\begin{aligned} V_2 = \quad &- \left(\frac{g_A}{2f_\pi}\right)^2 (\vec{\tau}_1 \cdot \vec{\tau}_2)(\vec{\sigma}_1 \cdot \vec{\nabla}_1)(\vec{\sigma}_2 \cdot \vec{\nabla}_2)Y(r_{12}) - (1 \leftrightarrow 2) \\ &+ \; 2(C_S + C_T\vec{\sigma}_1 \cdot \vec{\sigma}_2)\delta(\vec{r}_{12}) \cdots \end{aligned} \tag{10.70}$$

[§]Unless needed specifically we shall not indicate the scaled parameters. They can be easily implemented for the tree order while care is needed for higher orders.

where $Y(r) \equiv e^{-m_\pi r}/4\pi r$ and $C_{S,T}$ are unknown constants. The δ-function force summarizes effects of all the massive degrees of freedom consistent with chiral invariance that have been integrated out from the effective Lagrangian.

- *Many-body forces*

Next we consider three-body forces. The graph given in Fig. 10.3 is a Feynman diagram which is genuinely irreducible and hence is a bona-fide three-body force as defined above. However the middle vertex involving $NN\pi\pi$ is of the form $N^\dagger V_0 N \sim N^\dagger \vec{\tau} \cdot \vec{\pi} \times \dot{\vec{\pi}} N \sim O(Q^2/m_B)$. So it does not contribute to the leading order as it is suppressed to $O(Q^3/m_B^3)$. Next

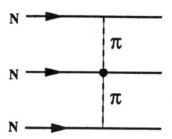

Fig. 10.3. Three-body forces suppressed up to $O(Q^2)$.

consider the Feynman diagrams Fig. 10.4. In terms of time-ordered graphs, some graphs are reducible in the sense defined above, so we should separate them out to obtain the contribution to the potential. A simple argument shows that when the reducible graphs are separated out in a way consistent with the chiral counting, *nothing* remains to the ordered considered, so the contribution vanishes provided static approximation in the one-pion exchange is adopted. This is easy to see as follows. In the Feynman diagram, each pion propagator of three-momentum $\vec{q}$ and energy q_0 is of the form $(q_0^2 - \vec{q}^2 - m_\pi^2)^{-1}$. The energy is conserved at the πNN vertex, so q_0 is just the difference of kinetic energies of the nucleon $p^2/2m_N - p'^2/2m_N$ which is taken to be small compared with the three-momentum of the pion $|\vec{q}|$. Therefore to the leading order, q_0 should be dropped in evaluating the Feynman graphs. This means that as long as the intermediate baryon is a nucleon, it is included in the iteration of the static two-body potential. Bona-fide three-body potentials arise only at higher orders, say,

$O(Q^3/m_N^3)$. ¶This means that if one does the static approximation as is customarily done in nuclear physics calculations, there are no many-body forces to the leading chiral order. *This result justifies the standard approximation in nuclear physics from the point of view of chiral symmetry and, in an indirect sense, of QCD.* Note however that to the order considered, calculating *two-body* processes – such as deuteron properties– would require q_0 in the two-body potential since there are no three-body graphs to cancel its effect.

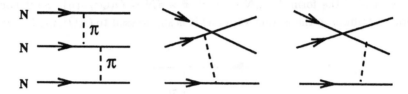

Fig. 10.4. Feynman diagrams for three-body amplitudes involving pairwise one-pion exchange that consist of both reducible and irreducible graphs.

We should remark at this point that if we were to include baryon resonances (in particular the Δ's) in the chiral Lagrangian as we will do later, there could be many-body contributions at the level of Fig. 10.4. The graph with a Δ in the intermediate state is a genuine three-body force in the space of nucleons and such graphs have been considered in the literature.

A problem with nuclear forces à la chiral symmetry is the number of unknown constants that appear if one goes beyond the tree order. To one loop and beyond, many counterterms are needed to renormalize the theory and to make a consistent chiral expansion. The contact four-fermion interactions will receive further contributions from the loop graphs and its number will increase as derivatives and chiral symmetry breaking mass terms are included. The contact terms are δ functions in coordinate space and describe high-energy degrees of freedom that are "integrated out" to describe low-energy sectors. To handle the zero-range interactions, we need to understand short-range correlations which we know from phenomenology, play an important role in nuclear dynamics. The scale involved here is not the small scale Q of order of pion mass but vastly greater. In order to have an idea as to whether – and in what sense – the chiral expansion

---

¶This simple argument is based on Weinberg.[46] For higher-order consideration, see [47].

is meaningful, we need to calculate graphs beyond the tree order. The pragmatic point of view one can take is that chiral symmetry can *constrain* phenomenological approaches in nuclear-force problem, but not really *predict* any new phenomena. Even so, remarkably enough, a recent calculation of the nucleon-nucleon interactions at one-loop order by van Kolck[48] who used a momentum cutoff implemented with form factors shows that the NN phase shifts can be understood up to about 100 MeV by one-loop potentials involving pions and nucleons alone (except perhaps for the tensor-force component which is presumably influenced by the ρ exchange).

One of the crucial questions in connection with the in-medium Lagrangian discussed above is: In view of the scaling nucleon mass in medium, at what point does the heavy-fermion approximation break down, say, in heavy nuclei? The point is that the smallness of many-body forces depends crucially on the largeness of the nucleon mass. Will the dropping mass not invalidate the low-order chiral expansion and thereby increase the importance of many-body forces in heavy nuclei?

(ii) Exchange currents

When a nucleus is probed by an external field, the transition is described by an operator that consists of one-body, two-body and many-body ones. It is usually dominated by a one-body operator and in some situations two-body operators play an important role. It has been known since some time that under certain kinematic conditions, the two-body operator can even dominate. The well-known example is the threshold electrodisintegration of the deuteron at large momentum transfers. Another case is the axial-charge transition in nuclei. In this section, we describe how one can calculate many-body currents from chiral symmetry point of view. First we have to introduce external fields into the effective Lagrangian. Denote the vector and axial-vector fields by

$$V_\mu = \frac{\tau^i}{2} \mathcal{V}_\mu^i, \quad A_\mu = \frac{\tau^i}{2} \mathcal{A}_\mu^i. \tag{10.71}$$

Then the coupling can be introduced by gauging the Lagrangian (10.37) and (10.38). This is simply effectuated by the replacement

$$D_\mu \rightarrow \mathcal{D}_\mu \equiv D_\mu - \frac{i}{2}\xi^\dagger \left(\mathcal{V}_\mu + \mathcal{A}_\mu\right)\xi - \frac{i}{2}\xi \left(\mathcal{V}_\mu - \mathcal{A}_\mu\right)\xi^\dagger,$$

$$A_\mu \rightarrow \Delta_\mu \equiv A_\mu + \frac{1}{2}\xi^\dagger \left(\mathcal{V}_\mu + \mathcal{A}_\mu\right)\xi - \frac{1}{2}\xi \left(\mathcal{V}_\mu - \mathcal{A}_\mu\right)\xi^\dagger. \tag{10.72}$$

The corresponding vector and axial-vector currents denoted respectively by $J^{i\mu}$ and $J_5^{i\mu}$ can be read off

$$
\begin{aligned}
\vec{J}_{5\mu} &= -f_\pi \left[\partial_\mu \vec{\pi} - \frac{2}{3f_\pi^2} \left(\partial_\mu \vec{\pi} \, \vec{\pi}^2 - \vec{\pi} \, \vec{\pi} \cdot \partial_\mu \vec{\pi} \right) \right] \\
&+ \frac{1}{2}\overline{B} \left\{ 2g_A S^\mu \left[\vec{\tau} + \frac{1}{2f_\pi^2} \left(\vec{\pi} \, \vec{\tau} \cdot \vec{\pi} - \vec{\tau} \, \vec{\pi}^2 \right) \right] \right. \\
&\left. + v^\mu \left[\frac{1}{f_\pi} \vec{\tau} \times \vec{\pi} - \frac{1}{6f_\pi^3} \vec{\tau} \times \vec{\pi} \, \vec{\pi}^2 \right] \right\} B + \cdots, \qquad (10.73)
\end{aligned}
$$

$$
\begin{aligned}
\vec{J}_\mu &= \left[\vec{\pi} \times \partial_\mu \vec{\pi} - \frac{1}{3f_\pi^2} \vec{\pi} \times \partial_\mu \vec{\pi} \, \vec{\pi}^2 \right] \\
&+ \frac{1}{2}\overline{B} \left\{ v^\mu \left[\vec{\tau} + \frac{1}{2f_\pi^2} \left(\vec{\pi} \, \vec{\tau} \cdot \vec{\pi} - \vec{\tau} \, \vec{\pi}^2 \right) \right] \right. \\
&\left. + 2g_A S^\mu \left[\frac{1}{f_\pi} \vec{\tau} \times \vec{\pi} - \frac{1}{6f_\pi^3} \vec{\tau} \times \vec{\pi} \, \vec{\pi}^2 \right] \right\} B + \cdots. \qquad (10.74)
\end{aligned}
$$

Here B is the "heavy" nucleon field defined before.

It is now easy to see what the chiral counting tells us[31]. Consider an amplitude involving four external (two incoming and two outgoing) nucleon fields interacting with a slowly-varying electroweak current. It can be characterized just as for nuclear forces with the exponent (10.47) with one difference: the vertices involving electroweak currents satisfy instead

$$
d_i + \frac{1}{2}n_i - 2 \geq -1. \qquad (10.75)
$$

Thus the leading order graphs are given by $L = 0$ with a single vertex probed by an electroweak current for which $d_i + \frac{1}{2}n_i - 2 = -1$ and πNN vertices satisfying $d_i + \frac{1}{2}n_i - 2 = 0$. One can easily see that such graphs are given only by one-pion exchange terms in which the current is coupled to a πNN vertex or hooks onto a pion line and that to this order, *there is no contribution from the four-nucleon contact interaction or from vector-meson exchanges.* Effectively then, there are only two soft-pion graphs for electromagnetic field (*i.e.*, the "pair" and "pionic" graphs in the jargon used in the literature[49]) and only one graph ("pair") for weak currents. These are given in Fig. 10.5. We should emphasize that there are no heavy-meson or multi-pion exchanges at this order, a situation that renders the calculation for currents markedly simpler than for forces. What this means is that when massive degrees of freedom are explicitly included, they can

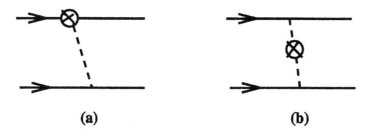

Fig. 10.5. The leading "soft-pion" graphs contributing to exchange electromagnetic currents ([a] and [b]) and to exchange axial charge ([a]). The circled cross here stands for a vector current V^μ or an axial-vector current A^μ.

contribute only at $O(Q^2)$ relative to the tree order. Put more precisely, if one includes one soft-pion (A_π), two or more pion $(A_{\pi\pi})$ and vector meson (A_V) exchanges, excited baryons (A_{N^*}) and appropriate form factors (A_{FF}) as one customarily does in exchange current calculations, consistency with chiral symmetry requires that even though each contribution could be large, the sum must satisfy

$$A_\pi + A_{\pi\pi} + A_V + A_{N^*} + A_{FF} \approx A_\pi(1 + O(Q^2)). \tag{10.76}$$

This is the "chiral filter" effect conjectured a long time ago[50]. Later we will give a simple justification of this result by one-loop chiral perturbation theory and give a numerical value for the $O(Q^2)$ term. The surprising thing is that the relation (10.76) holds remarkably well for the famous threshold electrodisintegration of the deuteron as well as in the magnetic form factors of ^{3}He and ^{3}H to large momentum transfers[51]. In the case of nuclear axial-charge transitions mediated by the time component of the isovector axial current which will be discussed below, one can construct a phenomenological model with the requisite chiral symmetry that fits nucleon-nucleon interactions which satisfies this condition well, with the $O(Q^2)$ contribution amounting to less than 10% of the dominant soft-pion contribution[52].

(iii) Axial-charge transitions in nuclei

We now illustrate how our formalism with the scaling works by applying it to what is known as axial-charge transition in nuclei. It involves the time component of isovector axial current $\bar{J}_5^{\mu\pm}$ inducing the transition $0^+ \leftrightarrow 0^-$ with $\Delta I = 1$. This transition in heavy nuclei is quite sensitive to the soft-pion exchange as well as to the scaling effect[53] and illustrates nicely the power of chiral symmetry in understanding subtle nuclear phenomena.

- *Tree order*

The axial-charge transition is described by a one-body ("impulse") axial-charge operator plus a two-body ("exchange") axial-charge operator described above‖ Three-body and higher many-body operators do not contribute at small energy-momentum transfer involved here. In the past[50], these operators were calculated without the scaling effect. Now from the foregoing discussion, it is obvious that we will get the operators of exactly the same form except that the masses m_i's are replaced by $m_i^\star$'s and the coupling constants g_i's are replaced by $g_i^\star$'s. This means that in the results obtained previously, we are to make the following replacements

$$m_N \to m_N^\star = m_N \Phi(\rho), \qquad f_\pi \to f_\pi^\star = f_\pi \Phi(\rho) \qquad (10.77)$$

with Φ a density-dependent constant to be determined elsewhere. In doing this, we are to preserve low-energy theorems as in matter-free space but with scaled parameters. At the mean-field level, the coupling constants $g_{\pi NN}$ and g_A are not affected by density as argued before. We shall come back to loop effects later. Evaluating the diagrams of Fig. 10.6, we have the axial-charge density one-body and two-body operators

$$J_5^{0\pm} = J_5^{(1)\pm} + J_5^{(2)\pm}, \qquad (10.78)$$

$$J_5^{(1)\pm} = -g_A \Phi^{-1} \sum_i \tau_i^\pm (\vec{\sigma}_i \cdot \vec{p}_i / m_N) \delta(\vec{x} - \vec{x}_i), \qquad (10.79)$$

$$J_5^{(2)\pm} = \frac{m_\pi^2 g_{\pi NN}}{8\pi m_N f_\pi \Phi} \sum_{i<j} (\tau_i \times \tau_j)^\pm [\vec{\sigma}_i \cdot \hat{r}\delta(\vec{x} - \vec{x}_j) + \vec{\sigma}_j \cdot \hat{r}\delta(\vec{x} - \vec{x}_i)] Z(r_{ij})$$

$$(10.80)$$

with

$$Z(r) = \left(1 + \frac{1}{m_\pi r}\right) e^{-m_\pi r} / m_\pi r \qquad (10.81)$$

where $r_{ij} = \vec{x}_i - \vec{x}_j$ and $\vec{p}$ the initial momentum of the nucleon making the transition. We have set the current momentum equal to zero. In writing

---

‖A much more spectacular case for the success of chiral perturbation theory in nuclei is the precise calculation of the process $n + p \to d + \gamma$ at thermal neutron energy[54]. The axial-charge transition in heavy nuclei offers a more interesting case where the scaling figures in addition to chiral symmetry.

(10.80), we have assumed that the pion mass does not scale. In addition, it will be shown later that due to short-distance effects, the ratio $g_A^\star/f_\pi^\star$ remains unscaled up to nuclear matter density.

Let us denote the one-body matrix element calculated with the scaling by $\mathcal{M}_1^*$ and the corresponding two-body exchange current matrix element by $\mathcal{M}_2^*$

$$\mathcal{M}_a^* \equiv \langle f | J_5^{(a)\pm} | i \rangle, \quad a = 1, 2. \tag{10.82}$$

Let $\mathcal{M}_a$ denote the quantities calculated with the operators in which no scaling is taken into account. For comparison with experiments, it is a common practice to consider the ratio

$$\epsilon_{MEC} \equiv \frac{\mathcal{M}^{observed}}{\mathcal{M}_1}. \tag{10.83}$$

This is a ratio of an experimental quantity over a theoretical quantity, describing the deviation of the impulse approximation from the observation. The prediction that the present theory makes is remarkably simple. It is immediate from (10.78) that

$$\epsilon_{MEC}^{th} = \frac{\mathcal{M}_1^* + \mathcal{M}_2^*}{\mathcal{M}_1} = \Phi^{-1}(1 + R) \tag{10.84}$$

where $R = \mathcal{M}_2/\mathcal{M}_1$.

- *One-loop corrections*

If one is interested in computing to $O(Q^2)$ relative to the soft-pion term in (10.76), there are two ways to proceed. One is to introduce vector mesons and saturate the correction by *tree graphs* and the other is to compute one-loop graphs with pions. Clearly the former is easier. However tree graphs with vector mesons are incomplete as we know from $O(\partial^4)$ in $\pi\pi$ scattering. There are nonanalytic terms coming from pion loops in addition to the counterterms saturated by the vectors. In this subsection, we shall compute the $O(Q^2)$ correction to R using pions and nucleons only. The Δ resonance does not contribute and the effects of the vector mesons ρ and a_1 are effectively included in the treatment through the loop effect. These are discussed in the work of Park *et al* [32,55].

Let the corrected ratio be

$$R \to (1 + \Delta)R \tag{10.85}$$

where Δ is given by the loops according to the counting rule given in (10.47). Δ can come from two sources. One is the loop correction to the soft-pion term of Fig. 10.5, essentially given by one-loop corrections to the vertex $J_5^0 N N \pi$ since there is no one-loop correction in HFF to the $\pi N N$ vertex other than renormalizing the $\pi N N$ coupling constant $g_{\pi N N}$. Now the one-loop correction to the $J_5^0 N N \pi$ is nothing but the one-loop correction to the isovector Dirac form factor F_1^V, so it is completely determined by experiments. We will call the corresponding correction $\Delta_{1\pi}$. The other class of diagrams that give rise to $\Delta_{2\pi}$ is two-pion exchange graphs given in Figure 10.6. There are many other graphs one can draw but they do not contribute to the order considered when the HFF is used. In particular the

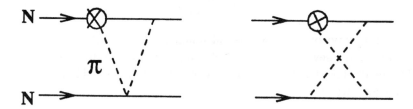

Fig. 10.6. The non-vanishing two-pion-exchange graphs of $O(Q^2)$ relative to those of Fig. 10.5. All possible insertions of the current are understood.

four-fermion contact interactions that appear in nuclear forces do not play any role in the currents. In evaluating these graphs [Fig. 10.6], counter-terms (there are effectively two of them) involving four-fermion field with derivatives are needed to remove divergences. They would also contribute to nuclear forces at the same chiral order. Now the finite counterterms that appear are the so-called "irrelevant terms" in effective theory (as they are down by the inverse power of the chiral scale parameter) and can only be probed at very short distance. Indeed the currents that come from such terms are δ functions and derivatives of the δ functions, so they should be suppressed if the wave functions have short-range correlations. Therefore the counterterms that appear here can be ignored. This can also be understood by the fact that such terms do not figure in phenomenological potentials. They could perhaps be seen if one measures small effects at very short distances. Thus *embedded in nuclear medium, the unknown counterterms do not play any role. This is the part of "chiral filtering" by nuclear matter of short-wavelength degrees of freedom.* A corollary to this "theorem" is that if the dominant soft-pion contribution is suppressed by kinematics or by selection rules, then the next order terms including $1/m_N$ corrections must enter. This is the other side of the same coin as predicted a long ago [50]. For instance, the s-wave pion production in

$$pp \to pp\pi^0 \tag{10.86}$$

is closely related to the matrix element of the axial charge operator. But because of an isospin selection rule, the soft-pion exchange current cannot contribute. Since the single-particle matrix element is kinematically suppressed, the contribution from higher chiral corrections will be substantial. This turns out to be the case[56]. The case with the space component of the axial current, namely, the coupling constant g_A in nuclei, is discussed below.

Apart from these counterterms which are rendered ineffective inside nuclei, there are no unknown parameters once the masses of the nucleon and the pion and constants such as g_A, f_π are fixed. We quote the results, omitting details. The quantity we want is

$$\Delta_{1\pi} = \langle \delta J_5^{(2)}(1\pi) \rangle / \langle J_5^{(2)}(soft) \rangle, \tag{10.87}$$

$$\Delta_{2\pi} = \langle \delta J_5^{(2)}(2\pi) \rangle / \langle J_5^{(2)}(soft) \rangle \tag{10.88}$$

where $J_5^{(2)}(soft)$ is given by the soft-pion term (10.80)

$$J_5^{(2)}(soft) = T^{(1)} \frac{d}{dr} \left[-\frac{1}{4\pi r} e^{-m_\pi r} \right], \tag{10.89}$$

with the matrix element to be evaluated in the sense of (10.82) and

$$
\begin{aligned}
\delta J_5^{(2)}(1\pi) &= \frac{m_\pi^2}{6} \langle r^2 \rangle_1^V \frac{m_\pi^2}{f_\pi^2} J_5^{(2)}(soft) \\
&\quad + \frac{T^{(1)}}{16\pi^2 f_\pi^2} \frac{d}{dr} \left\{ -\frac{1+3g_A^2}{2} \left[K_0(r) - \tilde{K}_0(r) \right] \right. \\
&\quad \left. + (2+4g_A^2) \left[K_2(r) - \tilde{K}_2(r) \right] \right\}, \tag{10.90}
\end{aligned}
$$

$$
\begin{aligned}
\delta J_5^{(2)}(2\pi) &= \frac{1}{16\pi^2 f_\pi^2} \frac{d}{dr} \left\{ -\left[\frac{3g_A^2 - 2}{4} K_0(r) + \frac{1}{2} g_A^2 K_1(r) \right] T^{(1)} \right. \\
&\quad \left. + 2g_A^2 K_0(r) T^{(2)} \right\}, \tag{10.91}
\end{aligned}
$$

where $n\pi$ in the argument of the δJ_5's stands for $n\pi$ exchange in the one-loop graphs and $\langle r^2 \rangle_1^V$ is the isovector charge radius of the nucleon. We have used the notations

$$T^{(1)} \equiv \vec{\tau}_1 \times \vec{\tau}_2 \, \hat{r} \cdot (\vec{\sigma}_1 + \vec{\sigma}_2), \tag{10.92}$$

$$T^{(2)} \equiv (\vec{\tau}_1 + \vec{\tau}_2) \hat{r} \cdot \vec{\sigma}_1 \times \vec{\sigma}_2 \tag{10.93}$$

and the functions K_i and $\tilde{K}_i$ defined by

$$K_0(r) = -\frac{1}{4\pi r} \int_0^1 dx \, \frac{2x^2}{1 - x^2} E^2 e^{-Er},$$

$$K_2(r) = -\frac{1}{4\pi r} \int_0^1 dx \, \frac{2x^2}{1 - x^2} \left(\frac{1}{4} - \frac{x^2}{12} \right) E^2 e^{-Er} \tag{10.94}$$

and

$$\tilde{K}_0(r) = \frac{1}{4\pi r} \int_0^1 dx \, \frac{2x^2}{1 - x^2} \frac{m_\pi^2}{m_\pi^2 - E^2} \left(m_\pi^2 e^{-m_\pi r} - E^2 e^{-Er} \right),$$

$$\tilde{K}_2(r) = \frac{1}{4\pi r} \int_0^1 dx \, \frac{2x^2}{1 - x^2} \left(\frac{1}{4} - \frac{x^2}{12} \right) \frac{m_\pi^2}{m_\pi^2 - E^2}$$
$$\times \left(m_\pi^2 e^{-m_\pi r} - E^2 e^{-Er} \right) \tag{10.95}$$

with $E(x) = 2m_\pi/\sqrt{1 - x^2}$. These are Fourier-transforms of McDonald functions and their derivatives, with δ function terms excised on account of short-range correlations.

For a qualitative estimate, we shall calculate the relevant matrix elements in Fermi-gas model with a short-range correlation function in the simple form (in coordinate space)

$$g(r) = \theta(r - d), \quad r = |\vec{r}_1 - \vec{r}_2|. \tag{10.96}$$

Figure 10.7 summarizes the result for various short-range cutoff d. The relevant d commonly used in the literature is 0.7 fm. It is difficult to pin down the cutoff accurately but this is roughly what realistic calculations give.

Numerically although $\Delta_{1\pi}$ and $\Delta_{2\pi}$ can be non-negligible individually, there is a large cancellation, so the total is small, $\Delta \lesssim 0.1$ leading to the chiral filter mechanism mentioned before. This cancellation mechanism which is also operative in nuclear electromagnetic processes mediated

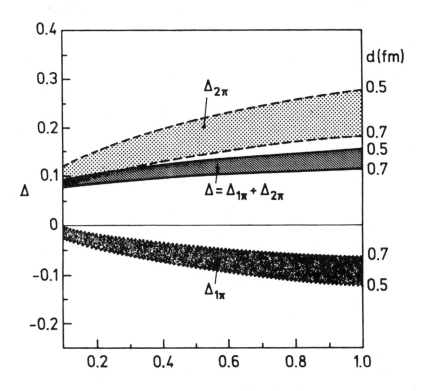

Fig. 10.7. $\Delta = \Delta_{1\pi} + \Delta_{2\pi}$ as function of density ρ/ρ_0 where $\rho_0 \approx \frac{1}{2}m_\pi^3$ is the nuclear matter density for given cutoff d.

by magnetic operators as in the electrodisintegration of the deuteron and the magnetic form factors of the tri-nucleon systems is characteristic of the strong constraints coming from chiral symmetry, analogous to the pair suppression in πN scattering.

- *Absence of multi-body currents*

For small energy transfer (say, much less than m_π), three-body and other multi-body forces do not contribute to $O(Q^2)$ relative to the soft-pion term. This can be seen in the following way.

Consider attaching axial-charge operator to the vertices in the graphs for three-body forces, Figures 10.3 and 10.4. We need not consider those graphs where the operator acts on any of the nucleon line. Now the argument that the sum of all such irreducible graphs cannot contribute to the order considered as long as the static pion propagator is used for two-body soft-pion exchange currents parallels exactly the argument used for the forces.

The situation is quite different, however, if the energy transfer is larger than, say, the pion mass as in the case where a Δ resonance is excited by an external electroweak probe. In such cases, there is no reason why three-body and higher-body currents should be small compared with one- and two-body currents. This phenomenon should be measurable by high intensity electron machines at multi-GeV region.

- *Comparison with experiments*

We shall estimate the size of the loop correction Δ in Fermi-gas model. This will give us a global estimate. Results of a more realistic treatment will be mentioned below. It turns out that the ratio R depends little on nuclear model and on mass number. In Fermi-gas model, it is given by[57]

$$R = \frac{g_{\pi NN}^2}{4\pi^2 g_A^2} \frac{p_F}{m_N} \left[1 + \lambda - \lambda(1 + \frac{\lambda}{2}) \ln(1 + \frac{2}{\lambda}) \right] \tag{10.97}$$

with $\lambda \equiv m_\pi^2/2p_F^2$, where p_F is the Fermi momentum (which is $(3\pi^2 \rho/2)^{1/3}$ for nuclear matter). Numerically this ranges from $R \approx 0.39$ for $\rho = \rho_0/2$ to $R \approx 0.54$ for $\rho = \rho_0$. Now in the same model with a reasonable short-range correlation, Δ comes out to be about 0.09, pretty much independent of nuclear density, so that we can take

$$R(1 + \Delta) \approx 1.09R. \tag{10.98}$$

The only undetermined parameter in this theory is the scaling factor Φ. Ideally one should determine it from experiments. Unfortunately it is a mean-field quantity which makes sense only for large nuclei and has not been pinned down yet. Once it is fixed in heavy nuclei, one could then use local-density approximation to describe transitions in light nuclei. At the moment it is known only from QCD sum-rule calculations of hadrons in medium, *e.g.*, the vector-meson mass[58]. For our purpose, it is good enough to take

$$\Phi(\rho) \approx 1 - 0.15\frac{\rho}{\rho_0}. \tag{10.99}$$

With eqs. (10.98) and (10.99), we predict

$$\epsilon_{MEC} \approx 1.6, 1.9 \text{ for } \rho/\rho_0 \approx 0.6, 1 \tag{10.100}$$

corresponding, roughly, to the average densities of ^{16}O and ^{208}Pb, respectively. These results are confirmed in more reliable shell-model calculations[55]. This prediction is in agreement with the empirical values[59]

$$\epsilon_{MEC}^{exp} = 1.61, 1.79 \sim 1.91 \text{ for } A = 16, 208. \tag{10.101}$$

(iv) g_A^* in Nuclei

As discussed above, at the tree level of the effective Lagrangian, the axial-vector coupling constant g_A is not renormalized in medium. Here we illustrate how the next order *in-medium* chiral correction can introduce a density dependence in the constant. Such a correction is actually required by the observation that the value of g_A measured in the neutron beta decay

$$g_A = 1.262 \pm 0.005 \tag{10.102}$$

is quenched to 1 in medium and heavy nuclei[60]. It is significant, however, that this quenching is *not* due to an influence of a modified vacuum, in other words, a "fundamental quenching" as often thought in the past. Of course at an asymptotic density, g_A should reach unity as a consequence of the melting of the quark condensate.

In nuclei, g_A^* is measured in Gamow-Teller transitions, namely the space component of the axial current. For this component, the exchange current given by the soft-pion theorem discussed above is suppressed. Therefore according to the corollary to the "chiral filter" argument, corrections to g_A^* must come from loop terms and/or higher-Q Lagrangian pieces or many-body currents involving short-distance physics. It is known that the dominant correction comes from the Landau g_0' interaction involving the excitation of Δ-hole states as given in Fig. 10.8. In the kinematical regime involved here, this effect can be incorporated as a "counterterm" with the Δ degree of freedom integrated out in the effective Lagrangian. To estimate its contribution, we extend the effective Lagrangian to decuplet (spin-3/2)

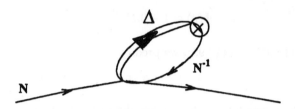

Fig. 10.8. The Δ-hole graph that plays the dominant role in quenching g_A in nuclei. The cross stands for the space component of the axial current, the thick solid line stands for the Δ isobar and N^{-1} for a nucleon hole.

baryons for three flavors. This can be done in the following way[61].

Introduce a Rarita-Schwinger field Δ^μ obeying the constraint $\gamma_\mu \Delta^\mu = 0$. This field carries $SU(3)$ tensor indices abc that are completely symmetrized to form the decuplet. The chirally symmetric Lagrangian involving the decuplet with its coupling to octet baryons and mesons equivalent to (10.32) is

$$\mathcal{L}^\Delta = i\bar\Delta^\mu \gamma_\nu \mathcal{D}^\nu \Delta_\mu - m_\Delta \bar\Delta^\mu \Delta_\mu \ + \ \alpha\left(\bar\Delta^\mu A_\mu B + h.c.\right)$$
$$+ \ \beta\bar\Delta^\mu \gamma_\nu \gamma_5 A^\nu \Delta_\mu, \qquad (10.103)$$

where α and β are unknown constants, B stands for the spin-1/2 octet baryons, Δ^μ the spin-3/2 decuplets Δ, Σ^*, Ξ^* and Ω and A^μ is the "induced" axial vector $A^\mu = \frac{i}{2}(\xi\partial^\mu\xi^\dagger - \xi^\dagger\partial^\mu\xi)$. Sum over all $SU(3)$ indices (not explicitly written down here) is understood. The covariant derivative acting on the Rarita-Schwinger field is defined by

$$\mathcal{D}^\nu \Delta^\mu_{abc} = \partial^\nu \Delta^\mu_{abc} + (V^\nu)^d_a \Delta^\mu_{dbc} + (V^\mu)^d_b \Delta^\mu_{adc} + (V^\mu)^d_c \Delta^\mu_{abd} \qquad (10.104)$$

with $V^\mu = \frac{1}{2}(\xi\partial^\mu\xi^\dagger + \xi^\dagger\partial^\mu\xi)$. Under chiral $SU_L(3) \times SU_R(3)$ transformation, the Rarita-Schwinger field transforms as

$$\Delta^\mu_{abc} \rightarrow g^d_a g^e_b g^f_c \Delta^\mu_{def}. \qquad (10.105)$$

As before, we have terms quadratic in baryon fields which we may write

$$\mathcal{L}_4^\Delta = \gamma \left(\bar{\Delta}^\mu C_{\mu\nu} B \bar{B} \mathcal{F}^\nu B + h.c. \right)$$
$$+ (\Delta\Delta BB) + (\Delta B \Delta B) + (\Delta\Delta\Delta\Delta) + \cdots, \quad (10.106)$$

where C and $\mathcal{F}$ are appropriate tensors allowed by Lorentz and isospin invariances. We have indicated explicitly only the ΔBBB components as they play – in particle-hole channel – the essential role for g_A^*.

As with the nucleon, we can use the heavy-baryon formalism for the Δ. Evaluating the diagram in Fig. 10.8 requires information on the axial-vector coupling constant g_A^Δ connecting to Δ and N and the contact interaction g'_Δ for $\Delta N \leftrightarrow NN$ in the *particle-hole* channel. The latter is analogous to the Landau-Migdal particle-hole interaction $g'_0 \vec{\tau}_1 \cdot \vec{\tau}_2 \vec{\sigma}_1 \cdot \vec{\sigma}_2 \delta(r_{12})$ governing spin-isospin interactions in nuclei and can be written in terms of the transition spin operator $\vec{S}$ and transition isospin operator $\vec{T}$ as[62]

$$g'_\Delta \vec{T}_1 \cdot \vec{\tau}_2 \vec{S}_1 \cdot \vec{\sigma}_2 \delta(r_{12}). \quad (10.107)$$

Neither g_A^Δ nor g'_Δ is known sufficiently well. Assuming that the axial-vector current couples to the Δ and N in the same way as the pion does, one usually takes

$$g_A^\Delta / g_A \approx 2. \quad (10.108)$$

We can then roughly estimate in small-density approximation that

$$g_A^* / g_A = 1 - a g'_\Delta \rho (m_\Delta - m_N)^{-1} + \cdots \quad (10.109)$$

with $a \approx (\frac{2}{3} m_\pi^{-1})^2$ (where the factor $(2/3)$ comes from the matrix element of the transition spin and isospin). It turns out that the acceptable empirical value (coming from pion-nucleus scattering) for g'_Δ is about $1/3$. Taking the ΔN mass difference $\sim 2.1 m_\pi$, we have

$$g_A^* / g_A \approx 0.8 \quad \text{for} \quad \rho = \rho_0. \quad (10.110)$$

This makes g_A^* approximately 1 in nuclear matter.

There are many other diagrams contributing at the same order but none of them is as important as Fig. 10.8 except possibly for core polarizations involving particle-hole excitations through tensor interactions. What one observes in nuclei is of course the sum of all. The point of the above result is that g_A is quenched in nuclei due to higher-order chiral effects of short-range character and not due to soft pions. An important consequence of this quenching of g_A in matter is that at least up to nuclear matter density, we expect that

$$\frac{g_A^\star}{f_\pi^\star} \approx \frac{g_A}{f_\pi}. \tag{10.111}$$

Since at an asymptotic density g_A must approach 1, this scaling must break down at some high density. It is not known where.

10.3.6. *Nuclei in the "swelled world"*

The case of axial-charge transitions (and the isovector magnetic dipole process which we did not discuss here) corresponds to small energy and momentum transfers $q_\mu \approx 0$. The scaling relations (10.55)

$$\frac{m_B^*}{m_B} \approx \frac{m_V^*}{m_V} \approx \frac{m_\sigma^*}{m_\sigma} \approx \frac{f_\pi^*}{f_\pi} \equiv \Phi(\rho) \tag{10.112}$$

can be applied to other nuclear processes of similar nature. For this relation to be useful, it is necessary that the process studied should be dominated by tree graphs, with next chiral order corrections suppressed as in the axial-charge example seen above. In phenomenological approaches in nuclear physics, this is usually the case. Let us therefore consider the cases where the scaling (10.112) makes a marked improvement in fitting experimental data. Here we discuss a few of such examples[37,63].

(i) Tensor forces in nuclei

At tree level, there are two terms contributing to tensor forces in nuclei. The first is the tensor force coming from one-pion exchange. As noted before, there is little scaling in the one-pion exchange potential, so this part will be left untouched by medium. The second is a contribution from one-ρ exchange and this part will be scaled. Effectively the tensor force has the form

$$V_T(\vec{q}, \omega) = -\frac{\bar{f}_\pi^2}{m_\pi^2} \left[\frac{\vec{\sigma}_1 \cdot \vec{q}\,\vec{\sigma}_2 \cdot \vec{q} - \frac{1}{3}(\vec{\sigma}_1 \cdot \vec{\sigma}_2)q^2}{(q^2 + m_\pi^2) - \omega^2} \right] \tau_1 \cdot \tau_2$$

$$+ \frac{f_\rho^{*2}}{m_\rho^{*2}} \left[\frac{\vec{\sigma}_1 \cdot \vec{q}\,\vec{\sigma}_2 \cdot \vec{q} - \frac{1}{3}(\vec{\sigma}_1 \cdot \vec{\sigma}_2)q^2}{(q^2 + m_\rho^{*2}) - \omega^2} \right] \tau_1 \cdot \tau_2 \tag{10.113}$$

with $q^2 \equiv \bar{q}^2$. Constants $\bar{f}$ (not to be confused with the f_π) and f_ρ^* are defined by the above formula. Now using

$$\frac{f_\rho^{*2}}{m_\rho^{*2}} \approx \frac{f_\rho^2}{m_\rho^2} \Phi(\rho)^{-2} \tag{10.114}$$

and the empirical observation

$$\frac{f_\rho^2}{m_\rho^2} \approx 2 \frac{\bar{f}_\pi^2}{m_\pi^2} \tag{10.115}$$

we see that the tensor force would vanish when

$$\frac{1}{2} \Phi^2 \frac{\langle (q^2 + m_\rho^{*2}) \rangle}{\langle (q^2 + m_\pi^2) \rangle} \approx 1 \tag{10.116}$$

where we have set $\omega \approx 0$ for static force. This happens when $\langle q^2 \rangle \approx 7.6 m_\pi^2$ for $\Phi \approx 0.8$ corresponding to nuclear matter density. This is within the range of momentum a meson carries between nucleons in nuclear matter. Such effects of a medium-quenched tensor force are observed in (e, e') and (p, p') processes on heavy nuclei at several hundreds of MeV: The value of Φ required is $m^*/m = 0.79 - 0.86$ which is consistent with the value needed to explain the enhanced axial charge transitions seen in Pb nuclei (see eq. (10.99)).

An interesting consequence of the quenched tensor force is that it suppresses the core-polarization quenching of $g_A^\star$ in heavy nuclei, so the main mechanism for leading to $g_A^\star = 1$ in nuclear matter is the Δ-hole effect discussed above. Of course for light nuclei, the tensor force is not suppressed much and hence the core-polarization quenching remains important. The fact that $g_A^\star \approx 1$ in also light nuclei can be understood by the combined effect of the core polarization and the short-range Δ-hole component.

(ii) Other processes

Other evidences are:

- Electroweak probes. In addition to the weak current discussed above, electromagnetic probes also provide an indirect evidence of the scaling effect. Specifically, the scaling of the vector meson and nucleon masses can be used to explain the longitudinal-to-transverse response function ratios in $(e, e'p)$ processes with such nuclei as ^{3}He, ^{4}He and

^{40}Ca. Experimentally there is an indication for quenching in the longitudinal response function that seems to increase with nuclear density while the transverse response remains unaffected. This feature can be economically (but perhaps not uniquely) explained by the scaling vector meson masses, although there is controversy as to whether the experiments do really show this phenomenon.

Some puzzles in deep-inelastic lepton processes in nuclei are simply explained by means of the BR scaling.[63] Further developments in this direction are expected as new machines such as the one at CEBAF go into operation.

- Hadronic probes. Some long-standing problems in hadron-nucleus scattering (*i.e.*, the kaon-nucleus and nucleon-nucleus optical potentials) are resolved by a quenched vector-meson mass. A stiffening of nucleon-nucleus spin-isospin response is predicted and is observed. These effects are manifested through the forces mediating the processes and the change in the range of such forces due to the scaling masses.

10.4. Hadronic Phase Transitions

So far we have been discussing response functions and two-body forces in terms of chiral Lagrangians. In this subsection we will turn to the bulk property of nuclear matter, particularly its structure under extreme conditions of high temperature and/or density. The question which remains largely unanswered up to date is whether or not effective Lagrangians can address the bulk property of hadronic matter. Our discussion given here is therefore very much tentative and could eventually be drastically modified.

10.4.1. *Nuclear matter*

Before discussing matter with density $\rho > \rho_0$ where $\rho_0 \approx m_\pi^3/2$ the ordinary nuclear matter density, we discuss briefly how the chiral Lagrangians that we shall use fare in describing ordinary nuclear matter.

It has been a disturbing point for chiral symmetry in nuclear physics that a Lagrangian with chiral symmetry treated at low order fails to describe correctly both nuclear matter and finite nuclei. Instead the Walecka model[64] with the scalar σ and the vector ω coupled to nucleons when treated at mean field is found to work fairly well both for nuclear matter and nuclei. This model however has no *apparent* chiral symmetry. Since the pion has no mean field, the only solidly established low-energy excitations of QCD (as seen in the preceding chapters), Goldstone bosons seem to play no role whatsoever in the ground-state property and low-energy excitations of nuclear systems. This has led many people to believe that chiral Lagrangians are useless for nuclear physics. This seemed to be in glaring conflict with the power of low-energy theorems and effective theories in some nuclear *response functions*.

This issue is revived and scrutinized from a novel perspective. One resolution of this paradox[39] involves a mean-field treatment of effective chiral Lagrangians with the Brown-Rho scaling described above, making the bridge to Walecka mean-field theory.

A related approach is to look for non-topological solitons or *"Q Balls"*[65] of an effective action calculated in chiral perturbation theory to high chiral orders with a Lagrangian consisting of the lowest-energy excitations, *i.e.*, nucleons and pions. Such an effective action would contain not only higher-order loops and higher-dimension counterterms but also higher-order fermion (baryon) fields as described by Weinberg. The ground state of the system for $A = \infty$ may then be described as a "chiral liquid" matter and finite nuclei as "chiral liquid" drops[66]. Here the pion field $\sqrt{\vec{\pi}^2}$ develops a mean field very much like Walecka's σ field and nonlinearity in the pion and nucleon fields induces the repulsion corresponding to Walecka's ω mean field. An interesting open problem is to relate this "chiral liquid" to the Landau-Migdal Fermi liquid and compute the parameters that figure in the latter in terms of the chiral constants that appear in the former. One valuable outcome of this relation would be the in-medium parameters of the theory determined in terms of the properties of the chiral liquid soliton. The power of this point of view is that one could pattern after the description in condensed matter physics of Fermi liquids and instabilities therefrom leading to phase transitions[29,34] in terms of effective field theories. Here phase transitions are driven by instabilities associated with "marginal" terms and some "irrelevant" terms kinematically enhanced in the effective Lagrangian absent in normal Fermi liquids.

The approaches given above are most probably related. We will essentially exploit both approaches in treating phase transitions in nuclear matter associated with the meson condensations treated below.

Finite temperature phase transitions associated with QCD variables will be discussed in Chapter 11.

10.4.2. *Goldstone boson condensation*

(i) Mean-field theory

At large density or temperature, there can be two basically different phase transitions seen from the point of view of effective chiral Lagrangians: Goldstone boson condensation and chiral/deconfinement phase transitions. In this section, we consider the density effect and discuss the first. We will focus on kaon condensation because pions are not likely to condense at low densities and hence are uninteresting. We shall first discuss a simple mechanism by which kaon condensation may occur, as suggested first by Kaplan and Nelson[67], at a density as low as a few times the nuclear matter density. We will see that the phenomenon is quite robust. In the next subsection, the simple result obtained here will be corroborated in a more realistic chiral perturbation calculation.

Consider infinite (nuclear) matter or more specifically the compact star matter. The Lagrangian we will consider is the sum $\mathcal{L}^U + \mathcal{L}_B + \delta\mathcal{L}_B$ of eq. (10.41) in flavor $SU(3)$ sector which we write with simplified coefficients,

$$
\begin{aligned}
\mathcal{L} &= \frac{f^2}{4}\,\mathrm{Tr}\partial_\mu U\partial^\mu U^\dagger + c\,\mathrm{Tr}\,(\mathrm{Tr}\mathcal{M}U + h.c.) \\
&+ \mathrm{Tr}B^\dagger i\partial_0 B + i\,\mathrm{Tr}B^\dagger[V_0, B] - D\,\mathrm{Tr}B^\dagger\vec{\sigma}\cdot\{\vec{A}, B\} - F\,\mathrm{Tr}B^\dagger\vec{\sigma}\cdot[\vec{A}, B] \\
&+ b\,\mathrm{Tr}B^\dagger\,(\xi\mathcal{M}\xi)\,B + d\,\mathrm{Tr}B^\dagger B\,(\xi\mathcal{M}\xi + h.c.) \\
&+ h\,\mathrm{Tr}B^\dagger B\,(\mathcal{M}U + h.c.) + \cdots.
\end{aligned}
\tag{10.117}
$$

The ellipsis stands for chiral symmetric terms quadratic in derivatives (A), terms linear in $\mathcal{M}$ and powers of derivatives (B), terms higher order in $\mathcal{M}$ (C) etc. The decuplet fields are not included here although they play a role. The class B and class C terms are not relevant at tree order that we will be considering here but the class A term is of the same order in the chiral counting as the terms linear in $\mathcal{M}$ given in (10.117). For the moment we will not worry about this class A terms but return to them later in discussing kaon-nuclear scattering at the same order.

Taking the Lagrangian (10.117) effectively as a Ginzburg-Landau form, we will treat it in tree order, so the parameters appearing in (10.117) can be fixed directly by experiments in free space. Weak leptonic pion decay gives $f = f_\pi \approx 93$ MeV. Nuclear β decay and semileptonic hyperon decay fix $F \approx 0.44$ and $D \approx 0.81$. The kaon and η masses give through the Gell-Mann-Oakes-Renner relation (at the lowest order in the mass matrix)

$$m_s c \approx (182\text{MeV})^4. \tag{10.118}$$

From baryon mass splittings, we can fix

$$bm_s \approx -67 \text{ MeV}, \tag{10.119}$$

$$dm_s \approx 134 \text{ MeV}. \tag{10.120}$$

The coefficient h is related to the KN sigma term and has a large uncertainty. For definiteness we shall use, following Politzer and Wise[68],

$$hm_s \approx -310 \text{ MeV} \tag{10.121}$$

which implies

$$m_s \langle p|\bar{s}s|p \rangle = m_s \frac{\partial m_N}{\partial m_s} = -2(d+h)m_s \approx 350 \text{ MeV}. \tag{10.122}$$

This large strangeness content of the proton is a consequence of the tree-order calculation and can be remedied at higher order in chiral perturbation theory as mentioned in Chapter 9. (The Callan-Klebanov model predicts a much smaller value.) In fact, the two extreme limits, $m_s \to 0$ and $m_s \to \infty$, indicate[68] that something is amiss with (10.122): In the former case, the s-quark contribution to the baryon mass is trivially zero since $\langle p|\bar{s}s|p \rangle$ is smooth in that limit. In the latter case, the s quark can be integrated out and one gets

$$m_s \frac{\partial m_N}{\partial m_s}\Big|_{m_s \to \infty} \to \frac{2}{9} m_n \approx 65 \text{ MeV}, \tag{10.123}$$

thus the two limits give a much smaller value. Note, however, that as mentioned already in Chapter 9, lattice gauge calculations[69] give a value as large as (10.122).

Let the Hamiltonian obtained from (10.117) be denoted as H. To ensure charge neutrality of the matter system, we consider

$$\tilde{H} = H + \mu Q \tag{10.124}$$

with Q the electric charge and μ the corresponding Lagrange multiplier (chemical potential). Let the expectation value of $\tilde{H}$ with respect to the lowest-energy state be denoted by $\tilde{E}$. Then the charge neutrality condition is

$$\frac{\partial \tilde{E}}{\partial \mu} = \langle Q \rangle = 0. \tag{10.125}$$

When this condition is satisfied, $\tilde{E} = E$, the quantity we wish to calculate. For reasons that will be obvious, we include the pion in addition to kaons and assume the ground-state expectation values

$$\langle \pi^- \rangle = e^{-i\mu t} e^{i\vec{p}_\pi \cdot \vec{x}}, \tag{10.126}$$

$$\langle \bar{K}^0 \rangle = e^{i\vec{q} \cdot \vec{x}} v_0, \tag{10.127}$$

$$\langle K^- \rangle = e^{-i\mu t} e^{i\vec{k} \cdot \vec{x}} v_K. \tag{10.128}$$

For the neutral kaons, the chemical potential is zero. As for the charged pions and kaons, the time dependence of the field is precisely given by the chemical potential with $p_{\pi 0} = k_0 = \mu$. In our consideration, the negatively charged mesons π^- and K^- are involved. The Hamiltonian mixes the neutron state to baryons with unit charge of strangeness $S = -1$ since we will be looking at the threshold for kaon condensation and hence limiting to quadratic order in kaon field, thus allowing $|\Delta S| = 1$. The states that can be admixed are Λ, Σ^0, and Σ^-. The proton will also be mixed through pion (π^-) condensation. The lowest such quasi-particle state can then be occupied up to the Fermi momentum $p_F = (3\pi^2 \rho)^{1/3}$ to form the ground state whose energy density to $O(1)$ in $1/m_B$ is then given by

$$\begin{aligned} \tilde{E}/V \equiv \tilde{\epsilon} &= \tilde{\epsilon}(0) + (\vec{q}^2 + m_K^2)|v_0|^2 + (\vec{k}^2 + m_K^2 - \mu^2)|v_K|^2 \\ &+ (\vec{p}_\pi^2 + m_\pi^2 - \mu^2)|v_\pi|^2 + \rho \Delta \epsilon \end{aligned} \tag{10.129}$$

where

$$\begin{aligned} \Delta \epsilon &= \left(\frac{\mu}{2f^2} - \frac{(F+D)^2}{2\mu f^2} \vec{p}_\pi^2 \right) |v_\pi|^2 \\ &+ \left[(4h + 2d + 2b) \frac{m_s}{2f^2} - \frac{(3F+D)^2}{8(d-2b)m_s f^2} \vec{q}^2 - \frac{(D-F)^2}{8dm_s f^2} \vec{q}^2 \right] |v_0|^2 \\ &+ \left[-\frac{\mu}{2f^2} + (4h+2d)\frac{m_s}{2f^2} - \frac{(D-F)^2}{(2dm_s - \mu)2f^2} \vec{k}^2 \right] |v_K|^2 \\ &+ \cdots . \end{aligned} \tag{10.130}$$

The $\Delta\epsilon$ given here is an *in-medium* one-loop term with the graphs contributing to (10.130) given in Fig. 10.9: The terms containing the coefficients b, d and h come from Fig. 10.9a, explicitly proportional to the chiral symmetry breaking quark mass matrix which are of s-wave KN interactions, the terms proportional to $\vec{q}^2$ and $\vec{k}^2$ come from Fig. 10.9b and are intrinsically from p-wave K-N interactions and finally the term linear in μ is from the vector field term V_μ coupled to the baryon current $\bar{B}\gamma_0 B$, included in Fig. 10.9a.

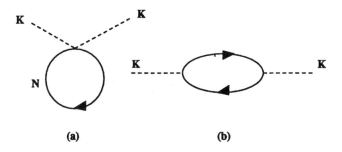

(a)   **(b)**

Fig. 10.9. One-loop graphs contributing to the energy density of dense matter proportional to v_K^2 for K condensation. The solid line stands for baryons and the dotted line for the kaon.

Within the approximation used here, the p-wave interaction is linear in the square of the kaon three-momentum. Thus for a density for which

$$\rho\left[\frac{(3F+D)^2}{8(d-2b)m_s f^2} + \frac{(D-F)^2}{8dm_s f^2}\right] > 1 \tag{10.131}$$

which corresponds to the density $\rho \>\sim 3.2\rho_0$, increasing $\vec{q}^2$ would bring in increasing attraction triggering a K^0 condensation. However in reality, there is a form factor associated with the vertex (which in the chiral counting would correspond to higher loops) that would prevent the arbitrary increase in the momentum. It is likely that the LHS of (10.131) is never greater than 1 in which case one would have $\vec{q}^2 = 0$. Let us assume this for the moment and ignore the K^0 mode. We now minimize the energy density ϵ with respect to the pion momentum and K^- momentum. From $\partial\tilde{\epsilon}/\partial\vec{p}_\pi = 0$, we get

$$\mu = (F+D)^2\frac{\rho}{2f^2}. \tag{10.132}$$

We will confine to s-wave KN interactions and set $\vec{k} = 0$. Then the charge neutrality $\partial \tilde{\epsilon}/\partial \mu = 0$ gives

$$p_\pi^2 = \left(\frac{\rho}{2f^2}\right)^2 (F+D)^2 \left\{ [2(F+D)^2 - 1] + [2(F+D)^2 + 1] |\frac{v_K}{v_\pi}|^2 \right\}.$$

(10.133)

We thus have

$$
\begin{aligned}
\tilde{\epsilon} =\ & \epsilon(v_\pi, v_K) = \epsilon(0,0) \\
& + |v_\pi|^2 \left\{ m_\pi^2 - (F+D)^2 \left(\frac{\rho}{2f^2}\right)^2 [(F+D)^2 - 1] \right\} \\
& + |v_K|^2 \left\{ m_K^2 - (F+D)^2 [(F+D)^2 + 1] \left(\frac{\rho}{2f^2}\right)^2 \right. \\
& \left. + \left(\frac{\rho}{2f^2}\right)(4h + 2d)m_s \right\} + \cdots.
\end{aligned}
$$

(10.134)

We know from nuclear beta-decay data – and the discussion given above – that the axial-vector coupling constant g_A is renormalized in medium from its free space value of 1.26 to about 1. This means that the main correction to the above formula would be to replace $(F + D)$ above by $g_A^* \approx 1$. Including nucleon-nucleon interaction effects generated through the contact interaction of (10.70) in the *particle-hole spin-isospin channel* will further reduce the effective g_A, as is well-known in the Landau-Migdal formalism. This makes the coefficient of $|v_\pi|^2$ always greater than zero to a much larger density than relevant here. Therefore we expect that

$$v_\pi \approx 0.$$

(10.135)

The vanishing VEV of the pion field in the presence of a nonvanishing kaon VEV would imply by (10.133) an arbitrarily large pion momentum. Such a situation would be prevented by form factor effects arising at higher loop order. Since there cannot be any kaon condensation without pions (condensed or not) (this can be easily verified by looking at the energy density in the absence of pion field and showing that the coefficient of $|v_K|^2$ is always greater than zero), it pays to have a little bit of pion VEV at the cost of a large pion momentum. *What matters is that the pion contribution to the energy density be as small as possible, not necessarily strictly zero.*

The critical density for kaon condensation ρ_K is

$$\rho_K \approx -f^2 m_s(2h+d)\left\{[1 + \frac{2m_K^2}{m_s^2(2h+d)^2}]^{\frac{1}{2}} - 1\right\} \qquad (10.136)$$

where we have set $g_A^* \approx 1$. Numerically this gives $\rho_K \approx 2.6\rho_0$ for $hm_s = -310$ MeV and $\rho_K \approx 3.7\rho_0$ for $hm_s = -140$ MeV (which is close to $hm_s = -125$ MeV corresponding to the extreme case with $m_s\langle p|\bar{s}s|p\rangle \approx 0$). Even when $(g_A^*)^2 = 1/2$, the critical density increases only to about 5.7 times ρ_0. This is of course too high a density for the approximations used to be valid. Nonetheless one observes here a remarkable robustness against changes in parameters, in particular with respect to the greatest uncertainty in the theory, namely the quantity h and also g_A^*. The insensitivity to the strangeness content of the proton, namely, the parameter $(h+d)$ makes the prediction surprisingly solid. Furthermore, while $(F+D) \to 1$ banishes the pion condensation, this affects marginally the kaon property. It seems significant that the critical density for kaon condensation is robust against higher-order corrections while pion condensation is extremely sensitive.

(ii) Kaon condensation in "nuclear star" matter

In the above discussion, pions played a crucial role in triggering kaon condensation. Here we present an alternative way for the kaons to condense in neutron star matter which does not require pion condensation. The key idea is that in neutron star matter, energetic electrons – reaching hundreds of MeV in kinetic energy in dense neutron star matter – can decay into kaons if the kaon mass falls sufficiently low in dense matter, as we will show below [70], via

$$e^- \to K^- + \nu. \qquad (10.137)$$

This process can go into chemical equilibrium with the beta decay

$$n \to p + e^- + \nu_e \qquad (10.138)$$

with the neutrinos diffusing out. This means that the chemical potentials satisfy

$$\mu_{K^-} = \mu_{e^-} = \mu_n - \mu_p . \qquad (10.139)$$

Let x denote the fraction of protons generated by (10.137) and (10.138), so

$$\rho_n = (1-x)\rho, \qquad \rho_p = x\rho \qquad (10.140)$$

and define

$$\rho \equiv u\rho_0 \tag{10.141}$$

with ρ_0 being the nuclear matter density. With the addition of protons, we gain in energy (*i.e.*, the symmetry energy) by the amount

$$E_s/V = \epsilon_s[(1 - 2x)^2 - 1]\rho_0 u = -4x(1 - x)\epsilon_s\rho_0 u \tag{10.142}$$

where we have assumed, for simplicity, that the symmetry energy depends on the density linearly with the constant ϵ_s determined from nuclei,

$$\epsilon_s \approx 32\text{MeV}. \tag{10.143}$$

More sophisticated formulas for symmetry energy could be used – and will be used later when we discuss loop corrections – but for the present purpose this form should suffice. In any event the symmetry energy above nuclear matter density is unknown, so this constitutes one of the major uncertainties in our discussion.

We shall ignore pions and neutral kaons, focusing on s-wave K^--nuclear interactions and make a drastically simplified discussion here since a refined higher order chiral perturbation calculation will be described shortly. Instead of (10.129), we have a simpler form for the energy density,

$$\tilde{E}/V \equiv \tilde{\epsilon} = \tilde{\epsilon}(0) + (m_K^2 - \mu^2)|v_K|^2 + \rho\Delta\epsilon \tag{10.144}$$

where $\tilde{\epsilon}(0)$ represents the sum of the kinetic energy density of the symmetric nuclear matter and the isospin-independent part of the nuclear interaction contribution and

$$\Delta\epsilon = -\frac{1}{2f^2}\left[\mu(1 - x) + 2\mu x - (4h + 2d)m_s\right]|v_K|^2$$
$$-4x(1 - x)\epsilon_s + x\mu + \cdots. \tag{10.145}$$

The first term in the square bracket of (10.145) is the neutron contribution through vector exchange (namely, the V_0 term in (10.117)) and the second the corresponding proton contribution. The factor of 2 for the proton contribution can be easily understood by the fact that the K^-p interaction with vector meson exchanges is twice as attractive as the K^-n interaction: This can be readily verified by looking at the isospin structure of the vector current V_μ. Defining

$$\hat{\mu} = \mu + \frac{\rho_0 u}{4f^2}(1 + x) \tag{10.146}$$

we can rewrite (10.144)

$$\tilde{\epsilon} = \tilde{\epsilon}(0) + \left[(m_K^*)^2 - \hat{\mu}^2 + \frac{\rho_0^2 u^2}{16 f^4}(1+x)^2 \right] |v_K|^2$$
$$-4x(1-x)\epsilon_s \rho_0 u + x\rho_0 u\mu \tag{10.147}$$

where

$$(m_K^*)^2 = m_K^2 + \frac{\rho_0 u}{2 f^2}(4h + 2d)m_s. \tag{10.148}$$

The charge neutrality condition is gotten by balancing the negative charge of the kaon against the positive charge of the proton, *i.e.*,

$$\hat{\mu} = x \frac{\rho_0 u}{2|v_K|^2}. \tag{10.149}$$

The condition (10.149) gives x as a function of $|v_K|^2$ since the proton and neutron chemical potentials are functions of x. Since by beta equilibrium $\partial \tilde{\epsilon}/\partial x = 0$ we have one more condition

$$\hat{\mu} \approx \frac{\rho u}{2 f^2}(1+x) - \frac{16 f^2}{|v_K|^2} x(1-2x)\epsilon_s. \tag{10.150}$$

Substituting (10.149) and (10.150) into eq.(10.147) with the electron contribution taken into account and using the numerical values $dm_s \approx 134$ MeV, $hm_s \approx -310$ MeV (consistent with lattice results discussed in Chapter 9), $\epsilon_s \approx 32$ MeV, $f = 93$ MeV, we find the critical density

$$\rho_K \approx 2\rho_0. \tag{10.151}$$

This critical density is comparable to that predicted for the most optimistic case of the kaon condensation triggered by pion condensation. The more sophisticated calculation described later using the same physical mechanism and chiral perturbation theory corroborates this simple estimate.

Formation of "nuclear stars"

As negatively charged kaons condense, neutrons convert to protons to neutralize the charge by exploiting the symmetry energy. At the critical point, the proton fraction x is of the order of 10 % and once in the condensed phase, x quickly increases reaching 50 % at a density not so much above the critical density. Thus the compact star is more a nuclear matter than neutron matter that one reads in standard astrophysics textbooks. Therefore it seems more appropriate to call the compact star a "nuclear star" rather than neutron star. This has a dramatic effect on the cooling of the stars. As pointed out recently by Lattimer *et al.*[71], the direct URCA process can occur in compact stars if the proton concentration exceeds some critical value in the range of $(11 - 15)\%$. The direct URCA can cool the compact stars considerably faster than any other known mechanism, be that meson condensation or quark-gluon plasma. For detailed discussions on this interesting topic, we refer to a review by Kubodera [72].

(iii) Higher-order chiral perturbation theory

The tree-order $O(Q^2)$ calculation discussed so far does not correctly describe kaon-nucleon scattering[73]. This can, however, be remedied simply by including the terms of $O(Q^2)$ missing from the Lagrangian (10.117). It turns out[74] that the terms so far left out go as ω^2 which at threshold contributes importantly but at the condensation point, $\omega = \mu$, plays an unimportant role and affects little the result obtained above.

One can do better by including loops that are required if one wants to go to $O(Q^3)$ in chiral expansion[76]. This corresponds to going to next-to-next-to-leading order in chiral perturbation theory. We discuss this here to show that the kaon condensation prediction is remarkably robust. One also learns an important lesson on effective theory approach to meson-condensation phenomena in analogy to condensed matter physics.

Since we want to do a calculation at one-loop order, it is necessary to write down the Lagrangian to $O(Q^3)$, the latter appearing as counterterms. In terms of the velocity-dependent octet baryon fields B_v, the octet meson fields $\exp(i\pi_a T_a/f) \equiv \xi$, the velocity-dependent decuplet baryon fields T_v^μ, the velocity four-vector v_μ and the spin operator S_v^μ ($v \cdot S_v = 0$, $S_v^2 = -3/4$), the vector current $V_\mu = [\xi^\dagger, \partial_\mu \xi]/2$ and the axial-vector current $A_\mu = i\{\xi^\dagger, \partial_\mu \xi\}/2$, the Lagrangian density to order Q^3, relevant for the low-energy s-wave scattering, reads

$$\mathcal{L} = \mathcal{L}^{(1)} + \mathcal{L}^{(2)} + \mathcal{L}^{(3)} \tag{10.152}$$

with

$$\begin{aligned}
\mathcal{L}^{(1)} =\ & \mathrm{Tr}\bar{B}_v(iv \cdot \mathcal{D})B_v + 2D\,\mathrm{Tr}\bar{B}_v S_v^\mu\{A_\mu, B_v\} + 2F\,\mathrm{Tr}\bar{B}_v S_v^\mu[A_\mu, B_v] \\
& -\bar{T}_v^\mu(iv \cdot \mathcal{D} - \delta_T)T_{v,\mu} + \mathcal{C}(\bar{T}_v^\mu A_\mu B_v + \bar{B}_v A_\mu T_v^\mu)
\end{aligned}$$

$$+2\mathcal{H}\bar{T}_v^\mu(S_v \cdot A)T_{v,\mu}, \tag{10.153}$$

$$\begin{aligned}
\mathcal{L}^{(2)} = \ & a_1 \operatorname{Tr}\bar{B}_v\chi_+ B_v + a_2 \operatorname{Tr}\bar{B}_v B_v\chi_+ + a_3 \operatorname{Tr}\bar{B}_v B_v \operatorname{Tr}\chi_+ \\
& +d_1 \operatorname{Tr}\bar{B}_v A^2 B_v + d_2 \operatorname{Tr}\bar{B}_v(v \cdot A)^2 B_v + d_3 \operatorname{Tr}\bar{B}_v B_v A^2 \\
& +d_4 \operatorname{Tr}\bar{B}_v B_v(v \cdot A)^2 + d_5 \operatorname{Tr}\bar{B}_v B_v \operatorname{Tr}A^2 + d_6 \operatorname{Tr}\bar{B}_v B_v \operatorname{Tr}(v \cdot A)^2 \\
& +d_7 \operatorname{Tr}\bar{B}_v A_\mu \operatorname{Tr}B_v A^\mu + d_8 \operatorname{Tr}\bar{B}_v(v \cdot A) \operatorname{Tr}B_v(v \cdot A) \\
& +d_9 \operatorname{Tr}\bar{B}_v A_\mu B_v A^\mu + d_{10} \operatorname{Tr}\bar{B}_v(v \cdot A)B_v(v \cdot A), \tag{10.154}
\end{aligned}$$

$$\begin{aligned}
\mathcal{L}^{(3)} = \ & c_1 \operatorname{Tr}\bar{B}_v(iv \cdot D)^3 B_v + c_2 \operatorname{Tr}\bar{B}_v(iv \cdot D)(i^2 D^\mu D_\mu)B_v \\
& +g_1 \operatorname{Tr}\bar{B}_v A_\mu(iv\cdot \overleftrightarrow{D})A^\mu B_v + g_2 \operatorname{Tr}\bar{B}_v A_\mu(iv\cdot \overleftrightarrow{D})A^\mu \bar{B}_v \\
& +g_3 \operatorname{Tr}\bar{B}_v v \cdot A(iv\cdot \overleftrightarrow{D})v \cdot A B_v + g_4 \operatorname{Tr}\bar{B}_v v \cdot A(iv\cdot \overleftrightarrow{D})v \cdot A\bar{B}_v \\
& +g_5 \left(\operatorname{Tr}\bar{B}_v A_\mu \operatorname{Tr}(iv\cdot \overrightarrow{D})A^\mu B_v - \operatorname{Tr}\bar{B}_v A_\mu(iv\cdot \overleftarrow{D}) \operatorname{Tr}A^\mu B_v \right) \\
& +g_6 \Big(\operatorname{Tr}\bar{B}_v v \cdot A \operatorname{Tr}B_v(iv\cdot \overrightarrow{D})v \cdot A \\
& \quad\quad - \operatorname{Tr}\bar{B}_v v \cdot A(iv\cdot \overleftarrow{D})v \cdot A \operatorname{Tr}B_v v \cdot A \Big) \\
& +g_7 \operatorname{Tr}\bar{B}_v[v \cdot A, [iD^\mu, A_\mu]]B_v + g_8 \operatorname{Tr}B_v[v \cdot A, [iD^\mu, A_\mu]]\bar{B}_v \\
& +h_1 \operatorname{Tr}\bar{B}_v\chi_+(iv \cdot D)B_v + h_2 \operatorname{Tr}\bar{B}_v(iv \cdot D)B_v\chi_+ \\
& +h_3 \operatorname{Tr}\bar{B}_v(iv \cdot D)B_v \operatorname{Tr}\chi_+ + l_1 \operatorname{Tr}\bar{B}_v[\chi_-, v \cdot A]B_v \\
& +l_2 \operatorname{Tr}\bar{B}_v B_v[\chi_-, v \cdot A] + l_3[\operatorname{Tr}\bar{B}_v\chi_-, \operatorname{Tr}B_v v \cdot A], \tag{10.155}
\end{aligned}$$

where the covariant derivative $\mathcal{D}_\mu$ for baryon fields is defined by

$$\begin{aligned}
\mathcal{D}_\mu B_v &= \partial_\mu B_v + [V_\mu, B_v], \\
\mathcal{D}_\mu T^\nu_{v,abc} &= \partial_\mu T^\nu_{v,abc} + (V_\mu)_a^d T^\nu_{v,dbc} + (V_\mu)_b^d T^\nu_{v,adc} \\
&\quad + (V_\mu)_c^d T^\nu_{v,abd}, \tag{10.156}
\end{aligned}$$

δ_T is the $SU(3)$ invariant decuplet-octet mass difference, and

$$\chi_\pm \equiv \xi\mathcal{M}\xi\pm\xi^\dagger\mathcal{M}\xi^\dagger. \tag{10.157}$$

To $O(Q^3)$, there are more terms involving the decuplets but they are not relevant for the present calculation, so we leave them out.

The Lagrangian (10.152) contains the even-parity octet and decuplet baryons but when protons are present as in our case, the odd-parity $\Lambda(1405)$ is known to play an important role in kaon-nucleon interaction. This state is, in the skyrmion description, a bound state of an $l = 0$ kaon and an

$SU(2)$b soliton, an odd-parity partner of the $\Lambda(1116)$ which is the Callan-Klebanov skyrmion described in Chapter 7. It may be described as a K^-p (weakly) bound state in a potential model. It is this $\Lambda(1405)$ which makes the K^-p scattering near threshold repulsive. As pointed out above, bound states are not amenable to chiral perturbation treatment, so the natural thing to do is to introduce the $\Lambda(1405)$ as an $SU(3)$ singlet elementary field. The relevant Lagrangian to the leading order can be written as

$$
\begin{aligned}
\mathcal{L}_{\Lambda_R} = {}& \bar{\Lambda}_R(iv \cdot \partial - (m_{\Lambda_R} - m_B))\Lambda_R \\
& + \left(\sqrt{2}\bar{g}_{\Lambda_R}\,\mathrm{Tr}(\bar{\Lambda}_R v \cdot AB_v) + h.\,c.\right).
\end{aligned} \tag{10.158}
$$

Finally there is a piece in the Lagrangian that involves multi-baryon fields contributing in many-body systems. We will restrict ourselves to four-Fermi interactions, although there is no a priori reason why six-Fermi interactions cannot contribute but the latter represents "irrelevant" terms in the sense of renormalization group theory and hence is expected to be suppressed. Now among four-Fermi interactions, we make one further simplification by assuming that all four-Fermi interactions that involve no strange baryons are subsumed in the parameters extracted from nuclear processes not involving strange particles. Then for s-wave kaon-nucleon interactions, we can reduce *all* four-Fermi interactions containing $\Lambda(1405)$ to one with two unknown parameters of mass dimension -2

$$
\mathcal{L}_{4f} = c_\Lambda \bar{\Lambda}_R \Lambda_R \mathrm{Tr}\bar{B}B + c_\Lambda^\sigma \bar{\Lambda}_R \sigma \Lambda_R \mathrm{Tr}\bar{B}\sigma B. \tag{10.159}
$$

Our Lagrangian consisting of (10.152), (10.158) and (10.159) looks daunting at first sight but it turns out that a considerable simplification is obtained in heavy-baryon formalism for the s-wave kaon-nuclear interactions. For instance, among the contributions coming from (10.152), the subleading terms (*i.e.*, terms of $O(Q^\nu)$ with $\nu \geq 2$) involving the spin operator S_μ do not contribute to the s-wave $K^\pm N$ amplitudes, since they are proportional to $S \cdot q$, $S \cdot q'$, or $S \cdot qS \cdot q'$, all of which vanish. Thus there is no contribution to the s-wave meson-nucleon scattering amplitude from one-loop diagrams in which the external meson lines couple to baryon lines through the axial vector currents. This reduces the number of diagrams from thirteen to six topologically distinct one-loop graphs for the s-wave meson-nucleon scattering apart from the usual radiative corrections in external lines.

- *Kaon-nucleon scattering near threshold*

We need first to determine the parameters of the Lagrangian – the constants in $\mathcal{L}^{(1,2)}$ and the counterterms in $\mathcal{L}^{(3)}$ – from the threshold data of kaon-nucleon scattering. Now given the empirical scattering lengths **

$$a_0^{K^+p} = -0.31 \, \text{fm}, \qquad a_0^{K^-p} = -0.67 + i\,0.63 \, \text{fm}$$
$$a_0^{K^+n} = -0.20 \, \text{fm}, \qquad a_0^{K^-n} = +0.37 + i\,0.57 \, \text{fm} \qquad (10.160)$$

and the theoretical expressions calculated to one-loop order (that is, to $O(Q^3)$), one can determine all the parameters of the Lagrangian for calculating the off-shell s-wave KN scattering amplitude needed for the equation of state for kaon condensation.

The predicted scattering lengths are as follows:

$$
\begin{aligned}
a_0^{K^\pm p} &= \frac{m_N}{4\pi f^2(m_N + M_K)} \Big[\mp M_K - \frac{\bar{g}_{\Lambda_R}^2 M_K^2}{m_B \mp M_K - m_{\Lambda_R}} \\
&\quad + (\bar{d}_s + \bar{d}_v)M_K^2 + [(L_s + L_v) \pm (\bar{g}_s + \bar{g}_v)]M_K^3 \Big] \\
a_0^{K^\pm n} &= \frac{m_N}{4\pi f^2(m_N + M_K)} \Big[\mp \frac{M_K}{2} + (\bar{d}_s - \bar{d}_v)M_K^2 \\
&\quad + [(L_s - L_v) \pm (\bar{g}_s - \bar{g}_v)] \Big].
\end{aligned} \qquad (10.161)
$$

It is not difficult to see where the terms in (10.161) come from[74]. The terms linear in M_K are the leading order $(O(Q))$ terms from $\mathcal{L}^{(1)}$ at tree order, the one proportional to $\bar{g}_\Lambda^2$ comes from the Born graph with $\Lambda(1405)$ in the intermediate state (with a complex mass when perturbative unitarity is invoked) contributing only to the proton target, the next terms of the form $(\bar{d}_s \pm \bar{d}_v)M_K^2$ – where $d_{s,v}$ are known linear combinations of the constants a_i and d_i in $\mathcal{L}^{(2)}$ – are the tree-order contribution from $\mathcal{L}^{(2)}$ and the last term multiplied by M_K^3 consists of crossing-even finite one-loop graphs $((L_s \pm L_v))$ calculated with $\mathcal{L}^{(1)}$ and crossing-odd one-loop graphs with counterterms grouped into $\bar{g}_{s,v}$ from $\mathcal{L}^{(3)}$. The four-Fermi interaction (10.159) plays no role here. Now from the experimental width and mass of the $\Lambda(1405)$, the coupling constant $\bar{g}_\Lambda$ is fixed

$$\bar{g}_\Lambda^2 \approx 0.25 \qquad (10.162)$$

---

*Data on K^+N scattering are not precise, so no error bars are quoted. To be consistent, no fine-tuning of the parameters of the Lagrangian will be made.

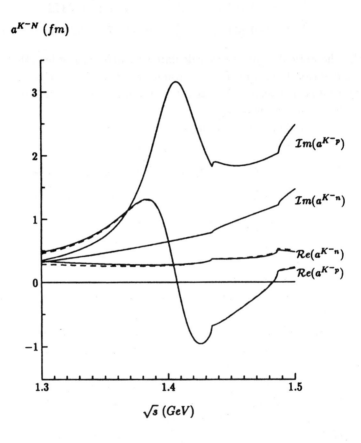

Fig. 10.10. The $K^- N$ amplitude as a function of $\sqrt{s} = \omega + m_N$. The quantity plotted a is related to the invariant amplitude T by $a = [4\pi(1 + \omega/m_N)]^{-1}T$.

and evaluating finite integrals gives

$$L_s M_K \approx -0.109 \,\text{fm},$$
$$L_v M_K \approx +0.021 \,\text{fm}. \tag{10.163}$$

The remaining parameters can then be determined from the experimental widths; they are

$$\bar{d}_s \approx 0.201 \,\text{fm}, \qquad \bar{d}_v \approx 0.013 \,\text{fm},$$
$$\bar{g}_s M_K \approx 0.008 \,\text{fm}, \qquad \bar{g}_v M_K \approx 0.002 \,\text{fm}. \tag{10.164}$$

Having fixed the parameters of the theory, we can calculate the s-wave forward scattering amplitude $a^{K^{\pm}N}(\omega)$ for $\omega \neq M_K$ to the chiral order $O(Q^3)$. This is the off-shell KN amplitude that will come in, together with other contributions, for kaon condensation. The result is given in Fig.10.10.

- *Critical density*

We shall now calculate the critical density for kaon condensation in nuclear star matter introduced above. The energy density is of the form

$$
\begin{aligned}
\tilde{\epsilon}(u, x, \mu, v_K) = & \frac{3}{5} E_F^{(0)} u^{\frac{5}{3}} \rho_0 + V(u) + u\rho_0 (1 - 2x)^2 S(u) \\
& - [\mu^2 - M_K^2 - \Pi_K(\mu, u, x)] v_K^2 + \mathcal{O}(v_K^3) \\
& + \mu u \rho_0 x + \tilde{\epsilon}_e + \theta(|\mu| - m_\mu)\tilde{\epsilon}_\mu
\end{aligned} \tag{10.165}
$$

where $E_F^{(0)} = \left(p_F^{(0)}\right)^2 / 2m_N$ and $p_F^{(0)} = (3\pi^2 \rho_0/2)^{\frac{1}{3}}$ are, respectively, the Fermi energy and momentum at nuclear density. The $V(u)$ is a potential for symmetric nuclear matter which we assume is subsumed in contact four-Fermi interactions (and one-pion-exchange – nonlocal – interaction) in the non-strange sector. It will affect the equation of state in the condensed phase but not the critical density, so we will drop it from now on. The nuclear symmetry energy $S(u)$ – also subsumed in four-Fermi interactions in the non-strange sector – does play a role as we know from above: Protons enter to neutralize the charge of the condensing K^-'s making the resulting compact star "nuclear" rather than neutron star.[75] We take the standard form

$$S(u) = \left(2^{\frac{2}{3}} - 1\right) \frac{3}{5} E_F^{(0)} \left(u^{\frac{2}{3}} - F(u)\right) + S_o F(u) \tag{10.166}$$

where $F(u)$ is the potential contribution to the symmetry energy which we take as before

$$F(u) = u. \tag{10.167}$$

It turns out that the choice of $F(u)$ does not significantly affect the critical density, so (10.167) is good enough for our purpose. The contributions from the filled Dirac seas of electrons and muons are included in (10.165).

One sees from the energy density written to quadratic order in v_K that the only ingredient needed at higher chiral order is the kaon self-energy Π_K. This means that we need to consider up to two-loop graphs in medium. The off-shell KN amplitude therefore contributes the linear density term depicted in Fig.10.11a. We shall denote it ρT. Put inside the matter, the one-loop graphs contributing to the KN scattering amplitude develop non-linear density dependence as depicted in Fig.10.11b-f. We shall denote it $\rho \delta T$. In addition to the above terms, the presence of $\Lambda(1405)$ contributes, through the four-Fermi interaction (10.159), the graphs Fig.10.11g-h and, through a kaon exchange, the graph Fig.10.11i. The sum of these contributions will be denoted Π_Λ. Putting all these together, we have

$$\Pi_K(\omega) = - \left(\rho_p T_{free}^{K^- p}(\omega) + \rho_n T_{free}^{K^- n}(\omega) \right)$$
$$- \left(\rho_p \delta T_{\rho N}^{K^- p}(\omega) + \rho_n \delta T_{\rho N}^{K^- n}(\omega) \right) + \Pi_\Lambda(\omega). \tag{10.168}$$

Contrary to its appearance, this does not involve the linear density approximation. Nonetheless, the only thing that is new here (in the sense of parameters involved) is the last term: The others are given entirely by the part of the Lagrangian determined from the free-space scattering data. One finds

$$\Pi_\Lambda(\omega) = -\frac{\bar{g}_\Lambda^2}{f^2} \bar{g}_\Lambda^2 \left(\frac{\omega}{\omega + m_B - m_\Lambda} \right)^2 \left\{ c_\Lambda \rho_p \left(\rho_n + \frac{1}{2} \rho_p \right) - \frac{3}{2} c_\Lambda^\sigma \rho_p^2 \right\}$$
$$- \frac{\bar{g}_\Lambda^4}{f^4} \rho_p \left(\frac{\omega}{\omega + m_B - m_\Lambda} \right)^2 \omega^2 \left(\Sigma_K^p + \Sigma_K^n \right) \tag{10.169}$$

where

$$\Sigma_K^N(\omega) = \frac{1}{2\pi^2} \int_0^{k_{F_N}} d|\vec{k}| \frac{|\vec{k}|^2}{\omega^2 - M_K^2 - |\vec{k}|^2}. \tag{10.170}$$

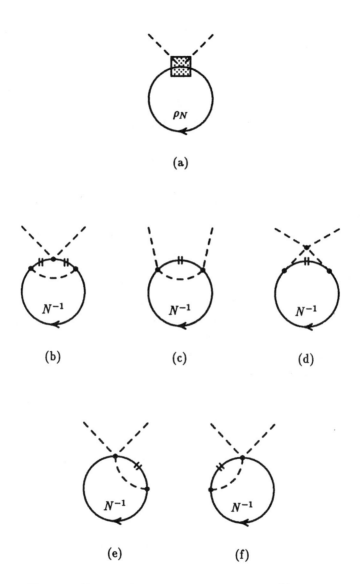

Fig. 10.11. The K^- self-energy diagrams: The square blob in (a) represents the free-space off-shell amplitude, the double slash in (b)-(f) stands for the in-medium nucleon propagator, the external dotted line for K^- and the internal dotted line for the octet pseudo-Goldstone bosons. The nucleon hole state is labeled as N^{-1}.

Evidently the constants $C_\Lambda^{S,T}$ cannot be extracted from on-shell kaon-nucleon data. They can however be determined from kaonic-atom data since the kaonic atom "sees" the off-shell kaon self-energy. In fact the relevant quantity is

$$\kappa(x) = c_\Lambda(1 - \frac{1}{2}x) - \frac{2}{3}c_\Lambda^\sigma x \qquad (10.171)$$

with $x \equiv \rho_p/\rho$. The presently available kaonic-atom data imply that for $x \approx 1/2$

$$\frac{3}{4}\kappa f_\pi^2 \approx 10 \qquad (10.172)$$

corresponding to an attraction of order of (180 ± 20) MeV at a density $u \approx 0.97$. This is rather crude, so we shall not adhere to this value too literally. We will just take it as a canonical value. It turns out that the critical density is *not at all* sensitive to this value once the condensation is triggered.

We should note that the above value is "natural" in the sense that the coefficient of the four-Fermi interactions is of $O(1)$.

The condensation process takes place when $x < \frac{1}{2}$, so the presently available kaonic-atom constraint leaves one parameter free. When isotopic effects are measured in kaonic atom, this freedom will be eliminated. At the moment the best one can do is to impose the "naturalness" condition on the free parameter, say, $c_\Lambda \approx O(1)$. The result turns out to be again quite insensitive to the precise value of this free parameter. The result is summarized in Fig.10.12. The predicted critical density is $u_c \sim (3-4)$. When the BR scaling of subsection (10.3.5) is suitably incorporated – which in the present framework corresponds to taking into account a selected set of higher-Fermi interactions in the Lagrangian, the resulting value comes out to be, independently of the parameters,

$$u_c \sim 2. \qquad (10.173)$$

This is also what was obtained in the mean-field chiral Lagrangian theory of [39].

- *Role of four-Fermi interactions*

One novel ingredient in the loop calculation given above is the presence of the diagrams Fig.10.11g, h that involve the $\Lambda(1405)$. Their contribution is crucial for attraction in kaonic atoms. On the other hand, the critical

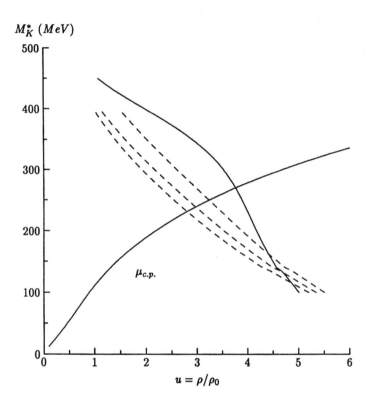

Fig. 10.12. The M_K^* vs density u for $(c_\Lambda - c_\Lambda^\sigma)f_\pi^2 = 10$ implied by kaonic-atom data and $c_\Lambda f_\pi^2 = 10, 5, 0$ in broken lines from left to right. The solid line is the linear density approximation. The point where the chemical potential $\mu_{c.p.}$ intersects these lines gives the critical point.

point for kaon condensation depends little on the strength of the four-Fermi interaction. Indeed the self-energy Π_Λ is quadratic in the kaon frequency ω and hence as ω runs down toward the regime where the condensation takes place, this term becomes "irrelevant," thereby explaining the negligible influence. A renormalization group flow analysis[77] confirms this result. This is unlike the four-Fermi interaction which becomes "marginally relevant" as in superconductivity leading to Cooper pair condensation. Nonetheless it is found that the absence of the attraction required for kaonic atoms would prohibit the condensation completely, so the four-Fermi interactions do play a crucial role in *triggering* the condensation process.

10.4.3. *Vector symmetry*

What happens when density is so high that there is a phase transition away from the Goldstone mode? Does it go into quark-gluon phase as expected at some density by QCD? Or does it make a transition into something else? Up to date, this question has not received a clear answer: It remains an open question even after so many years of QCD. Here we discuss a possible state of matter that is still hadronic but different from the normal hadron state and also from meson-condensed states – a state that realizes the "vector symmetry" of Georgi[78].

(i) Pseudoscalar and scalar Goldstone bosons

When one has the matrix element of the axial current

$$\langle 0|J_5^\mu|\pi\rangle = iq^\mu f_\pi \tag{10.174}$$

this can mean either of the two possibilities. The conventional way is to consider π's as Goldstone bosons that emerge as a consequence of spontaneous breaking of chiral $SU(3) \times SU(3)$ symmetry (with $N_f = 3$), with the current J_5^μ associated with the part of the symmetry that is broken (*i.e.*, the charge $\int d^3x J_5^0(x)$ being the generator of the broken symmetry). But there is another way that this relation can hold. If there are scalar bosons s annihilated by the vector current J^μ

$$\langle 0|J^\mu|s\rangle = iq^\mu f_s \tag{10.175}$$

with $f_\pi = f_s$, then the symmetry $SU(3) \times SU(3)$ can be unbroken. This can happen in the following way. Since J_5^μ and J^μ are part of $(8,1) +$

$(1, 8)$ of $SU(3) \times SU(3)$, all we need is that π's and s's are part of $(8, 1) + (1, 8)$. The $SU(3) \times SU(3)$ symmetry follows if the decay constants are equal. In nature, one does not see low-energy scalar octets corresponding to s. Therefore the symmetry must be broken. This may be understood if one imagines that the scalars are eaten up to make up the longitudinal components of the massive octet vector mesons. If this is so, then there must be a situation where the symmetry is restored in such a way that the scalars are liberated and become real particles, with the vectors becoming massless. This is the "vector limit" of Georgi. In this limit, there will be 16 massless scalars and massless octet vector mesons with the symmetry swollen to $[SU(3)]^4$ which is broken down spontaneously to $[SU(3)]^2$. In what follows, we discuss how this limit can be reached.

The full vector symmetry is

$$SU_L(3) \times SU_R(3) \times SU_{G_L}(3) \times SU_{G_R}(3). \tag{10.176}$$

This is a primordial symmetry which is dynamically broken down to

$$SU_{L+G_L}(3) \times SU_{R+G_R}(3). \tag{10.177}$$

As is well-known, the nonlinear symmetry

$$\frac{SU_L(3) \times SU_R(3) \times SU_{G_L}(3) \times SU_{G_R}(3)}{SU_{L+G_L}(3) \times SU_{R+G_R}(3)} \tag{10.178}$$

is equivalent to the linear realization

$$\{[SU_L(3) \times SU_{G_L}(3)]_{global} \times [SU_L(3)]_{local}\} \times \{L \to R\} \tag{10.179}$$

where the local symmetries are two copies of the hidden gauge symmetry. One can think of the two sets of hidden gauge bosons as the octet vectors ρ_μ and the octet axial vectors a_μ.

In order to supply the longitudinal components of both the vector and axial-vector octets of spin-1 fields, we would need 16 Goldstone bosons that are to be eaten up plus 8 pseudoscalars π to appear as physical states. Here we shall consider a minimal model where only the vector fields ρ are considered. The axials are purged from the picture. We can imagine this happening as an explicit symmetry breaking

$$SU_{G_L} \times SU_{G_R} \to SU_{G_L+G_R}. \tag{10.180}$$

There can be many different fates for the axials after the symmetry breaking. For instance, they could be banished to high mass $\gg m_\rho$.

(ii) The Lagrangian

To be more specific, we shall construct a model Lagrangian with the given symmetry. Since we are ignoring the axials, the Goldstone bosons can be represented with the transformation properties

$$\xi_L \quad \rightarrow \quad L\xi_L G^\dagger, \tag{10.181}$$

$$\xi_R \quad \rightarrow \quad R\xi_R G^\dagger \tag{10.182}$$

with ξ_L transforming as $(3, 1, \bar{3})$ under $SU_L(3) \times SU_R(3) \times SU_G(3)$ (with $G = G_L + G_R$) and ξ_R as $(1, 3, \bar{3})$. L, R and G are the linear transformations of the respective symmetries. The usual Goldstone bosons of the coset $SU_L(3) \times SU_R(3)/SU_{L+R}(3)$ are

$$U = \xi_L \xi_R^\dagger \tag{10.183}$$

transforming as $(3, \bar{3}, 1)$. In terms of the covariant derivatives

$$D^\mu \xi_L \quad = \quad \partial^\mu \xi_L - ig\xi_L \rho^\mu, \tag{10.184}$$

$$D^\mu \xi_R \quad = \quad \partial^\mu \xi_R - ig\xi_R \rho^\mu \tag{10.185}$$

the lowest derivative Lagrangian having the (hidden) local $SU(3)_{L+R}$ symmetry is

$$\frac{f^2}{2}\{ \operatorname{Tr}\left(D^\mu \xi_L D_\mu \xi_L^\dagger \right) + (L \rightarrow R)\} + \kappa \frac{f^2}{4} \operatorname{Tr}(\partial^\mu U \partial_\mu U^\dagger). \tag{10.186}$$

At the tree order with this Lagrangian, we have

$$f_s = f, \qquad f_\pi = \sqrt{1 + \kappa} f. \tag{10.187}$$

In the unitary gauge (*i.e.*, $\xi_L = \xi_R^\dagger = \sqrt{U}$), the vector meson mass is given by the KSRF relation

$$m_\rho = fg = \frac{1}{\sqrt{1 + \kappa}} f_\pi g. \tag{10.188}$$

Thus $f_s = f_\pi$ when $\kappa = 0$, *i.e.*, when the mixing LR vanishes. One can see that the standard hidden gauge symmetry result is obtained when

$\kappa = -1/2$. When $\kappa = 0$, we have an unbroken $SU(3) \times SU(3)$, but the vectors are still massive. The mass can disappear even if $f_\pi \neq 0$ if the gauge coupling vanishes. In the limit $g \to 0$ which can be attained at asymptotic density as described below, the vector mesons decouple from the Goldstone bosons, the scalars are liberated and the symmetry swells with the Goldstone bosons transforming

$$\xi_L \to L\xi_L G_L^\dagger, \tag{10.189}$$

$$\xi_R \to R\xi_R G_R^\dagger. \tag{10.190}$$

Note that the local symmetry G gets "recovered" to a *global* symmetry $G_L \times G_R$.

(iii) Hidden gauge symmetry and the vector limit

We shall now consider renormalization-group (RG) structure of the constants g and κ of the Lagrangian (10.186) reduced to $SU(2)$ flavor. The kinetic energy term of the ρ field should be added for completeness; we do not need it for this discussion. It is convenient to parametrize the chiral field as

$$\xi_{\mathrm{L,R}} \equiv e^{i\sigma(x)/f_s} e^{\pm i\pi(x)/f_\pi} \tag{10.191}$$

with $\sigma(x) = \frac{1}{2}\tau^a \sigma^a(x)$ and $\pi(x) = \frac{1}{2}\tau^a \pi^a(x)$. (Setting $\sigma(x) = 0$ corresponds to the unitary gauge with no Faddeev-Popov ghosts.) We are interested in what happens to hadrons described by this Lagrangian as matter density ρ and/or temperature T are increased[79]. For this, consider the vector meson mass in dense medium at zero temperature. It turns out that the mass formula (for $p^2 = 0$, *i.e.* off-shell) (10.188) is valid to all orders of perturbation theory[79,80]. We expect this to be true in medium, according to the argument given above. Therefore we need to consider the scaling behavior of the constants f_π, g and κ in medium. Now we know from above that f_π decreases as density is increased. This is mainly caused by the vacuum property, so we can assume that the effect of radiative corrections we are considering is not important. (Lattice gauge calculations show however that the f_π in heat bath changes little up to near the critical temperature $T_{\chi SR} \sim 140$ MeV.) Thus we have to examine how g and κ scale. The quantities we are interested in are the β functions for the coupling constants g and κ. These can be derived in a standard way with the Lagrangian

(10.186). At one-loop order, we have in dimensional regularization[79]

$$\beta_g(g_r) \equiv \mu\frac{dg_r}{d\mu} = -\frac{87 - a_r^2}{12}\frac{g_r^3}{(4\pi)^2}, \tag{10.192}$$

$$\beta_a(a_r) \equiv \mu\frac{da_r}{d\mu} = 3a_r(a_r^2 - 1)\frac{g_r^2}{(4\pi)^2} \tag{10.193}$$

where μ is the length scale involved (which in our case could be taken to be the matter density) and we have redefined

$$a = 1/(1 + \kappa). \tag{10.194}$$

To see how the factors in the beta function can be understood, let us discuss the first equation describing the way the coupling constant g runs. In terms of Feynman diagrams, the quantity on the right-hand side of (10.192) can be decomposed as

$$-\frac{1}{(4\pi)^2}\frac{87 - a_r^2}{12} = -\frac{1}{(4\pi)^2}\left[\frac{11n_f}{3} - (\frac{1}{2})^2\frac{1}{3} - (\frac{a_r}{2})^2\frac{1}{3}\right] \tag{10.195}$$

with the number of flavor $n_f = 2$. This is some analog to the beta function of QCD, $\beta_g^{\rm QCD} = -\frac{g^3}{(4\pi)^2}\left[\frac{11}{3}N_c - \frac{2}{3}n_f\right]$. The first term on the right-hand side of (10.195) is the vector meson loop which is an analog to the gluon contribution in QCD with the number of flavors n_f replacing the number of colors N_c (the vector meson ρ is the "gauge field" here), the second term is the σ loop contribution $g_{\rho\sigma\sigma}^2/3$ with $g_{\rho\sigma\sigma} = g/2$ and the last term the pion loop contribution $g_{\rho\pi\pi}^2/3$ with $g_{\rho\pi\pi} = a_r/2$.

As in QCD, the fact that the beta function is negative signals that the coupling constant runs down as the momentum (or possibly density or T) scale increases and at asymptotic momentum (or possibly at some density or T), the coupling would go to zero. Now the beta function is negative here as long as $a_r^2 < 87$ with the coupling constant running as

$$g^2(\mu) = \frac{8\pi^2}{(87 - a_r^2)\ln\frac{\mu}{\Lambda}} \tag{10.196}$$

with the HGS scale Λ defined as in QCD. The inequality $a_r^2 < 87$ is clearly satisfied as one can see from eq. (10.193) which says that there is an ultraviolet fixed point $a_r = 1$, so as the momentum increases the constant a would run toward 1. Of course one would have to solve the equations

(10.192) and (10.193) simultaneously to see the trajectory of a_r but it seems reasonable to assume that the a_r runs monotonically from the vacuum value $a_r = 2$ (or $\kappa = -1/2$) to the fixed point $a_r = 1$ (or $\kappa = 0$). The in-medium mass formula

$$m_\rho^\star = f^\star g^\star = \frac{1}{\sqrt{1 + \kappa^\star}} f_\pi^\star g^\star = 2\sqrt{1 + \kappa^\star} f_\pi^\star g_{\rho\pi\pi}^\star \qquad (10.197)$$

shows that apart from the condensate effect on $f_\pi^\star$, there are additional effects due to the increasing $(1 + \kappa^\star)^{1/2}$ factor and the decreasing $g^\star$. Asymptotically the falling $g^\star$ will win and hence the vector mass will go to zero as the coupling $g^\star$ vanishes, approaching the "vector limit" of Georgi.

(iv) Physics under extreme conditions

What happens to the hadronic matter governed by the HGS Lagrangian as density or temperature increases? At present, there is no direct prediction from QCD that the relevant limits can be reached by temperature or density. No model calculations purporting to indicate them are available. Nonetheless, the vector limit ideas are intuitively appealing and could play an important role in dense cold matter like in neutron stars and in hot dense matter like in heavy-ion collision. Some of these issues are discussed in the review by Brown and Rho[38].

In the $\kappa = 0$ limit, the vector mass is

$$m_\rho = 2 f_\pi g_{\rho\pi\pi} \qquad (10.198)$$

which in terms of physical values of f_π and $g_{\rho\pi\pi}$ is too big by roughly a factor of $\sqrt{2}$ whereas the current algebra result (KSRF) is $m_\rho = \sqrt{2} f_\pi g_{\rho\pi\pi}$, which is in good agreement with the experiment. This shows that at the tree order, the limit is rather far from reality as applied to the medium-free space. Loop corrections with the Lagrangian treating the κ term as a perturbation does improve the prediction[81], bringing it much closer to experiments. One possible scenario that emerges from the discussion given above is that density changes the state of matter from the phase of normal matter with $\kappa \neq 0$ and $g \neq 0$ to first $\kappa = 0$ and then to $g = 0$, resulting in a significant increase of light degrees of freedom below the critical point.

If we consider two flavors (ignoring the strange quark which is massive) in the chiral limit, then we can have three pions, three longitudinal components of the ρ meson, three longitudinal components of the axial vector a_1, the massless vectors ρ and a_1. Therefore in the vector limit with

$\kappa = g = 0$, we would have 21 massless degrees of freedom. This should be compared with $27\frac{1}{2}$ degrees of freedom expected from QCD. This would mean that there would be little increase of degrees of freedom from the "hadronic phase" to the quark-gluon phase. This scenario seems to be supported by lattice gauge calculations (on quark number susceptibility) and heavy-ion collisions ("cool" kaons) [38].

References

1. C. Coriano, R. Parwani, H. Yamagishi and I. Zahed, Phys. Rev. **D45**, 2542 (1992).
2. A.V. Manohar, Phys. Lett. **B336**, 502 (1994).
3. See for references, V.G. Makhanov, Y.P. Rybakov and V.I. Sanyuk, *The Skyrme Model*, Springer Series in Nuclear and Particle Physics, Springer-Verlag (Berlin 1993).
4. H. Yamagishi and I. Zahed, Mod. Phys. Lett. **A7**, 102 (1992).
5. A. Jackson, A.D. Jackson and V. Pasquier, Nucl. Phys. **A432**, 567 (1985).
6. R. Vinh Mau, M. Lacombe, B. Loiseau, W.N. Cottingham and P. Lisboa, Phys. Lett. **B150**, 259 (1985); E.M. Nyman and D.O. Riska, Int. J. Mod. Phys. **A5**, 1535 (1988).
7. J.J.M. Verbaarschot, T.S. Walhout, J. Wambach and H.W. Wyld, Nucl. Phys. **A468**, 520 (1987); V.B. Kopeliovich and B.E. Shtern, JETP Lett. **45**, 203 (1987); E. Braaten and L. Carson, Phys. Rev. **D38**, 3525 (1988).
8. N.S. Manton, Phys. Rev. Lett. **60**, 1916 (1988).
9. M.F. Atiyah and N.S. Manton, Commun. Math. Phys. **152**, 391 (1993).
10. W.Y. Crutchfield, N.J. Snyderman and V.R. Brown, Phys. Rev. Lett. **68**, 1660 (1992).
11. M. Bando, T. Kugo and K. Yamawaki, Phys. Repts. **164**, 217 (1988).
12. E. Braaten, S. Townsend and L. Carson, Phys. Lett. **B235**, 147 (1990).
13. M. Lacombe, B. Loiseau, J.M. Richard, R. Vinh Mau, J. Cote, D. Pires and R. DeTourreil, Phys. Rev. **C21**, 861 (1980).
14. T.S. Walhout and J. Wambach, Phys. Rev. Lett. **67**, 314 (1991).
15. H. Yamagishi and I. Zahed, Phys. Rev. **D43** 891 (1991).
16. V. Thorsson and I. Zahed, Phys. Rev. **D45** 965 (1992).
17. N.R. Walet and R.D. Amado, Phys. Rev. **C47**, 498 (1993).
18. R.D. Amado, B. Shao and N.R. Walet, Phys. Lett. **B314**, 159 (1993).
19. For a review, see M. Rho and G.E. Brown, Comments Part. Nucl. Phys. **10**, 201 (1981).
20. See, *e.g.*, Y. Nambu, in *Proceedings of the 1988 International Workshop on New Trends in Strong Coupling Gauge Theories*, ed. M. Bando *et al.* (World Scientific, Singapore, 1989) p. 2.
21. S. Weinberg, Physica (Amsterdam) **96A**, 327 (1979).
22. T.S. Park, D.P. Min and M. Rho, Phys. Rep. **233**, 341 (1993); G. Ecker, Prog.

Part. Nucl. Phys. **35** (1995); V. Bernard, N. Kaiser and U.G. Meissner, Int. J. Mod. Phys. **E4**, 193 (1995).

23. S. Coleman, J. Wess and B. Zumino, Phys. Rev. **177**, 2239 (1969); C. Callan, S. Coleman, J. Wess and B. Zumino, Phys. Rev. **177**, 2247 (1969).

24. For a nice exposition on this issue see Georgi's book, H. Georgi, *Weak Interactions and Modern Particle Theory* (The Benjamin/Cummings Publishing Co., Menlo Park, Calif. 1984).

25. J. Gasser, M.E. Sainio and A. Svarc, Nucl. Phys. **B307**, 779 (1988).

26. H. Georgi, Phys. Lett. **B240**, 447 (1990); E. Jenkins and A. Manohar, Phys. Lett. **B255**, 562 (1991); Phys. Lett. **B259**, 353 (1991); E. Jenkins, Nucl. Phys. **B368**, 190 (1991).

27. S. Weinberg, Phys. Lett. **B251**, 288 (1990); Nucl. Phys. **B363**, 3 (1991)

28. M. Luke and A. Manohar, Phys. Lett. **B286**, 348 (1992).

29. See J. Polchinski, "Effective field theory and the Fermi surface", in *Recent directions in particle physics*, ed. J. Harvey and J. Polchinski (World Scientific, 1993) p245ff.

30. S. Weinberg, Physica (Amsterdam) **96A**, 327 (1979); Phys. Lett. **B251**, 288 (1990); Nucl. Phys. **B363**, 3 (1991); Phys. Lett. **B295**, 114 (1992).

31. M. Rho, Phys. Rev. Lett. **66**, 1275 (1991).

32. T.-S. Park, D.-P. Min and M. Rho, Phys. Repts. **233**, 341 (1993).

33. J. Gasser and H. Leutwyler, Ann. Phys. (N.Y.) **158**, 142 (1984); Nucl. Phys. **250**, 465, 517, 539 (1985); Nucl. Phys. **B307**, 763 (1988).

34. See, *eg*, R. Shankar, Rev. Mod. Phys. **66**, 129 (1994).

35. V. Bernard, U.G. Meissner and I. Zahed, Phys. Rev. Lett. **59**, 966 (1987).

36. G.E. Brown and M. Rho, Phys. Rev. Lett. **66**, 2720 (1991)

37. C. Adami and G.E. Brown, Phys. Repts. **234**, 1 (1993).

38. G.E. Brown and M. Rho, Phys. Repts. in press (1995).

39. G.E. Brown and M. Rho, "From chiral mean fields to Walecka mean fields and kaon condensation," Nucl. Phys. **A** in press (1995).

40. T.-S. Park, D.-P. Min and M. Rho, "Chiral Lagrangian approach to vector exchange currents," Nucl. Phys. **A** in press (1995).

41. S. Weinberg, Phys. Rev. Lett. **65**, 1177 (1990).

42. S.R. Beane and U. van Kolck, Phys. Lett. **138**, 137 (1994).

43. T. Hatsuda and S.H. Lee, Phys. Rev. **C46**, R34 (1992).

44. X. Jin and D.B. Leinweber, "Valid QCD sum rules for vector mesons in nuclear matter," nucl-th/9510064.

45. R.J. Furnstahl, X. Jin and D.B. Leinweber, "New QCD sum rules for nucleons in nuclear matter," nucl-th/9511007.

46. S. Weinberg, Phys. Lett. **B295**, 114 (1992).

47. C. Ordóñez and U. van Kolck, Phys. Lett. **B291**, 459 (1992).

48. U. van Kolck, University of Texas thesis, 1993; L. Ray, C. Ordóñez and U. van Kolck, Phys. Rev. Lett. **72** 1982 (1994).

49. M. Chemtob and M. Rho, Nucl. Phys. **A163**, 1 (1971).

50. K. Kubodera, J. Delorme and M. Rho, Phys. Rev. Lett. **40**, 755 (1978).

51. B. Frois and J.-F. Mathiot, Comments Part. Nucl. Phys. **18**, 291 (1989).
52. I.S. Towner, Nucl. Phys. **A542**, 631 (1992).
53. K. Kubodera and M. Rho, Phys. Rev. Lett. **67**, 3479 (1991).
54. T.-S. Park, D.-P. Min and M. Rho, Phys. Rev. Lett. **74**, 4153 (1995).
55. T.-S. Park, I.S. Towner and K. Kubodera, Nucl. Phys. **A**, in press.
56. T.-H.H. Lee and D.O. Riska, Phys. Rev. Lett. **70**, 2237 (1993); C.J. Horowitz, H.O. Meyer and D.K. Griegel, Phys. Rev. **C49**, 1337 (1994).
57. J. Delorme, Nucl. Phys. **A374**, 541c (1982).
58. T. Hatsuda and S-H. Lee, Phys. Rev. **C46**, R34 (1992).
59. E.K. Warburton and I.S. Towner, Phys. Lett. **B294**, 1 (1992).
60. D.H. Wilkinson, in *Nuclear Physics with Heavy Ions and Mesons*, ed. R. Balian, M. Rho and G. Ripka (North-Holland, Amsterdam, 1978)
61. E. Jenkins and A.V. Manohar, Phys. Lett. **259**, 353 (1991).
62. G.E. Brown and W. Weise, Phys. Repts. **22C**, 279 (1975).
63. G.E. Brown, M. Buballa, Z.B. Li and J. Wambach, Nucl. Phys. **A593**, 295 (1995).
64. B. Serot and J.D. Walecka, Advances in Nuclear Physics, Vol. 16, ed. J.W. Negele and E. Vogt (Plenum, NY, 1986); B.D. Serot, Rep. Prog. Phys. **55**, 1855 (1992).
65. S. Coleman, Nucl. Phys. **B262**, 263 (1985), erratum-*ibid.* **B269**,744 (1986).
66. B.W. Lynn, Nucl. Phys. **B402**, 281 (1993).
67. D.B. Kaplan and A.E. Nelson, Phys. Lett. **B175**, 57 (1986).
68. H.D. Politzer and M.B. Wise, Phys. Lett. **B273**, 156 (1991).
69. M. Fukugita, Y. Kuramashi, M. Okawa and A. Ukawa, Phys. Rev. **D51**, 5319 (1995); S.-J. Dong and K.-F. Liu, "$\pi N \sigma$ term and quark spin content of the nucleon," Proc. Lattice '94, September 27-30, 1994, Bielefeld, Nucl. Phys. **B** (Proc. Suppl.) **42**, 322 (1995).
70. This part is based on G.E. Brown, K. Kubodera, V. Thorsson and M. Rho, Phys. Lett. **B291**, 355 (1992)
71. J.M. Lattimer, C.J. Pethick, M. Prakash and P. Haensel, Phys. Rev. Lett. **66**, 2701 (1991).
72. K. Kubodera, in *Proc. of Fifth Summer School and Symposium on Nuclear Physics*, Journal Korean Phys. Soc. **26**, S171 (1993).
73. G.E. Brown, C.-H. Lee, M. Rho and V. Thorsson, Nucl. Phys. **A567**, 937 (1994).
74. C.-H. Lee, H. Jung, D.-P. Min and M. Rho, Phys. Lett. **B326**, 14 (1994).
75. G. E. Brown and H. A. Bethe, Astrophys. J. **423**, 659 (1994).
76. C.-H. Lee, G.E. Brown and M. Rho, Phys. Lett. **B335**, 266 (1994); C.-H. Lee, G.E. Brown, D.-P. Min and M. Rho, Nucl. Phys. **A585**, 401 (1995).
77. H.K. Lee, M. Rho and S.-J. Sin, Phys. Lett. **B348**, 290 (1995).
78. H. Georgi, Nucl. Phys. **B331**, 311 (1990).
79. We will follow M. Harada and K. Yamawaki, Phys. Lett. **B297**, 151 (1992).
80. M. Harada, T. Kugo and K. Yamawaki, Phys. Rev. Lett. **71**, 1299 (1993).
81. P. Cho, Nucl. Phys. **B358**, 383 (1991).

CHAPTER 11

HADRONS AT FINITE TEMPERATURE

11.1. Introduction

The heavy-ion colliders in construction at RHIC and LHC have generated an intense activity in the physics of hot and dense hadronic matter. From asymptotic freedom, one expects that at sufficiently high temperatures and/or densities, hadronic matter in thermal and/or chemical equilibrium will behave as a weakly interacting system of quarks and gluons (quark-gluon plasma) [1,2,3]. As a result, the bulk thermodynamical quantities such as the energy, the pressure, the entropy, etc. should display black-body behavior. The plasma should screen for space-like momenta and display collective behavior for time-like momenta.

The low-temperature phase of QCD is dominated by the physics of pions and the constraints of the spontaneous breaking of chiral symmetry. It is therefore natural from the point of view of phenomenology to approach the higher temperature regime starting from effective hadronic theories discussed in the preceding chapters. The purpose of this chapter is to show how one can make use of the concepts introduced above to study the leading-temperature properties of hadrons in the chiral limit. We will also discuss some concepts related to the use of the sum rules at finite temperature, give an overview of the results obtained using instanton models, and comment on the importance of confinement. After reviewing the ideas behind the Hagedorn concept of a limiting temperature, we will give a brief discussion of the chiral critical point from the point of view of universality. A qualitative overview of the material based on intuitive understanding gained from experimental information available up to date has been recently reviewed[4]. One striking feature that emerged there was that as temperature (or density) goes across the transition point, effective hadronic degrees of freedom would cede in a completely smooth fashion to QCD degrees of freedom. (This kind of scenario was envisioned many years ago by Bég and Shei.[5]) In this chapter we wish to describe more precisely the current status of our understanding on the change of hadronic matter at high temperature.

The high-temperature phase of QCD is most likely dominated by the physics of quarks and gluons. After going over the main finite-temperature lattice results, we will argue that in the region of interest for most of the experimental analyses, namely $T_c < T < 3T_c$, the behavior of the quarks and gluons is not that of a free gas. More importantly, it is never that of a free gas when space-like correlations are probed. We will give a brief account of how ideas from dimensional reduction can be used to account for most of the lattice results.

11.2. Properties of Matter at Finite T

11.2.1. *Pion gas*

At low temperature the Gibbs averages are dominated by the lightest hadrons, *i.e.* pions. In the chiral limit, the role of the pion is further enhanced as all other hadrons are suppressed by Boltzmann factors. To understand these effects, let us consider the case where the vacuum is heated up. At low temperature, we produce a gas of pions. To leading order in the thermal pion density, the pressure $\mathbf{P}$ of the gas is given by the diagrams shown in Fig. 11.1. Specifically,

$$\mathbf{P} = \beta F = -\ln \mathrm{Tr}\left(e^{-\beta H}\, \mathbf{P}_*\right) = \mathbf{P}_\pi + \mathbf{P}_{2\pi} + \mathbf{P}_{3\pi} + \dots \qquad (11.1)$$

for one pion (a), two pion (b) and three pion (c) contributions, respectively. $\mathbf{P}_*$ in (11.1) is the projector onto physical states. Its presence is not benign as we will see further in this chapter. As the pions are on-shell in the heat bath, it follows that

$$\mathbf{P}_\pi = \int \frac{d\vec{k}}{(2\pi)^3}\, |\vec{k}|\, n_k = \frac{\pi^2}{30}T^4 \qquad (11.2)$$

where $n_k = 1/(e^{|k|/T} - 1)$ is the Bose distribution for massless pions. Equation (11.2) is just the black-body radiation contribution. The number of pions is just

$$\mathbf{n}_\pi = 3 \int \frac{d\vec{k}}{(2\pi)^3}\, n_k = \frac{3\zeta(3)}{\pi^2} T^3 \tag{11.3}$$

so that the average pion energy is $E_\pi = 3\mathbf{P}_\pi/\mathbf{n}_\pi \approx 2.7T$, valid for $T \geq m_\pi$. This is to be contrasted with the non-relativistic result $E_\pi = m_\pi + 3T/2$, valid for $T \leq m_\pi$. The pion mean separation is $d \sim 1/^3\sqrt{\mathbf{n}_\pi} \sim 1.4T$.

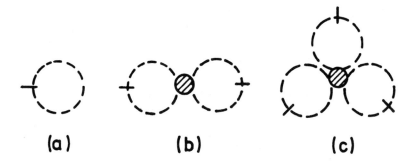

(a)   **(b)**   **(c)**

Fig. 11.1. (a) π-contribution to the pressure; (b) 2π-contribution; (c) 3π-contribution.

As the temperature is increased, the effects of interactions among the pions in the heat bath should be taken into account. From Fig. 11.1b, we have

$$\mathbf{P}_{2\pi} = \left(\prod_{i=1}^{2} \int \frac{d\vec{k}}{(2\pi)^3} \frac{n_{k_i}}{2|k_i|}\right) \left(\sum_{I=0}^{2} (2I+1)\mathbf{T}_I(k_1,k_2;k_1,k_2)\right) \tag{11.4}$$

where $\mathbf{T}_I$ is the forward $\pi\pi$-scattering amplitude in the $I = 1 \otimes 1 = (0,1,2)$ channels. At low temperature, the Bose distributions select energies $\omega \sim T$ which are small. Thus, the low-temperature part of (11.4) is dominated by the soft-pion physics discussed in the preceding chapters. Quite generally, if we think of $\mathbf{T}_I$ as a function of the Mandelstam variable s only for simplicity, then at low energies, we can write

$$\mathbf{T}_I(s) = \mathbf{T}_{I,2} + \frac{s}{f_\pi^2}\mathbf{T}_{I,4} + \frac{s^2}{f_\pi^4}\mathbf{T}_{I,6} + \cdots \tag{11.5}$$

with a threshold $s_\pi = (2m_\pi)^2 \to 0$. Because of Adler's condition (see

Chapter 4), at threshold,

$$\mathbf{T}_{I,2} = 16\pi m_\pi(a_0, a_1, a_2) = 16\pi m_\pi \left(\frac{7m_\pi}{16\pi f_\pi^2}, 0, -\frac{m_\pi}{8\pi f_\pi^2} \right) \to (0,0,0) \ (11.6)$$

as all the scattering lengths vanish in the chiral limit. There is no correction to the black-body contribution. This phenomenon carries over to the other parts of (11.5) as the forward scattering amplitude vanishes in the chiral limit. Explicitly,

$$
\begin{aligned}
\sum_{I=0}^{2}(2I+1)\mathbf{T}_{I,4} &= 1 \cdot \left(\frac{4s}{f_\pi^2} \right) + 3 \cdot \left(\frac{2(t-u)}{f_\pi^2} \right) + 5 \cdot \left(\frac{-2s}{f_\pi^2} \right) \\
&= \frac{6}{f_\pi^2}(t-u-s) = 0.
\end{aligned}
\tag{11.7}
$$

The first correction to the black-body limit comes out of the three-body parts Fig. 11.1c in the chiral limit. The result is [6]

$$\mathbf{P}_{3\pi} = \frac{4\pi^2}{30}T^4 \left(\frac{T^2}{12 f_\pi^2} \right)^2 \ln(\frac{\Lambda_P}{T}) \tag{11.8}$$

where $\Lambda_P \sim 275$ MeV. The correction to the black-body limit is of order T^8 and repulsive. This shows that in the temperature range of a 100 MeV or so, the pion gas is dilute and weakly interacting. The entropy of the gas is given by $\mathbf{S} = d\mathbf{P}/dT$.

11.2.2. *Pion mean-free path*

In heavy-ion collisions, the mean-free path of the pions in the central rapidity region plays an important role in the kinetic analysis. In hot matter, the pion dispersion relation is

$$\omega^2 = k^2 + m_\pi^2 + \mathbf{\Pi}(\omega, k) \tag{11.9}$$

where ω and k are the pion energy and momentum, respectively, and $\mathbf{\Pi}$ is the pion self-energy, which renormalizes to zero at $T = 0$. A complex $\mathbf{\Pi}$ causes the pions to attenuate in the medium. For real ω, (11.9) yields a complex wave-number, and hence a space attenuation. Alternatively, for real wave-numbers k, (11.9) yields a pole on the second Riemann sheet,

$$\omega = \sqrt{k^2 + m_\pi^2} - i\frac{\gamma(k)}{2} = \omega(k) - i\frac{\gamma(k)}{2} \tag{11.10}$$

with

$$\gamma(k) = -\frac{1}{\omega(k)} \operatorname{Im}\Pi(\omega(k), k). \tag{11.11}$$

Here $1/\gamma$ is the relaxation time of the pion wave in the medium. For a dilute gas, the imaginary part of Π is given by the cut of Fig. 11.2, modulo phase space. By the optical theorem, the cut is related to the total pion-pion scattering amplitude $\sigma_*(s)$ at tree level (*i.e.*, Weinberg term) [7],

$$\gamma(k_1) = \int \frac{d^3(k_2)}{(2\pi)^3} v_{\rm rel} \left(1 - \mathbf{v}_1 \cdot \mathbf{v}_2\right) \frac{1}{e^{\omega(k_2)} - 1} \sigma_*(s) \tag{11.12}$$

where $\mathbf{v}_{1,2}$ are the velocities of the incoming two pions, and $v_{\rm rel}$ the velocity of one relative to the the the other at rest. In the chiral limit $\sigma_*(s) = 5s/48\pi f_\pi^4$. The cross section rapidly grows with the energy. For $\sqrt{s} = 140$ MeV, the cross section is about 0.5 fm^2. Figure 11.3 shows the behavior of the mean-free path $\lambda = 1/v\gamma = k/\omega\gamma$ versus k in the chiral limit (dashed line) and for massive pions (solid line) at two temperatures, $T = 120$ MeV and $T = 150$ MeV [7]. In the chiral limit, there is no damping at large wavelengths. Zero energy pions do not interact. The mean-free path is pretty large at $T \sim 100$ MeV (about 10 fm), indicating that the free gas picture holds at these temperatures. The mean collision time was estimated by Goity and Leutwyler to be of order $\tau \sim 1/\langle\gamma\rangle_T \sim 12f_\pi^4/T^5$, since the cross section grows with T^2, and the pion density with T^3.

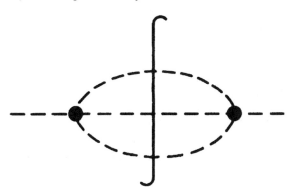

Fig. 11.2. Imaginary part of the pion self-energy contribution in the heat bath.

As the temperature is increased, the pion mean-free path drops dramatically, making the effects of interactions of pions among themselves and higher resonances important. While the effects of temperature alter considerably the structure of the ground state (vacuum), a quantitative understanding of these changes from a microscopic description is difficult. Below, we give some qualitative arguments to gain some understanding and present model calculations. First, however, we will derive a set of exact relations that hold in the low-temperature and chiral limit, independently of the model used.

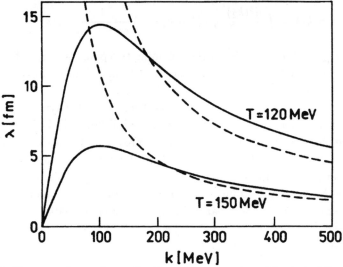

Fig. 11.3. Pion mean-free path as a function of its momentum. The dashed lines are for $m_\pi = 0$, and the solid lines for $m_\pi = 140$ MeV.

11.2.3. *Bulk parameters*

At zero temperature, the fermion condensate and the gluon condensate are the two key parameters of the vacuum state. We wish to know how they are affected by temperature. The fermion condensate at finite temperature follows from

$$\langle \overline{q}q \rangle_T = \text{Tr}\left(e^{-\beta(H-F)}\overline{q}q \; \mathbf{P}_* \right) = \langle \overline{q}q \rangle_0 + \langle \overline{q}q \rangle_\pi + \langle \overline{q}q \rangle_{2\pi} + \cdots \quad (11.13)$$

F is the free energy (11.1). The sum is over all physical states. At low temperature, the one-pion term is dominant

$$\langle \bar{q}q \rangle_\pi = \sum_{a=1}^{3} \int \frac{d^3 k}{(2\pi)^3} \frac{n_k}{2|k|} \langle \pi^a(k) | \bar{q}q | \pi^a(k) \rangle. \tag{11.14}$$

Using PCAC and the current-algebra commutation relations (see Chapter 4), it follows that

$$\langle \bar{q}q \rangle_\pi = - \left(\frac{T^2}{24 f_\pi^2} \right) \sum_{a=1}^{3} \langle 0 | [Q_5^a, [Q_5^a, \bar{q}q]] | 0 \rangle$$

$$= - \left(\frac{T^2}{12 f_\pi^2} \right) \frac{N_F^2 - 1}{N_F} \langle 0 | \bar{q}q | 0 \rangle \tag{11.15}$$

where N_F (=2) is the number of flavors. Thus

$$\langle \bar{q}q \rangle_T \approx \left(1 - \left(\frac{T^2}{12 f_\pi^2} \right) \frac{N_F^2 - 1}{N_F} \right) \langle 0 | \bar{q}q | 0 \rangle \tag{11.16}$$

to order T^2. This result was first derived by Leutwyler[8]. As the temperature is increased, the fermion condensate decreases. Thermal effects quench the symmetry breaking effects. This trend persists in higher pion contributions. Indeed, the two-pion contribution to (11.13) can be obtained in a similar way,

$$\langle \bar{q}q \rangle_{2\pi} = \sum_{a,b=1}^{3} \prod_{i=1}^{2} \int \frac{d^3 k_i}{(2\pi)^3} \frac{n_{k_i}}{2|k_i|} \langle \pi^a(k_1)\pi^b(k_2) | \bar{q}q | \pi^a(k_1)\pi^b(k_2) \rangle. \tag{11.17}$$

Successive use of the PCAC relation, the current-algebra commutation relation and the interaction between pions in the initial and final states give

$$\langle \bar{q}q \rangle_{2\pi} = - \frac{(N_F^2 - 1)}{2 N_F^2} \left(\frac{T^2}{12 f_\pi^2} \right)^2 \langle 0 | \bar{q}q | 0 \rangle. \tag{11.18}$$

The free part – coming from commutation relations – gives a positive contribution while the initial and final state interactions give a negative contribution. Since the $\pi\pi$-amplitude at threshold is of the order of m_π^2 and since the energy denominator is of order $1/m_\pi^2$, the effect of the $\pi\pi$ interaction in (11.17) is finite. For $N_F = 2$, the ratio of the former to the latter is $9/(-12)$[17]. This result is in agreement with the result derived from chiral

Lagrangians [6]. We refer to this work for a discussion on the contribution of higher resonances and pion masses at higher temperatures.

How does the gluon condensate change at finite temperature? To answer this question, we recall from Chapter 1 that in the chiral limit, the trace of the energy momentum tensor is anomalous

$$\langle 0|\Theta^\mu_\mu|0\rangle = \langle 0|\frac{2\beta}{g_R}\mathbf{G}^2|0\rangle. \tag{11.19}$$

Since there is no Lorentz covariant, non-perturbative regularization scheme for the unsubtracted energy-momentum tensor operator, we are forced to use the phenomenological definition of the gluon condensate from the QCD sum rules approach. In medium, we have

$$\langle\frac{\alpha_s}{\pi}G^2\rangle_T = \langle 0|\frac{\alpha_s}{\pi}G^2|0\rangle - \text{Tr}(e^{-\beta(H-F)}\Theta^\mu_\mu\,\mathbf{P}_*) \tag{11.20}$$

where the first term ensures that the energy momentum tensor has zero expectation value in the vacuum as required by Poincaré invariance and the second term is just $\mathcal{E} - 3\mathbf{P}$. For massless pions, the difference shows up at three-loop level (see above). The result is [6]

$$\langle\frac{\alpha_s}{\pi}G^2\rangle_T = \langle 0|\frac{\alpha_s}{\pi}G^2|0\rangle - \frac{\pi^2 T^8}{3240 f_\pi^2}N_F^2(N_F^2 - 1)\left(\ln(\frac{\Lambda_P}{T}) - \frac{1}{4}\right)(11.21)$$

to order T^8.

At finite temperature $\Theta^{00} \sim E^2 + B^2$ does not vanish. Thus E^2 and B^2 are two independent condensates. In the pion dominated phase, we expect

$$2\langle\Theta^{00}\rangle_T = \langle E^2 + B^2\rangle_T$$
$$\sim (N_F^2 - 1)b\int\frac{d\vec{k}}{(2\pi)^3}n_k\,|\vec{k}| = (N_F^2 - 1)\frac{bT^4}{90} \tag{11.22}$$

where $b = b(\mu \sim 1\text{GeV}) \sim 1.14$ [9] is a renormalization-dependent constant. It follows from (11.21) that to order T^4, the electric and magnetic condensates are the same. Thus

$$\langle E^2\rangle_T \sim \langle E^2\rangle_0 + (N_F^2 - 1)\frac{bT^4}{180},$$
$$\langle B^2\rangle_T \sim \langle B^2\rangle_0 + (N_F^2 - 1)\frac{bT^4}{180}. \tag{11.23}$$

The black-body contribution increases the expectation values of the condensates. The pion mass slightly affects this result.

This result is slightly counter-intuitive, as we expect the condensates to melt at finite temperature. While this result was derived in the chiral limit and at low temperature, an analysis of the lattice results suggests a somewhat different behavior. Indeed, lattice measurements of the pressure $(\mathcal{P})$ and energy density $(\mathcal{E})$ are amenable to the above condensates through [10,11]

$$\mathcal{E} = \alpha \langle E^2 + B^2 \rangle_{\text{lattice}} + \gamma \langle B^2 - E^2 \rangle_{\text{lattice}}$$
$$\mathcal{P} = \frac{\alpha}{3} \langle E^2 + B^2 \rangle_{\text{lattice}} - \gamma \langle B^2 - E^2 \rangle_{\text{lattice}} \qquad (11.24)$$

where the coefficients for pure glue are

$$\alpha = 0.5 - 0.08229 g^2 + \mathcal{O}(g^4),$$
$$\gamma = 0.0 - 0.01742 g^2 + \mathcal{O}(g^4). \qquad (11.25)$$

For QCD with three flavors, $\gamma \to 9\gamma/11$. In the lattice simulations, the zero-temperature part is subtracted so that (11.24) vanish at $T = 0$. The finite-temperature condensates follow from (11.24) by subtracting the perturbative black-body contribution, and adding back the $T = 0$ part of the condensates,

$$\langle E^2 \rangle_T = \langle E^2 \rangle_{\text{lattice}} - \langle E^2 \rangle_{\text{black-body}} + \langle E^2 \rangle_0,$$
$$\langle B^2 \rangle_T = \langle B^2 \rangle_{\text{lattice}} - \langle B^2 \rangle_{\text{black-body}} + \langle B^2 \rangle_0. \qquad (11.26)$$

The perturbative black-body contributions satisfy

$$\langle E^2 \rangle_{\text{black-body}} = \langle B^2 \rangle_{\text{black-body}} = \frac{1}{2}\mathcal{E}_{\text{black-body}}. \qquad (11.27)$$

Figure 11.4 shows the behavior of the electric (full) and magnetic (dashed) condensates versus T in GeV [10,11]. The zero-temperature condensates used there are

$$\frac{g^2}{\pi} \langle B^2 \rangle_0 = -\frac{g^2}{\pi} \langle E^2 \rangle_0 = (330 \pm 20 \text{ MeV})^4. \qquad (11.28)$$

The electric condensate increases at low T, while the magnetic condensate decreases. The latter behavior is not compatible with (11.23). The origin of this discrepancy may be due to the subtraction procedure adopted in defining these condensates at finite temperature, or inaccuracies in the lattice data. Since these data are fundamental to our understanding of the bulk transition in QCD, this discrepancy should be further investigated.

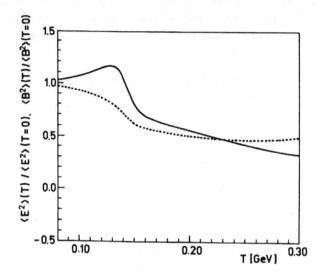

Fig. 11.4. Electric (full) and magnetic (dashed) condensates versus T in GeV [11].

11.2.4. *Hadron parameters*

Since the vacuum state is being changed by thermal effects, the next question is how we go about describing hadronic excitations in the heat bath. At low temperature, we can still think of hadrons as quasiparticles undergoing rescattering and absorption in a dilute environment of pions. This is reminiscent of the low-temperature electrons in normal metals, phonons and rotons in liquid ^{4}He. The processes of rescattering and absorption can be described using virial-type expansions. This approach is analogous to the description of collective potentials in nuclear many-body problems. [12] Here, we will again concern ourselves with exact results to leading order in the temperature.

Let us first consider the change in the pion mass m_π and decay constant f_π. Using the results of Chapter 5 with the vacuum average changed to a thermal average, we obtain

$$f_\pi^2(T) = f_\pi^2 \left(1 - \frac{2}{3} \frac{\langle \pi^2 \rangle_T}{f_\pi^2}\right),$$

$$m_\pi^2(T) = m_\pi^2 \left(1 + \frac{1}{6} \frac{\langle \pi^2 \rangle_T}{f_\pi^2}\right) \tag{11.29}$$

where

$$\langle \pi^2 \rangle_T = (N_F^2 - 1) \int \frac{d^4 p}{(2\pi)^4} \, 2\pi \, \delta(p^2) \, n_p = (N_F^2 - 1) \frac{T^2}{12} \tag{11.30}$$

is the matter part of the (isospin-averaged) pion propagator. Thus

$$f_\pi^2(T) = f_\pi^2 \left(1 - (N_F^2 - 1) \frac{T^2}{18 f_\pi^2}\right),$$

$$m_\pi^2(T) = m_\pi^2 \left(1 + (N_F^2 - 1) \frac{T^2}{72 f_\pi^2}\right) \tag{11.31}$$

in agreement with the result derived by Gasser and Leutwyler using effective Lagrangians [13]. The results (11.31) imply that the Gell-Mann-Oakes-Renner relation (see Chapters 2 and 4),

$$f_\pi^2(T) m_\pi^2(T) \approx -2m \langle \bar{q} q \rangle_0 \left(1 - \frac{N_F^2 - 1}{N_F} \frac{T^2}{12 f_\pi^2}\right) \approx -2m \langle \bar{q} q \rangle_T \tag{11.32}$$

holds to order T^2 in the heat bath. This relation is unlikely to hold to higher order, however, since the pion mass develops an imaginary part through absorption and the very definition of the mass becomes ambiguous. Similar relations can be sought for other mesons.

Consider now the case of a nucleon immersed in a heat bath of pions. To study the effects of the pion on the nucleon mass at low temperature, we can proceed in a number of ways. We can extract the mass by looking at the asymptotics of a nucleon-nucleon correlator in Euclidean space at finite temperature, or resum in real-time the rescattering effects around a propagating nucleon, or calculate the free energy at zero recoil. Because of the absorption caused by the rescattering of the nucleon in matter, these different ways give different results.

To illustrate these differences for the nucleon, consider the free energy of a nucleon immersed in a dilute pion gas. To leading order in the pion density, the free energy (minus the nucleon mass) is given by

$$\Delta F = \int_0^\infty \frac{dk}{\pi} \sum_{I,l} (2l+1) \delta'_{I,l}(k) \, \omega_k \, n_\omega \tag{11.33}$$

where $\omega_k = \sqrt{m_\pi^2 + k^2}$ is the pion energy, n_ω is the pion occupation number and $\delta_{l,I}$ is the phase-shift of a pion partial wave carrying angular momentum l and isospin I, expressed in the rest frame of the nucleon. Equation (11.33) is the thermal contribution to the zero-point energy in a spherical box of size R, with the boundary condition on the l-th partial wave

$$kR - l\frac{\pi}{2} + \delta_{l,I}(k) = n\pi \tag{11.34}$$

where n is an integer. From (11.34), the change in the number of pion states around the nucleon is $dn/dk = \delta'_{I,l}(k)/\pi$. Using the partial phase shifts $\delta_{I,l}$, the scattering amplitude $T_I(k)$ for a fixed isospin I is given by the spherical expansion

$$T_I(k) = i\frac{2\pi}{k} \sum_l (2l+1) \left(e^{2i\delta_{l,I}(k)} - 1 \right) P_l(\hat{k}). \tag{11.35}$$

Using (11.35) we can rewrite (11.33) in terms of the scattering amplitude, a procedure known from the work of Bethe and Uhlenbeck. The result is

$$\Delta F = \sum_I \int \frac{d^3k}{(2\pi)^3} \frac{\mathrm{Re} T_I(k)}{2\omega_k} \left(n_\omega - \frac{\omega_k}{T} e^{\omega_k/T} n_\omega^2 \right)$$
$$- \frac{i}{8\pi} \sum_I \int \frac{d^3k}{(2\pi)^3} \left(T_I T^{*\prime}_I - T_I^* T_I' \right)(k) \, \omega_k \, n_\omega. \tag{11.36}$$

The first term in (11.36) proportional to n_ω

$$\Delta M = \sum_I \int \frac{d^3k}{(2\pi)^3} \frac{\mathrm{Re} T_I(k)}{2\omega_k} n_\omega \tag{11.37}$$

contains the real part of the mass-shift that follows from the pole mass expansion of a nucleon rescattering off pions in the heat bath. This definition of the nucleon mass-shift has been used by Leutwyler and Smilga [14]. The second term in the first parentheses in (11.36) is an entropy-like term. The last term in (11.36) accounts for absorption, as unitarity requires in general a complex $\mathcal{T}$. Although the unitarity cuts are suppressed at low temperatures by powers of T/f_π, this is not the case of the Δ resonance which is $2m_\pi$ away from threshold. Equation (11.36) clearly shows the difference between a mass-shift that follows from the free energy and the real part of the pole mass given above.

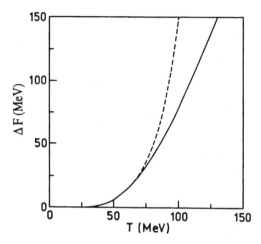

Fig. 11.5. Free energy (11.33) of a nucleon in a dilute gas of pions using Chew-Low theory (Yukawa model) (dashed line) and the Skyrme model[15] (solid line).

In Fig. 11.5, we show the behavior of ΔF versus T in MeV for a nucleon using phase shifts from Chew-Low theory (Yukawa model) and using a Skyrme model. In Fig. 11.6, we show the behavior of the pole mass-shift coming from (11.37) for the Yukawa model and the Skyrme model. While the free energy shift is positive for all temperatures (corresponding to a zero-point thermal energy), the pole-mass shift is negative at low temperature since we are below Δ. There is a change with increasing temperature, as we get through the Δ. This behavior was first observed by Leutwyler and Smilga [14] using the empirical phase shifts in (11.37).

Does the πNN-coupling constant change at finite temperature? Again, an answer to this question is only meaningful at low temperature, when the absorption is weak. The finite-temperature corrections follow from the thermal insertions shown in Fig. 11.7. The diagrams (a-b) renormalize the N and π mass and wave functions. The diagrams (c-d) renormalize the $g_{\pi NN}$ coupling. Because of the energy denominators, they are subleading in T.

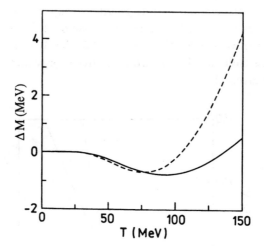

Fig. 11.6. Real part of the nucleon pole-mass shift (11.37) using Chew-Low theory (dashed line) and the Skyrme model[15] (solid line).

Order T^2 effects involve only thermal pions at tree level. Thus to order T^2, $g_{\pi NN}(T) \sim g_{\pi NN}(0)$.

At low temperature, the Goldberger-Treiman relation holds, as it involves only current conservation and the mass-shell condition, each of which is preserved by thermal pions to order T^2 (tree). Thus

$$\frac{g_A(T)}{f_\pi(T)} \approx \frac{g_A}{f_\pi} = \frac{g_{\pi NN}}{m_N} \tag{11.38}$$

to order T^2. This shows that g_A, much like f_π, is quenched by thermal effects

$$g_A(T) \approx g_A \left(1 - \frac{T^2}{6f_\pi^2}\right). \tag{11.39}$$

In the presence of a pion mass, these relations are expected to be slightly affected at low temperature. As the temperature is increased, the role of higher resonances cannot be ignored. More importantly, the entire concept of quasiparticles – and hence the use of "effective" parameters – is blurred when the absorption sets in. In this regime, it is more expedient and more reliable to resort to the concept of correlation functions.

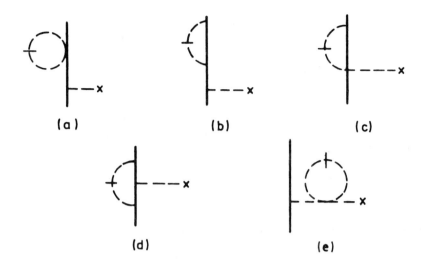

Fig. 11.7. Thermal pion insertions to the πNN-coupling to lowest order. The slash represents the thermal insertion and the dashed line with cross the external pion.

11.3. Correlators

11.3.1. *Hadron correlators*

To analyze the effects of hadronic correlations at finite temperature, consider the vector (V) and axial (A) correlators. In operator form, the time-ordered product reads

$$\mathbf{\Pi}_{\pm}(q) = i \int d^4x e^{-iq \cdot x} \mathrm{T}(A(x)A(0) \pm V(x)V(0)) \qquad (11.40)$$

where for simplicity we have dropped the spin indices. Other types of correlators (retarded, advanced, etc.) can be defined similarly. The Gibbs average of (11.40) reads

$$\Pi_{\pm}(q, T) = i \int d^4x e^{-iq \cdot x} \mathrm{Tr} \left(e^{-\beta(H-F)} \, \mathbf{P}_* \, \mathrm{T}(A(x)A(0) \pm V(x)V(0)) \right).$$

$$(11.41)$$

For a dilute pion gas and in the chiral limit, (11.41) is dominated by

$$\Pi_\pm(q,T) \sim \langle 0|\mathbf{\Pi}_\pm(q)|0\rangle + \int \frac{d^3k}{(2\pi)^3} \frac{n_k}{2|k|} \langle \pi(k)|\mathbf{\Pi}_\pm(q)|\pi(k)\rangle. \quad (11.42)$$

PCAC and current algebra to order $\mathcal{O}(1/f_\pi^4)$ give

$$\langle \pi|T\,(AA + VV)\,|\pi\rangle = 0,$$
$$\langle \pi|T\,(AA - VV)\,|\pi\rangle = -\frac{8}{f_\pi^2}\langle 0|T\,(AA - VV)\,|0\rangle. \quad (11.43)$$

Thus, to the same order,

$$\Pi_+(q,T) = (1 - 0)\,\Pi_+(q,0),$$
$$\Pi_-(q,T) = \left(1 - \frac{T^2}{3f_\pi^2}\right)\Pi_-(q,0). \quad (11.44)$$

In terms of the vector and axial correlators, (11.44) read

$$\Pi^V(T) = (1 - \epsilon)\Pi^V(0) + \epsilon\,\Pi^A(0),$$
$$\Pi^A(T) = (1 - \epsilon)\Pi^A(0) + \epsilon\,\Pi^V(0) \quad (11.45)$$

with $\epsilon = T^2/6f_\pi^2$, to order $\mathcal{O}(1/f_\pi^4)$. This result was first established by Dey, Eletsky and Ioffe [16]. It shows that to order T^2, the vector and axial correlators mix. The mixing follows from $a_1 \to \pi\rho \to a_1$ in the heat bath, a possible precursor to chiral symmetry restoration. These relations do not imply specific information on the spectral terms (thresholds, poles, resonances). The addition of a finite pion mass causes $\Pi_+(T) \neq \Pi_+(0)$, and gives rise to additional corrections of the type $e^{-m_\pi/T}$.

It is possible to understand these results from a quark point of view. Consider, for instance, (11.40) in Euclidean domain at high momentum (or short distance). Restricting to the currents

$$V_\mu = \frac{1}{2}\left(\overline{u}\gamma_\mu u - \overline{d}\gamma_\mu d\right),$$
$$A_\mu = \frac{1}{2}\left(\overline{u}\gamma_\mu\gamma_5 u - \overline{d}\gamma_\mu\gamma_5 d\right) \quad (11.46)$$

and using the operator product expansion in (11.40) we have ($Q^2 = -q^2 \geq 0$)

$$\Pi_{\mu,\nu,-}(q) = -(q_\mu q_\nu - q^2 g_{\mu\nu})\frac{2\pi\alpha_s}{Q^6}$$
$$\times \left(\overline{u}_L\gamma_\mu\lambda^a u_L - \overline{d}_L\gamma_\mu\lambda^a d_L\right)$$
$$\times \left(\overline{u}_R\gamma_\nu\lambda^a u_R - \overline{d}_R\gamma_\nu\lambda^a d_R\right) \quad (11.47)$$

to order $1/Q^6$ in the chiral limit. Here λ^a are $SU(3)$ color matrices. In the vacuum, the expectation value of the four-quark operator in (11.47) follows from vacuum dominance. After rewriting and taking trace over spin, we have, for instance,

$$\langle 0| \left(\bar{q}\gamma_\mu \lambda^a T^3 \frac{1+\gamma_5}{2} q \; \bar{q}\gamma_\mu \lambda^a T^3 \frac{1-\gamma_5}{2} q \right) |0\rangle$$

$$= -\frac{1}{4\cdot 9\cdot 16} \operatorname{Tr}_c(\lambda^a \lambda^a) \operatorname{Tr}_F(T^3 T^3) \operatorname{Tr} \left(\gamma_\mu \frac{1+\gamma_5}{2}\gamma_\mu \right) \qquad (11.48)$$

and

$$\langle 0|\bar{q}q\bar{q}q|0\rangle = -\frac{4}{9}\langle 0|\bar{q}q|0\rangle^2. \qquad (11.49)$$

In a dilute gas of pions, the vacuum expectation value is given by the forward one-pion matrix element times the thermal pion density $T^2/24$ in the chiral limit. The one-pion matrix element follows from PCAC and current algebra

$$\langle \pi| \left(\bar{q}\gamma_\mu \lambda^a T^3 \frac{1+\gamma_5}{2} q \; \bar{q}\gamma_\mu \lambda^a T^3 \frac{1-\gamma_5}{2} q \right) |\pi\rangle = -\frac{8}{f_\pi^2} \left(-\frac{4}{9}\langle 0|\bar{q}q|0\rangle^2 \right).$$

$$(11.50)$$

Thus

$$\Pi_-(T) \approx \left(1 + \frac{T^2}{24}\left(-\frac{8}{f_\pi^2} \right) \right) \Pi_-(0) = \left(1 - \frac{T^2}{3f_\pi^2} \right) \Pi_-(0) \quad (11.51)$$

in agreement with the result (11.44) derived on general grounds.

Most of the above results follow from current algebra and PCAC, since at low temperature the heat bath is dominated by weakly interacting pions. These results can in principle be extended to higher orders in pion physics. In fact, Eletsky and Ioffe have recently extended the analysis for the vector and axial correlators to T^4. [17] The analysis, however, is limited since the effects of higher resonances become important at higher temperatures, where the dilute pion gas approximation is no longer valid. A qualitative understanding of the effects of increasing temperature requires a better understanding of the ground state properties as well as the excitation spectrum in terms of the fundamental degrees of freedom. We shall now discuss some qualitative approaches that will help us to understand the effects of increasing temperature, and the departure from the chiral limit.

11.3.2. *Finite-T sum rules*

One way to investigate the effects of temperature beyond leading order has been to use finite-temperature QCD sum rules. The idea was developed by Bochkarev and Shaposhnikov [18] and others [19]. The starting point in the QCD sum rules is to make use of the analytical structure of a hadronic correlator via dispersion relations. For a given structure function $\Pi(\omega, |\vec{q}|)$, causality implies that Π is an analytic function in ω, except for cuts and poles. By Cauchy's theorem, it follows that

$$\text{Re}\Pi(\omega, |\vec{q}|) = \frac{1}{\pi} \int dz \, \frac{dz}{z - \omega} \, \text{Im}\,\Pi(z, |\vec{q}|). \qquad (11.52)$$

By the optical theorem, the imaginary part is related to the total cross section for the relevant channel. It reflects the time-like thresholds and is presently accessible by neither perturbation theory nor lattice simulations. The QCD sum rule strategy is to parametrize the right-hand-side (RHS) and fit it to the left-hand-side (LHS) using the operator product expansion (OPE) and general vacuum assumptions in the Euclidean domain.

Clearly, the present construction at finite temperature is not *a priori* limited to the low-temperature regime, and as such can be viewed as a way to go beyond the leading current-algebra results derived above. We will see, however, that the procedure has limitations of its own as well.

11.3.3. *OPE at finite T*

For a general approach to the OPE as applied to the QCD sum rules in the vacuum, we refer to the original work by Shifman, Vainshtein and Zakharov [20]. The essential idea proposed by these authors is that in order to define the OPE, it is essential to fix a renormalization scale μ. This scale is used to separate hard and soft momenta, the former contributing to the coefficient functions and the latter to matrix elements (vacuum averages or condensates). This means that there are *perturbative* contributions to the various condensates. In various toy models, these perturbative contributions can be calculated explicitly, and they do not have to be small compared to the nonperturbative contributions. The usual argument in the QCD sum rules is that the μ can be dialed such that the running coupling constant $\alpha_s(\mu)$ is small enough to allow a low-order perturbative calculation of the coefficient functions, with yet a small μ-dependence in the vacuum condensates (about 20% in the gluon condensate, for instance).

In the presence of a heat bath, there are additional (black-body) contributions to the matrix elements in the OPE that are perturbative and should be incorporated into the coefficient functions. To illustrate this basic point, we will derive the OPE at finite T for a free massless scalar theory

$$\mathcal{L} = \frac{1}{2}(\partial_\mu \phi)^2 - \frac{1}{2}m^2\phi^2. \tag{11.53}$$

Consider the retarded commutator

$$\Pi_R(x) = \theta(x^0)[: \phi^2(x) :, : \phi^2(0) :] \tag{11.54}$$

where :: denotes normal ordering with respect to the vacuum state. By successive commutation relations, one easily derives the following OPE for (11.54)

$$\Pi_R(x) = -2i\Delta_R(x)\tilde{\Delta}(x) + \sum_{\mu_1...\mu_n} x^{\mu_1}...x^{\mu_n} \frac{1}{n!} : \phi(0)\partial_{\mu_1}...\partial_{\mu_n}\phi(0) : \tag{11.55}$$

where $\tilde{\Delta}(x) = \langle 0|\{\phi(x), \phi(0)\}|0\rangle = -2\text{Im}\Delta_F(x)$. The first term corresponds to the graph given in Fig. 11.8, which is the free Feynman graph for the retarded commutator. Taking the Gibbs average of (11.55) amounts to taking Gibbs averages of all the operator matrix elements (here all twist-2 operators)

$$\langle : \phi(0)\partial_{\mu_1}...\partial_{\mu_n}\phi(0) : \rangle_T = \int \frac{d^3k}{(2\pi)^3} \frac{1}{k} n_k (ik_{\mu_1}, ..., ik_{\mu_n}) \tag{11.56}$$

where now n_k is the Bose distribution for the field ϕ. Substituting (11.56) into (11.55) and resumming yield [21]

$$\langle \Pi_R(x) \rangle_T = \int \frac{d^4k}{(2\pi)^4} e^{ik\cdot x} (1 + n_k) \text{Im}\Delta_F(k). \tag{11.57}$$

This result is identical to calculating the Feynman graph shown in Fig. 11.8b using Matsubara rules for finite-temperature field theory, and then analytically continuing the result to Minkowski space.

Thus by calculating the coefficient functions in the OPE (here the identity operator) using Euclidean T-perturbation theory, one effectively

Fig. 11.8. (a) Free Feynman graph for the retarded commutator. (b) Two-meson contribution to the $\phi^2 - \phi^2$ correlator.

resums order T (black-body) effects in the matrix elements of higher dimensional operators. [21] This is the same as reordering the operators in (11.54) around the black-body spectrum. Finally, we note that the lack of manifest Lorentz invariance in the heat bath implies that $\langle : \phi(0)\partial_{0_1}...\partial_{0_n}\phi(0) : \rangle_T$ is not related to $\langle : \phi(0)\partial_{x_1}...\partial_{x_n}\phi(0) : \rangle_T$. Thus the Lorentz constraint is to be relaxed in the OPE for finite-temperature calculations. The heat bath defines a specific frame.

11.3.4. *Hadrons in medium*

In medium, causal responses are related to the retarded correlator (as opposed to the Feynman propagator in the vacuum). Generically,

$$\Pi^V(q, T) = i \int d^4x \, e^{iq \cdot x} \, \theta(x^0) \, \mathrm{Tr} \left(e^{-\beta(H-F)} \, \mathbf{P}_* \, [V(x), V(0)] \right) \quad (11.58)$$

with $V(x)$ a local operator with specific J^{PC} assignment. The OPE for the commutator in (11.58) takes the form

$$[V(x), V(0)] = C_1(x, \mu) \mathbf{1} + \sum_n C_n(x, \mu) \, : \mathbf{O}_n(\mu) : \quad (11.59)$$

where μ is the subtraction scale. The commutator (left-hand-side) is μ-independent. The sum is over all operators consistent with the J^{PC} assignment, gauge invariance and $O(3)$ symmetry. Manifest $O(1,3)$ is broken down to $O(3)$ in the heat bath. Because of dimensionality, the coefficients $C_i(x)$ carry increasing powers of x so that the expansion converges at short distances. The typical contribution of (11.59) to (11.58) at short distances (or large Euclidean momenta) is of the form

$$\mathcal{C}_n(x,\mu)\,\mathrm{Tr}\left(e^{-\beta(H-F)}\mathbf{P}_* : \mathbf{O}_n(\mu) :\right) = \mathcal{C}_n(x,\mu)\langle: \mathbf{O}(\mu) :\rangle_T. \quad (11.60)$$

Being c-numbers, the Wilson coefficients are T-independent. All the temperature dependence is carried by the matrix elements. At very low temperature, the T-dependence of the matrix elements is in general unknown. A dilute pion gas system or lattice simulations at finite temperature may be used for an estimate. As the temperature is increased, a readjustment of the vacuum state is expected. In this case it is more efficient to reorder the operators in (11.59) and thus the thermal average (11.60) around the quark-gluon black-body spectrum as expected from duality. As a result, the Wilson coefficients acquire a T-dependence.

Since there is some controversy regarding the present formulations of the finite-T sum rule [22] we will not elaborate further on the formal aspects of the construction. We will simply quote some of the presently available results. Figure 11.9 shows the behavior in the 1^{--} meson channel of the ρ mass versus temperature as obtained by Adami, Hatsuda and Zahed [19]. The results by Hatsuda, Koike and Lee are qualitatively similar.[23] Below 100 MeV, nothing much happens to the ρ mass, while above 100 MeV, the ρ mass is found to drop appreciably. In this regime, however, corrections to the OPE becomes large. Clearly, more work is needed in these directions to clarify the formal aspects of the approach in medium, and also for assessing the reliability of the expansion in the critical region.

A dropping ρ mass with temperature is in agreement with earlier speculations by Pisarski [24] and with the scaling laws suggested by Brown and Rho and others [25] based on a sigma model implemented with scale anomaly. It is at variance with arguments presented by Pisarski and others, using vector dominance [26]. Lattice gauge calculations indicate that physical properties do not change appreciably in temperature up to $T \approx 0.93T_c$.[27]

A dropping ρ mass in medium – be that temperature-driven or density-driven – may have important consequences on the dilepton spectrum in heavy ion collisions at ultra-relativistic energies. If a mixed phase lives long enough, say 10 to 20 fms, then there is enough time for the thermal and compressed ρ's with a life-time of 1 fm to decay into dileptons, feeding copiously on the low-mass region of the ρ. Dilepton measurements from CERES experiment (as shown in Fig. 11.10) seem to require a source for

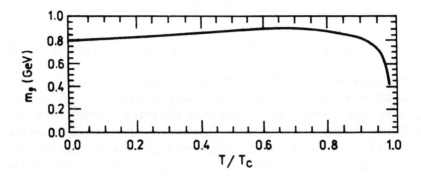

Fig. 11.9. Behavior of the ρ mass versus temperature using finite-temperature QCD sum rules as derived by Adami, Hatsuda and Zahed[19].

such dileptons at an invariant mass lower than that of free ρ mesons[28]. Calculations by Li, Ko and Brown [29] offer a surprisingly simple explanation of this observation with the dropping of the vector-meson masses at finite temperature and in a dense medium (as discussed in Chapter 10).

11.4. Chiral Models at Finite Temperature

Another way to go beyond the leading order effects in the temperature is to use at the mean-field level chiral effective models of the type discussed in the preceding chapters. The mean-field approach provides a meaningful starting point for detailed calculations. Whether it is reliable enough to be trusted is not really known. We observe, however, that a mean-field treatment of sigma models[30], instanton models[31,32,33], Nambu-Jona-Lasinio (NJL)[34] models, and other models of similar nature are all in qualitative agreement with each other in describing hadronic spectra and the character of the phase transition. This seems to indicate that what matters is the relevant symmetry, perhaps associated with universality as discussed below.

11.4.1. *Instantons at finite temperature*

The instanton model discussed in Chapter 2 captures essential features of the zero-temperature vacuum, and can be modified minimally to account for finite temperature [31,32,33,36]. The finite-temperature description amounts to looking at the system in an asymmetric (squeezed) box:

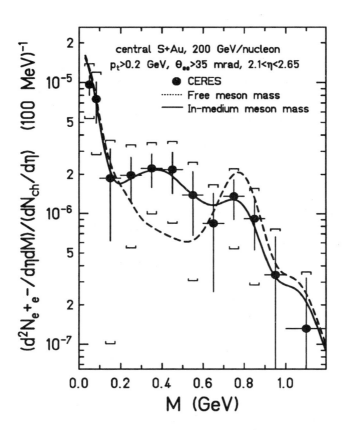

Fig. 11.10. Dilepton spectrum from the CERES experiment using S-Au at 200 GeV/n[28]. The dotted line represents the hadronic "cocktail" model with free-space masses and the solid line is the prediction of Li, Ko and Brown[29] with dropping vector-meson masses à la Brown-Rho scaling.

$1/T \times V_3$, with periodic conditions on the gauge fields and anti-periodic conditions for the quark fields. At low temperature, the system can be described either in terms of constituent quarks and classical gluons, or constituent quarks and mesons. Both descriptions will be presented in this section to illustrate their "duality."

(i) Constituent quarks and gluons

On a strip $1/T \times V_3$ the Yang-Mills equations admit periodic instanton solutions (calorons) [37]. It was noted earlier by Gross, Pisarski and Yaffe [36] that the temperature provides a natural cutoff to the large size instantons. At finite temperature and including fermions, the one-instanton density (2.23) becomes [36]

$$d(\rho, T) \sim d(\rho, 0) e^{-\frac{1}{3}(2N_c + N_f)(\pi \rho T)^2}. \qquad (11.61)$$

The temperature effects in (11.61) are induced by the hard thermal quarks and gluons treated here perturbatively to one-loop order. The exponential suppression induced by thermal effects may however be questioned[38]. The reason is that the low-temperature phase is pion-dominated. The pions are very ineffective in suppressing instantons as color singlets have difficulty in coupling to colored configurations. This point of view will be further discussed below, although it is a bit too radical since the system does not confine.

The number of instantons and anti-instantons at finite temperature in a fixed four-volume $1/T \times V_3$ follows from the extremum of [31,33]

$$Z = \sum_{N_+, N_-} \frac{1}{N_+! N_-!} \int \prod_I^N d\Omega_I \, d(\rho_I, T) \exp(-u_{I\bar{I}}/T) \det_{\text{soft}}(i\slashed{\nabla} + im). \qquad (11.62)$$

For a fixed number $N = N_+ + N_-$ of instantons and anti-instantons, the integration is over $4NN_c$ collective variables. $u_{I\bar{I}}$ is the interaction between instantons and anti-instantons in the $1/T \times V_3$ box, and the fermionic determinant runs over the soft (zero) modes. We refer the reader to Chapter 2 for further details. The soft part of the fermion determinant is given by

$$\det_{\text{soft}}(i\slashed{D} + im) \sim \prod_f \det(m_f^2 + \mathbf{T}^\dagger \mathbf{T}) \qquad (11.63)$$

where $\mathbf{T}$ is the overlap matrix element between the finite-temperature zero modes ϕ_I,

$$\mathbf{T}_{IJ} = \int_{1/T \times V_3} d^4x \, \phi_I(x) i \overleftrightarrow{\partial} \phi_J(x). \tag{11.64}$$

At finite temperature the instantons become calorons, and their respective zero modes are given by [37]

$$\phi_{R,L} = \frac{\sqrt{\Pi}}{2\pi\rho} \frac{1}{\Pi} \overleftrightarrow{\partial} (\frac{\Phi}{\Pi}) \epsilon_{R,L} \tag{11.65}$$

where

$$\Pi(t,r) = 1 + \frac{\pi T \rho^2}{r} \frac{\sinh(2\pi T r)}{\cosh(2\pi T r) - \cos(2\pi T t)},$$
$$\Phi(t,r) = (\Pi(t,r) - 1)\frac{\cos(\pi T t)}{\cosh \pi T r}, \tag{11.66}$$

and $\epsilon_{L,R}$ carry the spin and color structure. At finite temperature, the propagation of the quark zero mode along the spatial (radial) direction is damped by $e^{-\pi T r}$, while it oscillates with πT along the temporal direction.

The evaluation of the soft part of the fermion determinant (11.63) in an ensemble of instantons and anti-instantons at finite temperature requires a knowledge of how the quasiparticles redistribute themselves in the squeezed $1/T \times V_3$ box. Given the partition function (11.62) (with or without screening), one may in principle calculate the free energy density $\mathcal{F}$. The instanton density in the squeezed box follows from $\mathcal{F}$ at the minimum. Figures 11.11a and 11.11b show, respectively, the behavior of the free energy density and instanton density versus T when the screening factor in (11.61) is retained. For densities $n(T) \ll \Lambda^4$ [33],

$$\mathcal{F} = n(T)\left(c(T) + (\frac{N_f}{2} - 1)\log\frac{n(T)}{\Lambda^4}\right) \tag{11.67}$$

where $c(T)$ is a temperature-dependent constant. The behavior of the free energy density and the instanton density versus T, when the screening factor is suppressed, is shown in Figs. 11.12, respectively. Both calculations were carried out with $N_f = 3$.

The chiral condensate can be obtained either analytically [33] or using simulations [39] as we discussed in Chapter 2. The mean-field arguments

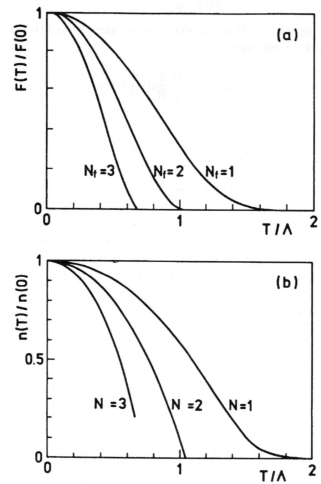

Fig. 11.11. Free energy (a) and instanton density (b) versus T, when the instantons are screened [33].

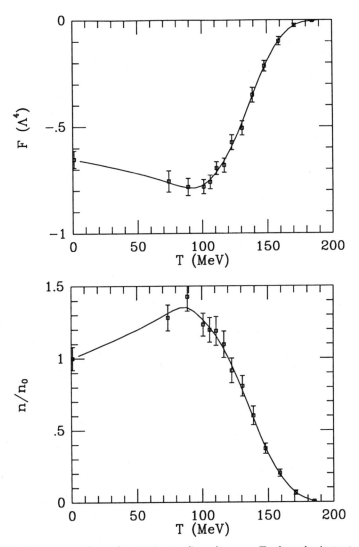

Fig. 11.12. Free energy (upper) and density (lower) versus T when the instantons are polarized (no screening).[39]

of Chapter 2, when extended to (11.62), yield a finite-temperature gap
equation

$$\int \frac{d^3k}{(2\pi)^3} T \sum_n \frac{M^2(\mathbf{k}, \omega_n)}{\mathbf{k}^2 + \omega_n^2 + M^2(\mathbf{k}, \omega_n)} = \frac{\epsilon\, n(T)}{4N_c}(1 - m\epsilon) \quad (11.68)$$

where the momentum-dependent mass is given by

$$M(\mathbf{k}, \omega_n) = \frac{\epsilon n(T)}{4N_c}(\mathbf{k}^2 + \omega_n^2)(\mathbf{k}^2 A_n^2 + \omega_n^2 B_n^2). \quad (11.69)$$

The sum in (11.68) is over the Matsubara frequencies $(2n + 1)\pi T$, and A_n
and B_n are Fourier coefficients related to the instanton and anti-instanton
zero modes [33]. Parameter ϵ is determined by solving (11.68) iteratively.
This fixes the value of the constituent mass $M(\mathbf{k}, 0)$ and the chiral condensate

$$\langle \bar{q}q \rangle = -4N_c\, T \int \frac{d^3k}{(2\pi)^3} \sum_n \frac{M(\mathbf{k}, \omega_n)}{\mathbf{k}^2 + \omega_n^2 + M^2(\mathbf{k}, \omega_n)}. \quad (11.70)$$

The behavior of (11.70) versus T is shown in Fig. 11.13 for $N_f = 1, 2, 3$, and
$m_s = 0$. For $N_f = 3$, the constituent mass is 370 MeV and the condensate
is $-(273\,\text{MeV})^3$ at zero temperature. The transition is first-order in this
case, with a critical temperature $T_c = 180$ MeV .[33] The transition was
found to be second-order for two flavors, and absent for one flavor.

Figure 11.14 shows the behavior of the chiral condensate in the case
where the exponential suppression is omitted [39]. The transition was characterized as weak first-order, with a critical temperature $T_c = 140$ MeV. This
is somehow lower than that found above. However, the chiral condensate
in the latter was adjusted to $-(216\,\text{MeV})^3$, which may explain part of the
discrepancy. Clearly both approaches seem to yield the same bulk transition, *albeit* with totally different physical assumptions on the instanton
content of the QCD vacuum. In the former approach, the instantons and
anti-instantons are squeezed (Debye screened) and disappear at T_c. In the
latter, the transition occurs when half of the instantons and anti-instantons
get polarized along the temporal direction. The polarization is favored because the screened hard quarks are dropped, while the screened soft quarks
are retained. The former are at the origin of the exponential suppression
in the instanton density to one-loop order, while the latter suppress the
polarization along the spatial directions.

The arguments presented in favor of molecular clustering implies that the semiclassical rules of calculations are not appropriate at finite temperature unless confinement is better understood. Indeed, thermal constituent quarks together with pions dominate the low-temperature phase of the instanton model, and contribute importantly to the vacuum pressure at finite temperature.

Which mechanism is favored in nature could only be decided by more dedicated lattice analyses. One potential and generic test of some of these ideas would be to carry out simultaneous lattice simulations, both quenched and unquenched. The polarization mechanism is absent in the quenched approximation.

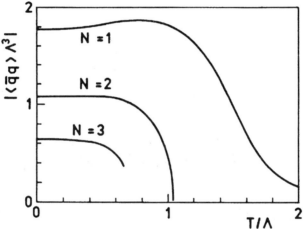

Fig. 11.13. Quark condensate versus T when the instantons are screened [33], for $N_f = 1, 2, 3$.

(ii) Susceptibilities

To probe the $U_A(1)$ content in the instanton vacuum at finite temperature, the topological charge correlator may be convenient. Indeed, for point-like instantons,

$$\chi(x - y) = \langle \sum_i Q_i \delta(x - x_i) \sum_j Q_j \delta_4(y - x_j) \rangle_T \qquad (11.71)$$

where $Q_\pm = \pm 1$ are the topological charges of (+) instantons, and (-) anti-

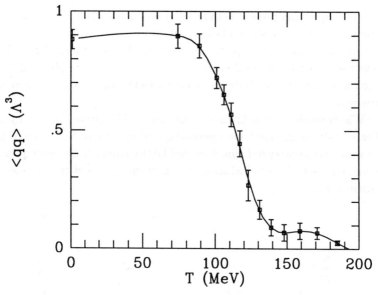

Fig. 11.14. Quark condensate versus T when the instantons are polarized [39], for $N_f = 3$.

instantons. A straightforward calculation using the results of Chapter 2 and the preceding discussion (with the $\eta - \eta'$ mixing) give [40]

$$\chi(x, y; T) = +\chi_*(T)\delta_4(x - y) - 2N_f \frac{\chi_*^2(T)}{f^2(T)} \frac{T}{2|\vec{x} - \vec{y}|} \sum_{n=-\infty}^{+\infty} e^{i\omega_n(x^0 - y^0)}$$

$$\times \left(\cos^2\theta(T)e^{-\sqrt{\omega_n^2 + m_{\eta'}^2(T)}|\vec{x} - \vec{y}|} + \sin^2\theta(T)e^{-\sqrt{\omega_n^2 + m_\eta^2(T)}|\vec{x} - \vec{y}|}\right) \quad (11.72)$$

where $\omega_n = 2\pi n T$ are the Matsubara frequencies, $\theta(T \sim 0) \sim -11^0$ [41], and

$$\chi_*(T) = \frac{f^2(T)}{2N_f}(m_{\eta'}^2 + m_\eta^2 - 2m_K^2)(T) \quad (11.73)$$

which is the finite-temperature version of the Witten-Veneziano formula (see Chapter 3). The static part $\chi(\omega = 0, \vec{x}; T)$ of (11.72) is dominated by the η' excitation. It provides a good way of measuring the η' mass at finite temperature. This static correlator offers a good probe for $U_A(1)$

restoration, and satisfies the following Ward identity

$$\int d^3x \; \chi(\omega = 0, \vec{x}, T) = \chi_*(T) \left(1 - \sum_{i=1}^{N_f} \frac{\chi_*(T)}{m_i \langle \overline{\psi}\psi \rangle(T)}\right)^{-1}. \quad (11.74)$$

It vanishes at the chiral transition point. This, however, does not necessarily mean that the $U_A(1)$ symmetry is restored. The restoration of the latter can be detected only through the large distance behavior of $\chi(0, \vec{x}, T)$. In the $U_A(1)$ broken phase, the correlator falls off exponentially, while in the symmetric phase it vanishes identically. This point is worth checking both on the lattice and using simulated instantons. For completeness, we note that the temporal asymptotics of (11.72) is free-field-dominated for temperatures $T \sim m_{\eta'}/2\pi \sim 150$ MeV $(0 < x^0 < 1/T)$. The compressibility $\sigma(T) \sim 2z(T)$, where $z(T)$ is the fugacity, and is expected to drop by $1/2$ at the chiral transition point, following the drop in the gluon condensate. Debye screening at high temperature causes the electric condensate to vanish. This would also mean in instanton models the vanishing of the magnetic condensate because of self duality.

(iii) Constituent quarks and mesons

To see how the above description translates into a description in terms of ordinary mesons at low temperature, we recall from Chapter 2 that the vacuum partition function in the bosonized form can be rewritten as [40]

$$Z[0] = Z_Q[0] \int d[K] \; e^{-W[K] + \int d^4x \; 2z \, \cos(\sqrt{2N_F} K^0/f)} \quad (11.75)$$

where $W[K]$ is the pseudoscalar effective action

$$W[K] = \int d^4x \left(\frac{1}{2}(\partial_\mu K^a)^2 - \frac{1}{2}\frac{\langle \overline{\psi}\psi \rangle}{f^2} \; \mathrm{Tr}(mK^2) + \mathcal{O}(K^4)\right) \quad (11.76)$$

with a mean fugacity $z \sim \langle N \rangle/2V_4$.* $Z[0]$ receives contribution from the one-loop constituent quark $Z_Q[0]$, as well as the light pseudoscalars and a self-interacting but heavy pseudoscalar singlet. An equivalent form for (11.75) is

$$Z[0] = Z_Q[0] \left(\int d'[K] e^{-W[K]}\right)$$

---

*For the sake of clarity, the octet-singlet mixing will be ignored in this section.

$$\sum_{N_\pm} \frac{z^{N_++N_-}}{N_+!N_-!} \prod_{i=1}^N \int d^4x_i \; e^{-N_F \frac{m_0^2}{f^2} \sum_{i,j} Q_i Q_j \left(\frac{1}{2\pi^2} \frac{K_1(m_0|x_i-x_j|)}{m_0|x_i-x_j|} \right)}$$

$$(11.77)$$

where m_0 is the singlet K^0 mass, and K_1 a modified Bessel (MacDonald) function. The prime in the path integral in (11.77) stands for the fact that singlet K^0 field is excluded from the mesonic measure. Equation (11.77) is the product of the vacuum partition function of the free massive-meson octet and a four-dimensional Coulomb gas with Yukawa interactions. The instanton-singlet mixing in the vacuum causes the instanton-anti-instanton pairs in the vacuum to be screened over distances of the order of $1/m_0$. [†]

At low temperature, the pressure is simply given by the one-loop effects following from (11.75).

$$\mathbf{P} + \mathbf{B}(T) = \mathbf{P}_\pi + \mathbf{P}_K + \mathbf{P}_\eta + \mathbf{P}_{\eta'} + \cdots \qquad (11.78)$$

where $\mathbf{B}(T)$ is the vacuum or "bag" pressure at finite temperature as defined in (11.22) (with the quark contribution), and $\mathbf{P}_a$ are the respective mesonic pressures. The ellipsis stands for the higher hadron resonances, in agreement with the Gibbs average. The mesons in (11.78) carry temperature-dependent masses, since the fermion condensate is temperature-dependent. At low temperature, the effects are small, and the pressures $\mathbf{P}_a$ in (11.78) are the usual black-body contributions. At low temperatures and in a confined phase, the vacuum pressure $\mathbf{B}$ (T) is T-independent [42] to leading order in N_c. This is not the case here and $\mathbf{B}$ (T) receives a contribution to order N_c from the free streaming constituent quarks. This term drives the phase transition in the preceding section.

We note that the instanton effects in (11.78) are only implicit, as they manifest themselves through an η' mass that is heavier than the η mass even at finite temperature (ignoring η' interactions). An equivalent way of describing the same physics is to rewrite (11.78) as follows

$$\mathbf{P} + \mathbf{B}(T) = \mathbf{P}_\pi + \mathbf{P}_K + \mathbf{P}_\eta + (\mathbf{P}_{K^0} + \mathbf{P}_I) + \cdots \qquad (11.79)$$

where $\mathbf{P}_{K^0}$ is the contribution to the pressure coming from a thermalized system of singlet mesons with mass m_0, and $\mathbf{P}_I$ the instanton contribution

---

[†]Note that as $x_i \to x_j$, the partition function diverges. This divergence can be smeared by providing the instantons with a core, which is tantamount to a renormalized fugacity.

following from the finite-temperature partition function [40]

$$Z_I[T] = \sum_{N_\pm} \frac{z^{N_+ + N_-}}{N_+! N_-!} \tag{11.80}$$

$$\times \prod_{i=1}^{N} \int_{R^3 \times 1/T} d^4 x_i \; e^{-N_F \frac{T}{f^2(T)} \sum_{i,j,n} Q_i Q_j \int \frac{d^3 q}{(2\pi)^3} \frac{e^{iq \cdot (x_i - x_j)}}{q_n^2 + m_0^2(T)}}$$

where $q_n = (2\pi n T, \vec{q})$, with $\mathbf{P}_{\eta'} = \mathbf{P}_{K^0} + \mathbf{P}_I$. The Coulomb-gas description is now manifest. This decomposition allows us to see how matter affects instantons. It also shows the artificial character of the decomposition in (11.79) compared to (11.78). At low temperature, the condensate (thus m_0) and the meson decay constant f do not change appreciably. The instanton screening-length remains about the same. At high temperature m_0 is expected to drop in this model [33], thereby weakening the screening. Our calculations break down when the screening becomes comparable to the interparticle distance, *i.e.* $m_0^4 \sim \langle N \rangle T / V_3$. Subleading terms in $1/N_c$ and the omitted scalars as well as higher-lying hadrons are important at high temperature.

11.4.2. *NJL models*

A large number of finite-temperature studies in the recent past have focused on the Nambu-Jona-Lasinio model (NJL) or extension thereof [34]. These models share much in common with the instanton and anti-instanton description given above. We highlight the similarities in a schematic way.

Since the NJL model, like the instanton model, does not confine, the low-lying spectrum is composed of constituent quarks and mesons. In the ground state, quarks condense and chiral symmetry is spontaneously broken. At high temperature the symmetry is restored. The restoration is mostly driven by entropy of the constituent quarks with temperature-dependent masses. To illustrate these points, consider the schematic Lagrangian in Euclidean space [‡]

$$\mathcal{L} = \psi^\dagger i\gamma \cdot \partial \psi + \frac{g^2}{2}\left((\psi^\dagger \psi)^2 + (\psi^\dagger i\gamma_5 \psi)^2 \right) \tag{11.81}$$

[‡]For simplicity we will present the argument for chiral $U(N_f) \times U(N_f)$, with $N_f = 1$. The mean-field results will be quoted for finite N_f.

or equivalently

$$\mathcal{L} = \psi^\dagger i\gamma \cdot \partial\psi + \psi_R^\dagger iP\psi_R + \psi_L^\dagger iP^\dagger\psi_L - \frac{1}{2g^2}PP^\dagger \qquad (11.82)$$

in the momentum range below a cutoff scale Λ. Here $P = \sigma + i\pi$ represents two auxiliary fields, and g a fixed coupling. The partition function associated to (11.82) follows by integrating over the fermion fields

$$Z[T] = \int dP\, e^{-\mathbf{S}_{\text{eff}}[P]} \qquad (11.83)$$

with the effective action $\mathbf{S}_{\text{eff}}[P] = \int_{\frac{1}{T} \times V_3} d^4x\, \mathcal{L}$ with

$$\mathcal{L} = +\frac{1}{2g^2}PP^\dagger - N_*T \int \frac{d^3k}{(2\pi)^3} \sum_{n=-\infty}^{+\infty} \ln\left(k^2 + \omega_n^2 + PP^\dagger\right) \qquad (11.84)$$

where $\omega_n = (2n+1)\pi T$ are the fermionic Matsubara energies, and $N_* = 4N_f N_c$ counts the quark degrees of freedom [§] The sum in (11.84) can be done readily. Up to an irrelevant constant, the result is

$$\mathcal{L} = +\frac{1}{2g^2}PP^\dagger$$
$$-N_* \int \frac{d^3k}{(2\pi)^3}\left(\sqrt{k^2 + PP^\dagger} + 2T\ln\left(1 + \exp(-\sqrt{k^2 + PP^\dagger}/T)\right)\right).$$
$$(11.85)$$

In the large N_c (number of colors) limit with $g^2 \sim 1/N_c$, (11.83) is dominated by the parity-even saddle point $P_* = m(T)$, following from the gap equation

$$\frac{1}{g^2} = 2N_*T \sum_{n=-\infty}^{+\infty} \int \frac{d^3k}{(2\pi)^3} \frac{1}{k^2 + \omega_n^2 + m^2(T)} \qquad (11.86)$$

or equivalently

$$\frac{1}{g^2} = N_* \int \frac{d^3k}{(2\pi)^3} \frac{1}{\sqrt{k^2 + m^2(T)}}\left(1 - 2\frac{1}{e^{\sqrt{k^2 + m^2(T)}/T} + 1}\right). \qquad (11.87)$$

[§] We have restored the flavor and color factors.

Since the constituent mass is momentum-independent (in contrast to the instanton case), the integral (11.87) is cut off at Λ for the results to be finite. The interpretation is straightforward. The gap equation (11.87) receives contribution from the vacuum $(+1)$ in (11.87) as well as the thermal bath made out of constituent quarks (this is because of no confinement). The (-2) indicates that the contribution comes from both quarks and anti-quarks, and the sign $(-)$ accounts for Pauli blocking. The fermion condensate associated to (11.85-11.87) reads

$$\langle \bar{\psi}\psi \rangle_T = -N_* \int_0^\Lambda \frac{d^3k}{(2\pi)^3} \frac{m(T)}{k^2 + m^2(T)} \times \left(1 - 2\frac{1}{e^{\sqrt{k^2+m^2(T)}/T}+1} \right).$$

$$(11.88)$$

The behavior of the quark condensate (11.88) versus T/T_c is shown in Fig. 11.15, for $\Lambda = 700$ MeV from Bernard *et al.* in [34]. Similar results were obtained by other authors.

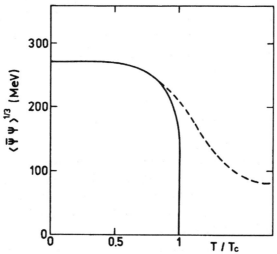

Fig. 11.15. Quark condensate versus T in the NJL model from Bernard *et al.* in [34]. The solid line is for current quark mass $m = 0$ and the dashed line for $m = 5$ MeV.

The T-dependent vacuum pressure following from (11.87) is solely given by the constituent quark loop (as indicated in the instanton case above)

$$\mathbf{P} = +\frac{T}{V_3}\ln\left(\frac{Z[T]}{Z[0]}\right)$$

$$= +2TN_*\int_0^\Lambda \frac{d^3k}{(2\pi)^3}\ln(1 + e^{-E/T})$$

$$+ N_*\int_0^\Lambda \frac{d^3k}{(2\pi)^3}(E(T) - E(0))$$

$$+ \frac{m^2(T) - m^2(0)}{2g^2} \tag{11.89}$$

where $E(T) = \sqrt{\mathbf{k}^2 + m^2(T)}$. Note that $\mathbf{P}$ scales like N_c. In a confining theory such as QCD, such terms are excluded from low-temperature phase, which should receive contributions solely from the color singlet mesons as we discussed above. This point will be emphasized below in the context of QCD_2. Here the same contributions follow by expanding $P = m(T) + \sigma + i\gamma_5\sigma$ around the saddle point, and by considering the quadratic fluctuations in σ and π. The inherited mesons carry a T-dependent mass. The T-dependence accounts for the mean-field interaction. Thermodynamics with dropping masses is a bit tricky since most of the conventional thermodynamical relations do not hold. They have to be rederived anew. Some of the preceding ideas have been used by Koch and Brown [11] to analyze the lattice QCD data for the energy, pressure and entropy.

11.4.3. *Random matrix models*

Chiral random matrix models, discussed broadly in Chapter 1, have proven to be a very economical framework for discussing the bulk aspect of the QCD vacuum. Starting from the assumption that chiral symmetry is spontaneously broken, a number of results on the QCD spectrum has been inferred from symmetry arguments alone and randomness. At zero temperature, the character of the Dirac spectrum is fundamental for understanding the mechanism behind the spontaneous breaking of chiral symmetry.

In this subsection we demonstrate that chiral random matrix models are schematic versions of Nambu-Jona-Lasinio models and their relatives, whether in vacuum or in matter.[35] We use then this analogy to emphasize the mean-field, universal character of the temperature driven phase transitions.

To illustrate these points we reconsider the schematic Lagrangian (11.82) in four Euclidean space,

$$\mathcal{L}_4 = +\psi_R^\dagger(i\gamma \cdot \partial + i\mu\gamma_4)\psi_L + \psi_L^\dagger(i\gamma \cdot \partial + i\mu\gamma_4)\psi_R$$
$$+ \psi_R^\dagger i(P + m)\psi_R + \psi_L^\dagger i(P^\dagger + m)\psi_L + \frac{1}{2g^2}PP^\dagger \qquad (11.90)$$

rewritten in the chiral basis $\psi = (\psi_R, \psi_L)$. We have added the current mass term and a chemical potential μ to be more general. Note that the Minkowski fields follow from the Euclidean fields through $(i\psi^\dagger, \psi) \to (\overline{\psi}, \psi)$. Equation (11.90) is defined on the strip $\beta \times V_3$ in Euclidean space, with $P(\tau + \beta, \vec{x}) = P(\tau, \vec{x})$, and $\psi(\tau + \beta, \vec{x}) = -\psi(\tau, \vec{x})$ and a 3-momentum cut-off Λ.

To compare this model to chiral random matrix models, further simplifications are needed. The anti-periodicity of the quark fields yields

$$\psi(\tau, \vec{x}) = \sum_{n=-\infty}^{+\infty} e^{-i\omega_n\tau}\, \psi_n^x \qquad (11.91)$$

where $\omega_n = (2n + 1)\pi T$ are the Matsubara frequencies ($T = 1/\beta$), and $x = 1, 2, ..., N$ label discrete points in space. Space is here a grid of dimension N, where each point contains a space-constant quark field of frequency n which is either left-handed (L) or right-handed (R). If we further assume that the auxiliary field P is constant in space and time, the action in (11.82) reduces dimensionally to 0-dimensional one with infinitely many Matsubara modes. The corresponding partition function can be readily found in the form

$$Z[T, \mu] = \int dP\, e^{-N\frac{\Sigma}{T}PP^\dagger}$$
$$\times \prod_{n=-\infty}^{+\infty} \det_2^N \beta \begin{pmatrix} i(m + P) & \omega_n + i\mu \\ \omega_n + i\mu & i(m + P^\dagger) \end{pmatrix} \qquad (11.92)$$

where we have rescaled (dimensionally reduced) the variables

$$q_n^x = \sqrt{V_3}\psi_n^x$$
$$\Sigma = V_3/2g^2 \qquad (11.93)$$

so *e.g.*, q_n^x are now dimensionless Grassmannian variables.

We can now rewrite the squares in (11.90) and integrate over the shifted auxiliary field. In terms of the rescaled fields, the result reads

$$\mathcal{L}_0 = q^\dagger((\Omega + i\mu)\gamma_4 + im)q + \frac{1}{N\Sigma}(q_R^\dagger q_R)\,(q_L^\dagger q_L) \qquad (11.94)$$

with $\Omega = \omega_n \mathbf{1}_n \otimes \mathbf{1}_x$. We note that a constant P implies that only certain combinations of the Matsubara modes are allowed to interact with each other. This is manifest when translating the local four-fermi interaction in (11.82) to frequency space (omitting space indices)

$$\int_0^\beta d\tau\,\left((\psi^\dagger\psi)^2 + (\psi^\dagger i\gamma_5\psi)^2\right) = 4 \sum_{n,m,k,l} \delta_{n+k,m+l}\,\psi_{Rn}^\dagger\psi_{Rm}\psi_{Lk}^\dagger\psi_{Ll}. \;(11.95)$$

A constant P implies the kinematic choice $n = m$ and $k = l$. Other choices are also possible, and they lead to different random matrix ensembles. At high temperature, all ensembles become the same since they are dominated by the two lowest Matsubara modes $\omega_0 = +\pi T$ and $\omega_{-1} = -\pi T$ (see below).

Consider now the new auxiliary matrix $\mathbf{A}_{n,m}^{x,y}$ with entries both in ordinary space x, y and frequency space n, m. Matrix $\mathbf{A}$ is doubly banded, with dimensions $(N \times N) \otimes (\infty \times \infty)$. Contrary to P, matrix A bosonizes the pairs of quarks of opposite chirality. For the lowest two Matsubara frequencies it is simply $(N \times N) \otimes (2 \times 2)$. In terms of $\mathbf{A}$, (11.95) becomes

$$\begin{aligned}\mathcal{L}_0 =&+q^\dagger((\Omega + i\mu)\gamma_4 + im)q \\ &+N\Sigma\mathrm{Tr}_{x,n}(\mathbf{A}\mathbf{A}^\dagger) + q_R^\dagger \mathbf{A} q_L + q_L^\dagger \mathbf{A}^\dagger q_R\end{aligned} \qquad (11.96)$$

where the trace in (11.96) is over x and n. The partition function associated to (11.96) is simply

$$Z[T,\mu] = \int d\mathbf{A}\; e^{-N\Sigma\mathrm{Tr}_{x,n}(\mathbf{A}\mathbf{A}^\dagger)}$$

$$\times \det{}_{2,x,n} \begin{pmatrix} im & \mathbf{A} + \Omega + i\mu \\ \mathbf{A}^\dagger + \Omega + i\mu & im \end{pmatrix}. \qquad (11.97)$$

This is an example of a chiral random matrix model. When restricted to $n = 0$ with $\omega_0 = 0$ and $\mu = 0$, this is just the chiral random matrix discussed in Chapter 1. For $\omega_0 = \pi T$ and $\mu = 0$ this is simply the chiral random matrix model discussed in [43]. Other versions are also possible[44]. For instance, if in (11.95) we choose instead $n = l$ and $m = k$, we will generate a chiral random matrix model discussed recently in [45] for $\mu = 0$. Although these models differ in their form, we will show below that they all lead to the same high temperature behavior that is controlled by a universal equation, with mean-field critical exponents. We would like to stress that despite the formal similarity between (11.92) and (11.97) the relation is non-trivial. In the high temperature limit, where only two lowest Matsubara frequencies contribute, the partition function (11.92) is saturated by the N-power of 2×2 determinant, whereas in the case of (11.97) we have the first power of determinant, built of $N \times N$ blocks of random matrix A shifted by lowest Matsubara frequency and/or chemical potential. Nevertheless, (11.92) and (11.97) are *equivalent* order by order in $1/N$.

We choose (11.92) to discuss the thermal properties of the random matrix model for large N with $\mathbf{n} = N/V_3$ fixed (thermodynamical limit). The effective potential associated with (11.92) is

$$
\begin{aligned}
H = N\Sigma\, PP^\dagger \\
-2NT \sum_{n=-\infty}^{+\infty} \ln\left(\beta^2(P+m)(P^\dagger+m) + \beta^2(\omega_n + i\mu)^2\right).
\end{aligned} \tag{11.98}
$$

The sum in (11.98) can be readily done (*cf.* (11.85))

$$
H = N\Sigma PP^\dagger - N\left(\omega + T\ln\prod_{\pm}(1 + e^{-(\omega \mp \mu)/T})\right) \tag{11.99}
$$

where $\omega = \sqrt{(P+m)(P^\dagger+m)}$. The first term is a simple version of the vacuum energy, while the second term combines the contribution from the "Dirac" sea and matter. In this schematic model, the Dirac spectrum is simplified to two-levels in frequency space $\pm\omega$ for each $x = 1, 2, ..., N$, and for each handedness L, R.[¶] There is no kinetic energy associated to the quarks in either (11.92) or (11.97). The pressure at high temperature is

$$
\mathbf{P} = T\ln Z = -F = 2\mathbf{n}\ln 2 - \mathbf{n}\Sigma PP^\dagger + \mathcal{O}(1/T) \tag{11.100}
$$

where the first term represents the thermal pressure of free constituent quarks, while the second term comes from the vacuum.

In large N, the extremum (denoted by star) of the energy (11.99) yields to a gap equation for P_* (we look for the solutions within the class

---

[¶]We have restricted the flavors to $N_F = 1$.

$P = P^\dagger$)

$$2\Sigma P_* = 1 - n - \bar{n} \tag{11.101}$$

where $n = (e^{\frac{\omega_* - \mu}{T}} + 1)^{-1}$ for particles and $\bar{n} = (e^{\frac{\omega_* + \mu}{T}} + 1)^{-1}$ for antiparticles. Using the rescaling (11.93), the constituent quark density is

$$\rho_Q = i\langle\psi^\dagger\gamma_4\psi\rangle = \mathbf{n}\,(n - \bar{n}) \tag{11.102}$$

while the quark condensate is

$$i\langle\psi^\dagger\psi\rangle = -2\mathbf{n}\,P_*\Sigma. \tag{11.103}$$

A non-vanishing P implies a non-vanishing quark condensate, hence a spontaneous breaking of chiral symmetry. The relations (11.101-11.103) are just schematic versions of the usual gap, density and condensate relations in the NJL models (11.86,11.89). We restrict our discussion to finite T but zero μ. We will later briefly mention the role of the chemical potential in this description.

For non-zero current mass m there is always a solution to the gap equation, *albeit* with $P \sim m$ for $T \sim T_c$ and $m \sim 0$. At high temperature limit (but with $T \sim T_c$) the gap equation (11.101) simplifies to

$$G(P_* + m)^3 + M^2(P_* + m) + h = \mathcal{O}(\frac{1}{T}) \tag{11.104}$$

with $h = -m$, $M^2 = (T - T_c)/T_c$ and $G = 1/(12T_c^2)$. The gap equation has the form of the cubic equation expected from mean-field treatments of chiral phase transitions [50] (see (11.124) with $g_3 = 0$). Therefore relation (11.104) provides the accurate description of the chiral phase transition consistent with universality.

Indeed, from (11.104) follows that at $T = T_c$, and for a weak external field h (current mass), $P \sim h^{\frac{1}{3}}$, hence the critical exponent $\delta = 3$. For $h = 0$ but $T \neq T_c$, $P \sim (T - T_c)^{\frac{1}{2}}$. Since the chiral condensate $i\langle\psi^\dagger\psi\rangle \sim P$, the critical exponent $\beta = \frac{1}{2}$. At $T \sim T_c$, the pressure is $\mathbf{P} \sim (T_c - T)$. As a result, the specific heat is $C = \partial E/\partial T \sim (T_c - T)^0$, with a critical exponent $\alpha = 0$.

The scalar quark susceptibility

$$\chi(T) = \langle(\overline{\psi}\psi - \langle\overline{\psi}\psi\rangle)^2\rangle \tag{11.105}$$

measures the correlation length produced by the exchange of a constant scalar in the random matrix model. It is saturated by the scalar mass, so $\chi \sim 1/m_\sigma^2$. This can be seen by considering the fluctuation $P = P_* + \sigma$ around the saddle point. In the Gaussian approximation, σ carries twice the constituent mass $m_\sigma(T) = 2m(T)$. Thus

$$\chi(T) \sim \frac{1}{m_\sigma^2} \sim \Sigma^2 \left(1 - \frac{T^2}{T_c^2}\right)^{-1}. \tag{11.106}$$

At the critical temperature $\chi \sim (T_c - T)^{-1}$, with a critical exponent $\gamma = 1$.

Since the random matrix model does not have any space-time dependence, it is not possible to use it to analyze the ν and η critical exponents. It is straightforward to understand that in the instanton model described above or in NJL models, all that was said above regarding critical exponents holds true. Moreover, the correlation functions at the critical point are dominated by the light pseudoscalars. In the pion channel, for instance, we expect that the static correlator approaches the asymptotics

$$C_\pi(x, T) \sim \frac{1}{|\vec{x}|} e^{-m_\pi(T) |\vec{x}|}. \tag{11.107}$$

Thus the spatial correlation length at T_c is

$$\xi \sim \frac{1}{m_\pi(T)} \sim \frac{1}{\sqrt{\langle \bar\psi\psi \rangle}} \sim (T_c - T)^{-\frac{1}{2}} \tag{11.108}$$

with critical exponents $\nu = \frac{1}{2}$ and $\eta = 0$. The set of all critical exponents derived so far $(\alpha, \beta, \gamma, \delta, \nu, \eta) = (0, \frac{1}{2}, 1, 3, \frac{1}{2}, 0)$ are known as the mean-field exponents. They are universal and characterize a phase transition that is second-order and of the type discussed by Landau and Ginzburg. This point will be further discussed below. Although our discussion focused on the $N_f = 1$ case, it is easily generalizable to several flavors, *albeit* with possibly non-Gaussian measures.

We end this subsection with the digression on the chemical potential. The role of the chemical potential in our description can be best seen from (11.101) by specializing to $T = 0$. In this case, $n = \Theta(\mu - \omega)$ and $\bar{n} = 0$, where Θ is a step function. Clearly, $\rho_Q \neq 0$ only when the chemical potential $\mu \geq \omega$, in which case $P = 0$ and hence a vanishing quark condensate. Our space independent quark modes have a Fermi surface that is sitting precisely at the level $+\omega$. Any finite quark density causes a restoration of chiral symmetry, since the latter is caused by the asymmetry in the spectrum obtained by populating the mode $-\omega$. Therefore to be predictive one has to account for the space-variation of the quark wave-function. That means a field-theory rather than a matrix model. In this case, we are back to the NJL descriptions or QCD itself. The situation $0 \leq \mu \leq \omega$ does not support any finite density (Fermi-level in the Dirac gap).

11.5. Finite Temperature and Confinement

In discussing the various models above, we have been ignoring the fact that the low-temperature part of the spectrum contains unconfined constituent quarks. In fact, the transitions we have observed so far are all driven by entropy, with most of the entropy carried by the constituent quarks. This is not the case in QCD. Indeed, below T_c, colored objects are barred from the heat bath by P_* in (11.1) from the start.

To contrast the thermodynamical arguments gotten from models with no confinement with those of a confining theory, we present here an argument due to Hansson and Zahed[42] in which the finite-temperature free energy in QCD_2 is analyzed in large N_c. Of course two-dimensional theory does not necessarily faithfully reproduce four-dimensional world, but one can gain some insight into the subtlety that may be involved.

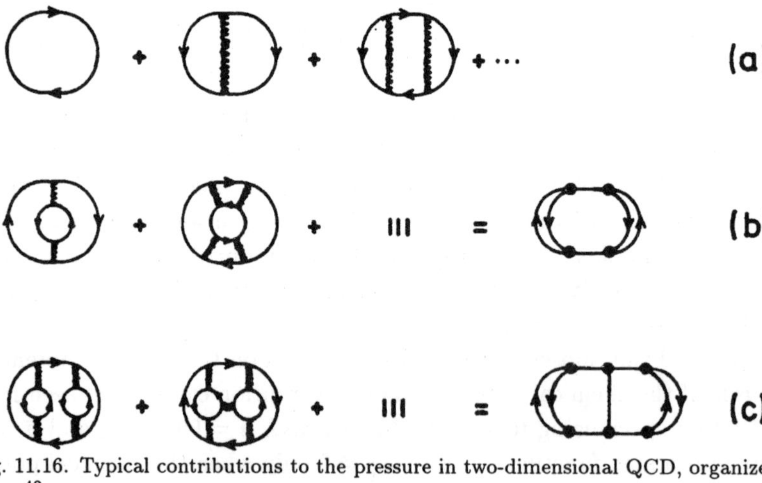

Fig. 11.16. Typical contributions to the pressure in two-dimensional QCD, organized in $1/N_c$.[42]

In two-dimensional QCD the gluons do not carry degrees of freedom.

The free energy can be organized in $1/N_c$, with typical contributions as shown in Fig. 11.16. The terms in (a) contribute to order N_c to the pressure (analog of $Z_Q[0]$ above), while the terms in (b) contribute to order N_c^0 to the pressure (analog of $Z_\pi[0]$ above), and so on. Below T_c – which is infinite in QCD_2, these terms should sum to *zero*, and the contribution to the pressure should be of order N_c^0, as we expect in a confining theory. To show this, we note that if the free energy density resulting from all the diagrams in (a) shown in Fig. 11.16 is denoted by $\mathcal{F}_1 = F_1(T, g^2)/V_1$ where V_1 is the volume, then [42]

$$g^2 \frac{d^2}{dg^2} \mathcal{F}_1 = \frac{1}{2} \mathrm{Tr}(S\Sigma), \tag{11.109}$$

where the trace is over color, spin and momentum. The sum over the momenta involves a sum over Matsubara energies. The constituent quark propagator $S(k)$ and the self-energy $\Sigma(k)$ are expressed in the light-cone gauge, $A_- = 0$, as discussed in Chapter 3 by (3.5) and (3.8), respectively. More explicitly

$$S(k) = \frac{-ik_-}{2k_+k_- - k_-\Sigma(k_-) - i\epsilon} \tag{11.110}$$

and

$$k_-\Sigma(k_-) = \frac{g^2}{\pi} \left(1 - \frac{|k_-|}{\lambda}\right) \tag{11.111}$$

where $\lambda \to 0$ is an infrared regulator. Note that for $\lambda \to 0$, the pole in the constituent quark is given by

$$E_p = -k + \frac{g^2}{\sqrt{2\pi\lambda}} \mathrm{sgn}(E - k) \tag{11.112}$$

where we have introduced the usual energy and momentum via $\sqrt{2}k_\pm = E \pm k$. Figure 11.17 shows the dispersion in energy of the constituent quark at zero temperature. Clearly, for any fixed λ, the constituent quarks get thermalized. However, for $\lambda \to 0$, the mass gap goes to infinity, and finite temperature cannot affect the spectrum. This is indeed the case. If we substitute eqs.(11.110-11.111) into (11.109) and subtract the temperature-independent part of the free energy $\mathcal{F}_1^{\mathrm{vac}}$, we have

$$g^2 \frac{d^2}{dg^2} (\mathcal{F}_1 - \mathcal{F}_1^{\mathrm{vac}})$$

$$= \frac{1}{2\pi} \int \frac{dk}{2\pi} \int_{-i\infty}^{+i\infty} \frac{dE}{2i\pi} \frac{1}{e^{E/T}+1} \mathrm{Ev} \left(\frac{1-|k_-|/\lambda}{2k_+ k_- - k_- \Sigma(k_-) - i\epsilon} \right)$$

(11.113)

where $\mathrm{Ev}[f(E)] = f(E)+f(-E)$. Performing the E-integration by contour, we pick up the pole at E_p given by (11.112). Thus the integral in (11.113) is suppressed by

$$\frac{1}{e^{E_p/T}+1} \sim e^{-\frac{1}{T\lambda}}.$$

(11.114)

Hence for any fixed temperature T, the integrand in (11.113) goes to zero in the limit when $\lambda \to 0$. There is no g-dependent $\mathcal{O}(N_c)$ contribution to the free energy below T_c. The is due to confinement.

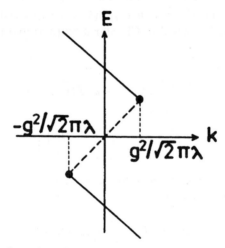

Fig. 11.17. Quark dispersion relation in light-cone gauge as discussed in the text.

    This result illustrates what may happen in four-dimensional QCD. There, we expect the pressure at low temperature to be dominated by pions. The strong infrared divergences in the quark and gluon sector will cause the order N_c^2 (gluonic) and $N_c N_f$ (fermionic) terms in the free energy to vanish identically, just as explained. The unconfined constituent quarks are forbidden below T_c. In four dimensions, however, we expect a phase transition to take place at $T_c \sim \Lambda_{QCD}$. As we approach this temperature from below, the mesonic interactions, though suppressed by $1/N_c$, will be dominant since the number of particles is likely to grow exponentially in a Hagedorn manner.

11.6. Hagedorn Temperature

Well before the advent of QCD, it was noted by Hagedorn [46] that in the narrow resonance approximation, the hadronic spectrum in nature grows exponentially leading to a limiting temperature T_H. Indeed, a fit to the experimentally known resonances suggests the following growth in the density of states

$$\rho(m) = A \frac{e^{m/T_H}}{m^B} \qquad (11.115)$$

with $A = .01 - .03$, $B \leq 3$ and $T_H \sim 160$ MeV. In the narrow resonance approximation, the partition function is given by

$$Z(T) \sim \int dm \, \rho(m) \, e^{-\frac{m}{T}}. \qquad (11.116)$$

As $T \to T_H$ the partition function diverges: The energy suppression is overcome by the entropy growth. From the hadronic point of view, T_H could be viewed as the maximum temperature.

The exponential growth in the spectral density is an essential property of strings. Indeed, the number of possible ways to draw a non-backtracking string of length $L = m/a$ on a discrete mesh of lattice size a is proportional to 5^L.[||] In the continuum, a Nambu string with fixed ends of length R in d dimensions carries an energy [47]

$$E_n^2 = (\sigma R)^2 - \frac{(d-2)}{24\alpha} + \frac{n}{\alpha} \qquad (11.117)$$

where the "Regge slope" is $\alpha = 1/(2\pi\sigma)$ and σ is the string tension. The first term in (11.117) is the classical string contribution, the second term is the zero-point contribution (regularized) from the transverse degrees of freedom,

$$\frac{(d-2)}{2} \left(\sum_{n=1}^{\infty} n \right)_{\text{reg}} = -\frac{1}{24}(d-2) \qquad (11.118)$$

[||]Here 5 is the number of ways that such a string can grow on a three-dimensional spatial lattice.

and the last term is the contribution from the n-th transverse mode. The lowest state in (11.117) is tachyonic. The density of transverse modes is given by [48]

$$\rho(n) \sim n^{-(d+1)/4} e^{2\pi \sqrt{(d-2)n/6}} \tag{11.119}$$

where $\alpha m^2 = n$ for large n. Using (11.118) the string partition function reads

$$Z(T) = \sum_{n=0}^{\infty} \rho(n) e^{-E_n/T}. \tag{11.120}$$

Using the asymptotic form of the density of states (11.119), we see that the statistical sum (11.120) diverges when

$$2\pi \sqrt{\frac{(d-2)n}{6}} \leq E_n \sim \frac{\sigma}{T} \left(\frac{2\pi n}{\sigma}\right)^{1/2}, \tag{11.121}$$

for $T \geq T_H$ with $T_H^2 = 3\sigma/\pi(d-2)$. Thus when the temperature exceeds the "Hagedorn temperature," the sum diverges. In the string description, the limiting temperature is caused by the tachyonic mode in (11.117). Indeed, for $T < T_H$ and fixed R, the statistical sum in (11.120) is dominated by the tachyonic mode (as all the other modes are Boltzmann suppressed), giving $Z(T) \sim e^{-\sigma(T)R}$ with a temperature-dependent string tension

$$\sigma(T) \sim \sigma(0) \left(1 - \frac{T^2}{T_H^2}\right)^{1/2}. \tag{11.122}$$

This behavior was first noted by Pisarski and Alvarez [49].

In nature, however, resonances are not sharp. At finite temperature, resonances are broadened by inelastic processes, and their contribution to a statistical sum is finite. Thus, while the concept of a limiting temperature is not physical, it does suggest the existence of a critical temperature for the hadronic world. Above this temperature, the hadronic basis is no longer optimal for describing hadronic interactions at finite temperature. It is more efficient to use a quark-gluon basis.

The fact that the string picture suggests that $\sigma(T) \to 0$ for $T \to T_H$ implies that the phase transition has a deconfining nature for a Yang-Mills theory. Indeed, if QCD string models describe the large distance properties of QCD with only heavy quarks, because of asymptotic freedom, we would expect that the string picture breaks down at some critical distance or temperature. As QCD strings have a finite transverse width, at high temperature, the string wraps and folds many times, touching itself at many places. In the folded regions space is like a big bag where quarks roam freely. There are strong indications from lattice simulations that such a phase transition takes place in the pure Yang-Mills theory.

11.7. Chiral Critical Point

When quarks are included there appears to be no simple way to distinguish between the confined and deconfined phases. However, for N_F flavors of massless quarks, we have seen in Chapters 1 and 2 that QCD is $SU_L(N_F) \times SU_R(N_F) \times U_V(1)$ symmetric, taking into account the $U_A(1)$ anomaly. We have also seen that this symmetry is spontaneously broken to $SU_{L+R}(N_F) \times U_V(1)$ in the vacuum. There are strong indications, mostly from numerical simulations, that for $N_F \geq 2$ there is a phase transition associated with the restoration of this symmetry. We note that for $N_F = 1$ the chiral symmetry is vacuous. This suggests that the chiral restoration in QCD (zero quark condensate) and the confinement-deconfinement transition in the Yang-Mills theory (zero string tension) are apparently distinct, as evidenced by the $N_F = 1$ gap.

11.7.1. *Universality class of chiral transition*

Using arguments based on universality, Pisarski and Wilczek [50] have suggested that at the critical point, QCD can be analyzed in terms of the Landau-Ginzburg model. A simple candidate for the order parameter for the chiral transition is the quark bilinear

$$\Phi_{ij} \sim \overline{q}_i(1 + \gamma_5)q_j \qquad (11.123)$$

which transforms as $(N_F, N_F^*) \oplus (N_F^*, N_F)$ under the chiral group. Other choices are also possible. The present choice (vector representation) is motivated by the success of the PCAC relations (see Chapter 4) and preliminary lattice simulations. Given this and according to the theory of critical phenomena, understanding the chiral transition amounts to finding out the universality classes of systems with the vector representation of the chiral group, assuming that the theory has an infrared fixed-point.

The most natural setup for analyzing the critical behavior of flavor QCD is the Landau-Ginzburg free energy. Setting the bare quark masses to zero (analogous to zero external magnetic field) we have [50]

$$\mathcal{F} = \frac{1}{2} \text{Tr} \left(\partial_i \Phi^\dagger \, \partial_i \Phi \right) + \frac{1}{2} m_\Phi^2 \, \text{Tr} \left(\Phi^\dagger \Phi \right) + g_1 \left(\text{Tr}(\Phi^\dagger \Phi) \right)^2$$
$$+ g_2 \, \text{Tr} \left(\Phi^\dagger \Phi \right)^2 + g_3 \left(\det \Phi + \det \Phi^\dagger \right) \qquad (11.124)$$

with m_Φ^2 the temperature-dependent renormalized (mass)^2, which is negative below and vanishes at the critical point, and g_1 and g_2 are two smooth
couplings at the critical point. Stability requires $g_2 > 0$ and $g_1 + g_2/N_F > 0$.
g is the quantum-coupling induced by instantons that breaks the $U_A(1)$
symmetry in the vacuum as discussed in Chapter 2.[**] The possible readjustment of the instantons in the vacuum may cause the coupling g_3 to
vary sharply at a fixed temperature T_* that may be distinct from the chiral
critical point [50]. This point will not be pursued here.

Depending on N_F, (11.124) may lead to a first- or second-order transition. Second-order phase transitions are characterized by infrared-stable
fixed points of the renormalization group. For $N_F \geq 3$ and $g_3 = 0$, an analysis of the renormalization group equations using the ϵ-expansion $(4 - \epsilon)$
shows that there is an infrared fixed-point (zero of the beta function) but
that it is not infrared-stable [51]. In other words, the transition could be a
fluctuation-induced first-order transition [50]. This is confirmed by numerical simulations [52]. For $g_3 \neq 0$ and $N_F = 3$, the determinantal interaction
is trilinear in Φ and renders the transition first-order. For $g_3 \neq 0$ and
$N_F = 4$ the determinantal interaction is quadrilinear, relevant but not relevant enough to generate an infrared-stable fixed point. For $N_F \geq 4$ the
determinantal interaction is an irrelevant operator at the critical point.

For $N_F = 2$ and a small g_3 (light η') the transition may turn first-
order with $SU_L(2) \times SU_R(2) \times U_A(1) \sim O(4) \times O(2)$ critical exponents. For
$N_F = 2$ and g_3 fixed (massive η') the transition may be second order with
$SU_L(2) \times SU_R(2) \sim O(4)$ critical exponents. This result has been confirmed
by simulations [52]. In this case it is more convenient to use the conventional
Gell-Mann-Lévy parametrization $\Phi = \sigma + i \vec{\tau} \cdot \vec{\pi}$ in (11.124), with $\sigma \sim \langle \bar{q}q \rangle$
and $\vec{\pi} \sim \langle \bar{q}\gamma_5 \vec{\tau} q \rangle$. Thus g_1 and g_2 in (11.124) are no longer independent. In
terms of the four-component field $\phi = (\sigma, \vec{\pi})$ the universality class is that
of the standard $O(4)$ invariant $n = 4$ Heisenberg magnet. The latter is
believed to describe a helical magnetic phase transition in dysprosium [53].

[**]This term is $Z_A(N_F)$ symmetric.

The empirical consequences of a second-order chiral phase transition with $O(4)$ critical exponents are many-fold: First, the specific heat scales like $C_V(T) \sim (T - T_c)^{-\alpha}$ with $\alpha = -0.19 \pm 0.06$, the negative α implying that the specific heat displays a cusp at the critical point; second, the free energy scales like $F(T) \sim (T - T_C)^{2-\alpha}$; third, the fermion condensate scales like $\langle \bar{q}q \rangle \sim (T - T_C)^{\beta}$ with $\beta = 0.38 \pm 0.01$; and fourth, the pion mass (inverse correlation length) scales like $m_\pi \sim (T - T_C)^{\nu}$ with $\nu = 0.73 \pm 0.02$ [54]. A number of lattice simulations have been carried out by various groups [55,56]. The results are however inconclusive due to the closeness of the $O(4)$ critical exponents to the $O(2)$ and mean-field (MF) critical exponents.

Finally, we note that for $N_F = 2$, the combination $(\sigma, \vec{\pi})$ belongs to the $(2, 2)$ multiplet of chiral $SU_L(2) \times SU_R(2)$. Other representations are of course possible provided that they include the pion multiplet (since they are massless below the transition in the chiral limit) and that they form an irreducible representation of the chiral group (since the symmetry is restored above the transition). The simplest alternative suggested by Brown and Rho[57] is the $(3, 1) \oplus (1, 3)$ representation of $SU_L(2) \times SU_R(2)$, which would imply that at the chiral critical point there would be six massless modes – the triplet of pions and their parity-doublet, isovector scalars – instead of the quartet $(\vec{\pi}, \sigma)$. These would be the π-modes and the longitudinal parts of the ρ-modes as described by Georgi's vector limit[58] discussed in Chapter 10. In this case long-range fluctuations would be observed in the vector-isovector channel (ρ-channel) instead of the scalar-isoscalar channel (quartet-channel). Lattice simulations to be discussed below, while consistent with the quartet scenario, cannot rule out the alternative vector-isovector scenario.

11.7.2. *Disoriented chiral condensate*

A second-order chiral phase transition would be followed by large-scale fluctuations in the chiral order parameter. In a hot-cold scenario such as the one entertained in heavy ion collisions, this may trigger the appearance of disoriented chiral condensates (DCC) as first suggested by Bjorken [59]. That is the manifestation of macroscopic regions of space with different orientations in flavor space, followed by massive isospin breaking and an enhancement in the soft-pion spectrum. The latter may be quenched by a finite pion mass, however. Indeed, the largest possible domain with chiral flavor orientation θ and volume V would be Boltzmann-suppressed by

$$e^{-Vm_\pi^2 f_\pi^2 \sin^2(\theta)/2T} \tag{11.125}$$

at temperature T. Thus $V^{1/3} \sim 2$ fm for $\theta \sim \pi/4$, suggesting a possible enhancement at $p_T \sim 1/V^{1/3} \sim 100$ MeV and not below. In the chiral limit, Rajagopal and Wilczek [54] have further suggested that massive breaking of isospin can be enhanced by off-equilibrium conditions (referred to as "quench").

This point has been critically re-examined by Gavin, Gocksch, and Pisarski [60]. Their main conclusion is that the growth of the DCC domains is characterized by the spinodal time $\tau_{sp} \sim 1/m_\sigma$, where m_σ is the spin-isospin singlet mass. After this time, the fields sense the non-linearities, hence the presence of a stable (cold) ground state, and the evolution shuts off. In strong coupling, the sigma mass is large, and the domains are found to be small, of the order of 1 fm, in agreement with the qualitative argument above.

11.7.3. *Mean-field scaling of chiral transition*

The universality arguments presented above have been challenged recently by Kogut and Kocić [61]. They have argued that although one expects the Gross-Neveu model to share the Ising critical exponents by symmetry arguments, this is not the case in $2 + 1$ dimensions. Through detailed lattice simulations they have concluded that the critical exponents in this case are mean-field and not Ising. Their point is that the standard argument based on dimensional reduction and universality breaks down in this case, since the fermions do not decouple at all temperatures. This is explicit at $T = 0$, where the model undergoes a phase change from weak to strong coupling in the presence of long-range forces. This critical behavior is borne out by large N calculations [62]. However, whether or not the Kogut-Kocić observation implies that the universality arguments given for QCD phase transitions are invalid is not clear. For instance, it is known that the large N arguments fail for the same model in $1+1$ dimensions, since they suggest a phase transition when there is none. To check this point, simulations were carried out for $N = 4$ and $N = 12$ for which the same conclusions were reached. Obviously one would like to do simulations for $N = 1$, but they seem to be numerically more difficult. It may be that what one is seeing in these simulations is the onset of the Ising domain. It is a well-known fact that most of the phase transitions are mean-field except in their Ginzburg window [63]. The latter is set by the ratio of the correlation length in the system to the mean-field length which is the size of the system. This window is where the true critical exponents are expected to dwell. The size of the window depends on the model used. If it is of order $1/N$ in the $2+1$ Gross-Neveu model, it would be hard to detect, unless N is set low. For normal superconductors this window is of order 10^{-14}, and reflects the fact that the Cooper pairs are loosely bound. Their size is about 1000 Å. In QCD, as discussed below, lattice simulations of space-like wave functions seem to indicate that the quarks are tightly bound all the way to T_c, suggesting that the Ginzburg window may be less narrow.

11.8. QCD Phase Diagram

Altogether, it is anticipated that QCD with two massless quarks undergoes a second-order transition with restoration of chiral $SU_L(2) \times SU_R(2)$ with possibly $O(4)$ critical exponents, and that QCD with three flavors or more undergoes a fluctuation-driven first-order transition. The addition of bare quark masses with a mass matrix M through $\text{Tr}(M\Phi)$ is equivalent to adding an external magnetic field. The latter may decrease the latent heat and weaken the character of the first-order transition.

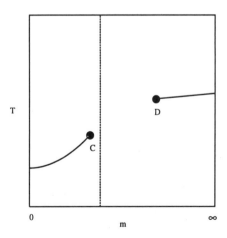

Fig. 11.18. QCD phase diagram for two light flavors of mass m and a fixed strange quark mass $m_s = 32m$. There are two critical points (m_C, T_C) and (m_D, T_D). QCD may lie in the cross-over region.[64]

Dedicated lattice simulations carried out by the Columbia group for three flavors are summarized in Fig. 11.18, in the $m - T$ plane where m is the common up and down quark mass, with a fixed strange quark mass $m_s = 32m$. There seems to be two critical temperatures, one for QCD with heavy quarks (D-line for deconfinement) that is close to the pure glue theory, and one for QCD with light fermions (C-line for chiral restoration). For fixed m, the transition across the solid lines versus T is first-order, and second-order elsewhere. The end points (C and D) may be tri-critical points. The important feature of this phase diagram is the appearance of a window $m_C < m < m_D$ for which there is a smooth crossover. QCD with realistic quark masses appears to lie within this window, perhaps close to the tri-critical point [50,64]. This important issue can only be settled by more detailed lattice simulations.

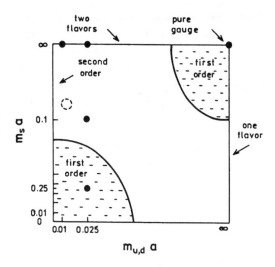

Fig. 11.19. Diagram of the QCD phase transitions as a function of the up (m_u) and strange (m_s) quark masses. The dashed circle is the experimental point.

Figure 11.19 shows the possible QCD phase diagram in the $m_u - m_s$ plane (with $m_u = m_d$) [65]. When all quark masses are heavy, the transition is first-order as in the pure $SU(3)$ Yang-Mills theory. The limit $m_s = \infty$ corresponds to the two-flavor case, where there is strong evidence for a second-order transition at $m_u = 0$. For $m_u = m_s$ the transition is first-order in the chiral limit. The dashed circle is a guess to where the hadronic world may lie.

11.9. Dirac Spectrum

Another aspect of the finite-temperature problem at the critical point is the global behavior of the Dirac spectrum near zero virtuality. Could it be that this behavior is universal, and if so what would be the relationship between this behavior and the universality classes discussed above? In other words, is it possible to marry the generic concept of randomness in the mesoscopic limit with those of universality, to reach model-independent statements on the bulk character of the transition? We do not know the answer. In the random phase, the "mean-free-path" of the quarks is what fixes the scale in the chiral condensate. At T_c, pion mean-free path is what determines the character of the transition. It is clear that the competition between these two scales is what determines the nature of the phase transition, with some important consequences on the structural aspects of the Dirac spectrum.

11.9.1. *Instantons*

In the screened version of the instanton and anti-instanton ensemble described above, the distribution of eigenvalues λ_n of the four-dimensional Dirac operator on the strip $1/T \times V_3$,

$$\nu(\lambda, T) = \langle \sum_n \delta(\lambda - \lambda_n) \rangle_T \qquad (11.126)$$

is a Wigner semi-circular distribution at finite but low temperature [33]

$$\nu(\lambda, T) = \frac{N}{\pi \kappa(T)} \left(1 - \frac{\lambda^2}{4\kappa^2(T)} \right)^{\frac{1}{2}} \qquad (11.127)$$

where $\kappa^2(T)$ is the temperature-dependent variance, which is proportional to the average of the trace of $\mathbf{T}^\dagger \mathbf{T}$, where $\mathbf{T}$ is the overlap matrix (11.64),

$$\kappa^2(T) = \frac{n(T)}{N_c} \, \alpha(T) \, \bar{\rho}^2 \qquad (11.128)$$

with

$$\alpha(T) = \frac{T}{\bar{\rho}^2} \int \frac{d^3 k}{(2\pi)^3} \sum_n (\mathbf{k}^2 + \omega_n^2)(\mathbf{k}^2 A_n^2 + \omega_n^2 B_n^2)^2. \qquad (11.129)$$

Here $\bar{\rho}$ is the mean size of the instanton at finite temperature [33]. The sum is over the fermionic Matsubara frequencies $2\pi(n+1/2)T$. The coefficients A_n and B_n are temperature-dependent and identical to the ones entering (11.69). For temperatures below the critical temperature $\bar{\rho} \sim \rho$,

$$\alpha(T) \approx \frac{6.6209}{1 + \frac{2}{3}\pi^2\bar{\rho}^2T^2}. \qquad (11.130)$$

As the temperature increases, the variance (11.128) shrinks, causing the distribution of eigenvalues to peak at the origin. This behavior is shown in Fig. 11.20. Chiral symmetry is restored, when $\kappa(T) \leq m$, where $m \to 0$ is a current-mass regulator (see Chapter 2). The restoration follows from the strong suppression of the instantons and anti-instantons due to screening. This spectrum reflects only the soft quark eigenvalues, and shows that chiral symmetry is restored through a re-localization of the quark zero modes due to the diluteness of the system.

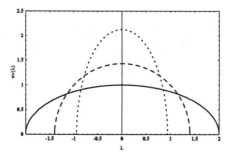

Fig. 11.20. Distribution of quark eigenvalues $\nu(\lambda, T)$ versus T, in a random but screened ensemble of instantons and anti-instantons.[33] The solid line represents the case $T = 0$. The dashed (dotted) line represents the case $T = 160$ MeV for 2(3) flavors, respectively. Normalization of the $T = 0$ distribution is arbitrary.

In the alternative description where the screening is ignored, a totally different distribution of eigenvalues for the Dirac operator has been noted numerically. We show in Fig. 11.21 the results of simulations for $N_f = 3$ and three temperatures. The spike in the middle is caused by the free streaming instantons and anti-instantons. The broadening of the spectrum away from $\lambda = 0$ is what causes the restoration of chiral symmetry, as discussed in Chapter 2. It is due to the localization of the quark zero modes in a molecular configuration (see Chapter 2).

11.9.2. *Random matrix*

In the context of the random matrix model discussed in Chapter 1, if we were to do the substitution $\Sigma \rightarrow \Sigma(T)$ with $\Sigma(T)$ following the T-dependent quark condensate, then in the large n limit, the spectral density would be semi-circular, with a variance that is temperature-dependent. This case is reminiscent of the screened instanton scenario.

If we use instead the random matrix ensemble discussed above [43], then the matrix structure in (11.97) shows that the πT terms do not allow a diagonalization of the integrand with the polar decomposition $A = U\Lambda V^{-1}$. This means that the density of eigenvalues does not follow from ordinary orthogonal polynomials. So far, there is no analytical form for (11.126) using the ensemble (11.97).

Detailed numerical studies show that the finite-temperature spectrum following from (11.97) has a double-hump structure, as shown in Fig. 11.22 for three temperatures $\pi T = 0.0, 0.4, 1.0$ in units where $\Sigma = 1$. This spectrum is reminiscent of the unscreened instanton calculations described above. When the temperature is increased, the random interaction is overcome by the πT term and a structural change in the spectrum results. The eigenvalues are peaked about $\lambda = 0.0, 0.4, 1.0$. This is also what happens in the molecular case, since only one Matsubara mode is selected by the molecule.

To help clarify these matters lattice simulations and studies of Dirac spectra below and about the critical temperature will be desirable. In their absence, we now turn to a qualitative discussion of some lattice simulations on bulk and correlation functions, and discuss their possible relevance to our understanding of the high-temperature phase.

11.10. Lattice Results

• By now, there is a large number of numerical simulations of finite-temperature QCD, where bulk as well as static correlators are investigated [66]. Since chiral symmetry is spontaneously broken in the vacuum, the first major question is: how is this symmetry affected by matter from first principles? Figure 11.23 shows the behavior of $\langle \bar{q}q \rangle$ versus temperature

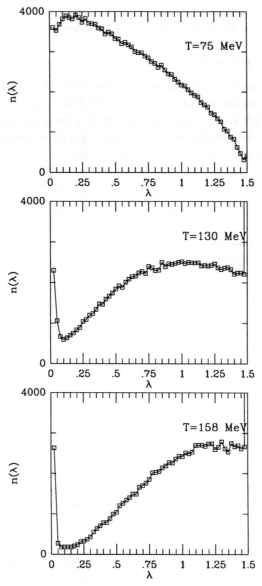

Fig. 11.21. Distribution of quark eigenvalues $\nu(\lambda, T)$ versus T, in a random but polarized (unscreened) ensemble of instantons and anti-instantons.[39]

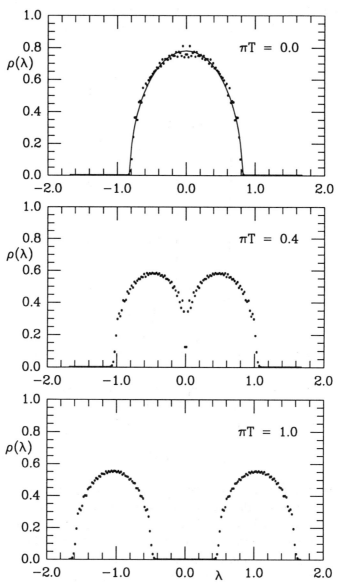

Fig. 11.22. Distribution of quark eigenvalues $\nu(\lambda, T)$ versus T, in a random matrix model.[43]

for four flavors. For temperatures of the order of 150 MeV, a substantial decrease in $\langle \bar{q}q \rangle$ is noted. Up to date, the compatibility of the low T-part of Fig. 11.23 with T-chiral perturbation theory as discussed above has not been investigated.

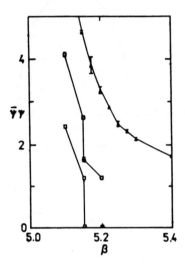

Fig. 11.23. Behavior of the fermion condensate as a function of $\beta = 6/g^2$ for four flavors and quark masses $m_q a = 0.025$ (triangles), 0.01 (circles) and 0.0 (squares).

- In the vacuum, color is expected to be confined at zero temperature, as we discussed in Chapter 1. At finite temperature, the order parameter for confinement is the Polyakov loop (Wilson line)

$$\langle P \rangle = \langle \, \mathrm{Tr} \, P e^{\int ig A_4 dx_4} \rangle \qquad (11.131)$$

where the average is a thermal average, A_4 is the imaginary time component of the gauge field and the trace is over one period $\beta = 1/T$ of the imaginary time. $\ln |\langle P \rangle|$ is usually assumed to be the free energy of an infinitely heavy quark. A sharp rise in the expectation value of the Polyakov loop is usually seen following the drop in the chiral condensate.

- Lattice measurements of the thermodynamical quantities in QCD show a rapid cross over in these quantities at about 150 MeV. Figure 11.24 shows the behavior of the energy density and pressure versus T/T_c, normalized to the black-body result [67]. In the regime $T_c < T < (2-3)T_c$, there are substantial deviations from the free massless gas limit, $\Theta^{00} = \mathcal{E} - 3\mathcal{P} \neq 0$. The fermionic susceptibilities (singlet and isotriplet) show also a rapid variation in the same temperature range as shown in Fig. 11.25 [68]. At low temperature the susceptibilities are exponentially suppressed by the mass of the hadronic modes.

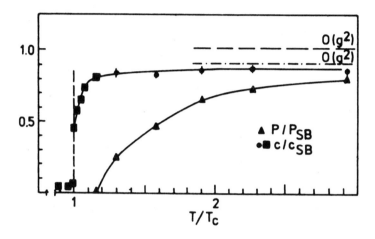

Fig. 11.24. Energy density and pressure normalized to the black-body limit on a finite lattice versus T/T_c. The dashed horizontal lines refer to the perturbative corrections to the black-body limit.

• The (connected) Polyakov-anti-Polyakov correlation function $\langle P(x) P^\dagger(0) \rangle$ has been measured both at low and high temperatures in the pure Yang-Mills theory. At low temperature, the correlation function displays an area-law behavior. Recent simulations at high temperature indicate a screening behavior. The screening range is estimated to be of the order of the inverse screening mass $1/gT$. The time-like Wilson loop shows an area-law behavior at low temperature, and a perimeter law at high temperature. This is to be contrasted with the space-like Wilson loop which displays an area-law behavior both at low and high temperatures. The space-like string tension at high temperature has been found to scale with T^2 [69].

• Lattice measurements of static hadronic correlation functions show that the screening lengths approach asymptotically $2\pi T$ (mesons) and $3\pi T$ (baryons) as shown in Fig. 11.26 [70] with the exception of the pion and the

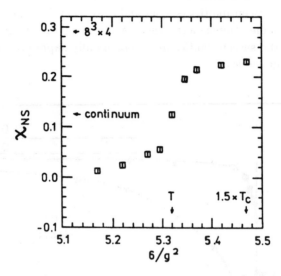

Fig. 11.25. The triplet susceptibility as a function of $\beta = 6/g^2$ for two flavors. The singlet susceptibility, not plotted here, is similar with however larger error bars.

sigma (which is scalar-isoscalar). The pattern displayed by the measured screening lengths suggests that the thermal state preserves chiral symmetry. The pion-sigma channel, however, suggests the presence of still large fluctuations possibly related with a chiral phase transition. The lattice measurements of the transverse correlations in the hadronic channels are shown in Fig. 11.27, for the pion and the ρ. [71] Almost no change is detected from the low to the high temperature regime. The correlations are expected to be absent in the free gas limit.

• Recent lattice measurements of the fermionic distribution around the Polyakov line (heavy quark) are shown in Fig. 11.28 [72]. At low temperature the fermionic distribution is localized around the heavy source in a range of the order of $1/\Lambda$ (where Λ is the QCD scale), and sums to -1. At high temperature, the distribution is considerably smeared.

The lattice measurements discussed so far are somewhat contradictory. On the one hand, they suggest that the bulk thermodynamical quantities such as the energy density, the pressure, the entropy, the susceptibilities, *etc.* when measured at high temperature are consistent with the

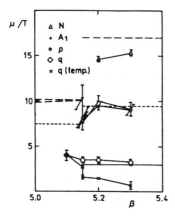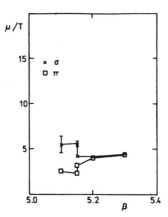

Fig. 11.26. (Left) Screening masses $\mu/T = m_H/T$ as a function of β for four flavors and $m_q a = 0.01$. $\beta = 5.15$ is the transition point. (Right) The same as (a) but for the pion and sigma correlator in the spatial directions.

black body limit. On the other hand, the space-like correlators, whether gluonic or hadronic, show strong evidence of correlations and thus an a priori non-black body limit.

We stress that all the lattice calculations performed to date reflect space-like physics. They have no bearing on time-like physics. It should be emphasized that the latter is what matters for rate calculations at high temperature.

11.11. Field Theories at High Temperature

11.11.1. *Hot* $O(N)$

To address the issues raised above, we shall follow the strategy along the line of reference[73]: First, we focus on space-like physics; second, we will work at $T = \infty$ and then expand in $1/T$. The rationale for this approach is the dimensional reduction (DR). We illustrate the approach with the $O(N)$ sigma model in two dimensions at high temperature. Similar work has been done using the Gross-Neveu model in large N in $2+1$ dimensions [74].

Consider the $O(N)$ model at finite temperature. The field $\vec{\phi}$ consists of N bosonic components constrained by $\phi \cdot \phi = 1/g^2$. The Lagrangian

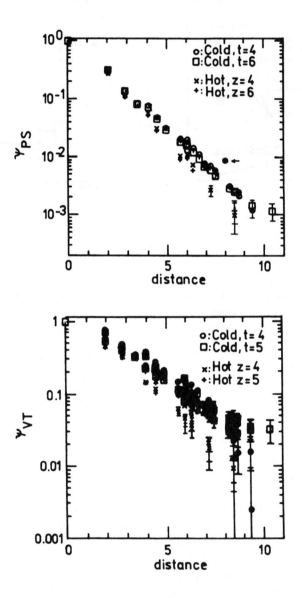

Fig. 11.27. The pion (PS) and ρ (VT) "spatial wave functions" at $T = 0$ and $T = 1.5$.

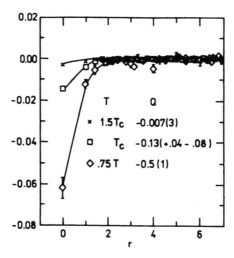

Fig. 11.28. Quark number density induced by a fixed quark at three temperatures.

density is

$$\mathcal{L} = \frac{1}{2}(\partial_\mu \vec{\phi}) \cdot (\partial^\mu \vec{\phi}). \tag{11.132}$$

The free energy of the system reduces to the free energy of $N-1$ free bosons, so to order g^0, we have $\mathcal{F}/V_1 \sim (N-1)\pi T^2/6$. The correlation function can be worked out in powers of g. To leading order and for large distances,

$$\langle \vec{\phi}(x) \cdot \vec{\phi}(0) \rangle = \frac{1}{g^2}\left(1 - \frac{1}{2}g^2(N-1)T|x| + \cdots\right) \sim \frac{1}{g^2}e^{-\frac{1}{2}g^2(N-1)T|x|}. \tag{11.133}$$

These results illustrate the points to be emphasized below. Indeed, while the free energy behavior reflects a free bosonic system to leading order, the correlation function has an exponential fall-off with a correlation length of the order of $1/(N-1)g^2T$. This correlation length is due to the long-wavelength modes in the $O(N)$ model which are sensitive to the curvature of the S^N manifold. Indeed, the Euclidean partition function reads

$$Z(T) = \int [d\vec{\phi}]\ \delta(\vec{\phi}^2 - \frac{1}{g^2})\ e^{-\int_0^{1/T} d\tau \int dx\ \frac{1}{2}(\partial_\mu \vec{\phi})^2}$$

$$= \int d\lambda (\det(-\nabla^2 + \lambda))^{N/2} e^{-\frac{1}{g^2} \int \lambda \, dx}. \tag{11.134}$$

The second equality follows from the integration over the $\vec{\phi}$ field after introducing a Lagrange multiplier λ. In the large N limit, the saddle point approximation to (11.134) gives

$$N \frac{T}{2\pi} \sum_{-\infty}^{+\infty} \int_0^\Lambda \frac{dk}{k^2 + (2\pi n T)^2 + \lambda} = \frac{1}{g^2} \tag{11.135}$$

which can be rearranged to give

$$\frac{NT}{2\sqrt{\lambda}} = \frac{1}{g^2(1 + \frac{N}{2\pi} g^2 \ln(\frac{T}{\Lambda}))} = \frac{1}{g^2(T)}. \tag{11.136}$$

The left-hand side of (11.136) is the $n = 0$ contribution in (11.135). The nonzero modes in (11.135) renormalize the coupling constant $g^2 \to g^2(T)$.

To enhance the physics of the nonzero modes it is more efficient to take the $T = \infty$ limit first, in which case the high-temperature problem for the $O(N)$ model reduces to a quantum mechanics problem (field theory in $0 + 1$ dimensions). Indeed, on the strip $[0, 1/T] \times R$, the bosonic fields are periodic,

$$\vec{\phi}(\tau, x) = \vec{\phi}_0(x) + \sum_{n \neq 0} e^{-i2\pi n T \tau} \vec{\phi}_n(x). \tag{11.137}$$

If we redefine $\vec{\phi}_0(x) \to \vec{x}(t)T$, then at high temperature only the zero modes contribute to $Z(T)$,

$$Z(T \to \infty) = \int d\vec{x} \, \delta(\vec{x}^2 - \frac{1}{g^2 T^2}) \, e^{-\int_0^{1/T} dt \, \frac{T}{2} \dot{x}^2}. \tag{11.138}$$

This is the partition function of a quantum-mechanical particle of mass T on a sphere of radius $R = 1/gT$. The Hamiltonian is $H = \vec{L}^2/2TR^2$ where $\vec{L}$ is the angular momentum in N dimensions. The eigenstates of the Hamiltonian are hyperspherical harmonics and the spectrum is given by $E_n = n(n + N - 2)/2TR^2$. There is a gap in the spectrum. The constant modes contribute zero to the free energy to leading order, but dominate the correlation function

$$T^2 \langle 0 | \vec{x}(t) \cdot \vec{x}(0) | 0 \rangle \sim e^{-E_1 t} \sim e^{-\frac{1}{2}(N-1)g^2 T t}. \tag{11.139}$$

11.11.2. *Hot three-dimensional QCD (QCD₃)*

We now proceed to discuss the high-temperature limit of QCD_3. The dimensional reduction scheme in QCD_3 is exact, in contrast to QCD_4 which has shortcomings beyond one loop. Many of the results derived in three dimensions can be carried over to four dimensions. Furthermore the three-dimensional theory has the advantage of being easier to analyze numerically. In this subsection, we will discuss in some details the reduction at high temperature for three-dimensional QCD, and briefly quote the main results for four-dimensional QCD.

Note that the pure Yang-Mills version of QCD_3 is also Z_N-symmetric. If the deconfinement phase transition is indeed related to the spontaneous breaking of Z_N at high temperature, then the three-dimensional theory should display also a phase transition that could be mapped onto the Z_N Potts model [75]. For both $N = 2, 3$ the transition could be second-order. In three dimensions parity could be spontaneously broken. If we assume that a condensate forms at low temperature and disappears at high temperature, then there is a possibility of a Z_2 transition related to parity restoration. If confirmed, this transition would be of the Ising type in two dimensions.

(i) Dimensional reduction

At high temperature, QCD_3 dimensionally reduces to QCD_2 plus a massive Higgs. To see this, consider three-dimensional QCD on the cylinder $[0, 1/T] \times R^2$

$$\mathcal{L} = \frac{1}{4}F_{\mu\nu}^2 + \psi^\dagger(-i\not{\partial} + g_3\,A)\psi \qquad (11.140)$$

where g_3 is a dimensionful coupling constant. (In three dimensions QCD is super-renormalizable.) The gauge fields obey periodic boundary conditions modulo periodic gauge transformations, and the fermions obey anti-periodic boundary conditions. Explicitly

$$A_\mu(\tau, x) = A_{\mu,0}(x) + \sum_{n\neq 0} e^{i2\pi nT\tau} A_{\mu,m}(x),$$

$$\psi(\tau, x) = \sum_\pm e^{\pm i\pi T\tau}\psi_\pm + \sum_{n\neq 0,-1} e^{i(2n+1)\pi T\tau}. \qquad (11.141)$$

At high temperature the theory truncates to R^2 with massless magnetic gluons A_i, massive electric gluons A_0 and "heavy fermions" with mass

$m = (2n + 1)\pi T$ [76],

$$\mathcal{L} = \frac{1}{4}F_{ij}^2 + \frac{1}{2}|(\partial_i - g_3\sqrt{T}A_i)\phi|^2 + \frac{1}{2}m_H^2\phi^2 + V_H(\phi)$$
$$+ \psi^\dagger(-i\partial\!\!\!/ + g_3\sqrt{T}A + \gamma^3 m + \gamma^3 g_3\sqrt{T}\phi)\psi. \qquad (11.142)$$

We have rescaled the fields to their canonical dimensions in two dimensions, $A_3 \to \sqrt{T}\phi$, $A_i \to \sqrt{T}A_i$ and $\psi \to \sqrt{T}\psi$. The screening mass m_H is infrared-sensitive in perturbation theory. Its self-consistent determination will be discussed below.

The effective potential V_H for the Higgs is induced by integrating over the magnetic gluons as shown in Fig. 11.29. At high temperature and for a fixed screening mass (to be specified below), the Higgs potential is subleading in $1/T$. Thus, to leading order, the Higgs field is massive and interacts solely with the magnetic gluons and the "heavy" fermions. Note that the fermions carry an energy m. However, through a γ_3 rotation of the spinors, this energy can be turned to a mass [76]. This will be assumed throughout.

Fig. 11.29. Induced Higgs potential V_H by dimensional reduction (DR) as discussed in the text. The wavy lines refer to the magnetic gluons, the dashed lines refer to the electric gluons.

(ii) Polyakov-Polyakov loop correlator

In the high-temperature limit, the Polyakov line simplifies to

$$P(x) = \text{Tr}\mathbf{P}e^{ig_3\int_0^{1/T} A_0(\tau,x)\,d\tau} \sim \text{Tr}e^{i\frac{g_3}{\sqrt{T}}\phi(x)}. \qquad (11.143)$$

Ignoring the Coulomb effects on the Polyakov sources, and to leading order in $1/T$, the connected part of the Polyakov-anti-Polyakov correlator is given by

$$\langle P(x)\, P^\dagger(0)\rangle_c \sim \frac{(g_3\sqrt{T})^4}{16T^4}\langle \phi^2(x)\, \phi^2(0)\rangle_c. \qquad (11.144)$$

This is the correlation function for two electric gluons. The correlator is dominated by the diagrams shown in Fig. 11.30. The asymptotic form of the correlator in Euclidean space is the tail of the closest-to-zero singularity in Minkowski space. This singularity corresponds to a bound state in $1+1$ dimensions. The bound-state equation is given by

$$-\frac{1}{m_H}\psi''(x) + \frac{1}{2}g_3^2 NT|x|\psi(x) = E\psi(x) \qquad (11.145)$$

in the regime where $m_H/T \ll 1$ (corresponding to the non-relativistic limit). The linear potential reflects the Coulomb potential in $1+1$ dimensions. The solutions to (11.145) are Airy functions. The spectrum follows from the zeros of the derivative of the Airy function $Ai'(-E_n) = 0$ (after proper rescaling). The lowest state

$$E_0 \sim \left(\frac{g_3^2 T}{\sqrt{m_H}}\right)^{2/3} \qquad (11.146)$$

determines the slope of the correlation function (11.145). Indeed, at high temperature we expect

$$\langle P(x)\, P^\dagger(0)\rangle \sim K_0((2m_H + E_0)|x|) \sim \frac{e^{-(2m_H+E_0)|x|}}{((2m_H + E_0)|x|)^{1/2}} \qquad (11.147)$$

which is to be contrasted with the free screening correlator (given by the free bubble of Fig. 11.30). The omitted Coulomb corrections amount to a self-energy correction.

$$\langle P(x)\, P^\dagger(0)\rangle \sim K_0^2(m_H|x|) \sim \frac{e^{-2m_H|x|}}{m_H|x|}. \qquad (11.148)$$

Note that at high temperature, the difference between (11.147) and (11.148) is in the pre-exponent and not the exponent, since $E_0/m_H \to 0$.

In QCD_3, the screening mass $m_H \sim g_3^2 T$ is infrared-sensitive and gauge-dependent in perturbation theory as indicated by Fig. 11.31. A self-consistent and gauge-independent analysis of the screening mass can be

Fig. 11.30. Leading contribution to the (connected) Polyakov-anti-Polyakov correlator in the dimensionally reduced theory to leading order.

performed from the loop expansion of the Polyakov-Polyakov correlator at high temperature. The two-loop contribution to (11.145) is shown in Fig. 11.32. The cross refers to the $-m_H^2$ insertion in the self-consistent (Hartree) definition of the electric mass, where the electric propagators carry a mass m_H^2. All the infrared and ultraviolet divergences cancel to two-loop order. The finite contributions from diagrams Fig. 11.32a and Fig. 11.32b are scale (μ) dependent: $g_3^2 NT(a + b\ln(m_H/\mu))$. The finite contribution of diagram 15c is also scale-dependent, $g_3^2 NT(\bar{a} + \frac{1}{2\pi}\ln(m_H/\mu))$. If we choose $\mu = m_H$, then the contribution from Fig. 11.32c can be made logarithmically large compared to the contributions from Fig. 11.32a and Fig. 11.32b in the large temperature limit. This large contribution can be reabsorbed in a self-consistent definition of the electric mass in a Hartree-type approximation[77],

$$-m_H^2 + g_3^2 NT \frac{1}{2\pi} \ln\left(\frac{T}{m_H}\right) = 0. \qquad (11.149)$$

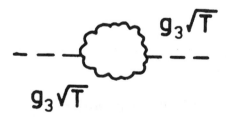

Fig. 11.31. One-loop contribution to the screening mass.

(iii) Spatial Wilson loops

That magnetic loops still obey an area law has a simple explanation

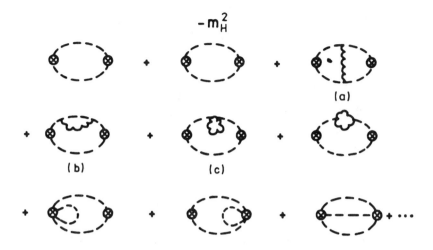

Fig. 11.32. Two-loop contribution to the (connected) Polyakov-anti-Polyakov correlator.

in high-temperature QCD in three dimensions. To leading order the static part of the magnetic gluon correlator in static-axial ($A_2 = 0$) gauge follows from the diagrams of Fig. 11.33,

$$\Delta_{ij} \sim \delta_{i1}\delta_{j1}\delta(x_1)\left(a|x_2| + \frac{c}{\sqrt{M x_2}}e^{-M|x|}\right) \qquad (11.150)$$

where $a = g_3^2 NT/24\pi m_H^2$, $b = g_3^2 NT/240\pi m_H^4$, c an arbitrary coefficient and $M^2 = (1 - a)/b$. In (11.150), only the leading temperature dependent part has been retained. The contribution of (11.150) to the space-like Wilson loop is due to the unscreened part of the magnetic gluon propagator. The result is $W_{xy} \sim e^{-\sigma_{xy} A}$, where A is the area in the xy-plane and $\sigma \sim a$ the temperature-dependent string tension to leading order. This result is gauge-invariant. On the other hand, note that the magnetic-magnetic correlator is exponentially screened. At large distances, it is dominated by the second term of (11.150) in the static-axial gauge

$$\langle B(x) \cdot B(0) \rangle \sim \frac{e^{-M|x|}}{\sqrt{M x_2}}. \qquad (11.151)$$

The linear term drops out in taking derivatives. This result is gauge-dependent. Note that an area-law behavior of the space-like Wilson loops is still consistent with an exponential fall-off in the correlation function of magnetic fields. We expect this result to extend to four dimensions as well.

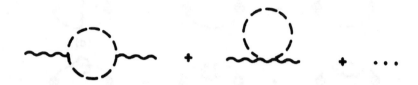

Fig. 11.33. Leading contribution to the magnetic polarization in the DR theory.

(iv) Static quark correlators

The static hadronic correlators can be analyzed in the dimensionally reduced theory by treating the fermions as infinitely heavy (or non-relativistic). To leading order in $1/T$

$$C_{\alpha,\alpha}(x) = \frac{1}{\beta^2} \langle \int_0^\beta d\tau \psi^\dagger \Gamma^\alpha \psi(\tau, x) \int_0^\beta d\tau' \psi^\dagger \Gamma^\alpha \psi(\tau', 0) \rangle$$
$$\sim \langle \psi^\dagger \Gamma^\alpha \psi(x) \, \psi^\dagger \Gamma^\alpha \psi(0) \rangle. \tag{11.152}$$

The fermions in (11.152) are restricted to the lowest Matsubara modes and carry a mass $m = \pi T$ after a γ_3-rotation. The dominant contribution to the hadronic correlator (11.152) follows from the diagram of Fig. 11.34. The Higgs and fermion insertions are subleading in the temperature. The large x-asymptotics of (11.152) is again given by the closest-to-zero singularity in Minkowski space. In Minkowski space the quark and anti-quark are massive (non-relativistic) and interact via a Coulomb potential in $1+1$ dimensions. The bound state equation is

$$-\frac{1}{m} \psi'' + \frac{e^2}{2} |x| \psi = E_\alpha \psi \tag{11.153}$$

where $e^2 = C_F g_3^2 T$ and C_F the value of the Casimir in the fundamental representation. The screening lengths are $m_\alpha = 2m + E_\alpha$, and characterize the correlator at long distance, $C_{\alpha\alpha}(x \to \infty) = e^{-m_\alpha |x|}$. For QCD$_3$ [76],

$$m_\alpha = 2\pi T + \frac{3}{2} \left(\frac{C_F^2 g_3^4 T}{4\pi} \right)^{1/3}. \tag{11.154}$$

The wave functions are Airy functions with a size of the order of $1/(g_3T)^{2/3}$. These wave functions are directly related to the transverse correlations. For four-dimensional QCD, the screening lengths can also be estimated [76]. The reduced wave functions directly relate to the correlations observed on the lattice as described above. The present discussion provides a simple explanation to the fact that while the screening lengths approach asymptotically "free" quark values, they are in fact reflecting strong space-like correlations. The correlations are subleading in the screening masses.

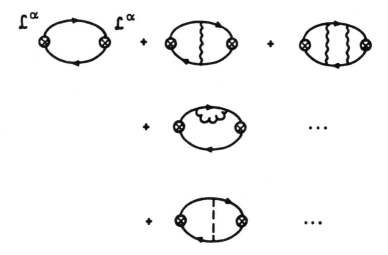

Fig. 11.34. Hadronic correlators in the DR theory.

(v) Quark number susceptibility

While the above considerations have provided some evidence for why the space-like physics at high temperature reflects correlations, it remains puzzling why the bulk properties of the high-temperature QCD phase are to some extent consistent with the black-body description. We will show now that this is in fact a consequence of the fact that the bulk properties sample *all* distance scales. Consider the isoscalar fermionic susceptibility in QCD_3

$$\chi(T) = \int_0^{1/T} d\tau \int d^2x \langle \psi^\dagger \gamma^3 \psi(\tau, x) \psi^\dagger \gamma^3 \psi(0,0) \rangle$$

$$\sim T \int d^2x \langle \psi^\dagger \sigma^3 \psi(x) \psi^\dagger \sigma^3 \psi(0) \rangle. \tag{11.155}$$

All fermionic modes contribute to $\chi(T)$ to leading order. Indeed [78]

$$\chi(T) = \frac{N_c T}{\pi} \sum_{H=-\infty}^{+\infty} \sum_{n=0}^{+\infty} \sin^2\left(\frac{n\pi}{2}\right) \frac{G_{nH}^2}{\mu_{nH}^2}. \tag{11.156}$$

The μ's are the invariant masses for bound quark-anti-quark states in $(1+1)$ dimensions with masses $m_F = (2F + 1)\pi T$, and G_{nH} the form factors following from the 't Hooft equation. Asymptotically, $\mu_{nH}^2 \to \pi\sqrt{2\pi n\sigma}/2\sigma$, $\sigma = g_3^2 NT$ and $G_{nH} \to \pi/\sqrt{2}$. The density of states per unit energy in (11.156) grows linearly with the energy $dn/dE \sim \sqrt{E^2 - 4m_F^2}/\pi\sigma$. The sums in (11.156) can be done exactly, leading to [78]

$$\chi(T) - \chi(0) \sim \frac{N}{\pi} \int \frac{dz}{e^{z/T} + 1} = N_c \left(\frac{\ln 2}{\pi}\right) T \tag{11.157}$$

which is the free gas result to leading order. We believe this result to extend to the bulk energy density, pressure and entropy as well.

11.11.3. Hot QCD_4

In four dimensions, the screening lengths and wave functions for the hadronic screening lengths follow from similar considerations [76,80]. The relevant non-relativistic Hamiltonian is obtained from Breit reduction. Keeping only the leading Coulomb interaction, we have

$$H = \sum_i \frac{\vec{p}_i^2}{2M} + \sum_{i<j} \frac{e^2}{2\pi} \left(\frac{1}{2} + \ln(M|r_i - r_j|)\right) \tag{11.158}$$

where $e^2 = g_3^2 C_F$ (quark-anti-quark) and $e^2 = g_3^2 C_F/2$ (quark-quark) with C_F the Casimir operator in the fundamental representation. To arrive at this expression one has to cancel the infrared singularities between self-energy and exchange contributions [76].

Non-asymptotic effects in the temperature may originate from a variety of terms. An example is the spin-spin interaction familiar from non-relativistic calculations which is due to one-gluon exchange, both temporal and spatial. The spin-spin interaction follows from the usual Breit construction. In our case, it is easy to note that in the dimensionally reduced theory, the expressions for the currents are similar to the ones encountered in the (3+1) dimensional case, except that the momenta in the 3-direction are identically zero. Since, in the non-relativistic approximation, the spin-dependent part of the current is $\chi^\dagger(\vec{p}')i\vec{\sigma} \times (\vec{p}' - \vec{p})\chi(\vec{p})$, the Pauli interaction depends only on the spin component in the 3-direction. A simple calculation gives,

$$H_{ss} = \sum_{i<j} \frac{e^2}{4M^2}(1 + 4S_3^i S_3^j)\ \delta^2(\vec{r}_i - \vec{r}_j). \qquad (11.159)$$

H_{ss} is subleading in the temperature and, as usual, it will only be considered as a perturbation. In the case of mesons, (11.159) simplifies to

$$H_{ss} = \frac{e^2}{2M^2}\ S_3^2\ \delta^2(\vec{r}_1 - \vec{r}_2) \qquad (11.160)$$

where S_3 is the total spin of the meson in the 3-direction. Different polarization components of the ρ (ρ_μ) acquire different screening masses. The components 1, 2 and 4 are measured on the lattice [79]. (In this work ρ_4 is mistakenly referred to as ρ_3.) Using (11.159) and the helicity assignment we have $m_\pi \simeq 2\pi T < m_{\rho_4} < m_{\rho_i}$ and

$$m_{\rho_i} - m_\pi = \frac{e^2}{M^2}|\Psi(\vec{0})|^2. \qquad (11.161)$$

The lattice measurements at temperatures not much above T_c do not exhibit this pattern but rather $\frac{3}{2}m_\pi \simeq m_{\rho_i} \simeq 2\pi T < m_{\rho_4}$. These results, however, are only good at subasymptotic temperatures which are still far away from the present lattice temperatures.

The baryonic screening masses follow essentially the same arguments. To leading order, $m_\alpha = 3M + E_\alpha$ is just the mass of three heavy quarks in a color-singlet configuration, interacting via Coulomb forces. In the ground state, the three quarks are in a symmetric spatial configuration, in which case the spin-spin interaction (11.159) simplifies to

$$H_{ss} = \frac{e^2}{2M^2}\ (S_3^2 + \frac{3}{4})\ \delta^2(\vec{r}_{12}), \qquad (11.162)$$

where S_3 again is the total spin along the 3-direction and $\vec{r}_{12}$ the relative two-body separation in the three-body system. This is to be compared with the spin-spin interaction (11.160) in the mesonic channel.

The ground-state baryon wave functions are antisymmetric in color and symmetric in spin, flavor and space. They can be estimated using variational techniques. Using Gaussian variational ansätze for the ground state wave function, we have

$$QCD_{2+1} : \psi(\rho, \eta) = \sqrt{\frac{8\alpha}{\pi}} e^{-\alpha(\rho^2 + \eta^2)}, \qquad (11.163)$$

$$QCD_{3+1} : \psi(\rho, \eta) = \frac{2\alpha}{\pi} e^{-\alpha(\rho^2 + \eta^2)}, \qquad (11.164)$$

where $\vec{\rho} = \frac{1}{\sqrt{2}}(\vec{r}_1 - \vec{r}_2)$ and $\vec{\eta} = \sqrt{\frac{2}{3}}(\vec{r}_1 + \vec{r}_2 - 2\vec{r}_3)$. Minimizing the expectation value of (11.164) with respect to α yields for QCD_{2+1}, $\alpha = \frac{1}{4}\left(\frac{9e^4 M^2}{\pi}\right)^{\frac{1}{3}}$ and for QCD_{3+1}, $\alpha = \frac{9Me^2}{16\pi}$. The corresponding baryonic screening masses are [81]

$$QCD_{2+1} : m_\alpha = 3M + \frac{9}{4}\left(\frac{3e^2}{M\pi}\right)^{\frac{1}{3}}, \qquad (11.165)$$

$$QCD_{3+1} : m_\alpha = 3M + \frac{9e^2}{8\pi}\left(2 - \mathbf{C} - \ln(\frac{e^2}{16\pi M})\right), \qquad (11.166)$$

where $\mathbf{C} \approx 0.577$ is Euler's constant. The spatial distribution of the baryonic correlators is Gaussian in both $\vec{\rho}$ and $\vec{\eta}$, and have widths $\sim 4\pi/e\sqrt{M}$.

These estimates are only qualitative, since space-like Wilson loops are known to obey an area-law behavior at high temperature [82]. Thus we expect the logarithmic potential to go over to a linearly confining potential, σr, where σ is the string tension, at large distances. Using a T-independent Coulomb plus linear potential, Koch et al. [80] have fitted the lattice simulations [82] as shown in Fig. 11.35. The finite-temperature $(T > T_c)$ space-like potential corresponds to the solid line, and the zero-temperature time-like potential to the dashed line. They used this potential to construct numerically the ρ "wave function" as shown in Fig. 11.36 for T=150 MeV (solid line), T= 250 MeV (short dashed line) and T = 350 MeV (long dashed line). The data are from Bernard et al. [71] for T=210 MeV. While the asymptotic form of the wave functions is drastically changed, the size and mass of the Coulomb bound state are only marginally affected by the confining forces, as can be inferred from the above discussion. More recent lattice simulations by Karkkainen et al. suggests a strong T^2 dependence in the string tension. With $\sigma = (24.8 - 44.0)T_c^2$ at a temperature of $(6 - 8)T_c$ and for

$QCD_4{}^{69}$, the above relations show that the size of the Coulomb bound state decreases by $\sim 2\%$, while its binding energy increases by $\sim 5\%$.

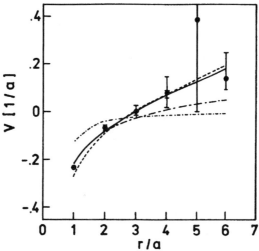

Fig. 11.35. Quark-anti-quark space-like potential at finite temperature [80] (solid line), compared to the zero-temperature time-like potential (dashed line).

As in the mesonic case, the degeneracy of the nucleon and Δ screening masses will be lifted by spin-dependent effects. At very high temperatures the splitting is due to the spin-spin interaction (11.162). This gives an upward shift to *both* the nucleon and the delta. The removal of the degeneracy is, however, only partial as the $\Delta(3/2, \pm 1/2)$ states remain degenerate with the $N(1/2, \pm 1/2)$ states. Specifically

$$m_N^{\pm 1/2} - m_\alpha = m_\Delta^{\pm 1/2} - m_\alpha = \frac{1}{3}\left(m_\Delta^{\pm 3/2} - m_\alpha\right) \qquad (11.167)$$

and the splitting is $\sim e^2/4M^2$. As already mentioned, the different polarizations of the vector and axial vector mesons are split, so the $j = \pm 1$ states are pushed up, while the $j = 0$ state is unaffected along with the pion and its scalar partner, the σ. At very high temperature where this effect should be dominant, we have the relation[81]

$$m_\Delta^{\pm 3/2} - m_\Delta^{\pm 1/2} = m_\Delta^{\pm 3/2} - m_N^{\pm 1/2}. \qquad (11.168)$$

The splitting is of the same order of magnitude and in the same *direction*

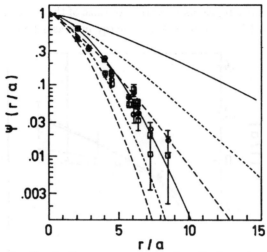

Fig. 11.36. Wave function for the ρ meson as a function of r for various temperatures T [80], and potentials. The lattice data are from [71].

as the meson splitting

$$m_\rho^{\pm 1} - m_\rho^0 = m_\rho^{\pm 1} - m_\pi. \qquad (11.169)$$

Since chiral symmetry is manifest at high temperature, these relations also hold for the chiral partners. A number of relations have been suggested for the gluonic correlators, together with a systematic discussion of the non-asymptotic effects at high temperature [81].

Finally, it should be clear from the lattice simulations as well as the arguments we presented in QCD_3 that the bulk properties in QCD_4 at high temperature should be sensitive to all distance scales in the plasma. In the thermal regime $\sim T$, the quarks and gluons behave as a free gas. In the Debye regime $\sim gT$, the quarks and gluons are screened [3,2,36]. Once the screening is taken into account, some sort of perturbation theory should be feasible as discussed by many authors[83,84,85]. In the magnetic regime $\sim g^2 T$, the quarks and gluons are no longer amenable to perturbation theory [86]. In this regime, both the bulk properties as well as correlation functions have to be analyzed non-perturbatively, much like what we have done here. Future experiments at RHIC and CERN will shed much light on these matters. A more thorough presentation of these theoretical issues goes beyond the scope of this book.

References

1. J.C. Collins and M.J. Perry, Phys. Rev. Lett. **34**, 1353 (1975).
2. A.M. Polyakov, Phys. Lett. **B72**, 477 (1978).
3. E. Shuryak, JETP **74**, 477 (1978).
4. G.E. Brown and M. Rho, "Chiral restoration in hot and/or dense matter," Phys. Repts., to appear.
5. M.A.B. Bég and S.-S. Shei, Phys. Rev. **D12**, 3092 (1975).
6. P. Gerbert and H. Leutwyler, Nucl. Phys. **B321**, 387 (1989).
7. J.L. Goity and H. Leutwyler, Phys. Lett. **B228**, 517 (1989).
8. H. Leutwyler, Nucl. Phys. **B4** (Supp.), 248 (1988).
9. V. Eletsky, P. Ellis and J. Kapusta, Phys. Rev. **D47**, 4084 (1993).
10. C. Adami, T. Hatsuda and I. Zahed, Phys. Rev. **D43**, 921 (1991).
11. V. Koch and G. E. Brown, Nucl. Phys. **A560**, 345 (1993).
12. E. Shuryak, Nucl. Phys. **A533**, 761 (1991).
13. J. Gasser and H. Leutwyler, Phys. Lett. **B184**, 83 (1987).
14. H. Leutwyler and A. Smilga, Nucl. Phys. **B342**, 302 (1990).
15. M. Kacir and I. Zahed, "Skyrmions at Finite Temperature", SUNY-NTG-95-54, to be published.
16. M. Dey, V. Eletsky and B. Ioffe, Phys. Lett. **B252**, 620 (1990).
17. V. Eletsky and B. Ioffe, Phys. Rev. **D51**, 2371 (1995).
18. A. Bochkarev and M. Shaposhnikov, Nucl. Phys. **B268**, 220 (1986).
19. R.J. Furnstahl, T. Hatsuda and S.H. Lee, Phys. Rev. **D42**, 1744 (1990); C. Adami, T. Hatsuda and I. Zahed; Phys. Rev. **D43**, 4312 (1991).
20. M. Shifman, A. Vainshtein and V. Zakharov, Nucl. Phys. **B147**, 385 (1979).
21. T.H. Hansson and I. Zahed, SUNY-NTG-90-22, unpublished; I. Zahed, in Proceedings "From fundamental fields to Nuclear Phenomena", Eds. J.A. McNeil and C.E. Price, World Scientific 1991.
22. V.L. Eletsky and B.L. Ioffe, Phys. Rev. **D47**, 3083 (1993); C.A. Dominguez and M. Loewe, UCT-TP-208-94, unpublished.
23. T. Hatsuda, Y. Koike and S.H. Lee, Nucl. Phys. **B394**, 221 (1993).
24. R.D. Pisarski, Phys. Lett. **B110**, 155 (1982).
25. G.E. Brown and M. Rho, Phys. Rev. Lett. **B178**, 2720 (1991).
26. R.D. Pisarski, Phys. Rev. **D52**, R3773 (1995); C. Gale and J. Kapusta, Nucl. Phys. **B357**, 65 (1991).
27. G. Boyd, S. Gupta, F. Karsch, E. Laermann, B. Petersen and K. Redlich, Phys. Lett. **B349**, 170 (1995).
28. G. Agakichiev, *et al.*, Phys. Rev. Lett. **75**, 1272 (1995).
29. G.Q. Li, C.M. Ko and G.E. Brown, Phys. Rev. Lett. **75**, 4007 (1995).
30. R.D. Pisarski and F. Wilczek, Phys. Rev. **D29**, 338 (1984); D. Metzger, H. Meyer-Ortmanns, and H.J. Pirner, Phys. Lett. **B321**, 66 (1994).
31. D.I. Diakonov and A.D. Mirlin, Phys. Lett. **B203**, 299 (1988).
32. E.M. Ilgenfritz and E.V. Shuryak, Nucl. Phys. **B319**, 511 (1989).
33. M.A. Nowak, J.J.M. Verbaarschot and I. Zahed, Nucl. Phys. **B325**, 581

(1989).

34. T. Hatsuda and T. Kunihiro, Phys. Rev. Lett. **55**, 158 (1985); V. Bernard, U.G. Meissner and I. Zahed, Phys. Rev. **D36**, 819 (1987); M. Lutz, S. Klimt and W. Weise, Nucl. Phys. **A542**, 521 (1992); G. Ripka and M. Jaminon, Ann. Phys. **218**, 51 (1992).

35. M.A. Nowak and I. Zahed, "Thermodynamics in Chiral Random Matrix Models", SUNY-NTG-96-2, to be published.

36. D.J. Gross, R.D. Pisarski and L.G. Yaffe, Rev. Mod. Phys. **53**, 43 (1981).

37. B. J. Harrington and H. K. Shepard, Phys. Rev. **D17**, 2122 (1978).

38. E.V. Shuryak and M. Velkovsky, Phys. Rev. **D50**, 3323 (1994); T. Schafer, E.V. Shuryak and J.J.M. Verbaarschot, Nucl. Phys. **B412**, 143 (1994).

39. T. Schäfer and E.V. Shuryak, SUNY Preprint, SUNY-NTG-95-22.

40. I. Zahed, Nucl. Phys. **B427**, 561 (1994).

41. R. Alkofer, M.A. Nowak, J.J.M. Verbaarschot and I. Zahed, Phys. Lett. **B233**, 205 (1989).

42. T.H. Hansson and I. Zahed, Phys. Lett. **B309**, 385 (1993).

43. A.D. Jackson and J.J.M. Verbaarschot, SUNY preprint, SUNY-NTG-95-26.

44. T. Wettig, A. Schafer, and H.A. Weidenmüller, hep-ph/9510258.

45. M. Stephanov, hep-lat/9601001.

46. R. Hagedorn, Nuov. Cim. Supp. **3**, 147 (1965).

47. P. Olesen, Nucl. Phys. **B267**, 539 (1986).

48. K. Huang and S. Weinberg, Phys. Rev. Lett. **25**, 895 (1970).

49. R.D. Pisarski and O. Alvarez, Phys. Rev. **D26**, 3735 (1982).

50. R.D. Pisarski and F. Wilczek, Phys. Rev. **D29**, 338 (1984).

51. A.J. Peterson, Nucl. Phys. **B190** [FS3], 188 (1981).

52. H. Gausterer and S. Sanilielevici, Phys. Lett. **B209**, 533 (1988).

53. G. Malmstrom and D. Geldart, Phys. Rev. **B21**, 1133 (1980).

54. K. Rajagopal and F. Wilczek, Nucl. Phys. **B399**, 395 (1993).

55. F. Karsch, Phys. Rev. **D49**, 3791 (1994); F. Karsch and E. Laermann, Phys. Rev. **D50**, 6954 (1994).

56. C. Detar, Nucl. Phys. **B42** (Proc. Suppl.), 73 (1995).

57. G.E. Brown and M. Rho, Phys. Lett. **B338**, 301 (1994).

58. H. Georgi, Nucl. Phys. **B331**, 311 (1990).

59. J.D. Bjorken, Phys. Rev. **D27**, 140 (1983); J.D. Bjorken, K.L. Kowalski and C.C. Taylor, SLAC preprint, SLAC-6109-93.

60. S. Gavin, A. Gocksch and R.D. Pisarski, Phys. Rev. Lett. **72**, 2143 (1994).

61. A. Kocić and J. Kogut, Phys. Rev. Lett. **74**, 3109 (1995).

62. B. Rosenstein, B. Warr and S. Park, Phys. Rev. **D39**, 3088 (1989).

63. V. Privman, P. Hohenberg and A. Aharony, in Phase Transitions and Critical Phenomena, **14**, Eds. C. Domb and J. Lebowitz, Academic Press (1991).

64. R.D. Pisarski, BNL preprint, hep-ph/9503330.

65. F.R. Brown, *et al.*, Phys. Rev. Lett. **65**, 2491 (1990).

66. S. Gottlieb, *et al.*, ANL-HEP-PR-92-57, UCSBTH-92-34.

67. F. Karsch, Nucl. Phys. (Supp.) **9B**, 357 (1989).

68. S. Gottlieb, *et al.*, Phys. Rev. Lett. **59**, 2247 (1987).
69. L. Karkkainen, *et al.*, Phys. Lett. **B312**, 173 (1993).
70. R.V. Gavai, *et al.*, Phys. Lett. **B241**, 567 (1990).
71. C. Bernard, *et al.*, Phys. Rev. Lett. **68**, 2125 (1992).
72. C. Detar, Nucl. Phys. (Supp.) **30B**, 66 (1993).
73. I. Zahed, Act. Phys. Pol. **B25**, 99 (1994).
74. S. Huang and M. Lissia, MIT preprint, hep-ph/9509360.
75. F. Wilczek, talk in "Quark Matter 1993", Bergen, Sweden.
76. T.H. Hansson and I. Zahed, Nucl. Phys. **B374**, 177 (1992).
77. E. D'Hoker, Nucl. Phys. **B201**, 401 (1982).
78. M. Prakash and I. Zahed, Phys. Rev. Lett. **69**, 3282 (1992).
79. C. Detar and J. Kogut, Phys. Rev. **D36**, 2828 (1987).
80. V. Koch, E.V. Shuryak, G.E. Brown and A.D. Jackson, Phys. Rev. **D46**, 3169 (1992).
81. T.H. Hansson, M. Sporre and I. Zahed, SUNY-NTG-93-12.
82. E. Manousakis and J. Polonyi, Phys. Rev. Lett. **58**, 587 (1987).
83. E. Braaten and R.D. Pisarski, Phys. Rev. Lett. **64**, 1338 (1990)
84. J.-P. Blaizot, Nucl. Phys. **A566**, 333C (1994).
85. R. Jackiw, "Nonperturbative results for high T QCD," hep-th/9502052.
86. A. Linde, Phys. Lett. **B96**, 289 (1980).

INDEX